Symbol	Meaning*	Symbol	Meaning*
k	Number of treatments (10.8)	Π_1, Π_2	Population proportions in populations 1 and 2 (7.7)
l	Width of class interval containing the median (2.4)	Π_i^0	Specified value of ith probability (9.2)
L_m	Lower limit of class interval containing the median (2.4)	$\hat{\Pi}$	Estimate of population proportion (7.8)
λ	Mean number of events per unit of time (Appendix 5.1)	$P(A)$	Probability of event A (3.3)
M	Median (2.4)	$P(A\ or\ B)$	Probability of event A, or event B, or both (3.4)
M	Number of times an experiment occurs (3.3)	$P(A\ and\ B)$	Probability of both event A and event B (3.4)
M_0	Population median if null hypothesis is true (9.10)	$P(A\,|\,B)$	Probability of event A, given that event B occurs (3.5)
m	Number of times a particular event occurs (3.3)	$P(x)$	Probability that random variable X equals x (4.3)
μ	Population mean (2.4)	$P(not\ A)$	Probability that event A does not occur (3.4)
μ_o	Population mean if null hypothesis is true (8.4)	p	Sample proportion (6.6)
μ_1, μ_2	Means of populations 1 and 2 (7.7)	p_1, p_2	Sample proportions from populations 1 and 2 (7.7)
μ_i	Mean of ith population (10.5)	$Pr\{a<X<b\}$	Probability that random variable X lies between a and b (7.5)
$\mu_{y.x}$	Mean of Y, given the value of X (11.4)	$Pr\{A\}$	Probability of event A (5.9)
N	Number of observations in the population (2.4)	Q_1	Lower quartile (Appendix 2.2)
n	Number of observations in the sample (2.4)	Q_3	Upper quartile (Appendix 2.2)
n	Number of Bernoulli trials (4.6)	R	Sample multiple correlation coefficient (12.6)
n	Number of blocks (10.8)	R	Interquartile range (Appendix 2.2)
n_1, n_2	Sample sizes from populations 1 and 2 (7.7)	R_1	Sum of ranks in Mann-Whitney test (9.11)
n_1, n_2	Numbers of each type in runs test (9.12)	R^2	Sample multiple coefficient of determination (12.6)
υ	Number of degrees of freedom of the χ^2 distribution (9.3)	RSS	Block sum of squares (10.8)
υ_1, υ_2	Numbers of degrees of freedom of the F distribution (10.4)	r	Sample correlation coefficient (11.11)
Π	Population proportion (7.4)	r^2	Sample coefficient of determination (11.10)
Π	Probability of success on each Bernoulli trial (4.6)	r	Number of rows (9.6)
Π_0	Population proportion if null hypothesis is true (8.5)		

(Glossary continues inside back cover)

*Number in parenthesis indicates the section where the symbol is introduced.

BASIC STATISTICS
With Applications

EDWIN MANSFIELD
UNIVERSITY OF PENNSYLVANIA

TO ACCOMPANY THE TEXT:

Basic Statistics: Problems, Exercises, and Case Studies

BASIC STATISTICS

WITH APPLICATIONS

W·W·NORTON & COMPANY·NEW YORK·LONDON

Copyright © 1986 by W. W. Norton & Company, Inc.

Printed in the United States of America.

FIRST EDITION

The text of this book is composed in Baskerville, with display type set in Delphian. Composition by NETS. Manufacturing by The Murray Printing Company. Book design by Nancy Dale Muldoon.

Library of Congress Cataloging in Publication Data
Mansfield, Edwin.
   Basic statistics with applications
   1. Statistics.   I. Title.
QA276.12.M33   1986   519.5   85-17338

ISBN 0-393-95393-9

W. W. Norton & Company, Inc., 500 Fifth Avenue, New York, N.Y. 10110
W. W. Norton & Company Ltd., 37 Great Russell Street, London WC1B 3NU
1 2 3 4 5 6 7 8 9 0

TO TWO EDMs:
Cornell 1, Penn 1

Contents

PART II
Probability and Probability Distributions   85

PART III

Sampling, Estimation, and Hypothesis Testing

215

PART IV

Chi-Square Tests, Nonparametric Tests, and the Analysis of Variance   357

PART V
Regression and Correlation   447

Preface

The field of statistics offers insights for a wide variety of disciplines, and statistical techniques are valuable tools for many professions. Why then do students complain that the subject is dull, abstract, and difficult to master? Today's students are more practical and career minded than any other generation in recent memory; yet they often fail to see the great utility of statistics for their own areas of study and for their future lives. How can this be?

I believe there is a major need—as yet unmet—for a text that conveys the power of statistics in the real world, that addresses important issues in psychology, sociology, political science, and economics, and that shows how statistical methods are used in professions like business, education, engineering, and nursing. My purpose here is to introduce students to statistics in a practical, clear, and interesting fashion, using an approach that is consistent with the abilities and backgrounds of typical students, but still reasonably rigorous and modern.

This book is written for first courses in statistics for undergraduates and graduate students. Over the past thirty years, I have taught such courses at the University of Pennsylvania, at Harvard University, and at Carnegie-Mellon University. Despite the fact that many good books exist in this field, I felt the need for a text that was more practical, broader in scope, and more flexible than those currently available. Gradually, building on materials used in the classroom, the present volume evolved.

There are several features that distinguish this book from other introductory texts:

1. *There is a continual emphasis on the practical applications of statistics.* In each chapter, there are many worked examples indicating how statistical techniques have been used to help solve important

problems in the real world. Also, in each chapter, attention is focused on case studies based on the actual experience of a particular person or organization. Some of these cases, ranging from NASA's problems with the reliability of the Apollo space mission to the controversies over the relationship between smoking and lung cancer, are presented for solution by the student. Also, there are 1-page inserts asking the student to spot errors in statistical procedures, as well as inserts dealing with topics such as the educational attainment of children in Japan, Taiwan, and the United States. In my judgment, the examples, the case studies, and the inserts add significantly to the effectiveness of the text, because students are motivated to learn material much more thoroughly when they are convinced that it has important practical uses.

2. *More attention is paid to the role of the computer here than in most other statistics books at this level.* The computer has enabled students (and seasoned practitioners) to perform statistical analyses more quickly and effectively. For the many instructors who want to introduce their students to the use of computers, I have provided brief general descriptions of relevant computer packages and their application. Also, in practically all chapters, there are a few exercises that ask for the interpretation of computer output or the use of programs like Minitab. Particular attention is devoted to the role of the computer in regression analysis.

At the same time, I have taken pains to insure that this material can easily be skipped in classes where no attention is paid to computers. The passages in the text on computer applications are brief and have been confined to a relatively few inserts and sections that are clearly marked, and are optional. As for the relevant exercises, they can be ignored as well. Put briefly, the material on computers is available for those instructors who want to use it, but it can readily be omitted in classes where it is not appropriate.

3. *This book provides a broader menu of topics than is included in most other texts at this level.* I have included optional sections and appendixes dealing with box plots, the Poisson distribution, the hypergeometric distribution, and other topics frequently neglected in books of this sort. These are topics which many instructors would like to touch on.

4. *If used in a one-semester (or one-quarter) statistics course this book provides more flexibility than most other texts.* It is designed so that several quite different types of one-semester courses can be taught from it. For example, some instructors may wish to focus considerable attention on analysis of variance and experimental design, while others may wish to emphasize regression techniques. The material is presented in sufficient depth—and enough of it is optional—so that a variety of types of one-semester courses can be pursued effectively.

5. *I have tried to assemble an unusually abundant and varied selec-*

tion of problems, problem sets, and supplementary materials for use by both the student and the instructor. In addition to the examples, case studies, and inserts, each chapter contains approximately 30 to 40 problems and problem sets. Further, *Basic Statistics: Problems, Exercises, and Case Studies,* which accompanies this text, includes numerous additional problems and questions as well as a detailed and realistic case study pertaining to each chapter. There is also a *Solutions Manual* for instructors, which includes solutions to all problems and case studies in the text. These supplements illustrate and help to illuminate the various topics in the text.

No mathematical background beyond high school algebra is required for an understanding of *Basic Statistics with Applications.* Mathematical derivations are generally not included in the text. The emphasis here is on providing students with solid and effective evidence concerning the power and applicability of modern statistical methods, on making sure that they can use these techniques, and on indicating the assumptions underlying these techniques, as well as their limitations. To accomplish these purposes, it is neither necessary nor appropriate to deluge students with mathematics.

This book is an adaptation of my *Statistics for Business and Economics,* which has been adopted by hundreds of colleges and universities in the United States and abroad. In writing this book, I have benefited from the comments and suggestions of many colleagues and students. The manuscript of this book was improved by the helpful comments of David Burdick, San Diego State University; Victor DeGhett, State University College at Potsdam, N.Y.; Kenneth Land, University of Texas, Austin; Neil Macmillan, Brooklyn College; Shelton Schmidt, Union College; W. Robert Stephenson, Iowa State University; Jessica Utts, University of California, Davis; Dennis Wackerly, University of Florida; and James Wright, University of Massachusetts. In addition, thanks go to the following teachers who have commented in detail on parts of the manuscript: Gordon Antelmen, University of Chicago; Arnold Barnett, Massachusetts Institute of Technology; Wallace Blischke, University of Southern California; Warren Boe, University of Iowa; Pat Donnelly, Drexel University; Judd Hammock, California State University at Los Angeles; Arthur Hoerl, University of Delaware; D. L. Mars, Louisiana State University; John Mattila, Wayne State University; Robert B. Miller, University of Wisconsin; Don R. Robinson, Illinois State University; Robert Sandy, Indiana University-Purdue University; Robert Schiller of Yale University; Stanley Steinkamp, University of Illinois; Donna Stroup, University of Texas at Austin; Robert J. Thornton, Lehigh University; Bruce Vavrichek, University of Maryland; Albert Wolinsky, California State University at Los Angeles; and Gordon P. Wright, Tuck School, Dartmouth College. Professor Boe supervised the checking of many of the solutions to the exercises. I also thank John Hawkins and Donald S. Lamm, who did their usual good job editing the work and guiding it through to publication.

I am indebted to the Biometrika Trustees for allowing me to reproduce material from the *Biometrika Tables for Statisticians;* and to the Rand Corporation for permission to reproduce material from its tables of random numbers. Further, I am grateful to the literary executor of the late Sir Ronald A. Fisher, F.R.S.; to Dr. Frank Yates, F.R.S.; and to Longman Group Ltd., London, for permission to reprint part of Table 3 from their book *Statistical Tables for Biological, Agricultural, and Medical Research* (6th edition, 1974). Also, thanks go to Thomas Ryan and Minitab, Inc. for permission to reproduce Minitab printouts. Finally, special thanks go to my wife, Lucile, and to my two children who helped in countless ways with the preparation of this book.

Philadelphia, 1986   E.M.

PART II

Descriptive Statistics

OBJECTIVES

Statistics is one of the most important and useful subjects taught in the nation's college and universities. The overall purpose of this chapter is to acquaint you with the types of problems that statistics can help you to solve. Many basic concepts, such as probability, population, and sample, are introduced. Among the specific objectives are:

1   To introduce you to basic statistical techniques like survey sampling and experimental design.

2   To indicate how a frequency distribution can be constructed from a set of numbers.

3   To show how a histogram, frequency polygon, and stem-and-leaf diagram can be used to represent a body of data.

What Is Statistics?

1.1 Descriptive and Inferential Statistics

According to one of America's leading playrights, the late George S. Kaufman, "The trouble with incest is that it gets you involved with relatives." According to some students, the trouble with statistics is that it gets you involved with figures. However, whether or not you feel a strong affinity for numbers, there is no reason why you cannot understand and make good use of statistical methods. Moreover, regardless of whether you are a sociologist or an engineer, a nurse or a psychologist, a teacher or a political scientist, it is essential that you have at least a basic knowledge of statistics.

The field of statistics has two aspects: descriptive statistics and inferential statistics. *Descriptive statistics* is concerned with summarizing and describing a given set of data. For example, every ten years, the Bureau of the Census gathers basic data concerning the number, age, and sex distribution (as well as occupational and educational composition) of the American people. Since the amount of raw data gathered is immense, the Census Bureau must condense and interpret this information to make it useful. If not properly handled, this process may distort the results. It is very important that it be done properly, or the entire census may result in misleading and incorrect conclusions.

Inferential statistics consists of a host of techniques that help people to arrive at rational decisions. Generally, the data available to a psychologist, engineer, government official, business executive, or physician are incomplete, in the sense that they pertain to only part of the population in which he or she is interested. Thus, social scientists continually have to test their theories on the basis of relatively small samples (since it would be too expensive to use larger samples). And government officials often rely on public opinion polls which include only a small fraction of the population. A central function of inferential statistics is to specify ways in which decisions based on incomplete information of this sort can be made as effectively as possible.

It is important to note at the outset that statistics is concerned with *the ways in which* data should be gathered (and *whether* data should be gathered at all), as well as with *how* a particular set of data should be analyzed once it has been collected. *What one can legitimately conclude from a particular set of data depends on how the data have been collected.* If a sociologist were to collect data concerning the tippling habits of the American people by sampling the inhabitants of nursing homes run by the Temperance Union, you would pay no attention to the results— and rightly so. But as we shall see, far less apparent—and thus more dangerous—mistakes can be made in data gathering with unfortunate and costly results both to those who collect and analyze the data and those who use them.

1.2 Sample Surveys

Population vs. Sample

The previous section mentioned two terms, *population* and *sample,* which must now be defined. *The **population** consists of the total collection of observations or measurements that are of interest to the statistician or decision maker.* It is difficult to give a more specific definition at this point, but in subsequent sections we shall provide many concrete examples of populations. *A **sample** is a subset of measurements taken from the population in which the statistician is interested.* As pointed out in the previous section, social scientists, hospitals, firms, government agencies, and other organizations continually draw samples to obtain needed information because it would be too expensive and time-consuming to try to obtain complete data concerning all relevant units. Thus, the federal government's unemployment statistics are based on a sample of persons rather than on a complete count of all workers without jobs and looking for work. Television ratings, which have so strong an influence over which programs survive and which are canceled, are based on the TV-viewing decisions of a sample of families. It is easy to find examples of the use of sampling in all areas of social and government activity.

Consider the case of the Gallup Poll, which has used sampling for many years to help predict election results.[1] To predict whether Walter Mondale or Ronald Reagan would win the presidency in1984, the Gallup Poll chose a sample of about 3,000 eligible voters. To choose this sample of voters, Gallup began by selecting a sample of geographical areas in the United States. Then a sample of precincts within each of the selected areas was chosen. Finally, a sample of households in each selected precinct was selected, and some members of each of these households were interviewed.

The interviewers asked each person a number of questions, some of which were designed to indicate whether the person was likely to vote. In this way, Gallup hoped to screen out those people who were unlikely to vote. To determine each person's preferences, the interviewer asked each one to fill out a secret ballot which contained the names of each of the candidates. The interviewer instructed the person to indicate on the ballot the candidate he or she would vote for if the election were held at the time of the interview.

How accurate has this sampling procedure been? Table 1.1 compares the Gallup Poll's prediction with the actual outcome of each presidential election since 1952. As you can see, the predicted percentage of the vote going to the winning candidate was generally in error by less than 2 percentage points. (Indeed, in almost half of the elections, the predicted percentage was in error by less than 1 percentage point.) This does not mean that all sampling procedures are as accurate as the Gallup Poll or that it is always better to sample than to carry out a complete count. What this example does illustrate is that, *if correct sta-*

Year	Winner	Percentage of vote Predicted	Actual	Error (percentage points)
1952	Eisenhower	51	55.4	+4.4
1956	Eisenhower	59.5	57.8	−1.7
1960	Kennedy	51	50.1	−0.9
1964	Johnson	64	61.3	−2.7
1968	Nixon	43	43.5	+0.5
1972	Nixon	62	61.8	−0.2
1976	Carter	49.5	51.1	+1.6
1980	Reagan	47	50.8	+3.8
1984	Reagan	59	59	0.0

Table 1.1
Results of Gallup Poll, Presidential Elections, 1952–84

Source: The Gallup Organization.

[1] George Gallup, "Opinion Polling in a Democracy," in J. Tanur et al. (eds.), *Statistics: A Guide to the Unknown* (San Francisco: Holden-Day, 1972).

tistical principles are known and observed, it is generally possible to design samples that will be sufficiently accurate for the purposes at hand. In some cases, pinpoint accuracy is required; in other cases, rough estimates will do. As we shall see in subsequent chapters, the nature of any given sampling procedure should depend on the accuracy required, as well as on the type of population from which the sample is drawn.

KINDS OF POPULATIONS AND SAMPLES

Let's return now to the concept of the population. What was the population in the case of the Gallup poll? It was the set of observations indicating which candidate each eligible voter in the United States supported. Since there are millions of eligible voters in the United States, this population contained millions of observations. Many populations do not contain nearly so many observations. For example, suppose that the U.S. Department of Labor wants to estimate the percentage of unemployed individuals in the labor force 16 years old and over in a particular small town. In this case, what is the population? It is a set of observations indicating whether each person in the labor force 16 years old or over does or does not have a job. To simplify, suppose that there are in this town only 25 individuals in the labor force of age 16 or over, and that the employment status of each one is as shown in Table 1.2. (That is, the first person's employment status is given by the first word in the first column, the second person's employment status is given by the second word in the first column, and so on, until the last person's employment status is given by the last word in the fifth column.) Then the observations in Table 1.2 would constitute the population.

Table 1.2
Hypothetical Population
of Whether or Not
Employed, 25 People

Unemployed	Employed	Employed	Employed	Employed
Employed	Employed	Unemployed	Employed	Employed
Employed	Employed	Employed	Employed	Employed
Employed	Employed	Employed	Employed	Employed
Employed	Employed	Employed	Employed	Employed

Using a quite different example, suppose that a psychologist is interested in the characteristics of famous scientists. She administers a Wechsler IQ test to all of the (living) Nobel Prize winners in a particular field. (According to Lubin, Larsen, and Matarazzo[2], the Wechsler Adult Intelligence Scale is the most frequently used psychological test.)

[2] B. Lubin, R. Larsen, and J. Matarazzo, "Pattern of Psychological Test Usage in the United States: 1935–82," *American Psychologist,* April 1984.

In this case, the population consists of the set of scores achieved by the Nobel Prize winners. To simplify, suppose that there were only 20 such Nobel Prize winners and that the score of each one is as shown in Table 1.3; then the population consists of the 20 numbers in Table 1.3.

151	161	163	157	160
143	148	158	142	151
139	137	152	141	142
132	129	147	138	144

Table 1.3
Hypothetical Population of Scores of 20 Nobel Prize Winners on Wechsler IQ Test

When carrying out sample surveys, *a listing of all the elements or units in the population is often called a **frame.*** In the case of the Gallup Poll, the frame is a list of all eligible voters. In the case of the unemployment rate, the frame is a listing of all persons in the labor force 16 years old or more. *A **census** is a survey that attempts to include all the elements or units in the frame.* For example, the Census Bureau's decennial Census of Population attempts to include each person in the United States. If the frame is very large, it is generally impossible to include all the elements or units; nevertheless, surveys that do so with reasonable success are called censuses.

Populations are of various types. Some consist of *quantitative* information, whereas others consist of *qualitative* information. In the Nobel Prize winners example, the population consists of quantitative measurements; that is, ones that can be expressed numerically. Each observation or measurement in this population is the score that a particular Nobel Prize winner achieved on the Wechsler IQ test. Each such score is a number. In the unemployment case, the population consists of qualitative (nonnumerical) information. Each observation or measurement in this population is a statement of whether or not the person in question has a job. Other qualitative characteristics of a person include his or her sex and education.

Quantitative vs. Qualitative Data

Populations also can be *finite* or *infinite*. If a population contains a finite number of members, it is called a *finite population*. Some finite populations, such as the hypothetical ones in Tables 1.2 and 1.3, contain relatively few members; others have a very large number of members. If a population contains an unlimited number of members, it is called an *infinite population*. Typically, infinite populations are conceptual constructs. In flipping a coin again and again, the sequence of heads and tails is an infinite population if the flipping is conceived of as continuing indefinitely. In some cases, as we shall see in subsequent chapters, it is convenient to treat a finite population as if it were infinite if the size of the sample is a small proportion of the size of the population.

As indicated above, a sample is a subset of measurements taken from the population.[3] For example, suppose that we choose the individuals in the third column of Table 1.2 as our sample; then the sample consists of "employed," "unemployed," "employed," "employed," and "employed." Or suppose that we choose as a sample the Nobel Prize winners in the first column in Table 1.3; then the sample consists of 151, 143, 139, and 132. A sample can be drawn in a variety of ways. Among those discussed in subsequent chapters are simple random sampling, stratified random sampling, systematic sampling, and cluster sampling. In order to use sample data intelligently, it is essential that you have some familiarity with each of these methods.

1.3 Experimental Design

Unlike the case of the Gallup Poll, many statistical investigations are not aimed simply at estimating an unknown percentage or an unknown total. Instead, their purpose is to estimate the effect of one or more factors on some dependent variable. For example, a psychologist may be interested in the relationship between the number of categories into which unrelated words are grouped and the number of words that subjects can recall.[4] Or a political scientist may be interested in estimating the effect of the closeness of past elections on the proportion of citizens registered to vote.[5]

Frequently, scientists, engineers, and policy makers carry out experiments to help shed light on such questions. For example, Imperial Chemical Industries (ICI), the huge British chemical firm, carried out the following experiment to estimate the effect of a chlorinating agent on the abrasion resistance of a certain type of rubber.[6] Each of 10 pieces of this rubber was cut in half, and one-half of each piece was treated with the chlorinating agent, while the other half was untreated. Then the abrasion resistance of each half-piece was evaluated on a machine, and the difference between the abrasion resistance of the treated half-piece and the untreated half-piece was computed. Table 1.4 shows the 10 differences (one corresponding to each of the pieces in the sample).

---

[3] Even if a subset of measurements is taken from a population in which the statistician or decision maker is *not* interested, it is a sample; but it is a sample from the wrong population.

[4] H. Gleitman, *Psychology* (New York: W. W. Norton, 1981), p. 289.

[5] See E. Tufte, "Registration and Voting," in J. Tanur et al. (eds.) *Statistics: A Guide to the Unknown.* More will be said on this score in Chapter 12.

[6] See O. Davies, *The Design and Analysis of Industrial Experiments* (London: Oliver and Boyd, 1956), p. 13.

These 10 differences can be viewed as a sample of 10 from the infinite population of differences which would result if pieces of rubber were subjected to this test indefinitely. Viewed in this way, the sample enabled ICI to estimate the average effect of the chlorinating agent on the abrasion resistance of the rubber. Also, it enabled ICI to *test certain hypotheses* about this effect. For example, some of ICI's personnel were interested in testing the hypothesis that on the average, this chlorinating agent had no effect on the abrasion resistance of the rubber. As we shall see in subsequent chapters, statistical methods can be used to test this hypothesis, based on the sample in Table 1.4.

Piece	Difference*
1	2.6
2	3.1
3	−0.2
4	1.7
5	0.6
6	1.2
7	2.2
8	1.1
9	−0.2
10	0.6

Table 1.4
Difference in Abrasion Resistance (Treated Part Minus Untreated Part), 10 Pieces of Rubber

Source: O. Davies, *The Design and Analysis of Industrial Experiments,* p. 13.
* The units in which these differences are expressed are given in the source and need not concern us here.

Just as it is not easy to design a sample survey effectively, so it is not easy to create the proper design for an experiment. The time to worry about how an experiment should be designed is before, not after, the experiment is carried out. Too often, an investigator finds that he or she cannot draw useful conclusions from an experiment because the experiment was improperly designed. *Before carrying out an experiment (or a sample survey) it is essential that the objectives of the experiment (or sample survey) be defined precisely.* Without such a statement of objectives, it is impossible to formulate a design that will obtain the desired information at reasonable cost. With such a statement of objectives, the statistician can provide useful and time-tested guidance for conducting the experiment and obtaining the desired information at minimum or close-to-minimum cost.

1.4 The Importance of Probability in Statistics

As indicated above, a central feature of inferential statistics is its use of information concerning a sample to make inferences about the nature of the population from which the sample is drawn. For example, the Gallup Poll used sample data to make an inference concerning the proportion of all voters who would vote for a particular presidential candidate. And Imperial Chemical Industries used sample data to make an inference concerning the true effect of a chlorinating agent on the abrasion resistance of rubber. Any conclusion based on a sample must be subject to a certain amount of uncertainty. Thus, even if ICI finds that the chlorinating agent increases abrasion resistance for 150 pieces of rubber, there is always a chance that these results are somehow a fluke and that, if the sample size were increased to 1,000 or to 10,000, the results would be reversed.

How much certainty can you attach to a particular statement about a population that is based on the results of a sample? Intuitively, you are likely to feel that the answer depends on the size of the sample. The bigger the sample, the more confidence you are likely to have in the sample results; the smaller the sample, the less confidence you are likely to have in the sample results. (Subsequent chapters will show that the sample size is indeed one determinant of how much confidence you can put in a sample result, but it is by no means the only determinant.) Why do you feel that increases in sample size increase the amount of confidence you should have in the sample results? Because, as the sample size increases, the ***probability*** that the sample result departs greatly from the corresponding result for the entire population becomes smaller and smaller. Thus, as the Gallup Poll takes a bigger and bigger sample of voters, it seems logical that we can place more and more confidence in the sample estimate of the proportion of votes that each candidate will receive.

Probability

But what exactly do we mean by a probability? And how can we measure a particular probability? These questions are important to the statistician and to the user of statistics alike. To go beyond vague, intuitive notions about the degree of accuracy of a particular sample result, we must draw on probability theory, a branch of mathematics distinct from, but closely related to, statistics. (To delve deeply into probability theory, one needs a considerable mathematical background. However, no mathematics beyond elementary high-school algebra is needed to understand the elements of probability theory required for an introductory course in statistics.) Until we present a more adequate definition of a probability in Chapter 3, we shall treat probability intuitively. That is, if a particular event has a 50–50 chance of occurring, we shall say that the probability of its occurring is 0.5. Similarly, if a particular event has one chance in four of occurring, we shall say that the probability of its occurring is 0.25. Or if a particular event has one chance in ten of occurring, we shall say that the probability of its occurring is 0.1.

1.5 Bias and Error

Because the field of statistics attempts to make inferences from a sample concerning a population, statistics must necessarily be concerned with **error.** Any result based on a sample is likely to depart in some measure from the corresponding result for the population. For example, in Table 1.1 we saw that the Gallup Poll's predicted percentage of the vote going to the winning candidate has often been in error by one percentage point or more. When the Gallup Poll takes a sample in the next election, it too will probably be in error, perhaps by more than one percentage point, perhaps by less. Similarly, the results of ICI's sample (shown in Table 1.4) are almost certainly in error to some extent. That is, if ICI were to test 10,000 pieces of rubber rather than 10 pieces, the results would probably differ from those in Table 1.4. To repeat, all results based on a sample are likely to be in error, since the sample does not contain information concerning all items in the population.

Error

  The error in a particular sample result consists of two parts: ***experimental or sampling error*** (sometimes called ***random error***) and ***bias.*** Experimental or sampling errors occur because of a large number of uncontrolled factors, which we can subsume under the term *chance.* For example, if the Gallup Poll draws 10 samples of voters at random[7] (each sample containing 3,000 voters and each being selected at the same time), the results will differ from one sample to another. Why? Because of the luck of the draw. Or in the case of ICI's experiment, the average difference in abrasion resistance between treated and untreated half-pieces will differ from one sample of 10 pieces of rubber to another sample of 10 pieces of rubber. Why? Because the instruments that measure abrasion resistance contain small errors, because human beings make mistakes in reading these instruments, because the pieces of rubber sometimes are not homogeneous, and so on. Errors of this kind are experimental or sampling errors.

  The essential characteristic of experimental or sampling errors is that

---

[7] The concept of drawing a sample at random is more sophisticated and technical than can be appreciated or explained adequately at this point. For present purposes, it is sufficient to regard *at random* as meaning that each unit in the population has the same chance of being drawn in the sample. Much more will be said about this in later chapters.

Bias

they can reasonably be expected to cancel each other out over a period of time or over a large number of experiments or samples. Many uncontrolled factors operate to cause the results of a sample to be in error, sometimes on the high side and sometimes on the low side. However, if the sample or the experiment is repeated a large number of times, the errors tend to offset each other when the results are averaged. In contrast, *bias consists of a systematic and persistent type of error which will not tend to cancel out.* For example, the Gallup Poll, instead of choosing voters at random, might select its samples in such a way that Republicans are more likely to be included than Democrats. If this were the case, there would be a systematic and persistent tendency for the sample estimates to overstate the percentage of votes received by the Republican candidate, and this error would not tend to cancel out if the sample were repeated many times and the results were averaged.

To summarize, an important reason for distinguishing between experimental or sampling error and bias is that *increases in a sample's size tend to reduce experimental or sampling errors, while such increases do not reduce bias.* Thus, although (as pointed out previously) the accuracy of the results of a sample tends to increase with the sample's size, not all errors can be eliminated by increasing sample size. If there are serious biases, the results may be considerably in error even if the sample is huge.

TWO IMPORTANT CAUSES OF BIAS

Bias can be particularly dangerous when, like termites or dry rot, its presence is undetected. Two types of methodological errors are commonly causes of serious bias. The first arises when *a sample is taken from a population that differs in an important way from the population that the statistician or decision maker is really interested in.* For example, suppose that you want to estimate the proportion of people in Chicago who are college graduates. To do so, you go to three Chicago suburbs and pick a sample of their residents. Such a sample is likely to result in substantial bias because you are sampling from the wrong population. Rather than sampling from the population of all Chicago residents, you are sampling from the population of Chicago suburban residents. And since the educational level is likely to be higher in the suburbs than in the inner city, the result is likely to be biased.[8]

---

[8] Of course, one should not be overly dogmatic about this. Sometimes it is not possible to obtain an adequate frame for the population one is interested in, and the only possibility is to do the best one can with a frame which is only approximately what one would like to use. If this frame is close enough to what is really needed, little harm will result. Sometimes, however, it is not close enough, even though the analyst believes it to be.

A second methodological mistake which can be the source of serious bias arises when *the effect of the variable one wants to measure is mixed up inextricably with the effect of some other factor.* In this case, if one finds that the variable in question seems to have a certain effect, this estimated effect may be biased because it also reflects the effect of another factor. Table 1.5 presents part of the results of a nationwide test of the effectiveness of the Salk antipolio vaccine.[9] These data seem perfectly adequate to measure the effects of the vaccine on the incidence of polio in children, but in fact they contain a major bias because only those second-graders who received their parents' permission were given the vaccine, whereas all first- and third-graders were used to indicate what the incidence of polio would be without the vaccine. The bias arises because the incidence of polio was substantially lower among nonvaccinated children who did not receive permission than among those who did. Why? Because children of higher-income parents were more likely than children of lower-income parents to receive permission—and they were also more likely to contract polio. (Surprising as it may seem, children who grow up in less hygienic conditions are less likely to contract this disease!)

School grade	Treatment	Number of children	Number afflicted with polio	Number afflicted per 100,000 children
Second	Salk vaccine	222,000	38	17
First and third	No vaccine	725,000	330	46

Table 1.5
Incidence of Polio in
Two Groups of
Children

Source: W. S. Youden, "Chance, Uncertainty, and Truth in Science," *Journal of Quality Technology,* 1972.

Table 1.6 shows the results of a more adequate experimental design in which this bias is eliminated. In this sample, children who received permission are differentiated from those who did not. Those with permission were assigned at random to either the group receiving the vaccine or the group receiving the placebo (something similar in appearance to the vaccine, but of no medical significance). As you can see, the results indicate that the reduction in the number afflicted with polio (per 100,000 children) due to the vaccine is much larger than in Table 1.5. Biases of the sort contained in Table 1.5 and in the Chicago educational survey can result in distorted information which can lead

[9] See W. S. Youden, "Chance, Uncertainty, and Truth in Science," *Journal of Quality Technology,* 1972.

Table 1.6
Modified Experiment
with Salk Vaccine

Permission given	Treatment	Number of children	Number afflicted with polio	Number afflicted per 100,000 children
Yes	Salk vaccine	201,000	33	16
Yes	Placebo	201,000	115	57
No	None	339,000	121	36

Source: See Table 1.5.

decision makers to make costly and embarrassing mistakes. It is worth taking some trouble to avoid the methodological errors that result in these biases—unless, of course, you don't mind making serious mistakes.

EXERCISES

1.1 A Boston psychologist, interested in the characteristics of children of drug-addicted mothers, wants to determine the percentage of such children in Boston that have a particular form of learning disability.
(a) What is the population?
(b) How can a frame be obtained?
(c) Is the population finite or infinite?
(d) Does the population consist of qualitative or quantitative information?

1.2 To obtain an estimate of how well his students understand the material, an instructor asks each of the students in the first row of his class a question. Six out of the seven students in the first row come up with the right answer. Do you think that such a sample contains a bias? If so, what is it?

1.3 The public school teachers in Philadelphia went on strike in the fall of 1981. Suppose that a newspaper asked each of the teachers on picket lines whether or not the strike would be settled on terms unfavorable to the union.
(a) Would the results be a sample of the opinions of all public school teachers in Philadelphia?
(b) Would the population be finite or infinite?
(c) Would the population contain quantitative or qualitative information?
(d) What possible biases are likely to exist in the results?

1.4 The mayor of a seaside vacation town must decide whether or not to support an effort to close down one of the town's elementary schools. To determine the extent to which public opinion favors such an action, she hires several interviewers to ask individuals selected at random on the town's boardwalk from 10 A.M. until noon each day during the second week in August whether they feel that an elementary school with less than 200 students should remain in operation.

(a) What is the population from which this sample is drawn?

(b) Is this population finite or infinite?

(c) Does this population consist of qualitative or quantitative measurements?

(d) What deficiencies can you see in this sample design? What possible biases are likely to exist in the results?

(e) What improvements in the sample design would you suggest?

1.5 Another seaside vacation resort is the scene of considerable controversy over whether or not bars should be allowed to stay open after midnight. The local newspaper, which favors the existing arrangements whereby bars must close at midnight, points out that when a neighboring community allowed bars to stay open after midnight the crime rate increased.

(a) What are the weaknesses in the newspaper's argument?

(b) Do you think that an experiment could be run to resolve this type of controversy? If so, what sort of an experiment should it be?

(c) Do you think that an analysis of data concerning past changes in crime rates in this and other communities might help resolve this controversy? If so, what kind of data should be examined, and how might they be analyzed?

1.6 A major private university mails a questionnaire to a sample of 10 percent of its alumni. This questionnaire asks whether the recipient of the questionnaire is for or against the university's policies with respect to curriculum, faculty hiring, coed dormitories, and a variety of other topics. About 30 percent of the questionnaires are filled out and returned. Do you think that there is likely to be bias as well as sampling error in the results? Explain.

1.6 The Frequency Distribution

At the start of this chapter, we pointed out that the field of statistics can be divided into two parts: descriptive statistics and inferential statistics. One of the central tools of descriptive statistics is the frequency distribution. In this section, we describe the nature of a frequency distribution, how it can be represented graphically, and some guidelines for constructing frequency distributions.

CLASS INTERVALS

Although one seldom is provided with the entire population in which one is interested, let's start by assuming that you undertake a study for which you do have the entire population. Specifically, you are interested in determining the voter turnout rate (that is, the percent of voting age population that voted) in the 51 states (including the District of Columbia) in 1980. The numbers in Table 1.7 constitute the popula-

Table 1.7
Percent of Voting Age
Population That Voted
in 1980 Elections, by
State (Including
District of Columbia)

State	Percent voting	State	Percent voting
Maine	64.6	North Carolina	43.5
New Hampshire	57.2	South Carolina	40.6
Vermont	57.8	Georgia	41.3
Massachusetts	59.0	Florida	48.6
Rhode Island	58.6	Kentucky	50.0
Connecticut	61.0	Tennessee	48.8
New York	47.9	Alabama	48.8
New Jersey	54.9	Mississippi	52.0
Pennsylvania	51.9	Arkansas	51.4
Ohio	55.4	Louisiana	53.1
Indiana	57.7	Oklahoma	52.3
Illinois	57.8	Texas	44.9
Michigan	59.9	Montana	65.1
Wisconsin	67.2	Idaho	68.0
Minnesota	70.1	Wyoming	53.2
Iowa	63.0	Colorado	56.0
Missouri	58.7	New Mexico	51.0
North Dakota	64.8	Arizona	44.7
South Dakota	67.6	Utah	64.6
Nebraska	56.8	Nevada	41.2
Kansas	56.8	Washington	57.5
Delaware	54.6	Oregon	61.5
Maryland	50.0	California	49.0
District of Columbia	35.5	Alaska	57.8
Virginia	47.6	Hawaii	43.6
West Virginia	52.9		

Source: Statistical Abstract of the United States.

tion. Given these numbers, how can we summarize and describe them? (That is, how can we put them into a more compact and comprehensible form?) The need to condense and summarize the data is not as great in this example as in many other cases, since there are only 51 numbers, in contrast to the hundreds or thousands of numbers in some other populations. However, this example can illustrate the essential principles on which we shall focus.

As a first step toward summarizing these data, it is convenient to establish certain *class intervals,* which we define as *classes or ranges of values that the observations or measurements can assume.* For example, Table 1.8 shows eight class intervals for these voter turnout rates, the first being 35.0–40.0 percent, the second 40.0–45.0 percent, and so on. In other words all values greater than or equal to 35.0 percent and

Turnout rate (percent)	Number of states
35.0 and under 40.0	1
40.0 and under 45.0	7
45.0 and under 50.0	6
50.0 and under 55.0	12
55.0 and under 60.0	14
60.0 and under 65.0	6
65.0 and under 70.0	4
70.0 and under 75.0	1
Total	51

Table 1.8
Frequency Distribution of 1980 Voter Turnout Rates, by State (Including District of Columbia)

less than 40.0 percent are included in the first class interval, and all values greater than or equal to 40.0 percent and less than 45.0 percent are included in the second class interval, and so on. Each class interval must have a *lower limit* and an *upper limit.* An observation or measurement falls in a particular class interval if it is greater than or equal to this class interval's lower limit and less than its upper limit. The lower limit of the first class interval is 35.0, the lower limit of the second class interval is 40.0, and so on. The upper limit of the first class interval is 40.0, the upper limit of the second class interval is 45.0, and so on. The *width* of a class interval is equal to its upper limit minus its lower limit.

Given these class intervals, the next step is to determine how many observations or measurements in the population fall into each class interval. Using the data in Table 1.7, we find that one state falls into the first class interval in Table 1.8, seven fall into the second class interval, six fall into the third class interval, and so on. The resulting data, shown in Table 1.8, are called a *frequency distribution. A **frequency distribution** shows the number of observations or measurements that are included in each of the class intervals.* Thus, a frequency distribution, if properly formulated, can summarize a set of numbers very effectively. For example, the frequency distribution in Table 1.8 makes it much easier to see how high the voter turnout rates were in 1980. Clearly, only a small number of states (five) had voter turnout rates greater than or equal to 65.0 percent, and only a small number (eight) had voter turnout rates of less than 45.0 percent. The bulk of the states had voter turnout rates between 45.0 and 65.0 percent. This frequency distribution shows the salient features of the raw data in Table 1.7.

If the population consists of qualitative measurements, the classes in the frequency distribution are not ranges of numerical values, but possible qualitative observations. For example, if the population con-

Frequency Distribution

Table 1.9
Frequency Distribution
of Employment Status
of People in the
Civilian Labor Force,
United States,
December 1983

Employment status	Number of persons
Employed	102,941,000
Unemployed	9,195,000
Total	112,136,000

Source: Economic Report of the President, February 1984.

sists of observations concerning whether various people in a certain area are employed or unemployed, the frequency distribution shows the number of persons in the relevant area who are employed and the number unemployed. Table 1.9 constitutes such a frequency distribution for the United States in December 1983. Often, frequency distributions of this sort are presented graphically in the form of bar charts like that in Figure 1.1. Each bar in Figure 1.1 has a length representing the number of people employed or unemployed.

GRAPHICAL REPRESENTATIONS OF FREQUENCY
DISTRIBUTIONS

When a population consists of quantitative measurements (such as the voter turnout rates of the 51 states), a useful way to represent the population's frequency distribution is to construct a *histogram.* The

Figure 1.1
Bar Chart Showing
Frequency Distribution
of Employment Status
of People in the
Civilian Labor Force,
December 1983

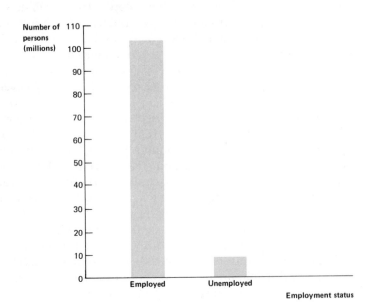

histogram has a horizontal axis and a vertical axis, as shown in Figure 1.2. The horizontal axis is scaled to display all values of the measurements in the population (in this case the voter turnout rates of the 51 states). The horizontal axis is divided into segments corresponding to the class intervals, and a vertical bar is erected on each of these segments. In Figure 1.2 the segment from 35.0 percent to 40.0 percent on the horizontal axis is the first class interval, and on this segment a vertical bar is erected. Similarly, a vertical bar is erected on the segment of the horizontal axis corresponding to each of the other class intervals.

Histogram

A histogram's vertical axis shows the number of observations (measurements) in the population which fall into any particular class interval. The key point to understand about a histogram is that *the area of each vertical bar is proportional to the number of observations in the relevant class interval.* Thus, if the width of each class interval is the same, the height of the vertical bar for each class interval equals the number of observations (measurements) in the population that fall within this class interval. For example, in Figure 1.2 the height of the vertical bar for the first class interval (from 35.0 percent to 40.0 percent on the horizontal axis) is 1 unit because the voter turnout of 1 out of 51 states falls into this class interval. Clearly, the histogram in Figure 1.2 is a simple and effective way to present the frequency distribution in Table 1.8 in graphical form. (More will be said about histograms in Chapter 2, where we take up the case where the class intervals have unequal widths.)

An alternative way of representing a frequency distribution graphically is to construct a *frequency polygon.* Like the histogram, the frequency polygon has a horizontal axis showing possible values of the measurements in the population, and a vertical axis showing the number of observations or measurements for each particular class in-

Frequency Polygon

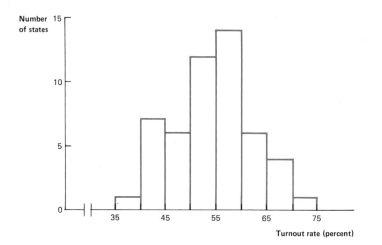

Figure 1.2
Histogram of Population of 1980 Voter Turnout Rates of 51 States (Including District of Columbia)

ACCESS TO COMPUTERS IN THE PUBLIC SCHOOLS

Microcomputers are becoming increasingly evident in the nation's public schools. They are used for drill practice and testing, for games and word processing, and for classes in computer literacy and computer science. One issue that is being raised is whether the educational system is providing equitable computer opportunities to the poor, women, minorities, and disabled persons. In 1983, one study found that the 12,000 wealthiest schools are four times as likely to have microcomputers as are the 12,000 poorest schools. There is concern about unequal access to computer learning as a consequence of a student's social and economic status.

Basic statistical methods of the sort discussed in this chapter have been used to throw light on this topic. For example, Ronald Anderson, Wayne Welch, and Linda Harris,[10] reporting on the 1982 National Assessment in Science, provided the following bar chart to show how the percentage of students (of age 13) that use computer terminals varies by type of school—rural, ghetto, urban/rich, and others. (Unlike the bar chart in Figure 1.1, there are four bars, not two, because there are four types of schools. Also, this bar chart shows the *percentage* of students, whereas Figure 1.1 shows the *number* of persons. A bar chart can be used to represent data in either percentage or ordinary numerical form.) Clearly, the percentage is lower in the rural and ghetto schools, and the chart is an effective way to portray this fact.

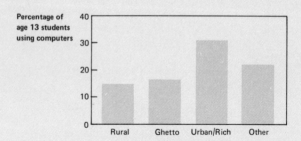

[10] R. Anderson, W. Welch, and L. Harris, "The Inequities in Opportunities for Computer Literacy," *The Computing Teacher,* April 1984.

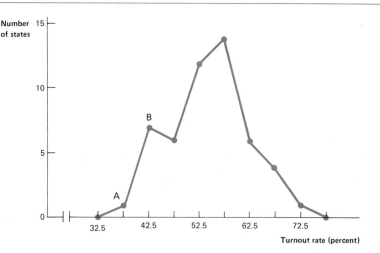

Figure 1.3
Frequency Polygon of
Population of 1980
Voter Turnout Rates of
51 States (Including
District of Columbia)

terval. To construct a frequency polygon, we first find the *class mark* for each class interval. (The class mark is the point in a particular class interval midway between the interval's upper and lower limit.) We then plot the number of observations or measurements within a particular class interval against the class mark for that interval. In Figure 1.3, if we plot the number of state voter turnout rates in the first class interval against the class mark for this interval (37.5 percent), we get point *A*. If we plot the number of state voter turnout rates in the second class interval against the class mark for this interval (42.5 percent), we get point *B*. When we connect all resulting points, the resulting geometric figure is a frequency polygon. (Since there are zero firms in the 30.0–35.0 percent class interval and in the 75.0–80.0 percent class interval, the points for these class intervals lie on the horizontal axis.) Like the histogram, the frequency polygon is useful for portraying a frequency distribution graphically.

1.7 How to Construct A Frequency Distribution

The point of constructing a frequency distribution is to condense the mass of data contained in an array of population values into a table that is more readily appraised and comprehended. To do this, the observations in the population are grouped into class intervals, as we have seen. There are no simple, cut-and-dried rules to insure that one will construct a proper frequency distribution. Moreover, there is no single frequency distribution which is the best representation of a given set of data; ordinarily, a number of different frequency distributions will do well enough. But in constructing a frequency distribution, one would be wise to consider the following guidelines.

DEFINITION OF CLASS INTERVALS. In defining class intervals, it is essential that each measurement in the population fall into some class interval. Thus, it would be wrong to define class intervals in Table 1.8 so that voter turnout rates above 70 percent fell into no class interval. Such a definition of class intervals would result in a frequency distribution which distorts, at least to some extent, the nature of the population. Equally important, none of the class intervals should overlap. Thus, in Table 1.8 it would be wrong to define one class interval as 35.0–40.0 percent and another as 37.0–42.0 percent. If class intervals were defined in this way, it would be possible for a given measurement in the population to fall into more than one class interval.

WIDTH OF CLASS INTERVALS. In general, statisticians prefer to construct class intervals of equal widths because this makes it easier to compare the number of observations in different class intervals and to carry out a variety of computations which are described in the next chapter. However, it is sometimes not practical to use class intervals of equal widths. For example, in constructing a frequency distribution of the annual income of American families, we might want the width of class intervals to be $4,000 up to an income level of $40,000, and to be $10,000 from an income level of $40,000 to $80,000. Families with incomes of $80,000 or more might be included in a single class interval with no finite upper limit. (Such a class interval is called *open-ended.*) In this case, the reason for preferring class intervals of unequal widths is that the bulk of the nation's families have incomes of under $40,000, and it is important in this income range to have the width of class intervals small enough so that the frequency distribution portrays the distribution of income with a reasonable degree of accuracy. However, it would be inappropriate to use the same width of class interval for the very affluent, since so many class intervals would result that the frequency distribution would be very unwieldy. For example, if $4,000 were used as the width of each class interval for incomes ranging from zero to $1 million, there would be 250 intervals!

NUMBER OF CLASS INTERVALS. Although there are no hard-and-fast rules for how many class intervals to use, it is generally recommended that there be no fewer than 5 and no more than 20. If data are lumped into too few class intervals, there is a major loss of information because much of the variation cannot be determined from the frequency distribution. However, if too many class intervals are used, the frequency distribution does little more than reproduce the array of data. (If there are enough class intervals, each interval may contain only 1 or zero observations.) What is needed is something in-between, and the answer is inevitably a matter of judgment.

RELATIONSHIP BETWEEN NUMBER AND WIDTH OF CLASS INTERVALS. When deciding on the number and width of class intervals to use, it is important to realize that interval width is related to the number of class intervals one picks. If all class intervals are to have equal widths, the following formula provides a reasonable estimate of how wide they should be:

$$\text{width of class interval} = \frac{\text{largest value} - \text{smallest value}}{\text{number of class intervals}}$$

where "largest value" is the largest value of an observation in the population, and "smallest value" is the smallest value of an observation in the population. What this formula does is indicate how wide each class interval must be in order to cover the distance between the largest and smallest values, given the chosen number of class intervals. Thus, if we want nine class intervals of equal widths for the data in Table 1.7,

$$\text{width of class interval} = \frac{70.1 - 35.5}{9} = 3.84,$$

since the largest voter turnout rate is 70.1 percent and the smallest is 35.5 percent. According to this result, the width of the class interval should be about 3.84 percentage points.

POSITION OF CLASS INTERVALS. If possible, one should try to set up class intervals so that the midpoint or class mark of each interval is close to the average of the observations included in the class interval. For example, suppose that the data to be summarized are monetary amounts that tend to concentrate on multiples of $1 ($1; $2; $3; and so on). Then it would be better to use class intervals with multiples of $1 as class marks (class intervals of $0.50 and under $1.50; $1.50 and under $2.50; and so on) than ones with class marks that are quite different from the actual concentration of observations in the class intervals. The reason is that in calculating averages and other summary measures from frequency distributions, we assume that the class mark is representative of the values in each class interval. (More will be said about this in the next chapter.)

WHAT
COMPUTERS CAN
DO

OBTAINING A HISTOGRAM

Modern computer technology has had an important impact on statistics, as well as on other fields. A variety of computer packages are available to carry out the calculations described in this book. Four of the most popular packages are: Minitab, SAS, SPSS, and BMDP. It is relatively simple to use these packages, and they take a great deal of the drudgery out of statistical calculation.[11]

To illustrate how these packages can be used, suppose that we want to construct a histogram showing the frequency distribution of states by crime rate (number of offenses known to the police per 1,000 population). The crime rate in each state (excluding the District of Columbia) is as follows:

Maine	4.2	North Dakota	3.0	Louisiana	5.3
New Hampshire	4.3	South Dakota	3.0	Oklahoma	4.8
Vermont	5.1	Nebraska	4.2	Texas	6.0
Massachusetts	5.8	Kansas	5.4	Montana	5.0
Rhode Island	5.9	Delaware	6.7	Idaho	4.5
Connecticut	5.8	Maryland	6.6	Wyoming	5.1
New York	6.9	Virginia	4.7	Colorado	7.4
New Jersey	6.2	W. Virginia	2.6	N. Mexico	6.2
Pennsylvania	3.7	N. Carolina	4.5	Arizona	7.6
Ohio	5.4	S. Carolina	5.3	Utah	5.8
Indiana	4.5	Georgia	5.6	Nevada	8.6
Illinois	5.0	Florida	8.0	Washington	6.7
Michigan	6.9	Kentucky	3.5	Oregon	7.0
Wisconsin	4.8	Tennessee	4.3	California	7.6
Minnesota	4.7	Alabama	4.9	Alaska	6.6
Iowa	4.7	Mississippi	3.5	Hawaii	6.5
Missouri	5.4	Arkansas	3.8		

To use Minitab to construct such a histogram, all that one has to do (after logging on to the computer) is to type the following:

[11] For a description of each of these packages, see J. Lefkowitz, *Introduction to Statistical Computer Packages*, Boston: Duxbury Press, 1985. For a manual instructing students how to use Minitab, see T. Ryan, B. Joiner, and B. Ryan, *Minitab Student Handbook*, Boston: Duxbury Press, 1976 and 1985.

```
-SET C1
4.2 4.3 5.1 5.8 5.9 5.8 6.9 6.2 3.7 5.4 4.5 5.0 6.9 4.8 4.7 4.7 5.4
3.0 3.0 4.2 5.4 6.7 6.6 4.7 2.6 4.5 5.3 5.6 8.0 3.5 4.3 4.9 3.5 3.8
5.3 4.8 6.0 5.0 4.5 5.1 7.4 6.2 7.6 5.8 8.6 6.7 7.0 7.6 6.6 6.5
-HISTOGRAM OF COLUMN C1
```

The first line says that the numbers should be put in column C1 of the worksheet that Minitab maintains in the computer. The next three lines contain the numbers themselves. The last line commands the computer to formulate a histogram.

After you type the above five lines, the computer automatically prints out the frequency distribution and the histogram, which are shown below:

```
        MIDDLE OF    NUMBER OF
        INTERVAL     OBSERVATIONS
          2.5          1    *
          3.0          2    **
          3.5·         3    ***
          4.0          3    ***
          4.5          8    ********
          5.0          7    *******
          5.5          6    ******
          6.0          7    *******
          6.5          5    *****
          7.0          3    ***
          7.5          3    ***
          8.0          1    *
          8.5          1    *
```

Note that the histogram is printed sideways; that is, the bars (consisting of asterisks) are horizontal, not vertical.

Our aim here is not to teach you how to use these packages, but to acquaint you with their general nature and usefulness. Specialized manuals are available to teach you the details. The important point for present purposes is that computer packages exist which enable you to do statistical calculations simply, rapidly, and cheaply.

THE CUMULATIVE FREQUENCY DISTRIBUTION

In some cases, statisticians and decision makers are interested in the number of measurements in the population that lie below or above a certain value. Thus, a political scientist might want to know how many states had voter turnout rates below 45.0 percent in 1980. In cases of this sort, it is useful to construct a cumulative frequency distribution, showing *the number of measurements in the population that are less than particular values.* Table 1.10 shows the cumulative frequency distribution for the state voter turnout rates. The difference between the *cumulative* frequency distribution in Table 1.10 and the *ordinary* frequency distribution in Table 1.8 is that the cumulative frequency distribution shows the number of states with voter turnout rates *less than* those given in Table 1.10, whereas the ordinary frequency distribution in Table 1.8 shows the number of states in *each* of the class intervals.

Table 1.10
Cumulative Frequency Distribution of Population of 1980 Voter Turnout Rates of 51 States (Including District of Columbia)

Turnout rate (percent)	Number of states
Less than 35.0	0
Less than 40.0	1
Less than 45.0	8
Less than 50.0	14
Less than 55.0	26
Less than 60.0	40
Less than 65.0	46
Less than 70.0	50
Less than 75.0	51

If one has the frequency distribution for a set of data, it is easy to construct a cumulative frequency distribution. To construct a cumulative frequency distribution for the state voter turnout rates, we proceed as follows. The number of states with voter turnout rates of less than 40.0 percent is 1 (the figure in the lowest class interval in Table 1.8); the number of states with voter turnout rates of less than 45.0 percent is 8 (the sum of the figures in the lowest two class intervals); the number of states with voter turnout rates of less than 50.0 percent is 14 (the sum of the figures in the lowest three class intervals); and so on. To obtain the number of states with voter turnout rates of less than the upper limit of a particular class interval, we add up (or *cumulate,* which

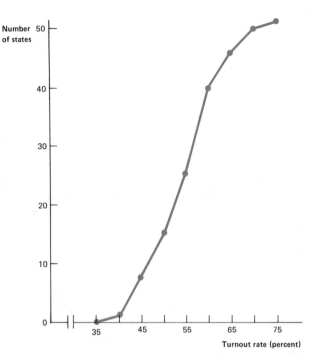

Figure 1.4
Ogive Showing
Cumulative Frequency
Distribution of
Population of 1980
Voter Turnout Rates of
51 States (Including
District of Columbia)

accounts for the name of the cumulative frequency distribution) the number of states in this and all lower class intervals.

Just as the ordinary frequency distribution can be portrayed graphically, so can the cumulative frequency distribution. To construct such a graph one needs only to plot the number of observations in the population with values less than a certain number against the number itself. For example, since there is one state with a voter turnout rate less than 40.0 percent, we would plot 1 on the vertical axis against 40.0 percent on the horizontal axis. Since there are eight states with voter turnout rates less than 45.0 percent, we would plot 8 on the vertical axis against 45.0 percent on the horizontal axis. The results of plotting and connecting all such points with straight lines are shown in Figure 1.4. The resulting curve is called an *ogive*. Such curves are encountered   *Ogive* frequently in the social sciences, business, engineering, medicine, and elsewhere.[12]

---

[12] In this section we have considered "less than" cumulative frequency distributions and ogives. It is also possible to construct "greater than" cumulative frequency distributions and ogives, which show the number of measurements that *exceed* particular values. To obtain the number of states with voter turnout rates *greater than* the lower limit of a particular class interval, we add up the number of states in this and all higher class intervals.

EDUCATIONAL
ACHIEVEMENT IN
JAPAN, TAIWAN,
AND THE UNITED
STATES

The future of any nation depends on the educational achievement of its children. In 1983, the National Commission on Excellence in Education reported that the scholastic performance of American children was low relative to their counterparts in other countries. Studies by psychologists, sociologists, and others associated with the Center for Advanced Study in the Behavioral Sciences also indicated that Japanese and Taiwanese children seemed to outperform American children. These reports and studies received an enormous amount of attention in the media, and were discussed by politicians, scientists, educators, and others.

Relatively simple statistical methods played an important role in these studies. To illustrate the sorts of techniques that were used, consider the work carried out at the Center for Advanced Study in the Behavioral Sciences.[13] Three cities were chosen: Minneapolis (in the U.S.), Sendai (in Japan), and Taipei (in Taiwan). All are large, prosperous cities. Within each city, 10 schools were chosen, and two first-grade and two fifth-grade classrooms were chosen at random from each school. Then 6 boys and 6 girls were chosen at random from each classroom. These 1,440 children (240 first graders and 240 fifth graders from each city) were studied intensively. Each child was given a mathematics test and a battery of cognitive tasks. The mother of each child was interviewed, and each child's teachers filled out a questionnaire.

A comparison of the mathematics scores of the American, Japanese, and Taiwanese children indicated that the Americans tended to perform poorly. "Among the 100 top first graders in mathematics, there were only 15 American children. And only one American child appeared among the top 100 fifth graders. In whatever way the data are summarized, the poor performance of American children is evident."[14]

According to the researchers, one important reason for this is that the Japanese and Taiwanese children spend appreciably more time in school than their American counterparts. While the American school year is only about 178 days, the school year in Taipei and Sendai is about 240 days. Furthermore, as shown by

[13] Harold Stevenson, "Making the Grade: School Achievement in Japan, Taiwan, and the United States," *1983 Annual Report,* Stanford: Center for Advanced Study in the Behavioral Sciences, 1983, pp. 41–51.
   [14] Ibid., p. 43.

the bar charts presented below, American children spend a much smaller proportion of their time in school on mathematics. Among first graders, for example, about 25 percent of classroom time was spent on mathematics in Japan, whereas less than 15 percent was spent on it in the United States.

Also, American children spend less of their time outside school on homework. "On weekdays, American first-grade children were estimated to spend an average of 14 minutes a day on homework, and in fifth grade, 46 minutes a day. In Japan, the corresponding averages were 37 and 57 minutes. In Taiwan, first graders were estimated to spend an average of 77 minutes a day on homework, and fifth graders 114 minutes."[15]

Based on these simple statistical methods, the researchers conclude that the low mathematical achievement of American children is due partly to their spending less time in school, to their spending less time outside school on homework, and to their spending a smaller proportion of their school time on mathematics than is the case in the other two countries. Of course, this is only part of the story, but it may be an important part. According to these researchers, the United States should consider policies to increase the length of time children spend in school and the amount of homework. Needless to say, this is a controversial recommendation.

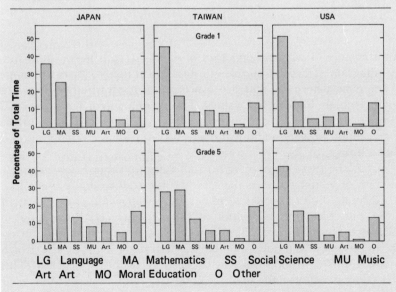

LG Language   MA Mathematics   SS Social Science   MU Music
Art Art   MO Moral Education   O Other

[15] Ibid., p. 46.

FREQUENCY DISTRIBUTIONS FOR SAMPLE DATA

Finally, in setting up an actual frequency distribution, we usually do not have all the measurements in the population. If, as is usually the case, we have data concerning only a sample from the population, a frequency distribution (constructed according to the principles discussed above) is a useful way to describe and summarize the data. For example, suppose that we had the voter turnout rates from some, not all, of the states. Under these circumstances, it still would be useful to formulate a frequency distribution to summarize the data, and the same principles apply. Thus, our discussion in previous parts of this section applies to both sample and population data.

But two important points should be recognized. First, *the frequency distribution of a sample is different from the frequency distribution of the population from which the sample is drawn.* Because the sample contains only a portion of the measurements in the population, the two frequency distributions are different and should not be confused. Second, *although we ordinarily do not have the data to construct a frequency distribution for the entire population, such a frequency distribution exists or can be imagined. It is important that we be able to visualize and think about this frequency distribution, since the questions that statistical investigations attempt to answer often are questions about this frequency distribution.*

1.8 The Stem-and-Leaf Diagram

Another way of summarizing a set of data is to construct a stem-and-leaf diagram, which is a simple graphical technique to show the range of the data, where the data are concentrated, and whether there are any extremely high or low values. The guidelines for constructing a frequency distribution set forth on pages 21–25 do not apply to the construction of a stem-and-leaf diagram. Instead, each observation is characterized by its *stem* and its *leaf.*

For the three-digit voter turnout rates in Table 1.7, we can use the first or leading digit as the *stem* and the *trailing* digits as the *leaf.* The leading digit or stem determines the row in which a voter turnout rate is put. The first observation in Table 1.7 (which is 64.6) is put in the fourth row of Figure 1.5, since its leading digit or stem is 6 (the figure to the left of the horizontal bar). The trailing digits or leaf (which equals 4.6) is written in this row to the right of the vertical bar. This procedure is followed for all of the observations in Table 1.7, the result being shown in Figure 1.5.

Clearly, this stem-and-leaf diagram is a useful representation of the data. It can be constructed quickly, and it has the advantage over the histogram that one loses no information concerning the value of each observation. To make the diagram somewhat neater, one can order the data within a row from lowest to highest, as shown in Figure 1.6. Also, if there seem to be too few rows, one can use two lines per stem; for example, leaf digits 0, 1, 2, 3, and 4 can be put on the first line of the stem, and leaf digits 5, 6, 7, 8, and 9 can be put on the second line. Thus, in Figure 1.6, we could use two lines for each stem. To illustrate, take the case where the stem is 6. The first line for this stem would include 1.0, 1.5, 3.0, 4.6, 4.6, and 4.8. The second line would include 5.1, 7.2, 7.6, and 8.0.

If the data consist of numbers with more than three digits, we might use the first two digits as the stem, and the second two digits as the leaf. The remaining digits might be ignored. Thus, the stem of 204,640 might be 20, and the leaf might be 46.

Figure 1.5
Stem-and-Leaf Diagram for Voter Turnout Rates, by State (including District of Columbia)

3	5.5
4	7.9 7.6 3.5 0.6 1.3 8.6 8.8 8.8 4.9 4.7 1.2 9.0 3.6
5	7.2 7.8 9.0 8.6 4.9 1.9 5.4 7.7 7.8 9.9 8.7 6.8 6.8 4.6 0.0 2.9 0.0 2.0 1.4 3.1 2.3 3.2 6.0 1.0 7.5 7.8
6	4.6 1.0 7.2 3.0 4.8 7.6 5.1 8.0 4.6 1.5
7	0.1

Source: Table 1.7.

Figure 1.6
Stem-and-Leaf Diagram for Voter Turnout Rates, Within-Row Data Ordered from Lowest to Highest

3	5.5
4	0.6 1.2 1.3 3.5 3.6 4.7 4.9 7.6 7.9 8.6 8.8 8.8 9.0
5	0.0 0.0 1.0 1.4 1.9 2.0 2.3 2.9 3.1 3.2 4.6 4.9 5.4 6.0 6.8 6.8 7.2 7.5 7.7 7.8 7.8 7.8 8.6 8.7 9.0 9.9
6	1.0 1.5 3.0 4.6 4.6 4.8 5.1 7.2 7.6 8.0
7	0.1

Source: Table 1.7.

EXERCISES

1.7 A psychologist is interested in the accuracy with which 100 children can estimate the width of an object which in fact is 240 inches wide. The errors vary from 0 to almost 20 inches. The psychologist wants to formulate a frequency distribution of the size of the errors. Determine upper and lower limits and the class mark for the first and last class intervals. (Assume that there are 10 class intervals.)

1.8 The math SAT score of the entering freshmen at Old Ironsides University varies from 520 to 719. Determine upper and lower limits and the class mark for the first and last class intervals. (Assume that there are 10 class intervals.)

1.9 The amount of life insurance carried by 40-year-old males in a Midwestern town varies from zero to $400,000. Construct a table with about ten classes into which these data might be grouped. What is the class mark in each interval?

1.10 The monthly salary of secretaries at the McNair Company varies from $930 to $1,975. Construct a table with seven classes of equal width into which these data might be grouped. What is the class mark in each interval?

1.11 A group of 40 schizophrenic adults takes an examination to test manual dexterity. Their scores are as follows:

81	62	76	81	61	80	42	53
83	93	78	86	75	82	76	60
78	98	92	74	73	81	78	65
76	51	63	79	71	43	71	95
74	75	50	71	69	58	72	98

(a) Group these numbers into a frequency distribution.
(b) Construct the corresponding histogram.

1.12 The cost per day (in dollars) of a room in 30 hospitals in Canada is as follows:

114	100	40	65	130	110
110	80	100	140	150	100
83	70	100	120	180	80
25	50	50	100	170	160
40	82	30	120	110	190

(a) Group these numbers into a frequency distribution.
(b) Construct the corresponding frequency polygon.

1.13 (a) Given the following frequency distribution of the weights of 2,000 soldiers, construct the cumulative frequency distribution.

Weight (pounds)	120 and under 140	140 and under 160	160 and under 180	180 and under 200	200 and under 220	220 and under 240
Number of soldiers	205	371	403	523	312	186

(b) Draw the ogive.

1.14 The frequency distribution of the ages of secretaries at the McNair Company is as follows:

Age (years)	18 and under 24	24 and under 30	30 and under 36	36 and under 42	42 and under 48	48 and under 54	54 and under 60
Number of secretaries	2	1	4	5	3	2	1

(a) Construct the cumulative frequency distribution.
(b) Draw the ogive.

1.15 There are 1,712 male students and 1,024 female students at Old Ironsides University. Among the males, 785 come from the eastern U.S., the rest from the western U.S. Among the females, 302 come from the eastern U.S., the rest from the western U.S.
(a) Draw a bar chart showing the distribution of students by sex.
(b) Draw a bar chart showing the distribution of students by region.

1.16 A psychologist in Washington, D.C., sends 112 packages containing psychological reports by express mail to New York City during 1986. Four of these packages do not arrive on the next day (as promised by the express mail service).
(a) Draw a bar chart showing the distribution of packages by whether or not they were late.
(b) One-half of the packages that arrived on the next day (and none of those that did not arrive on the next day) did *not* contain checks. Draw a bar chart showing the distribution of packages containing checks by whether or not they were late.

1.17 Use a stem-and-leaf diagram to summarize the data in exercise 1.11.

1.18 The age at which each of the first 39 presidents of the United States was inaugurated was as follows:

57	68	56	54	60
61	51	46	42	62
57	49	54	51	43
57	64	49	56	55
58	50	50	55	56
57	48	47	51	61
61	65	55	54	52
54	52	55	51	

Use a stem-and-leaf diagram to summarize these data.

The Chapter in a Nutshell

1. The field of statistics consists of two parts, descriptive statistics and inferential statistics. *Descriptive statistics* is concerned with summarizing and describing a set of data. *Inferential statistics* consists of techniques which help decision makers come to rational decisions under uncertainty. Statistics is concerned with whether data should be gathered at all, with how data should be gathered, and with how a particular set of data should be analyzed once it has been collected.

2. Government agencies, firms, natural scientists, and social scientists are continually engaged in *sampling* in order to obtain needed information, because it would be too expensive and time-consuming to try to obtain complete data concerning all relevant units. A *population* consists of the total collection of observations or measurements that are of interest to the statistician or decision maker in solving a particular problem. (A listing of all the elements or units in the population is called a *frame.*) A population can consist of quantitative or qualitative information, and it may be finite or infinite. A *sample* is a subset of measurements taken from the population.

3. Just as statistics is useful in designing a sample survey, it is also useful in designing an experiment. In both a survey and an experiment the objective should be to obtain the desired information at minimum cost. To promote this objective, it is important that the purposes of the experiment or sample survey be defined precisely before data are collected.

4. Modern statistical techniques are useful in promoting more rational decisions under uncertainty. Almost all decisions are made under uncertainty because it is seldom possible for the decision maker to forecast accurately the consequence of each alternative course of action. Statisticians use the concept of *probability* to measure the amount of confidence that one can have in various sample results.

5. Because statistics attempts to make inferences from a sample concerning a population, it must be concerned with error. After all, any sample result is likely to depart in some measure from the corresponding result for the total population. The error in any particular sample result is composed of two parts: *experimental or sampling error,* and *bias.* Experimental or sampling error is due to a large number of uncontrolled factors which we subsume under the shorthand expression *chance. Bias* consists of a persistent, systematic sort of error. Increases in a sample's size tend to reduce experimental or sampling error, but not bias.

6. To summarize a body of data, it is useful to construct a *frequency distribution,* which is a table showing the number of measurements or observations that fall into each of a number of class intervals. To establish a frequency distribution, one must set up certain well-defined class intervals, each interval being defined by a lower limit and an upper limit. Frequency distributions are often presented in graphical as well as tabular form. A *histogram* is composed of a series of bars or rectangles; the bottom of each bar is the line segment on the

horizontal axis corresponding to the interval from the class interval's lower limit to its upper limit. The area of each bar is proportional to the number of cases in the class. A *frequency polygon* is another type of graphical representation of a frequency distribution. Still another valuable graphical means of representing a body of data is the *stem-and-leaf diagram*.

Chapter Review Exercises

1.19 A psychologist studying techniques of child-rearing interviews each of the parents of a 4-year-old boy and gives each parent a battery of objective tests. To determine whether the father or mother will be interviewed first, the psychologist flips a coin.
(a) What is the probability that the father will be interviewed first?
(b) If the psychologist carries out this procedure with a sample of 100 4-year-old boys, will the proportion of cases where the father will be interviewed first equal the probability in (a)?

1.20 To determine the relative safety of today's aircraft and those of 40 years ago, a researcher compares the number of aircraft accidents in 1985 with that in 1945. Finding that the number of accidents has increased, the researcher concludes that today's airplanes are not as safe as the earlier ones. Do you agree? Why, or why not?

1.21 A frequency distribution of bills issued in a particular year by a psychiatrist contains ten class intervals, each with a width of $100. Does it follow that the largest bill is $1,000 (that is, 10 times $100) greater than the smallest bill? Why or why not?

1.22 How can one calculate the class mark of an open-ended class interval? Is it possible?

1.23 A psychologist obtains information from 30 18-year-old alcoholics concerning the age at which each first consumed any form of alcoholic beverage. The results (in years) are as follows:

9.1	17.2	14.8	12.3	13.6
11.2	15.7	13.7	11.2	11.5
12.7	11.0	11.4	10.1	10.9
10.5	10.9	9.2	9.9	9.2
14.3	6.2	8.0	15.8	8.1
16.8	9.2	15.1	14.7	7.0

(a) Construct a frequency distribution of these data.
(b) Construct a histogram based on these data.
(c) Construct a frequency polygon based on these data.

1.24 According to the Department of Agriculture, the total net income per farm in each of 10 midwestern states in 1973 was:

State	Income per farm (dollars)
Illinois	13,224
Indiana	11,282
Iowa	19,685
Kansas	17,018
Michigan	6,171
Minnesota	19,456
Nebraska	17,790
North Dakota	35,631
Ohio	5,212
South Dakota	22,928

(a) Are there too few states to construct a frequency distribution of their incomes per farm?

(b) If not, what frequency distribution would you suggest?

1.25 A supermarket inspects the work of one of its clerks at the checkout counter. For each customer, it determines what the customer's bill should be and what this clerk calculates it to be. For 50 customers the frequency distribution of the difference between the latter and the former is as follows:

Error	Number of customers
−$1.00 and under −$0.75	1
−$0.75 and under −$0.50	2
−$0.50 and under −$0.25	4
−$0.25 and under −$0.00	30
$0.00 and under $0.25	6
$0.25 and under $0.50	2
$0.50 and under $0.75	2
$0.75 and under $1.00	2
$1.00 and under $1.25	1

(a) What are the class intervals in this frequency distribution?

(b) Are the class intervals of equal width? (That is, is the difference between the upper and lower limit the same for each class interval?)

(c) If the error in a particular customer's bill equals −$0.25, into which class interval would this item fall?

(d) Construct a cumulative frequency distribution for the data.

(e) Plot the ogive for the cumulative frequency distribution for the data.

1.26 (a) Given only the frequency distribution in Table 1.8 (not the original data in Table 1.7), which of the following questions would you be able to answer?

(i) How many of these states had voter turnout rates equal to 60.0 percent or more?

(ii) How many had voter turnout rates of less than 47.5 percent?

(iii) How many had voter turnout rates of less than 50.0 percent?

(iv) How many had voter turnout rates of 57.5 percent or more?

(v) How many had voter turnout rates of at least 60.0 percent but less than 75.0 percent?

(vi) How many had voter turnout rates of at least 62.5 percent but less than 70.0 percent?

(b) If the width of each class interval in Table 1.8 were cut in half, how many of the questions in parts (i) to (vi) above would you be able to answer? What are some of the advantages of relatively narrow class intervals? What are some of their disadvantages?

2

OBJECTIVES

One of the fundamental purposes of statistics is to describe a body of data effectively. The overall purpose of this chapter is to present various measures of central tendency (such as the mean, median, and mode) and various measures of dispersion (such as the range and standard deviation), as well as to indicate ways that descriptive statistics sometimes are misused. Among the specific objectives are:

1   To define and indicate the advantages and disadvantages of the mean, median, and mode.

2   To define and indicate the advantages and disadvantages of the range, variance, and standard deviation.

3   To show how inappropriate comparisons, very small and obviously biased samples, improper choice of averages, neglect of the variation about an average, and misleading graphs can lead to serious errors.

Summary and Description of Data

2.1 Introduction

As indicated in the previous chapter, descriptive statistics is concerned with the summary and description of a body of data. While this concise definition may make descriptive statistics seem cut-and-dried and perhaps a bit dull, this is by no means true. The proper summary of a body of data involves much more than arithmetic. It entails avoiding a variety of pitfalls (many of which are discussed in this chapter) that can lead the unwary analyst or decision maker to false conclusions. Data can also be distorted intentionally by unscrupulous individuals and organizations to mislead others. To avoid being misled, a knowledge of descriptive statistics is essential.

2.2 Summary Measures

In Chapter 1, we showed how a frequency distribution can be constructed to summarize and describe a set of data. However, in many situations where a frequency distribution would be too detailed and cumbersome, a few summary measures can present concisely the salient features of the data. Although summary measures provide much less information than the frequency distribution, in many situations the

39

lack of a certain amount of information is not crucial, and the greater conciseness of the summary measures makes them more useful than the frequency distribution. In general, two types of summary measures are most frequently used: measures of central tendency and measures of dispersion.

MEASURES OF CENTRAL TENDENCY. Often, one wants a single number to represent the "average level" of a set of data. In other words, a number is needed that will indicate where the frequency distribution is centered. This number should tell us what a "typical" value

Table 2.1
Percent of Voting Age Population That Voted in 1980 Elections, by State (Including District of Columbia)

State	Percent voting	State	Percent voting
Maine	64.6	North Carolina	43.5
New Hampshire	57.2	South Carolina	40.6
Vermont	57.8	Georgia	41.3
Massachusetts	59.0	Florida	48.6
Rhode Island	58.6	Kentucky	50.0
Connecticut	61.0	Tennessee	48.8
New York	47.9	Alabama	48.8
New Jersey	54.9	Mississippi	52.0
Pennsylvania	51.9	Arkansas	51.4
Ohio	55.4	Louisiana	53.1
Indiana	57.7	Oklahoma	52.3
Illinois	57.8	Texas	44.9
Michigan	59.9	Montana	65.1
Wisconsin	67.2	Idaho	68.0
Minnesota	70.1	Wyoming	53.2
Iowa	63.0	Colorado	56.0
Missouri	58.7	New Mexico	51.0
North Dakota	64.8	Arizona	44.7
South Dakota	67.6	Utah	64.6
Nebraska	56.8	Nevada	41.2
Kansas	56.8	Washington	57.5
Delaware	54.6	Oregon	61.5
Maryland	50.0	California	49.0
District of Columbia	35.5	Alaska	57.8
Virginia	47.6	Hawaii	43.6
West Virginia	52.9		

Source: Statistical Abstract of the United States.

of the measurements might be. To illustrate, let's return to the data concerning the voter turnout rates in the 51 states (including the District of Columbia), which are shown in Table 2.1. If these data were collected to determine how large a percentage of the potential voters in our country actually exercise their right to vote, it might be sensible to ask: "What is the typical level of the 1980 voter turnout rates in the 51 states (including the District of Columbia)?" As we shall see, there are several types of averages or measures of central tendency that can be used to help answer this question; and the choice among these depends on the purposes of the investigator and the nature of the data.

MEASURES OF DISPERSION. In addition to knowing the "average level" of a set of data, it is important to know the degree to which the individual measurements vary about this average. In other words, we need to know whether a frequency distribution is tightly packed around its average or whether there is a great deal of scatter about it. In the case of voter turnout rates in Table 2.1, an important question is: "Regardless of what the average level of these turnout rates may be, to what extent does the voter turnout rate vary from state to state?" Statisticians have devised a number of measures of dispersion, which will be described in subsequent sections. As in the case of measures of central tendency, the choice among measures of dispersion depends on the purposes of the investigator and the nature of the data.

2.3 Parameters and Statistics

If we have all the measurements in a given *population,* we can calculate summary measures for that population as a whole. Such summary measures are called *parameters.* For example, if we calculate a particular kind of average of the voter turnout rates in Table 2.1, the resulting average is a parameter since it is calculated from all the measurements in the relevant population. (Recall from Chapter 1 that these 51 observations are regarded as the population.) Or if we calculate a particular kind of measure of dispersion (again using the turnout rates in Table 2.1), the result is a parameter since the calculation is based on all the measurements in the relevant population. As has been pointed out, we seldom have all the measurements in an entire population, but this does not mean we are not interested in the parameters of the population. On the contrary, *much of inferential statistics is designed to draw inferences from a sample concerning the value of a population parameter.*

If we have only *sample* data, we can calculate summary measures for the sample: such summary measures are called **statistics.** For example, if a political scientist chooses a sample of 10 states and calculates a particular kind of average of the voter turnout rates in these states, the resulting average is a statistic since it is calculated from the *measurements in a sample.* Or if the political scientist calculates a particular kind of measure of dispersion, using the voter turnout rates in these 10 states, the result is also a statistic since the calculation is based on the *measurements in a sample.* As we shall see in subsequent chapters, a statistic from a sample is often used to estimate the analogous parameter of the entire population from which the sample is drawn. Thus, the proportion of voters in the Gallup Poll's sample that favors a particular presidential candidate is used to estimate the proportion of voters favoring him or her in the population.

2.4 Measures of Central Tendency

THE ARITHMETIC MEAN

There are several important types of measures of central tendency. The one used most frequently is the arithmetic mean. Like Molière's character in *Le Bourgeois Gentilhomme*, who was surprised to learn that he had been speaking prose all his life, you may be surprised to learn that you have been using the arithmetic mean for a long time (although you probably have not been calling it by that name). *The* **arithmetic mean** *is the sum of the numbers included in the relevant set of data divided by the number of such numbers.* Let N denote how many numbers there are in a population; thus in Table 2.1, $N = 51$. If we order these numbers from 1 to N, X_1 being the first number, X_2 being the second number, and so on up to X_N, which is the Nth number, then the population mean is

Formula for μ

$$\mu = \frac{X_1 + X_2 + X_3 + \cdots + X_N}{N},$$ *(2.1)*

where μ is the Greek letter *mu.* In particular, if we order the numbers in Table 2.1 from the top down (beginning with the left-hand column), X_1 being 64.6, X_2 being 57.2, and so forth, with X_{51} being 43.6, then in the case of this population,

$$\mu = \frac{64.6 + 57.2 + 57.8 + 59.0 + \cdots + 57.8 + 43.6}{51}$$

$$= \frac{2777.8}{51}$$

$$= 54.47.$$

Consequently, the arithmetic mean of the population turns out to be 54.47 percent.

If the arithmetic mean is calculated for a sample rather than for the whole population, it is designated as $\overline{X}$ rather than μ. Whereas N stands for the number of measurements in the population, n stands for the number of measurements in the sample. Thus, the sample mean is defined as

$$\overline{X} = \frac{X_1 + X_2 + X_3 + \cdots + X_n}{n}.$$

(2.2)   *Formula for* $\overline{X}$

This expression is sometimes written in the following form:

$$\overline{X} = \frac{\sum_{i=1}^{n} X_i}{n},$$

where Σ is the mathematical *summation sign*. What does ΣX_i mean? It   *Summation Sign*
means that the numbers to the right of the summation sign (that is, the values of X_i) should be summed from the lower limit on i (which is given below the Σ sign) to the upper limit on i (which is given above the Σ sign). Thus, in this case it means that X_i is to be summed from $i = 1$ to $i = n$. In other words, ΣX_i means the same as $X_1 + X_2 + \cdots + X_n$. (For further discussion of the summation sign and its uses, see Appendix 2.1.)

Whether the set of data is a sample or a whole population, it is sometimes necessary to calculate the arithmetic mean from grouped data—that is, from a frequency distribution. For example, according to a study in the *American Psychologist*, the age at which 101 17-year-old American smokers first tried smoking is given by the frequency distribution in Table 2.2.[1] Suppose that we want to calculate the average age at which the people in this sample first tried smoking. Since the age at

---

[1] See the publication cited in Table 2.2 for the nature and sources of these data; for pedagogical reasons, we assume here that it pertains to 101 people. This is an innocuous assumption, adopted to help us to illustrate the use of the statistical techniques discussed here.

Table 2.2
Frequency Distribution of Sample of 101 17-Year-Old Smokers, by Age When They First Tried Smoking*

Age (years)	Number of people
6 and under 8	6
8 and under 10	9
10 and under 12	11
12 and under 14	25
14 and under 16	30
16 and under 18	20
Total	101

* See note 1.

Source: J. Matarazzo, "Behavorial Health's Challenge to Academic, Scientific, and Professional Psychology," *American Psychologist,* January 1982. The lower limit of the first class interval is assumed to be 6 years. For present purposes, this assumption is innocuous.

which each person in the sample first tried smoking is not given in Table 2.2, we cannot use equation (2.2) to calculate $\overline{X}$. However, we can approximate the sum of the X_i by assuming that *the midpoint (class mark) of each class interval can be used to represent the value of the measurements in that class interval.* Thus, the sample mean can be approximated by

$$\overline{X} = \frac{f_1 X_1' + f_2 X_2' + \cdots + f_k X_k'}{n} = \frac{\sum\limits_{j=1}^{k} f_j X_j'}{n}, \tag{2.3}$$

where f_1 is the number of measurements in the first class interval, X_1' is the midpoint of the first class interval, f_2 is the number of measurements in the second class interval, X_2' is the midpoint of the second class interval, and so on.

Applying equation (2.3) to the data in Table 2.2, we find that the sample mean can be approximated by

$$\overline{X} = \frac{6(7) + 9(9) + 11(11) + 25(13) + 30(15) + 20(17)}{101}$$

$$= \frac{1359}{101} = 13.46.$$

Thus the sample mean is about 13.5 years. If the mean of the measurements in each class interval is close to the midpoint of the class interval, this approximation should entail only a small amount of error.[2] Even if data are available for all the measurements, calculating the mean from a frequency distribution of the data may be easier and less expensive.

THE WEIGHTED ARITHMETIC MEAN

In some cases the measurements in a sample or a population should not be weighted equally, as in equations (2.1) and (2.2). Since some states are much more populous than others, it might be argued that a state's voter turnout rate should be weighted according to its population in determining the average level of voter turnout rates. If w_i is the weight attached to the ith measurement in a sample, the weighted arithmetic mean is

$$\overline{X}_w = \frac{\sum_{i=1}^{n} w_i X_i}{\sum_{i=1}^{n} w_i}. \tag{2.4}$$

For example, suppose that we have a sample of three states' voter turnout rates: 50 percent, 55 percent, and 60 percent. The state with the 50 percent voter turnout rate has a voting age population of 10 million, whereas the other two states have a voting age population of 5 million each. If a state's voting age population is used to weight its voter turnout rate, the weighted arithmetic mean of the voter turnout rates of these three states is

$$\overline{X}_w = \frac{10(50) + 5(55) + 5(60)}{10 + 5 + 5} = \frac{1075}{20} = 53.75.$$

Thus, the weighted mean is 53.75 percent.

Comparing equation (2.3) with equation (2.4) shows that the arithmetic mean based on grouped data is a type of weighted arithmetic mean: It is a weighted mean of the midpoints of the class intervals, the weight attached to each particular midpoint being the number of measurements falling within that class interval.

---

[2] At this point, you should understand more fully why we recommended in the previous chapter that class intervals be constructed so that the midpoint of each class interval is close to the average of the observations included in it. If this is done, the error in using equation (2.3) will be very small.

THE MEDIAN

Other than the mean, the most widely used measure of central tendency is the median, which is defined as the *middle value* of the relevant set of data. In other words, the median is the *value that divides the set of data in half, 50 percent of the measurements being above (or equal to) it and 50 percent being below (or equal to) it.* Let's consider the voter turn-out rates of the 51 states (including the District of Columbia) again. If we list these 51 states in the order of their voter turnout rates (from lowest to highest), the voter turnout rate of the 26th state must be the middle value. In other words, as many states have voter turnout rates exceeding this value as have voter turnout rates falling below it. According to this method of listing, we find that the *median* voter turnout rate of the 51 states (including the District of Columbia) is 54.9 percent. (To verify this, make such a listing of the states in Table 2.1, and see for yourself.)

If there is an even number of observations in a frequency distribution, *none* of the observations can be the middle value. For example, among the four numbers 2, 4, 6, and 8 there can be no "middle" number or median because the middle lies *between* two of the numbers —specifically, between 4 and 6. To resolve this difficulty, convention dictates that *if there is an even number of observations, the mean of the middle pair of observations is regarded as the median.* (In this case the median would be the mean of 4 and 6, or 5.)

Just as the mean frequently must be approximated from a frequency distribution, so the median also must often be approximated in this way. The first step in calculating the median from a frequency distribution is to find the class interval that contains the median. To do this, we start with the lowest class interval, cumulate the number of measurements in one, two, three, and subsequent class intervals, stopping with the interval where the cumulated number of measurements first exceeds or equals $n/2$ if the measurements are a sample or $N/2$ if they are the whole population. This particular class interval contains the median.

To calculate the median from the frequency distribution of the 101 teenagers in Table 2.2, we cumulate the number of teenagers with initial smoking ages less than the upper limit of each class interval. Thus, as shown in the third column of Table 2.3, the number of teenagers that first tried smoking before the age of 8 years was 6, the number that began before the age of 10 years was 15, and so on. As specified in the previous paragraph, we must keep on cumulating until we reach the first interval where the cumulated number of measurements exceeds $n/2$ (50.5 in this case, because $n = 101$). Since the cumulated number is 26 for the third class interval and 51 for the fourth

Age (years)	Number of people	Cumulated number of people*
6 and under 8	6	6
8 and under 10	9	15
10 and under 12	11	26
12 and under 14	25	51
14 and under 16	30	81
16 and under 18	20	101

Table 2.3
Finding the Class Interval in Which the Median Age When Teenagers First Tried Smoking Is Situated, 101 People

* This is the number of people with ages *less than* the *upper limit* of the relevant class interval.

class interval in Table 2.3, it is clear that the fourth class interval (12 and under 14 years) is the first where the cumulated number exceeds 50.5. Thus, this is the class interval in which the median is located.

To estimate where in this class interval the median is situated, we assume that the interval's measurements are *spaced evenly* along its width of 2 years. If so, the 25 measurements are 12.04; 12.12; 12.20; ··· ; 13.80; 13.88; and 13.96. The median must be the highest measurement, since it must be larger than 24 of the measurements in this interval. Thus, the median must be 13.96.

A general expression for finding the median in grouped data of this sort is

$$M = \left(\frac{n/2 - c}{f_m}\right) l + L_m, \qquad (2.5)$$

where c is the number of measurements in class intervals below the one containing the median; f_m is the number of measurements in the class interval containing the median; l is the width of the class interval containing the median; and L_m is the lower limit of the class interval containing the median. Since (as explained in footnote 3) $n = 101, c = 26$, $f_m = 25$, $l = 2$, and $L_m = 12$,

[3] It is clear from Table 2.3 that the number of measurements in class intervals below the one containing the median is 26, since the "12 and under 14 years" class interval includes the median. The width of this class interval equals 2 years. The number of observations in this class interval is 25, and the lower limit of this class interval is 12. Thus, $c = 26, l = 2, f_m = 25$, and $L_m = 12$.

$$M = \left(\frac{50.5 - 26}{25}\right) 2 + 12$$

$$= \left(\frac{24.5}{25}\right) 2 + 12$$

$$= 13.96.$$

Thus, this formula results in precisely the same answer (13.96) as was obtained in the previous paragraph. This will always be true.

USES OF THE MEAN AND THE MEDIAN

Both the mean and the median are important and useful measures of central tendency (that is, of the "average level" of a set of data). In some circumstances the mean is a better measure than the median, and in others the reverse is true. The following factors are among the most important determinants of whether the mean or the median should be used.

SENSITIVITY TO EXTREME OBSERVATIONS. The median is often preferred over the mean when the latter can be influenced strongly by extreme observations. For example, how do we go about computing the average income of the families in an apartment building containing 19 families, 8 of which earn $10,000 per year, 10 of which earn $12,000 per year, and 1 of which earns $1 million per year? (The latter presumably has the penthouse.) The mean income of the 19 families equals

$$\frac{8(\$10,000) + 10(\$12,000) + 1(\$1,000,000)}{19} = \frac{\$1,200,000}{19}$$

$$= \$63,157.$$

However, this figure is not a very good description of the yearly income level of the majority of the families in the building. A better measure might be the median, which in this case is $12,000 per year. The median is much less affected by the one extreme point (the millionaire), which raised the mean very considerably.

OPEN-ENDED CLASS INTERVALS. As you will recall from our discussion of class intervals in Chapter 1, it is not unusual for frequency distributions to have open-ended class intervals—that is, class intervals with no finite upper or lower limits. For example, in a frequency distribution of the annual income of American families, two class intervals might be "less than $1,000" and "$100,000 and more." Each of these class intervals is open-ended.[4] If one needs to calculate an average from a frequency distribution with one or more open-ended class intervals, there may be no alternative but to use the median, since calculation of the mean requires a knowledge of the sum of the measurements in the open-ended classes.

MATHEMATICAL CONVENIENCE. The mean rather than the median is often the preferred measure of central tendency because it possesses convenient mathematical properties that the median lacks. For example, the mean of two combined populations or samples is a weighted mean of the means of the individual populations or samples. On the other hand, given the medians of two populations or samples, there is no way to determine what the median of the two populations combined or two samples combined would be.

EXTENT OF SAMPLING VARIATION. As pointed out earlier, sample statistics such as the sample mean or the sample median are often used to estimate the population mean. A major reason for preferring the mean to the median is that the sample mean tends to be more reliable than the sample median in estimating the population mean. In other words, the sample mean is less likely than the sample median to depart considerably from the population mean. This is a very important consideration which will be more fully appreciated in Chapter 7, where we shall cover this topic in greater detail.

THE MODE

Another often-used measure of central tendency is the *mode*, which is defined as the *most frequently observed value of the measurements in the relevant set of data.* For example, if an industrial psychologist were to ask people to indicate which of several colors they preferred on a particular piece of machinery, and if 100 people preferred red, 50 preferred green, and 20 preferred yellow, then the mode would be at the color red. Or if there were 19 families in your apartment building, 8 of which earned $10,000 per year, 10 of which earned $12,000 per year, and 1 of which earned $1 million per year, the mode would be at $12,000. Why? Because $12,000 is the most frequently observed value of family income in the building.

[4] At first glance, you may think that the "less than $1,000" class interval has a finite lower limit—namely, zero. This is incorrect because some people have negative incomes, and there is no limit on how large losses of this sort can be.

Figure 2.1
Mode of a Frequency
Polygon

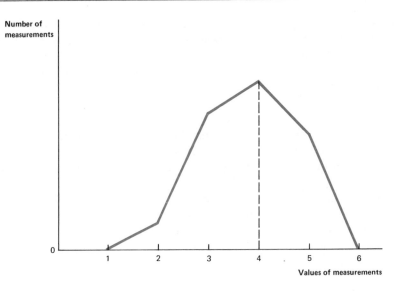

When data are presented in the form of a frequency distribution, the mode can be estimated as *the midpoint or class mark of the class interval containing the largest number of measurements.* The mode of the ages in Table 2.2 is 15 years. Why? Because the largest number of measurements (30) is contained in the class interval "14 and under 16 years," and the midpoint of this class interval is 15 years. The class interval containing the largest number of measurements is called the modal class. Thus, the modal class in Table 2.2 is "14 and under 16 years."

Based on a graphical representation of a frequency distribution such as a frequency polygon, it is easy to find the mode of a body of data. We need only find the value along the horizontal axis where the frequency polygon achieves its *maximum* vertical height. For example, the mode of the frequency distribution portrayed in Figure 2.1 equals 4. Some frequency distributions (like the one shown in Figure 2.2) have more than one mode. If a frequency distribution has more than one mode, it is called ***multimodal;*** if it has two modes, it is called ***bimodal.*** Frequency distributions with more than one mode often arise because two or more quite different types of measurements or observations are included. If we were to form a frequency distribution of the heights of American adults, we might find two modes, one at the modal height for men and one at the modal height for women. Great care must be exercised in constructing and interpreting measures of central tendency for multimodal distributions since *measures like the mean or the median may fall between the modes and be unrepresentative of*

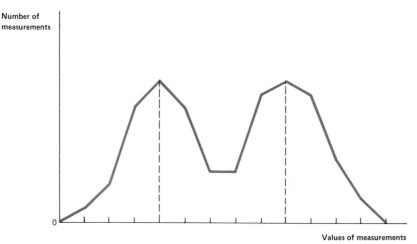

Number of measurements

0

Values of measurements

Figure 2.2
Frequency Polygon of
a Bimodal Frequency
Distribution

the bulk of the measurements lying near the separate modes. In cases where
a multimodal frequency distribution arises because two or more quite
different types of measurements or observations are included, it often
is wise to construct a *separate* frequency distribution for each type
rather than combine them. In the example above, one frequency distribution might be constructed for men's heights and another frequency distribution for the heights of women.

RELATIONSHIPS AMONG THE MEAN, MEDIAN, AND MODE

Having discussed the three principal measures of central tendency (the mean, median, and mode), we must describe how these
three measures are related to one another. If the frequency distribution of a set of data has a single mode and is *symmetrical,* as in panel A of
Figure 2.3, the mean, median, and mode coincide. (Symmetrical
means that if we were to "fold" the distribution at its mean, the part of
the distribution to the left of the mean would be a perfect match for the
part to the right of the mean.) Many frequency distributions are not
symmetrical, but are **skewed to the right** (as in panel B of Figure 2.3) or   *Skewness*
skewed to the left (as in panel C of Figure 2.3). A frequency distribution
that is skewed to the right has a long tail to the right, whereas one that
is skewed to the left has a long tail to the left. As shown in Figure 2.3, if
the frequency distribution is skewed to the right, the mean generally
exceeds the median, which in turn exceeds the mode. If the frequency
distribution is skewed to the left, the mode generally exceeds the median, which in turn exceeds the mean (also shown in Figure 2.3).

Figure 2.3
Relationship of Mean, Median, and Mode

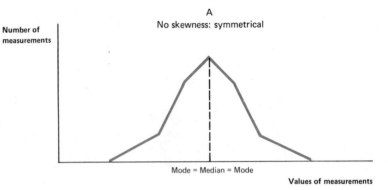

A
No skewness: symmetrical

Number of measurements

Mode = Median = Mode

Values of measurements

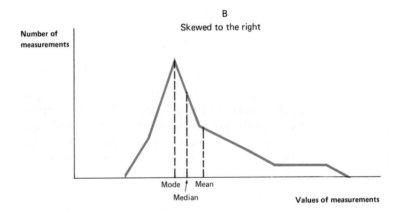

B
Skewed to the right

Number of measurements

Mode   Mean
Median

Values of measurements

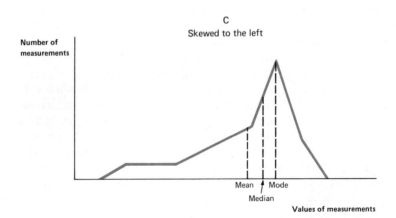

C
Skewed to the left

Number of measurements

Mean   Mode
Median

Values of measurements

EXERCISES

2.1 A psychology professor obtains data concerning the number of classes per month that 18 students missed during their four years in college. The results are as follows:

1.14	1.33	1.04	1.51	1.68	1.72	2.04	1.36	1.77
1.21	1.46	1.85	1.73	1.46	1.77	1.83	1.86	1.56

(a) Determine the mean number of classes per month that these students missed.
(b) Determine the median number.
(c) Is the above set of numbers a sample or a population?
(d) Are the mean and median you determined in (a) and (b) parameters or statistics?

2.2 A hospital wants to determine the average age of its nurses. It chooses ten nurses (out of the 289 that work for the hospital), and finds that their ages (in years) are:

46	49	32	30	27	49	62	53	37	39

(a) Find the mean age.
(b) Find the median age.
(c) Is the above set of numbers a sample or a population?
(d) Are the mean and median you determined in (a) and (b) parameters or statistics?

2.3 A government agency has two divisions, one with 4,000 employees, the other with 8,000 employees. The mean hourly wage of the employees is $6.05 in the first division and $7.39 in the second division. What is the mean hourly wage of all the agency's employees?

2.4 A group of 400 female employees and 600 male employees were given an IQ test. The mean IQ for all employees was 115, and the mean IQ for the female employees was 117. What was the mean IQ for the male employees?

2.5 The frequency distribution of the daily cost (in dollars) of commuting back and forth to work by 100 employees of a hospital is:

Cost	0 and under $1	$1 and under $3	$3 and under $5	$5 and under $7
Number of employees	29	32	29	10

(a) Find the mean daily cost of commuting.
(b) Find the median daily cost of commuting.

2.6 The frequency distribution of the number of years of schooling of 100 workers at a hospital is:

Number of years	0 and under 8	8 and under 12	12 and under 16	16 and over
Number of workers	1	12	49	38

(a) Find the mean years of schooling (if possible).
(b) Find the median years of schooling (if possible).
(c) What are the relative advantages here of the mean and median?

2.7 A clinical psychologist administers a test to 16 brain damaged 14-year-old children, with the following results:

88	89	89	98	98	90	92	95
98	98	98	91	93	94	96	97

(a) What is the mode?
(b) Is the frequency distribution skewed? If so, is it skewed to the left or to the right?
(c) Which is largest: the mean, median, or mode?

2.8 An Amtrak official obtains data on a particular day concerning the length of time (in minutes) that the metroliners leaving New York take to reach Philadelphia, with the following results:

93	90	91	93	89	90	94	115	88	88

(a) What is the mean?
(b) What is the median?
(c) What are the advantages here of the median over the mean?

2.9 According to the *Statistical Abstract of the United States,* the annual number of successful earth orbiting space payload launchings by the United States during 1971–81 was as follows:

1971	45	1975	30	1979	18
1972	33	1976	33	1980	16
1973	23	1977	27	1981	18
1974	27	1978	34		

(a) Calculate the mean number of annual successful launchings during 1971–76.
(b) Calculate the mean number of annual successful launchings during 1977–81.
(c) Interpret and compare the results of (a) and (b) of this question.
(d) Calculate the median number of annual successful launchings in 1971–1976. Calculate the median number of annual successful launchings in 1977–81. How big is the difference between these two medians? Interpret your results.

2.10 In a township in Virginia, all lots are 1/4 acre, 1/2 acre, 1 acre, or 2
acres. The frequency distribution of lot sizes for all residential property
in this township is:

Size of lot (acres)	*Number of lots*
1/4	100
1/2	500
1	50
2	20

What is the mode of this frequency distribution? Is the mode bigger than
the mean size of lot? Is it bigger than the median size of lot?

2.5 Measures of Dispersion

IMPORTANCE OF DISPERSION

In the previous section we saw that measures of central tendency
provide useful summary information concerning the general level or
average value of a body of data. However, this obviously does not mean
that such measures alone can provide a complete or adequate description of the data. A case that is close to home can illustrate the limitations of such measures. Suppose that you have taken an examination
and that the instructor, after grading the exam, announces to the class
that the mean grade is 75. Then he or she hands back the exams, and
you find that your paper has received a grade of 80. Clearly, it is hard
to interpret this grade on the basis of information concerning only the
average grade. The variability about the average is very important,
too. If the grades are highly variable (as in panel A of Figure 2.4), then a
large number of your classmates may have received a grade higher
than yours. On the other hand, if there is little variability in the grades
(as in panel B of Figure 2.4), then you may have received close to the
highest grade.

If there is enough variability about an average, the average may
not mean much. If Mr. Rich, one of the world's wealthiest men, is
driven to the airport by his chauffeur, what is the average income level
of the two occupants of the car? Mr. Rich's income is $10 million a year
and his chauffeur's income is $10 thousand a year; thus, the mean in-
come of the two is $5,005,000 per year. This is a misleading figure,
however, since it vastly overstates the chauffeur's income and vastly
understates the income of Mr. Rich. In cases of this sort, then, we see
that an average can be quite misleading. (More will be said about this
later in the chapter.)

Figure 2.4
Histogram of Examina-
tion Grades

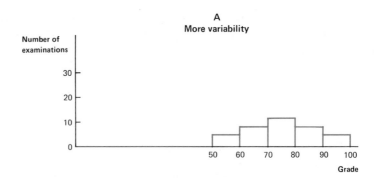

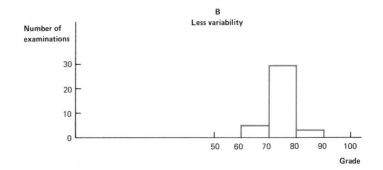

DISTANCE MEASURES OF DISPERSION

Both summary measures of dispersion and summary measures of central tendency are important in describing a body of data, but neither alone will suffice. There are two types of summary measures of dispersion: distance measures and measures of average deviation. ***Distance measures*** *describe the variation in the data in terms of the distance between selected measurements.* The most frequently used distance measures are the range and the interquartile range.

Range. Perhaps the simplest measure of variability is the *range,* which is *the difference between the highest and lowest values in the body of data.* In the example of the exam grades discussed above, if the lowest grade in the class is 50 and the highest is 99, then the range is $99 - 50 = 49$. Although the range is a popular measure of variability, particularly because it is so easy to compute, it has the important disadvantage of being unaffected by the values of all observations other than the highest and the lowest. For example, the range of the exam grades would be the same (49) if the grades were distributed evenly from 50 to 99, or if all grades other than the highest or lowest fell between 70 and 80. Yet the variability certainly is less in the latter case.

Interquartile Range. Another common measure of variability is the *interquartile range*, which is defined as *the difference between the third quartile and the first quartile.* The *third quartile* is the value such that 75 percent of the observations lie below it; the *first quartile* is the value such that 25 percent of the observations lie below it. Thus, the interquartile range measures *the spread bounding the middle 50 percent of the values of the observations.* (Note that the *second quartile,* which is the value such that 50 percent of the observations lie below it, is another name for the median.) One advantage of the interquartile range is that it can be calculated from frequency distributions with open-ended class intervals, while the range cannot.

Quartiles

Measures Based on Percentiles. In addition to the range and the interquartile range, other distance measures of dispersion can be based on percentiles. The *Xth percentile* is defined as the value that exceeds X percent of the observations. Thus, the 90th percentile is the value such that 90 percent of the observations lie below it. The difference between the 90th percentile and the 10th percentile is a possible measure of dispersion; so is the difference between the 99th percentile and the 1st percentile. The former measures the spread bounding the middle 80 percent of the values of the observations. The latter measures the spread bounding the middle 98 percent of them.

Percentiles

THE VARIANCE AND STANDARD DEVIATION

Although the distance measures of dispersion are sometimes used, they are not as important as measures of average deviation, the most significant of which are the variance and the standard deviation. The *variance* of the measurements in a population, denoted by σ^2 (sigma squared), is defined as *the arithmetic mean of the squared deviations of the measurements from their mean.* Thus if $X_1, X_2, \ldots, X_N$ are the measurements in the population,

$$\sigma^2 = \frac{(X_1 - \mu)^2 + (X_2 - \mu)^2 + \cdots + (X_N - \mu)^2}{N}$$

$$= \frac{\sum_{i=1}^{N} (X_i - \mu)^2}{N}$$

(2.6)

Formula for σ^2

The variance is a measure of dispersion, but it is expressed in units of squared deviations or squares of the values of the measurements rather than in the same units as the measurements. The *standard deviation* is a measure of dispersion which is expressed in the same units as the measurements. The standard deviation of the measurements in a population is denoted by σ (sigma), which is the positive square root of the variance. In other words, the standard deviation is

Formula for σ

$$\sigma = \sqrt{\frac{\sum_{i=1}^{N}(X_i - \mu)^2}{N}}$$

(2.7)

Since the standard deviation is so important in statistics, its definition is worth discussing in greater detail. To begin with, note that if X_i is the ith measurement in the population, then the difference between this measurement and the population mean equals $(X_i - \mu)$. This is the deviation of the ith measurement from the population mean, and the square of this deviation obviously equals $(X_i - \mu)^2$. Next, let's find the mean of these squared deviations. Since there are N of these squared deviations, this means equals

$$\frac{\sum_{i=1}^{N}(X_i - \mu)^2}{N}$$

Recall from our earlier discussion that Σ is the summation sign, which means that the numbers to its right—that is, $(X_i - \mu)^2$—should be summed from the lower limit on i (given below the summation sign) to the upper limit on i (given above the summation sign). Thus, in this case, $(X_i - \mu)^2$ is to be summed from $i = 1$ to $i = N$. (In other words, the squared deviations are to be summed for all observations.) Then, to obtain the mean of the squared deviations, the resulting sum is divided by N. Finally, we must find the square root of this mean, the result being the standard deviation. Equation (2.7) shows this complete procedure.

Intuitively, it seems clear that *the more dispersion there is in a body of data the bigger the standard deviation will be.* If there is no dispersion at all, every observation will equal the population mean, with the result that every one of the deviations will equal zero. (Thus, the standard deviation will equal zero.) As the dispersion in the data increases, the deviations of the observations from the population mean will tend to increase as well, and so will the mean of the squared deviations. (Thus, the standard deviation will also increase.) For this reason, if one knows that the standard deviation of the measurements in one population is higher than the standard deviation of the measurements in another population, this indicates that there is more dispersion in the former population than in the latter.[5]

If the body of data is a sample rather than a population, the formulas for the variance and the standard deviation are somewhat different from those used for the entire population. Specifically, the sample variance, denoted by s^2, is defined as

$$s^2 = \frac{\sum_{i=1}^{N} (X_i - \overline{X})^2}{n - 1}.$$

(2.8)   *Formula for s^2*

And the sample standard deviation, denoted by s, is defined as

$$s = \sqrt{\frac{\sum_{i=1}^{n} (X_i - \overline{X})^2}{n - 1}}.$$

(2.9)   *Formula for s*

There are three differences between these formula and those given for the population variance and standard deviation. First, the *sample* mean $\overline{X}$ is substituted for the *population* mean μ. Second, the squared deviations from the mean are summed over all measurements in the *sample,* not the *population.* Third, the sum of the squared deviations from the mean is divided by $(n - 1)$, not by N.

Generations of statistics students have been puzzled over why the denominator of the sample variance is $(n - 1)$, while the denominator of the population variance is N. Basically, the reason is that the sample variance, if its denominator were n, would tend to underestimate the population variance. (More will be said about this in later chapters.)

CALCULATING THE STANDARD DEVIATION

To illustrate the computations involved in calculating the population standard deviation, let's take two cases in which we have all the data in the population and thus are able to compute the population standard deviation. The first case is the population of incomes in a house where there are two families, one with an income of $20,000, the other with an income of $30,000 per year. Since there are only two

[5] However, if the means of the two populations differ, a better procedure for comparing their dispersion may be to calculate the coefficient of variation in each population. (See exercise 2.38.)

numbers in this population ($20,000 and $30,000), the population mean is $25,000 and the deviations from the mean (in thousands of dollars) are

$$(X_1 - \mu) = 20 - 25 = -5,$$

$$(X_2 - \mu) = 30 - 25 = 5.$$

Thus, the mean of the squared deviations (the variance) equals

$$\frac{\sum_{i=1}^{2} (X_i - \mu)^2}{2} = \frac{(-5)^2 + (5)^2}{2} = 25,$$

and the standard deviation is[6]

$$\sqrt{\frac{\sum_{i=1}^{2} (X_i - \mu)^2}{2}} = \sqrt{25} = 5 \text{ (thousands of dollars)}$$

Next, let's take a somewhat more complicated case—that of the voter turnout rates of the 51 states (including the District of Columbia). What is the standard deviation of these voter turnout rates? The first column of Table 2.4 shows each state's voter turnout rate, and the second column shows the amount by which it deviates from the mean (that is, $X_i - \mu$, where, as we found in an earlier section, $\mu = 54.47$). In the third column, the square of this deviation—that is, $(X_i - \mu)^2$—is calculated for each state. The sum of these squared deviations is shown to equal 3,142.2739:

$$\sum_{i=1}^{51} (X_i - \mu)^2 = 3,142.2739.$$

Thus,

$$\frac{\sum_{i=1}^{51} (X_i - \mu)^2}{N} = \frac{3,142.2739}{51} = 61.613$$

---

[6] In this very simple case, the standard deviation equals the deviation of each observation from the mean. For populations containing more than two observations, this generally will not be the case.

and

$$\sigma = \sqrt{\frac{\sum\limits_{i-1}^{51} (X_i - \mu)^2}{N}} = \sqrt{61.613} = 7.8.$$

Consequently, the standard deviation of the voter turnout rates in the 51 states (including the District of Columbia) is 7.8 percentage points.

Having calculated the standard deviation for the population in two cases, let's turn now to the calculation of the sample standard deviation; at the same time, we will illustrate how the standard deviation can be calculated from a frequency distribution. Specifically, we will estimate the standard deviation of the ages when the 101 teenagers in

X_i	$X_i - \mu$	$(X_i - \mu)^2$	X_i	$X_i - \mu$	$(X_i - \mu)^2$
64.6	10.13	102.6169	43.5	−10.97	120.3409
57.2	2.73	7.4529	40.6	−13.87	192.3769
57.8	3.33	11.0889	41.3	−13.17	173.4489
59.0	4.53	20.5209	48.6	− 5.87	34.4569
58.6	4.13	17.0569	50.0	− 4.47	19.9809
61.0	6.53	42.6409	48.8	− 5.67	32.1489
47.9	− 6.57	43.1649	48.8	− 5.67	32.1489
54.9	0.43	0.1849	52.0	− 2.47	6.1009
51.9	− 2.57	6.6049	51.4	− 3.07	9.4249
55.4	0.93	0.8649	53.1	− 1.37	1.8769
57.7	3.23	10.4329	52.3	− 2.17	4.7089
57.8	3.33	11.0889	44.9	− 9.57	91.5849
59.9	5.43	29.4849	65.1	10.63	112.9969
67.2	12.73	162.0529	68.0	13.53	183.0609
70.1	15.63	244.2969	53.2	− 1.27	1.6129
63.0	8.53	72.7609	56.0	1.53	2.3409
58.7	4.23	17.8929	51.0	− 3.47	12.0409
64.8	10.33	106.7089	44.7	− 9.77	95.4529
67.6	13.13	172.3969	64.6	10.13	102.6169
56.8	2.33	5.4289	41.2	−13.27	176.0929
56.8	2.33	5.4289	57.5	3.03	9.1809
54.6	0.13	0.0169	61.5	7.03	49.4209
50.0	− 4.47	19.9809	49.0	− 5.47	29.9209
35.5	−18.97	359.8609	57.8	3.33	11.0889
47.6	− 6.87	47.1969	43.6	−10.87	118.1569
52.9	− 1.57	2.4649			
			Total		3,142.2739

Table 2.4
Calculation of $\Sigma(X_i - \mu)^2$ for the Population of Voter Turnout Rates of the 51 States (Including the District of Columbia)

the sample in Table 2.2 first tried smoking. As in the calculation of the arithmetic mean from a frequency distribution in equation (2.3), we assume that the *midpoint of each class interval (that is, the class mark) can be used to represent each value of the measurements in that class interval.* This means that we assume that the ages of the teenagers in the "6 and under 8 years" class interval can be approximated by 7 years, that the ages of the teenagers in the "8 and under 10 years" class interval can be approximated by 9 years, and so on. Thus the sample standard deviation can be approximated by

$$
s = \sqrt{\frac{\sum_{j=1}^{k} f_j(X_j' - \overline{X})^2}{n - 1}} \tag{2.10}
$$

where f_j is the number of measurements in the jth class interval, X_j' is the midpoint of the jth class interval, and k is the number of the class intervals.

Table 2.5 shows how the formula for the sample standard deviation in equation (2.10) can be applied to the frequency distribution of the ages when teenagers first tried smoking. The first column of Table 2.5 provides the class intervals of this frequency distribution. The second column gives the number of measurements in each class interval (the value of f_j). The third column shows the midpoint of each class interval (the values of X_j'). Since we know from our earlier discussion (on page 44) that the sample mean is 13.46 years, the values of $(X_j' - \overline{X})$ are as shown in the fourth column, and the values of $(X_j' - \overline{X})^2$ are as shown in the fifth column. Finally, the sixth column shows the product of f_j and $(X_j' - \overline{X})^2$. Summing up the figures in the sixth column, we have the sum of the squared deviations from the sample mean, which is 823.0516. Dividing this sum by 100, which is $(n - 1)$, we get 8.2305, the sample variance. Taking the square root of 8.2305, we get 2.9, the sample standard deviation.

SHORTCUTS IN CALCULATING THE VARIANCE AND STANDARD DEVIATION

Modern electronic computers are often used to compute the variances and standard deviations of populations and samples. The advent of computers has transferred a great deal of drudgery from human beings to these mechanical aids. However, since many computations of this sort are still done by hand, it is useful to note that modifications of

Calculation of the Standard Deviation of the Ages When the Sample of 101 Teenagers First Tried Smoking

Age (years)	Number (f_j) of teenagers	Midpoint (X'_j) of class interval	$X'_j - \overline{X}$	$(X'_j - \overline{X})^2$	$f_j(X'_j - \overline{X})^2$
6 and under 8	6	7	−6.46	41.7316	250.3896
8 and under 10	9	9	−4.46	19.8916	179.0244
10 and under 12	11	11	−2.46	6.0516	66.5676
12 and under 14	25	13	−0.46	.2116	5.2900
14 and under 16	30	15	1.54	2.3716	71.1480
16 and under 18	20	17	3.54	12.5316	250.6320
					823.0516

Using equation (2.10), we find that

$$s = \sqrt{\frac{\sum\limits_{j=1}^{k} f_j (X'_j - \overline{X})^2}{n-1}} = \sqrt{\frac{823.0516}{100}} = 2.9.$$

Using the shortcut in equation (2.13),

$$s = \sqrt{\frac{\sum\limits_{j=1}^{k} f_j X'^2_j - \frac{1}{n}\left(\sum\limits_{j=1}^{k} f_j X'_j\right)^2}{n-1}}$$

Since

$$\sum\limits_{j=1}^{k} f_j X'^2_j = 6(7^2) + 9(9^2) + 11(11^2) + 25(13^2) + 30(15^2) + 20(17^2)$$
$$= 19{,}109$$

and

$$\sum\limits_{j=1}^{k} f_j X'_j = 6(7) + 9(9) + 11(11) + 25(13) + 30(15) + 20(17) = 1{,}359$$

$$s = \sqrt{\frac{19{,}109 - \frac{1}{101}(1{,}359)^2}{100}} = \sqrt{\frac{19{,}109 - 18{,}285.95}{100}} = 2.9$$

Thus, the answer is the same as for equation (2.10). Except for the effects of rounding errors, these answers must always be identical.

equations (2.8) and (2.9) often simplify the calculations. Specifically, these modifications are

$$s^2 = \frac{\sum\limits_{i=1}^{n} X_i^2 - \frac{1}{n}\left(\sum\limits_{i=1}^{n} X_i\right)^2}{n-1} \tag{2.11}$$

and

$$s = \sqrt{\frac{\sum\limits_{i=1}^{n} X_i^2 - \frac{1}{n}\left(\sum\limits_{i=1}^{n} X_i\right)^2}{n-1}}. \tag{2.12}$$

Similarly, if you are calculating the sample standard deviation based on a frequency distribution, it is often quicker and easier to use the following modification of equation (2.10) to carry out the calculations:

$$s = \sqrt{\frac{\sum\limits_{j=1}^{k} f_j X_j'^2 - \frac{1}{n}\left(\sum\limits_{j=1}^{k} f_j X_j'\right)^2}{n-1}}. \tag{2.13}$$

These modifications do not alter the answers given by the formulas in equations (2.8), (2.9), and (2.10). They are merely different ways of obtaining the same result. To verify this, see the continuation of Table 2.5 on page 63 which shows how the formula in equation (2.13) is applied to the frequency distribution of ages when teen-agers first tried smoking. Clearly, the result is the same as when the formula in equation (2.10) was used, but the calculations are easier and less time-consuming.

INTERPRETATION OF THE STANDARD DEVIATION

The standard deviation is the most important summary measure of dispersion. If the frequency distribution of a population conforms to the so-called normal distribution (to be discussed in detail in Chapter 5), then we know the percentage of measurements in the population that fall within 1, 2, or 3 standard deviations of the population mean. Specifically, 68.3 percent of the measurements lie within ± 1 standard deviation of the mean, 95.4 percent of the measurements lie within ± 2 standard deviations of the mean, and 99.7 percent of the measurements lie within ± 3 standard deviations of the mean.

Thus, if we know that the population of verbal SAT scores of the entering freshmen at a particular university conforms to the normal

How Often Are Psychologists' Numbers Wrong?

In every scientific discipline, observations sometimes are wrong because the people who record the observations are fallible, and mistakes are made. In 1978, Robert Rosenthal of Harvard University[7] reviewed the results of a number of published experiments in psychology, and determined from the basic data whether the experimenter had misrecorded any of his or her observations. For each of 15 experiments, the percent of the observations that were incorrectly recorded was as follows:

1.13	0.23	0.95
0.67	4.17	2.50
1.69	1.59	0.00
0.72	0.82	0.69
3.17	0.41	0.62

As you can see, the incidence of such errors is generally very small.

(a) Construct a frequency distribution of the percentage of observations that were incorrectly recorded.

(b) Do a stem-and-leaf diagram of the data.

(c) According to Rosenthal, "Getting *all* the errors out is probably not possible or even desirable . . ."[8] Why not? Isn't it always better to eliminate errors of this sort?

[7] R. Rosenthal, "How Often Are Our Numbers Wrong?" *American Psychologist*, November 1978, pp. 1005–8.

[8] Ibid., p. 1007.

distribution with a mean of 550 and a standard deviation of 30, it follows that 68.3 percent of the entering freshmen will have scores of 520 to 580, that 95.4 percent will have scores of 490 to 610, and that 99.7 percent will have scores of between 460 and 640. This is useful information. For example, if entering freshmen with scores below 490 or above 610 are to be put in remedial or advanced English classes, it follows that 4.6 percent of them will be put in such classes.

Turning to an actual case, consider the study by Sandra Hofferth and Kristin Moore of childbearing by young women in the United States.[9] The study included 1,268 women who were 27 years old sometime between 1971 and 1975 and who had had a child. According to

[9] S. Hofferth and K. Moore, "Early Childbearing and Later Economic Well-Being," *American Sociological Review*, October 1979.

their results, the mean age at the birth of a first child was 21.58 years for whites and 19.57 years for blacks; the standard deviation was 3.29 years for whites and 3.02 years for blacks. Thus, in this sample, white women tended to have their first child at a later age than black women, and their age at first birth tended to be somewhat more variable than that of black women.

EXERCISES

2.11 A sample is composed of the following five weights (in pounds): 1.8, 1.9, 2.1, 2.3, 2.0. Calculate the variance and standard deviation.

2.12 A finite population consists of the seven integers: 3, 4, 5, 6, 7, 8, 9. Compute the variance and standard deviation.

2.13 A sample of five tires is chosen, the diameters of the tires being 30.01, 30.02, 30.03, 30.03, and 30.02 inches.
(a) Calculate the variance and standard deviation.
(b) Subtract 30 inches from each observation in the sample, and calculate the variance and standard deviation.
(c) Explain why your results in (a) and (b) are or are not equal.

2.14 Use the shortcut formulas in equations (2.11) and (2.12) to compute the variance and standard deviation of the ages in Exercise 2.2.

2.15 Based on the frequency distribution in Exercise 2.5, use the shortcut formula in equation (2.13) to calculate the standard deviation of the daily cost of commuting, assuming that the 100 workers are a sample. If the 100 workers were the population, how would equation (2.13) have to be altered, and what would be the standard deviation?

2.16 Based on the data in Exercise 2.1, what is the range of the number of classes per month missed by the 18 students?

2.17 Mary Malone takes standardized test A, which has been given to 10,000 people. For these people, the mean score was 80 and the standard deviation was 10. Annie Laurie takes standardized test B, which has been given to 12,000 people. For these people, the mean score was 75 and the standard deviation was 12. If Mary scores 95 and Annie scores 94, which of the two does better relative to others who have taken the same test?

2.18 The mean and standard deviation of the prices charged by movie theaters in Tucson and Seattle for admission to a particular movie are:

City	Mean	Standard deviation
Tucson	$4.96	$0.38
Seattle	5.22	0.45

The Bijou Theater in Tucson charges $5.50, and the Biloxi Theater in Seattle charges $5.75. Which is more expensive relative to the other theaters in the same city?

2.19 Which of the following statements is true?
(a) The median equals one-half of the sum of the first and third quartiles.
(b) The first quartile equals the 25th percentile.
(c) The third quartile equals the 75th percentile.

2.20 In an Ohio township, the frequency distribution of the sizes of all residential lots is given below:

Size of lot (acres)	Number of lots
1/4	300
1/2	400
1	200
2	100

(a) What is the standard deviation of the lot sizes?
(b) What is the variance of the lot sizes?
(c) What is the range of the lot sizes?

2.21 Based on the data in Exercise 2.9 of this chapter, (a) what is the range of the annual numbers of successful launchings in 1971–76? (b) What is the range in 1977–81? (c) What is the range in 1971–81? (d) Can the range for 1971–76 or for 1977–81 be greater than for 1971–81?

2.22 A salesman made 100 visits to customers. The frequency distribution of the amount of commission he made per visit is as follows:

Amount of commission (dollars)	Number of visits
0 and under 20	60
20 and under 40	30
40 and under 60	10

Can you determine the range of the amount of commission earned per visit by the salesman? Can you obtain upper and/or lower bounds for this range? Explain.

2.6 How Descriptive Statistics Can Be Misused

The famous British prime minister Disraeli once said that there were lies, damned lies, and statistics. You are no doubt already on guard against many kinds of misuses of statistics. However, it is important for you to develop as much skill in rooting out statistical fallacies and chicanery as possible, since the world is full of pitfalls for the statistically unwary. This section covers five kinds of errors that frequently occur in descriptive statistics. These are errors that can result in costly mistakes, and even trained statisticians occasionally fall prey to them.

INAPPROPRIATE COMPARISONS. Suppose you read in the newspaper that the crime rate in your area is 2 percent lower than last year. Before accepting this conclusion, it would be wise to question whether the *definition* of crime on which these statistics were based has remained the same from last year to the present. In other words, might the apparent decrease in the crime rate be due to the fact that some types of behavior were defined as crimes last year which are no longer classified as such? Also, do the figures pertain to the same kind of population during the two periods, or could the apparent drop in the crime rate be due to the fact that the figures for the most recent period pertain to a different set of people than the earlier figures? For example, the later figures might represent all individuals, including young children, whereas the earlier ones might pertain only to individuals over 10 years old. Furthermore, are the later and earlier periods really comparable? Perhaps the later figures pertain to a portion of the year when crime is always relatively low, whereas the earlier figures pertain to a full year. Unless you can be reasonably certain that each of these possible discrepancies is nonexistent or relatively minor it is difficult, if not impossible, to interpret this kind of statistical item in the newspaper.

VERY SMALL OR OBVIOUSLY BIASED SAMPLES. Suppose you receive a report stating that a new device will increase the number of miles an automobile can travel per gallon of gasoline by 5 percent. On the surface this certainly sounds good, but this figure of 5 percent might be based on the effect of the device on the mileage of only three cars. If so, the figure of 5 percent may not apply to your car, because so small a sample may not be at all representative of the entire population of automobiles. Therefore, it is generally wise to ask about the size of the sample on which a particular statistical result is based. This is a good question to ask even in areas like medicine and the natural sciences, since despite the scientific and precise nature of the work, results sometimes are based on very small samples.

Some organizations and people use statistics as a drunk uses a lamppost: for support, not illumination. Thus individuals or firms sometimes carry out studies on one small sample after another until eventually and by chance one of the samples indicates what they want to prove about the product they sell. Then they publicize the results of this small sample, hoping thereby to influence their customers or potential customers. The makers of the new device to increase mileage per gallon may have tested the effect of their invention on one three-car sample after another. (They may also have discarded the results of many earlier small samples that showed the device had no effect!) Finally, when one sample indicated, by chance, a 5 percent increase in mileage per gallon, they may have publicized the results. Clearly, statistics based on such improperly selected samples can be worse than valueless.

IMPROPER CHOICE OF AVERAGE. For a given set of data, a person can use one measure of central tendency to indicate one thing and another measure of central tendency to indicate another. A real estate statesman trying to impress a potential customer with the high income level in a particular suburb quotes the *mean* family income in this suburb as $40,000 per year. His statement is perfectly true, but it is also misleading because the income *distribution* in the suburb happens to be highly skewed to the right. Perhaps 2 percent of the suburb's residents earn about $750,000 per year, and the rest earn between $20,000 and $30,000 per year. Thus, the mean of $40,000 is not a very representative figure.

On the other hand, the mayor of this particular suburb is trying to impress the state government with the *low* income level of the area (and hence, with its need for state aid). In contrast to the real estate salesman, the mayor chooses as a measure of central tendency the *mode* of the frequency distribution of families by income level. Because the income distribution is highly skewed to the right, the mode is likely to be considerably less than the mean and somewhat less than the median. The mode of the income distribution in the community may be only $24,000. Thus, as a measure of central tendency, the real estate salesman quotes $40,000, while the mayor quotes $24,000. Both are right; the difference is due to the fact that different measures of central tendency are being used. The moral, of course, is clear: Whenever someone quotes an "average" or a measure of central tendency, be sure to find out what kind of measure it is and how representative it is likely to be.

NEGLECT OF THE VARIATION ABOUT AN AVERAGE. Even if the proper kind of average is chosen, the average may be surrounded by so much variation that it alone may be misleading. It is sometimes argued that home builders, focusing their attention on the average number of people in the American family, build too many medium-sized homes and too few small and large ones. In other words, if the average family size in the United States is 3.6 persons, home builders may tend to neglect the considerable variation about this average, and build too many homes for families of 3 persons or 4 persons. (According to some observers, this has in fact occurred.)[10]

[10] See Darrell Huff and Irving Geis, *How to Lie with Statistics* (New York: Norton, 1954), part of which is contained in E. Mansfield, *Statistics for Business and Economics: Readings and Cases* (New York: Norton, 1980). Also, see Oskar Morgenstern's article in the latter book.

Another case of this sort is encountered in ads which announce that a certain toothpaste will result, on the average, in a such-and-such percent reduction in dental cavities. Besides the fact that some of these claims may be subject to some of the other problems discussed above, they may be misleading because there may be such great variation about the average. What is important to a particular consumer is whether the toothpaste will reduce his or her cavities. Because the variation among individuals in the effects of the toothpaste may be so great, the chance that it will have a beneficial effect on one particular person may be scarcely better than 50–50, even though on the average it may offer some protection against tooth decay.

MISINTERPRETATION OF GRAPHS AND CHARTS. Sometimes graphs and charts are presented so as to give a misleading impression. For example, a real estate salesman wants to run an ad showing the increase over time in the average price of houses in the suburb where he works. He decides to use a *pictogram,* a graph in which the size of the object in the picture indicates the relative size of the thing the object represents. The salesman uses the pictogram in Figure 2.5 to show that the average price of a house (represented by the size of the deed) doubled between 1965 and 1986. This chart certainly makes it appear that real estate prices have risen, and indeed, it is misleading in this respect. Why? Because the salesman doubled both the width and the height of the deed in Figure 2.5, thus *quadrupling* the area of the deed. Thus, based on the *area* of the deed, the chart makes it appear that real estate prices have quadrupled, not doubled.

Figure 2.5
Pictogram Showing the
Exaggerated Increase
in Average Price of
Houses, 1965–1986

Another frequently encountered error occurs in the construction of histograms where the class intervals are not of equal width. The frequency distribution of voter turnout rates of the 51 states (including the District of Columbia) is reproduced in Table 2.6. If we combine the first three class intervals in Table 2.6, we obtain the frequency distribution in Table 2.7. We then construct bars on the line segments of the horizontal axis corresponding to the classes of the frequency distribution, and we make the height of each bar equal to the number of cases in each class interval. The result is shown in panel A of Figure 2.6.

THE CASE OF THE NEW TOOTH POWDER

MISTAKES TO AVOID

Some years ago, a new brand of tooth powder was introduced in the United States. According to its advertisements, studies showed that it had "considerable success" in improving the health of a person's teeth. Suppose that these studies were designed in the following way. The manufacturer of this tooth powder chose three samples of seven people; the first sample agreed to use the new brand, the second sample used Brand X, and the third sample used Brand Y. After six months, a dentist examined each person's teeth and scored the health of his or her teeth on a scale of 0 (poorest possible score) to 100 (highest possible score). The results were as follows:

New brand: 65, 71, 53, 55, 34, 82, 77
Brand X   : 58, 60, 63, 90. 95, 89, 62
Brand Y   : 54, 38, 43, 61, 94, 96, 82

The manufacturer of the new tooth powder says that the average score for the new brand is 2 points higher than for Brand X and 4 points higher than for Brand Y. Would you agree with the manufacturer's advertisements? If not, what mistakes have been made?

SOLUTION: There are at least four errors or problems. (1) The data pertain to the health of a person's teeth, not to the extent of its improvement. (2) The manufacturer seems to be using the median, rather than the mean. One suspects that this choice was prompted by the fact that the median for the new brand (65) is higher than for Brand X (63) or Brand Y (61). The mean for the new brand (62.4) is less than for Brand X (73.9) or Brand Y (66.9). Unless some good reason is given for preferring the median over the mean, the results are suspect. (3) The samples are small, and the differences among the averages may be due merely to chance. (Ways of testing whether this is the case will be discussed in Chapter 8.) (4) The respondents who used the new brand "agreed" to do so, which suggests that the sample was not random and that the people who used the new brand may differ systematically in dental health from those who used the other brands. Advertisements based on statistical methods of this sort may be misleading and costly both in terms of consumers' pocketbooks and their health (if the advertised products are in fact less effective than other products).

Table 2.6
Frequency Distribution
of 1980 Voter Turnout
Rates, by State
(Including the District
of Columbia)

Turnout rate (percent)	Number of states
35.0 and under 40.0	1
40.0 and under 45.0	7
45.0 and under 50.0	6
50.0 and under 55.0	12
55.0 and under 60.0	14
60.0 and under 65.0	6
65.0 and under 70.0	4
70.0 and under 75.0	1
Total	51

Table 2.7
Frequency Distribution
of Voter Turnout
Rates, Unequal Width
of Class Intervals

Turnout rate (percent)	Number of states
35.0 and under 50.0	14
50.0 and under 55.0	12
55.0 and under 60.0	14
60.0 and under 65.0	6
65.0 and under 70.0	4
70.0 and under 75.0	1
Total	51

Comparing panel A of Figure 2.6 with Figure 1.2, it is clear that panel A gives a distorted picture. The reason is simple: In the class interval from 35.0 to 50.0 percent, there are 14 states, which means that on the average there are 4⅔ states for every five percentage points in this class interval. Thus, to be comparable with the other vertical bars, the height of the bar for this class interval should be 4⅔, not 14. Put differently, the *area* of each bar in a histogram should be proportional to the number of observations in the relevant class interval. (Recall Chapter 1.) Thus, since area equals width times height, the height of the bar in the 35.0–50.0 percent class interval should be 14 divided by 3 because this class interval is three times as wide as the others. The resulting histogram, corrected in this way, is shown in panel B of Figure 2.6. Put bluntly, panel A is wrong, panel B is right.

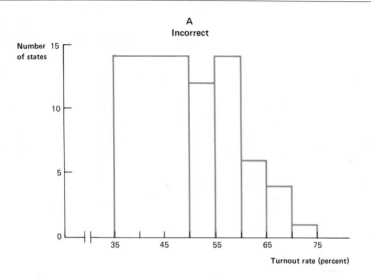

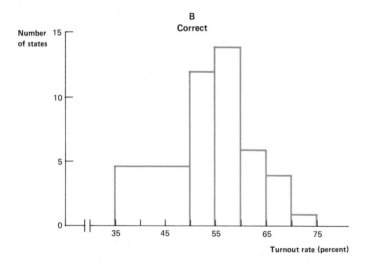

Figure 2.6
Incorrect and Correct
Histograms of Voter
Turnout Rates,
Unequal Widths of
Class Intervals

2.7 Removing Snow from the Big Apple

An actual crisis that faced the city of New York illustrates how some of the statistical techniques described in this and the previous chapter have been used to help solve important real problems. On February 9, 1969 a major snowstorm hit New York and paralyzed the city for days. Mayor John Lindsay, faced with a consequent political crisis, asked a group of statisticians and systems analysts to carry out a thorough study to find out what, if anything, was wrong with the city's snow-removal procedures.

Clearly, the first question the study faced was: How unusual was this snowstorm? That is, was it much more severe than those typically experienced in New York? In order to answer this question, the statisticians obtained U.S. Weather Bureau records of the depth of each snowfall in New York City in recent years. Based on these records, the statisticians constructed the frequency distribution shown in Table 2.8. The results indicate that the 1969 storm, with a snowfall of over 15 inches, was unusual indeed. (From 1948 to 1967 only three storms resulted in more than 15 inches of snow.)

The next question was: How much work has to be done to cope with a snowstorm? The three basic activities that must be performed are spreading salt, plowing, and snow removal; and the amount of each of these activities depends on the number of street-miles to be serviced. The Department of Sanitation divides all streets into three priority classes: primary, secondary, and tertiary (the clearance of primary streets being the highest priority and the clearance of tertiary streets the lowest). The analysts further subdivided the primary streets into emergency streets (ones of the very highest priority) and other primary streets. Based on this classification, the statisticians determined the number of miles of New York's streets in each of these classes. The resulting frequency distribution is shown in Table 2.9.

Next, the study had to investigate the city's existing capacity for spreading salt, plowing, and snow removal. The main types of equipment used to cope with snowstorms are spreaders and plows. Based on data regarding the frequency of downtime on such equipment, it was

Table 2.8
Frequency Distribution of Depths of Snowstorms, 1948–67, New York City

Depth (inches)	Number of snowstorms during 20-year period (1948–67)
1 and under 3	67
3 and under 5	26
5 and under 7	13
7 and under 9	4
9 and under 11	3
11 and under 13	3
13 and under 15	2
15 and under 17	3

Source: E. Savas, "The Political Properties of Crystalline H_2O: Planning for Snow Emergencies in New York," *Management Science*, October 1973.

Priority class	Miles of streets
Primary	
Emergency	1600
Other	930
Secondary	1978
Tertiary	1331
Total	5839

Table 2.9

Miles of New York City Streets, by Snow-Clearance Priority

Source: See Table 2.8.

assumed that about 40 percent of the spreaders and plows would be out of commission at the onset of a storm. As a result, it was concluded that about 134 spreaders and 1,050 plows would be usable. Based on estimates of the time taken to reach various streets and the length of rest breaks, meal breaks, and refueling time, as well as the speed at which spreaders and salters could move along the the snowy streets, it was possible to determine how long it would take the existing number of spreaders and plows to service the city's street network. The results indicated that "there is sufficient equipment available, in the aggregate, to plow *every* mile of *every* street in the city in only six hours, and to plow the high-priority streets in less than two hours."[11]

This finding, based on the simplest kind of statistical analysis, was startling, since it indicated that the city had enough plowing capability to do the required work in a very short period of time. Why, then, had it taken so long to clean up the snowstorm of February 9? The analysts reasoned that the distribution of snow plows might have been the problem and to find out they obtained data concerning the distribution of plows among the various boroughs of New York. The results, shown in Table 2.10, indicate clearly that relative to their number of miles of primary streets, Manhattan, Brooklyn, and the Bronx had a relatively large number of plows, whereas Queens and Richmond had relatively few. It was therefore no wonder that snow removal was relatively slow in Richmond, which with over 9 percent of the city's mileage of primary streets, had only about 4 percent of the city's plows.

Why were the plows so poorly allocated, and how could the allocation be improved? The reason for the misallocation was linked to the fact that snow plows are merely sanitation trucks fitted with plows. The city had allocated these vehicles among geographical areas in accord with their primary function, refuse collection. Thus, densely popu-

[11] E. Savas, "The Political Properties of Crystalline H_2O: Planning for Snow Emergencies in New York," *Management Science,* October 1973, p. 142.

Table 2.10
Allocation among
Boroughs of Plows and
Primary Street
Mileage, New York
City

Borough	Borough's percentage of total mileage of primary streets	Borough's percentage of city's snow plows
Manhattan West	8.6	9.2
Manhattan East	6.5	10.7
Bronx West	6.6	8.9
Bronx East	9.9	8.4
Brooklyn West	9.6	11.0
Brooklyn North	7.5	11.5
Booklyn East	6.8	9.8
Queens West	14.7	12.8
Queens East	20.8	13.5
Richmond	9.2	4.0
Total*	100.00	100.0

Source: See Table 2.8.
* Due to rounding errors, the percentages do not sum to exactly 100.0.

lated areas like Manhattan and Brooklyn, which produce much more refuse per street-mile than other areas, received a relatively large number of these vehicles. To improve the allocation of plows, the analysts recommended that plows be mounted on vehicles other than refuse trucks and that these other vehicles be distributed so as to make the plowing capability of each borough proportional to its mileage of primary streets.[12]

The results of the snow-crisis study were presented to Mayor John Lindsay in June 1969. Since this was a time when Lindsay was running for reelection, and since the snow-removal crisis of the previous February figured prominently in the campaign, it is not surprising that the mayor called for immediate adoption and implementation of the study's recommendations. During the following winter New York's new snow-emergency plan went into operation and, based on experience to date, seems to have been a success. During the next four years, only one big snowstorm occurred, and this time the snow was cleaned up within a few hours.

EXERCISES

2.23 "There were more civilian than military amputees during the war. During the period of the war, 120,000 civilians suffered amputations, but

[12] Other recommendations were made as well. This brief sketch cannot do justice to the study. For further details, see E. Savas, op. cit.

only 18,000 military personnel."[13] Does this prove that civilians were more likely to suffer amputations? If not, where does the fallacy lie?

2.24 A national magazine once published a story saying that farmers lead other groups in the consumption of alcohol. As evidence, it pointed to the fact that a rehabilitation center in rural Illinois treated more farmers than other occupational groups for alcoholism. Do you regard this evidence as unbiased? Why, or why not?

2.25 Data have been published which indicate that the more children a couple has, the less likely the couple is to get a divorce. Does this indicate that increases in the number of children are related causally to the likelihood of divorce? Why, or why not?

2.26 Based on the data in Exercise 2.10, a real estate agent says that the typical size of a lot in the relevant township is .54 acres. What sort of average is the agent using? How many lots are of this "typical" size?

2.27 In 1973, the total net income per farm was $19,685 in Iowa and $19,456 in Minnesota. Based on these data, a television commentator maintains that Iowa farmers were better off than Minnesota farmers in 1973. Do you consider this statement to be very meaningful? What sort of pitfall is present here? What would be a better way of interpreting the data concerning these two states?

2.28 "Patents are of little value since the Supreme Court invalidates most of the patents that come before it."[14] Do you agree with this statement? If not, in what way does it represent a misuse of statistics?

2.29 Based on the data in Table 2.8, a newspaper reporter writes that the average snowfall in New York is less than 2 inches and consequently the expense of preparing for snowfalls of more than 10 inches is far in excess of the potential benefits. Do you agree? If not, in what way is this statement a misuse of statistics? What sort of pitfall is present here?

2.30 The following is a distribution of the prices charged for a particular type of psychological test by 100 psychologists in Los Angeles.

$80 and under $90	41
$90 and under $100	9
$100 and under $110	6
$110 and under $120	44
	100

(a) A UCLA psychology professor says that the median price for such a test is about $100. Is this correct?
(b) Is $100 a good measure of central tendency? Why or why not?

---

[13] W. A. Wallis and H. Roberts, *Statistics* (Glencoe: Free Press, 1956), p. 91.
[14] Ibid., p. 98.

The Chapter in a Nutshell

1. There are several frequently used *measures of central tendency:* the *mean,* the *median,* and the *mode.* The *mean* is the sum of the numbers contained in the body of data divided by how many numbers there are. The *median* is a figure which is chosen so that one-half of the numbers in the body of data are below it and one-half are above it. The *mode* is the number that occurs most often. These three kinds of measures may differ substantially from one another. For example, if there are a few extremely high observations (as in most income distributions) the mean will be considerably higher than the median.

2. *Measures of variability or dispersion* tell us how much variation there is among the numbers in a body of data. Perhaps the simplest measure of variability is the *range,* which is defined as the difference between the highest and the lowest number in the body of data. However, the most important measure of dispersion is the *standard deviation,* which is defined as the square root of the mean of the squared deviations of the observations from their mean. The square of the standard deviation is called the *variance.*

3. There are many misuses of descriptive statistics. Figures are sometimes presented in such a way that they seem comparable when in fact they are based on different definitions, concepts, time periods, areas, and so forth. Sometimes figures are presented which are based on very small or obviously biased samples. Individuals and organizations sometimes choose the type of average that supports their case best, even if this information is misleading. An average is sometimes presented which has so much variability about it that the average alone is misleading. In addition, graphs and charts (such as pictograms) are sometimes presented so as to give a misleading impression. Be on your guard against improper statistical procedures of this sort.

Chapter Review Exercises

2.31 Do you think that the median amount paid by Americans in income tax in 1985 was less than the mean amount paid? Explain.

2.32 (a) A Medical Aptitude Test has been given to 6,000 males and 4,000 females who were premedical students at U.S. colleges and universities. Given the mean score for males and the mean score for females, describe a simple procedure for obtaining the mean score for all people who have taken the test.

(b) Given the median score for males and the median score for females, can you determine the median score for all people who have taken the test from this information alone?

2.33 (a) For *any* set of measurements, what is the sum of the deviations of these measurements from their mean?

(b) A set of measurements has a symmetrical distribution, and the median is 3. If there are 1,000 measurements, can you calculate their sum?

2.34 A salesman made 100 visits to customers. The frequency distribution of the amount of commission he earned per visit is as follows:

Amount of commission (dollars)	Number of visits
0 and under 20	60
20 and under 40	30
40 and under 60	10

(a) Calculate the mean amount of his commission per visit. Estimate the total commission he earned for all 100 visits.

(b) Calculate the median amount of commission earned by the salesman per visit. Based on information solely concerning this median, can you tell whether the total commission earned for all 100 visits exceeded $2,000? Why, or why not?

2.35 Use the data in Exercise 2.9 to calculate the standard deviation of the annual numbers of successful launchings (a) in 1971–76, (b) in 1977–81, (c) in 1971–81. (Assume in each case that you are dealing with the entire population.)

2.36 Based on the data in Exercise 2.34, is the distribution of commissions earned by the salesman skewed to the left? To the right?

2.37 The standard deviation is a measure of *absolute* dispersion or variability, and it is affected by the units of measurement. To illustrate this fact, calculate the standard deviation of the amount of commission per visit in Exercise 2.34. Then express the commissions in cents (not dollars) and calculate the standard deviation. What is the ratio of the latter standard deviation (for commissions expressed in cents) to the former standard deviation (for the commissions expressed in dollars)? Why?

2.38 To obtain a measure of dispersion or variability that is normalized for the units of measurement, statisticians often use a measure of *relative* variability such as the *coefficient of variation*. The coefficient of variation equals the standard deviation as a percentage of the mean. In other words, for a population it equals

$$V = \frac{\sigma}{\mu} \cdot 100.$$

(a) Prove that the coefficient of variation of the commissions in Exercise 2.34 is the same whether the commissions are measured in cents or dollars.

(b) From the data in Exercise 2.9 calculate the coefficient of variation of the annual number of successful launchings in 1971–76.

(c) From the data in Exercise 2.9 calculate the coefficient of variation of the annual number of successful launchings in 1977–81.

(d) Based on the coefficient of variation, were the annual numbers of successful launchings more variable in 1971–76 than in 1977–81?

2.39 In the fall of 1973, 8,442 men and 4,321 women applied for admission to the Graduate Division of the University of California, Berkeley. About 44 percent of the men and 35 percent of the women were admitted.[15]

(a) Was this evidence of a sex bias in admissions?

(b) Admissions were made separately for each major. Upon closer examination, it turned out that, major by major, the percentage of women accepted was about the same as the percentage of men accepted. Is this consistent with the above facts? Does there still appear to be a sex bias in admissions?

2.40 According to Sheldon and Eleanor Glueck,[16] a large percentage of juvenile delinquents are middle children (not the first or last born).

(a) Does this imply that being a middle child contributes to delinquency?

(b) Studies have shown that there is a strong direct relationship between family size and delinquency. Can this help to explain the Gluecks' results?

* 2.41 The life expectancy (in years) in 30 countries is as follows:[17]

United States	73	Germany	72	Soviet Union	70
Argentina	65	Greece	72	Spain	73
Australia	74	India	47	Sri Lanka	64
Bangladesh	46	Indonesia	46	Sweden	75
Belgium	71	Italy	73	Thailand	61
Brazil	60	Japan	76	Turkey	57
Canada	73	Mexico	60	United Kingdom	73
Chile	62	Nigeria	41	Venezuela	63
Egypt	54	Pakistan	48	Yugoslavia	70
France	73	Poland	71	Zaire	39

* Exercise requiring computer package, if available.

[15] D. Freedman, R. Pisani, and R. Purves, *Statistics* (New York: Norton, 1978).

[16] S. and E. Glueck, *Unraveling Juvenile Delinquency* (Cambridge: Harvard University Press, 1950).

[17] *Statistical Abstract of the United States, 1982–83* (Washington, D.C.: Bureau of the Census 1982), p. 862.

Use Minitab (or some other computer package) to

(a) construct a histogram of these data;

(b) calculate the mean of these data;

(c) calculate the standard deviation of these data (assuming that they are a sample).

If no computer package is available, do the calculations by hand.

Appendix 2.1

RULES OF SUMMATION

In this chapter, we encountered Σ, the mathematical summation sign. Since this sign will be used frequently in later chapters, it is worthwhile to summarize some of the rules of summation. We know from this chapter's discussion that

$$\sum_{i=1}^{n} X_i = X_1 + X_2 + \cdots + X_n.$$

From this fact, we can establish the validity of the following three rules.

The first rule is:

$$\sum_{i=1}^{n} aX_i = a \sum_{i=1}^{n} X_i,$$

where a is a constant. To prove that this rule is correct, note that

$$\sum_{i=1}^{n} aX_i = aX_1 + aX_2 + \cdots + aX_n$$

$$= a(X_1 + X_2 + \cdots + X_n)$$

$$= a \sum_{i=1}^{n} X_i,$$

which proves the rule.

The second rule is

$$\sum_{i=1}^{n} a = na.$$

To prove that this rule is correct, note that

$$\sum_{i=1}^{n} a = a \sum_{i=1}^{n} 1$$

$$= a(\underbrace{1 + 1 + \cdots + 1}_{n \text{ terms}})$$

$$= na,$$

which proves the rule.

The third rule is

$$\sum_{i=1}^{n} (X_i + Y_i) = \sum_{i=1}^{n} X_i + \sum_{i=1}^{n} Y_i.$$

To prove that this rule is correct, note that

$$\sum_{i=1}^{n} (X_i + Y_i) = X_1 + Y_1 + X_2 + Y_2 \cdots + X_n + Y_n$$

$$= (X_1 + X_2 + \cdots + X_n) + (Y_1 + Y_2 + \cdots + Y_n)$$

$$= \sum_{i=1}^{n} X_i + \sum_{i=1}^{n} Y_i,$$

which proves the rule.

Finally, let's also consider the concept of double summation. The expression

$$\sum_{i=1}^{n} \sum_{j=1}^{m} X_i Y_j$$

means the sum of the products of X and Y where X takes its first, second, $\cdots$, nth values and Y takes its first, second, $\cdots$, mth values. For example,

$$\sum_{i=1}^{3} \sum_{j=1}^{2} X_i Y_j = X_1 Y_1 + X_1 Y_2 + X_2 Y_1 + X_2 Y_2 + X_3 Y_1 + X_3 Y_2.$$

The following example illustrates the use of these rules of summation.

EXAMPLE 2.1 If $X_1 = 4$, $X_2 = 6$, and $X_3 = -3$, evaluate the following sums:

(a) $\sum\limits_{i=1}^{3} X_i$

(b) $\sum\limits_{i=1}^{2} X_i^2$

(c) $\sum\limits_{i=1}^{3} 3X_i$

SOLUTION: (a) $\sum\limits_{i=1}^{3} X_i = 4 + 6 - 3 = 7$

(b) $\sum\limits_{i=1}^{2} X_i^2 = 4^2 + 6^2 = 52$

(c) $\sum\limits_{i=1}^{3} 3X_i = 3 \sum\limits_{i=1}^{3} X_i = 3(7) = 21$

Appendix 2.2

THE BOX PLOT

Another graphical technique which is used to show the characteristics of a frequency distribution is the box plot. A box plot consists of a box, which extends from the lower (i.e., first) quartile of the frequency distribution to the upper (i.e., third) quartile of the distribution.[18] Thus, in Figure 2.7, the lower quartile of the distribution is 200, and the upper quartile is 300. This is obvious because the box extends from 200 to 300. Also, a solid line is drawn across the box at the location of the median. Thus, in Figure 2.7, it is evident that the median of the distribution is 240, since this is where the solid line is drawn.

A quick glance at a box plot can tell us a great deal about the location, dispersion, and skewness of a frequency distribution. In Figure 2.7, we can readily determine the median, since (as just indicated) it is at the point where a solid line is drawn across the box. The median provides us with a measure of

[18] More precisely, the box plot uses the *hinges* of a distribution, which may depart slightly from the quartiles. For details, see L. Ott, R. Larson, and W. Mendenhall, *Statistics: A Tool for the Social Sciences,* third edition (Boston: Duxbury Press, 1983).

Figure 2.7
A Box Plot

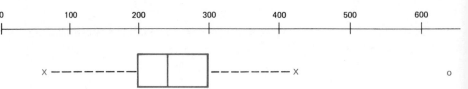

central tendency. To measure dispersion, we can calculate the interquartile range from the graph, since the width of the box equals the interquartile range. Information concerning the skewness of the distribution can be gathered by seeing whether the median is closer to the lower or upper quartile.

Besides the box, a box plot contains a dashed line between each quartile and its *adjacent value.* (The lower adjacent value is the most extreme observation in the distribution between the lower quartile and $Q_1 - 1.5R$, where Q_1 is the lower quartile and R is the interquartile range. The upper adjacent value is the most extreme observation in the distribution between the upper quartile and $Q_3 + 1.5R$, where Q_3 is the upper quartile and R is the interquartile range.) In addition, each adjacent value is marked with an x, and *extreme outliers* (observations less than $Q_1 - 3R$ or greater than $Q_3 + 3R$) are marked with the symbol °, as shown in Figure 2.7. The location of the dashed line and of these symbols tells us a great deal about the presence and number of observations that are far from the average.

Probability and Probability Distributions

3

OBJECTIVES

To understand statistical methods, you must have some familiarity with probability theory. The overall purpose of this chapter is to define what is meant by a probability and to indicate how one can calculate probabilities. Among the specific objectives are:

1　To present the frequency and subjective concepts of probability.

2　To state and apply the addition rule for probabilities.

3　To define and apply the concepts of conditional probability and independence.

4　To state and apply the multiplication rule for probabilities.

Introduction to Probability

3.1 Introduction

Statistical theory and practice rest largely on the concept of probability since, as stressed in Chapter 1, any conclusion concerning a population based only on a sample is subject to a certain amount of uncertainty. The concept of probability is by no means unfamiliar. For example, based on the past examination questions given by a professor, you make rough judgments as to the probability that he or she will include certain kinds of questions in a forthcoming test. Or based on the past performance of certain football teams, you make rough estimates of the probability that one team will defeat another. This chapter will provide an introduction to probability theory; in subsequent chapters, we shall build on this foundation and go deeper into probability theory.

3.2 Experiments, Sample Spaces, and Events

EXPERIMENTS AND SAMPLE SPACES

Any probability pertains to the results of a situation which we call an experiment. *An experiment is any process by which data are obtained through the observation of uncontrolled events in nature or through controlled procedures in a laboratory.* If you roll a die, this is an experiment which may have one of six outcomes, depending on which number comes up. Or if you flip a coin, this is also an experiment, which has as possible outcomes either a head or a tail. Or if you take a statistics course, this too is an experiment (although you may not prefer to think of it that way), which has as possible outcomes your passing or your not passing.

Sample Space

Any experiment can result in various outcomes. For example, one possible outcome of your experiment with the die is that a 4 may come up. And one possible (and hopefully very likely) outcome of your experiment with the statistics course is that you will pass it. *The sample space is the set of all possible outcomes that may occur as a result of a particular experiment. For example, the sample space for the experiment in which you rolled the die is*

$$S = \{1, 2, 3, 4, 5, 6\}. \qquad (3.1)$$

In other words, S is a set composed of the numbers that can come up when a die is thrown; that is, it is a set composed of $1, 2, \ldots, 6$. The symbol S is conventionally used to designate the sample space, and the outcomes in this set are called *elements* of the sample space.

It is important to recognize that the outcomes in a sample space need not be numbers. In the case of your experiment with the statistics course, there are two possible outcomes: You pass, or you don't pass. In this case, the sample space is

$$S = [\text{pass, not pass}].$$

Sometimes it is convenient in cases of this sort to designate the elements of the sample space as 0 and 1, where 0 stands for "pass" and 1 stands for "not pass." Thus, the sample space becomes

$$S = \{0, 1\}.$$

It frequently is useful to represent the sample space visually. One way to do this is to express the possible outcomes of the experiment as points on a graph. For example, if two dice are rolled simultaneously, then a graph showing the sample space can be drawn where the number coming up on the first die is located along the horizontal axis, and the number coming up on the second die is located along the vertical axis. This is shown in Figure 3.1. Each of the 36 points in the graph represents a possible outcome of the roll of the dice.

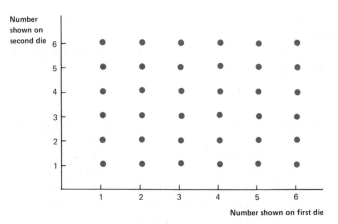

Figure 3.1
Sample Space for
Simultaneous Roll of
Two Dice

Another way to represent the sample space, particularly when an experiment is carried out in stages, is to use a *tree diagram*. Each fork in such a diagram shows the possible outcomes that may occur at a certain stage of the experiment. For example, if you take examinations two days in a row, on the first day there are two possibilities: You pass, or you don't pass. These possibilities are represented by the first fork in Figure 3.2. On the second day there are again two possibilities: Either you pass or you don't pass. These possibilities are represented by the two forks to the right of the first fork. The upper fork represents the outcomes if you passed on the first day, whereas the bottom fork represents the outcomes if you did not pass on the first day. At the right-hand end of the tree diagram we wind up with four points representing the four elements in the sample space, which are (1) passing on neither the first day nor the second; (2) passing on the first day but not on the

Tree Diagram

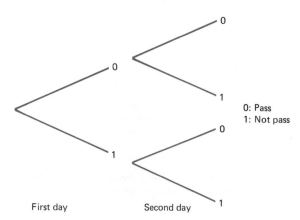

0: Pass
1: Not pass

Figure 3.2
Sample Space for Two
Days of Examinations

second; (3) passing on the second day but not on the first; and (4) passing on both days.

The following two examples should help illustrate the important concept of sample space.

EXAMPLE 3.1 Cornell University's Medical College published a study in 1981 of second opinions by physicians concerning elective surgery.[1] This is an important topic because it often is alleged that considerable unnecessary surgery is performed. Each patient, after receiving word from his or her physician as to whether or not surgery was advisable, contacted another physician and received a second opinion. What was the sample space for each patient?

SOLUTION. We can depict this sample space as four points on a graph, as shown in the figure below. In this graph, a 0 on the horizontal axis means that the first physician advised against surgery, while a 1 means that he or she recommended surgery. On the vertical axis, a 0 means that the second physician advised against surgery, while a 1 means that he or she recommended surgery.

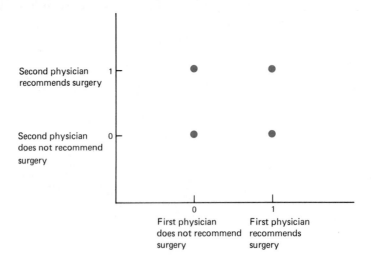

[1] M. Finkel, H. Ruchlin, and S. Parsons, *Eight Years' Experience with a Second Opinion Elective Surgery Program* (Washington, D.C.: Health Care Financing Administration, 1981).

EXAMPLE 3.2 Each year, Americans (and particularly New Yorkers) wonder whether there will be a massive electric power failure in New York City (such as occurred in July 1977). Next year, there may or may not be a failure; and in the year after next, there may or may not be one. (Only God and perhaps Consolidated Edison, the local power company, know for sure.) Use a tree diagram to depict the sample space for the next two years.

SOLUTION: We can depict this sample space by constructing the graph below, which shows a fork for the first year leading to a 0 (if a failure occurs) or a 1 (if a failure does not occur). Whether or not there is a failure in the first year, another such fork is given for the second year. Thus, at the right-hand end of the graph, we arrive at the four points in the sample space.

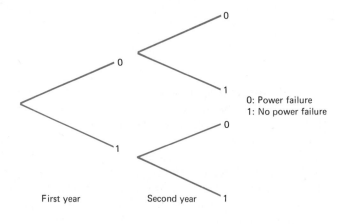

0: Power failure
1: No power failure

First year   Second year

EVENTS

One of the most important functions of probability theory is to enable us to calculate the probability of an event. *An **event** is a subset of a sample space.* A *subset* is any part of a set (including the whole set, which has all the elements in the sample space, and the empty set, which has no elements at all). Put less formally, an event can be defined as a group of zero, one, two, or more outcomes of an experiment. For example, if the experiment is rolling a particular die, one event is that the number that comes up is odd. Another event is that the number that comes up is a 3 or a 6. Still another event is that the number that comes up is a 5. There are many practical reasons (other than preparing for trips to Las Vegas or Atlantic City) why statisticians need to be able to calculate the probability of an event.

3.3 Probabilities

WHAT IS A PROBABILITY?

Frequency Definition of Probability

The probability of an event is the proportion of times that this event occurs over the long run if the experiment is repeated many times under uniform conditions. Thus, the probability that a particular die will come up a 1 is the proportion of times this will occur if the die is thrown many, many times; and the probability that the same die will come up a 2 is the proportion of times this will occur if the die is thrown many, many times. And so on.

In general, *if an experiment is repeated a very large number of times M, and if event A occurs m times, the probability of A is*

$$P(A) = \frac{m}{M} \qquad\qquad (3.2)$$

Thus, if a die is "true" (meaning that each of its sides is equally likely to come up when the die is rolled), the probability of its coming up a 1 is 1/6, because if it is rolled many, many times, this will occur one-sixth of the time. Moreover, even if the die is not true, this definition can be applied. Suppose that a local mobster injects some loaded dice into a crap game, and that one of the players (who is suspicious) asks to examine one of them. If he rolls this die, what is the probability that it will come up a 1? To answer the question, we must imagine the die in question being rolled again and again. After many thousands of rolls, if the proportion of times that it has come up a 1 is 0.195, then this is the probability of its coming up a 1.

Based on our definition of a probability, the following three fundamental propositions must be true:

1. *The probability of an impossible event must be zero.* This follows from the definition of a probability in equation (3.2), because if an event is impossible, the number of times the event occurs (that is, m) must equal zero.

2. *The probability of an event that is certain must equal 1.* This also follows from the definition of a probability in equation (3.2), because if an event is certain, the number of times the event occurs (m) must equal the number of times the experiment takes place (that is, M). Note that this implies that the probability that some element of the sample space occurs is 1, since the sample space includes all possible outcomes.

3. *The probability of any event must be no less than zero and no*

greater than 1. This, too, follows from the definition of a probability in equation (3.2). Since the number of times any event occurs (*m*) cannot be negative, its probability cannot be less than zero. Since the number of times any event occurs cannot exceed the number of times the experiment takes place (*M*), its probability cannot exceed 1.

CALCULATING THE PROBABILITY OF AN EVENT

The probability that an event will take place is the sum of the probabilities of the outcomes that comprise the event. For example, if a true die is rolled, what is the probability that it comes up with an odd number? The event in which the die comes up with an odd number can be broken down into three outcomes: (1) the die comes up a 1; (2) the die comes up a 3; and (3) the die comes up a 5. In other words, these three outcomes comprise this event. Since the probability of each of these outcomes is 1/6, the probability of this event is $1/6 + 1/6 + 1/6 = 1/2$. This is true because (1) if any of the outcomes comprising the event occurs, the event itself occurs, and (2) more than one outcome cannot occur simultaneously.

A somewhat more complex illustration involves the opinions of physicians concerning elective surgery. Suppose that two patients, Jonathan and Theresa, each ask the opinions of three physicians. Under these conditions, the sample space showing the number of physicians that recommend surgery for each patient is shown in Figure 3.3. For simplicity, let's assume that each of the 16 elements of this sample space is equally likely, which means that the probability of each point in Figure 3.3 is 1/16. (This implies, of course, that there is considerable dis-

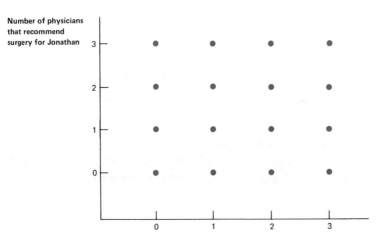

Figure 3.3
Sample Space for
Opinions of Physicians

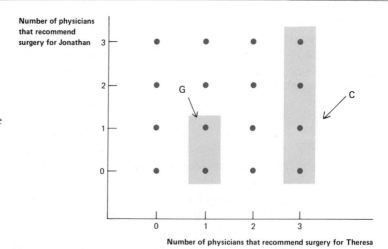

Figure 3.4
Subset of Sample Space
Corresponding to Two
Events

agreement over whether surgery is necessary.) Given these probabilities of the outcomes, we can readily calculate the probabilities of various other events.

To illustrate, what is the probability that three physicians recommend surgery for Theresa and that three or less recommend surgery for Jonathan? This event includes the 4 elements in the sample space designated as subset C in Figure 3.4. Thus, the probability of this event is the sum of the probabilities of these 4 outcomes, or $1/16 + 1/16 + 1/16 + 1/16 = 1/4$. What is the probability that one physician recommends surgery for Theresa and that less than two recommend surgery for Jonathan? This event includes the 2 elements in the sample space designated as subset G in Figure 3.4. Thus, the probability of this event is the sum of the probabilities of these 2 outcomes, or $1/16 + 1/16 = 1/8$.

PROBABILITY OF EITHER EVENT A OR EVENT B OR BOTH

Composite events

A composite event is an event that is defined by using combinations of other events. One type of composite event occurs if *either* or *both* of two other events occur. For example, if A is the event that you pass your psychology course and B is the event that you pass your statistics course, your graduation may depend on *either* or *both* of these events occurring. If so, the composite event that you graduate will occur only if *either* event A or event B *(or both)* occurs. Turning to the illustration involving physicians' opinions concerning surgery, consider the composite event that *either* (1) less than two physicians recommend

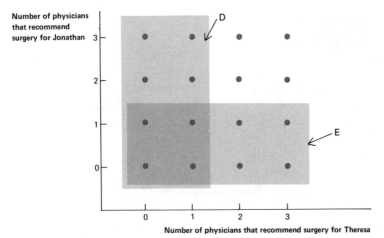

Figure 3.5
Subset of Sample Space
Corresponding to
Either of Two Events

surgery for Theresa; or (2) less than two physicians recommend surgery for Jonathan; or (3) *both* (that is, less than two physicians recommend surgery for both). Subset *D* of the sample space in Figure 3.5 is the set of outcomes where less than two physicians recommend surgery for Theresa (event 1). Subset *E* is the set of outcomes where less than two physicians recommend surgery for Jonathan (event 2). *The set of outcomes where either or both of these events occurs is composed of all outcomes in either subset D or E.* This set consists of the 12 points in the shaded area in Figure 3.5. Since the probability of each of these points is 1/16, the probability of this composite event is 12/16.

PROBABILITY OF BOTH EVENT A AND EVENT B

A second type of composite event occurs *only* if *both* of two other events occur. Thus, if *A* is the event that you pass your psychology course and *B* is the event that you pass your statistics course, your receiving a good job offer may be dependent on the occurrence of *both*. If so, the event that you receive a good job offer will occur only if *both* events *A* and *B* occur. Once again, take the case of the physicians' opinions concerning surgery. What is the probability that *both* Jonathan and Theresa receive recommendations for surgery from less than two of the physicians? This event occurs only if *both* of the following events occur: (1) Theresa receives less than two such recommendations, and (2) Jonathan receives less than two such recommendations. Since the set of outcomes comprising the first event is subset *D* of the

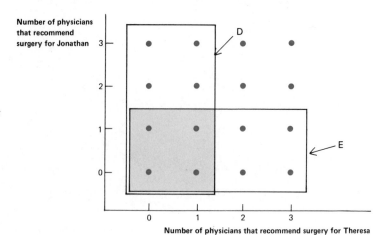

Figure 3.6
Subset of Sample Space
Corresponding to Both
of Two Events

sample space in Figure 3.6 and the set of outcomes comprising the second event is subset *E* of the sample space in Figure 3.6, *the set of outcomes comprising both of these events is composed of all outcomes in both subsets D and E.* In other words, it is the set of 4 points in the shaded area in Figure 3.6 where subsets *D* and *E* overlap. Since the probability of each of these points is 1/16, the probability of this composite event is 4/16.

EXERCISES

3.1 List all of the outcomes if a coin is flipped three times. If it is a fair coin, what is the probability that each of these outcomes will occur?

3.2 A box contains two white balls and one black ball. Two balls are to be taken from this box. After the first ball is selected, it will not be put back in the box. List all possible outcomes of this experiment. What is the probability that each of these outcomes will occur?

3.3 A true die has its 4-spot changed to a 2-spot. When tossed, what is the probability of obtaining (a) 3, (b) 2, (c) 4, (d) 2 or less, (e) 5 or more?

3.4 A school counselor says that the odds are 2 to 1 that John Cicero, a junior, will not be able to finish high school successfully. He also says that the probability that John will neither be able to finish high school successfully nor to finish college is 0.70. Are his two statements consistent? Explain.

3.5 A gambler makes a single roll with a pair of true dice. What is the probability that each of the following numbers comes up?
(a) Either a 7 or an 11
(b) Either a 2, a 12, or both

3.4 Addition Rule

CASE OF TWO EVENTS

As pointed out in the previous section, statisticians and decision makers frequently must calculate the probability that at least one of two events occurs. For instance, if a fair coin is flipped twice and we want to calculate the probability that it comes up heads at least once, we know that this will occur if (1) the coin comes up heads on the first flip; or (2) the coin comes up heads on the second flip. The probability of the first event is $1/2$, and the probability of the second event is $1/2$. The probability of both events (that is, that the coin will come up heads *both* times) is $1/4$. Can we use these probabilities to determine the likelihood that the coin will come up heads at least once in the two flips?

To solve problems of this sort, statisticians have devised the so-called addition rule, which is as follows.

ADDITION RULE: *If* A *and* B *are two events, and the probability of* A *is denoted by* P(A) *and the probability of* B *is denoted by* P(B), *then the probability of either* A *or* B *(or both), denoted by* P(A *or* B), *equals* P(A) + P(B) − P(*A and* B), *where* P(*A and* B) *is the probability that both* A *and* B *will occur.*

Formula for P(A or B)

Let's define *A* as the event that the coin comes up heads on the first flip, and *B* as the event that the coin comes up heads on the second flip. Then *P(A)* is the probability of heads on the first flip, which is $1/2$. And *P(B)* is the probability of heads on the second flip, which is $1/2$. And *P(A and B)* is the probability of heads on both the first and second flips, which is $1/4$. Thus, the probability of at least one heads is

$$P(A \text{ or } B) = P(A) + P(B) - P(A \text{ and } B) = 1/2 + 1/2 - 1/4 = 3/4.$$

This result accords with Figure 3.7, which shows the relevant sample space. There are four possible outcomes in this sample space. The outcomes correspond to
   (1) heads/heads (heads first flip, heads second flip)
   (2) heads/tails (heads first flip, tails second flip)
   (3) tails/heads (tails first flip, heads second flip)
   (4) tails/tails (tails first flip, tails second flip)

Each outcome has a probability of $1/4$. Since at least one heads comes up in three of these four outcomes, the probability of at least one heads is $3/4$, which accords with the result obtained from the addition rule.

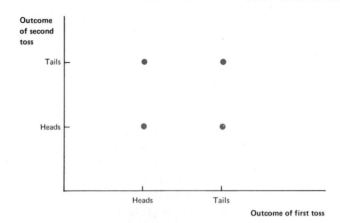

Figure 3.7
Sample Space for Two
Flips of a Coin

This example indicates why *P(A and B)* must be subtracted from the sum of *P(A)* and *P(B)* to obtain *P(A or B)*. If *P(A and B)* were not subtracted, the outcome where both events *A* and *B* occur (that is heads/heads) would be counted twice, since it is included in the outcomes where *A* (heads on the first flip) occurs, as well as in the outcomes where *B* (heads on the second flip) occurs. If it were not subtracted, the answer would be incorrectly given as 1, not 3/4.

MUTUALLY EXCLUSIVE EVENTS

In some cases, two events are **mutually exclusive;** that is, they cannot occur together. For example, if you roll a die once, it is impossible for the die to come up with *both* a 4 *and* a 5, since only one number can appear at a given time. Thus, the event that "the die shows a 4" and the event that "the die shows a 5" are mutually exclusive. *If two events are mutually exclusive, the probability that they both occur is zero; consequently, if the events,* A *and* B, *are mutually exclusive* P(A and B) = 0, with the result that we have the following addition rule for mutually exclusive events.

> ADDITION RULE FOR MUTUALLY EXCLUSIVE EVENTS: *If* A *and* B *are two mutually exclusive events, the probability of either* A *or* B, *denoted by* P(A *or* B), *equals* P(A) + P(B).

Clearly, this addition rule follows directly from the addition rule given earlier in this section since, if *P(A and B)* = 0, the earlier addition rule implies that this rule will be true. To illustrate the present addition rule, let's define *A* as "a die shows a 4" and *B* as "a die shows a 5." As we have seen, these two events are mutually exclusive (if only one die is thrown only once), so $P(A$ or $B) = P(A) + P(B) = 1/6 + 1/6 = 1/3$. Thus, the probability of either a 4 or a 5 is $1/3$.

CASE OF MORE THAN TWO EVENTS

Both the general addition rule and the special addition rule for mutually exclusive events can be extended to cases where more than two events are considered. For present purposes it is not necessary to give the extension of the more general addition rule, but it is important to give the extension of the addition rule for mutually exclusive events, which is as follows.

ADDITION RULE FOR ANY NUMBER OF MUTUALLY EXCLUSIVE EVENTS: *If* E_1, $E_2 \cdots E_n$ *are* n *events, the probability of* E_1 *or* E_2 *or* $E_3 \cdots$ *or* E_n, *denoted by* $P(E_1$ *or* E_2 *or* $E_3 \cdots$ *or* $E_n)$, *equals* $P(E_1) + P(E_2) + P(E_3) + \cdots + P(E_n)$, *if these events are mutually exclusive.*

To illustrate the use of this rule, let's define E_1 as "a die shows a 1," E_2 as "a die shows a 2," E_3 as a "die shows a 3," and E_4 as "a die shows a 4." Then since these four events are mutually exclusive (so long as only one die is thrown once), $P(E_1$ or E_2 or E_3 or $E_4) = P(E_1) + P(E_2) + P(E_3) + P(E_4) = 1/6 + 1/6 + 1/6 + 1/6 = 2/3$. Thus, the probability of a 1, a 2, a 3, or a 4 is $2/3$.

COMPLEMENT OF AN EVENT

The **complement** of an event occurs when the event itself does not occur. Thus, since the event or its complement is sure to occur, and since the event and its complement are mutually exclusive, the addition rule implies that if $P(A)$ is the probability that event A will occur, and if P (not A) is the probability that event A will not occur, then

$P(A) + P$ (not A) = 1,

or

$P(A) = 1 - P$ (not A).

This result is useful because, if it is easier to determine P (not A) than $P(A)$, then $P(A)$ can be obtained by deducting P (not A) from 1. Suppose that we want to determine the probability that if two dice are thrown, the number that comes up *will not equal* 2. The complement of this event is that the number *will equal* 2. Since the probability of this complementary event is $1/36$, the probability that the number will not equal 2 must be $1 - 1/36$, or $35/36$.

APPLICATIONS

The addition rule has a host of important applications. The following two examples illustrate ways in which it is used.

EXAMPLE 3.3 A new chemical product is about to be introduced commercially, and a great many chemical firms are competing to be first to put this product on the market. An industrial analyst believes that the probability is 0.30 that DuPont will be first, 0.15 that Dow will be first, 0.15 that Monsanto will be first, and 0.15 that Union Carbide will be first. Based on the analyst's beliefs, what is the probability that any one of these four firms will be first, assuming that a tie does not occur?

SOLUTION: Let E_1 be the event that DuPont is first, E_2 be the event that Dow is first, E_3 be the event that Monsanto is first, and E_4 be the event that Union Carbide is first. If a tie is not allowed, these events are mutually exclusive, since more than one firm cannot be first. Thus, $P(E_1$ or E_2 or E_3 or $E_4) = P(E_1) + P(E_2) + P(E_3) + P(E_4) = 0.30 + 0.15 + 0.15 + 0.15 = 0.75$. Consequently, the probability that any one of these four firms will be first is 0.75.

EXAMPLE 3.4 The Congressional Budget Office (CBO) gives summer jobs to two political science students, Mary Carp and John Minelli. The head of CBO hopes that at least one of these students will decide to go to work for CBO upon graduation. Based on past experience, it is believed that the probability that each student will decide to work for CBO is 0.3, and the probability that both will decide to work for CBO is 0.1. What is the probability that the hopes of the head of CBO will be fulfilled?

SOLUTION: Let E_1 be the event that Mary Carp will decide to go to work for CBO, and E_2 be the event that John Minelli will decide to do so. Since $P(E_1) = 0.3$, $P(E_2) = 0.3$, and $P(E_1$ and $E_2) = 0.1$,

$$P(E_1 \text{ or } E_2) = 0.3 + 0.3 - 0.1 = 0.5.$$

Thus, the probability that the hopes of the head of CBO will be fulfilled equals 0.5.

EXERCISES

3.6 Forty percent of the members of Congress favor bill *A*, and 20 percent favor bill *B*. Ten percent favor both bills *A* and *B*. What is the probability that a member of Congress favors either bill *A* or bill *B* (or both)?

3.7 A psychologist presents a subject with a list of nonsense syllables, the list is removed, and the subject is asked to recall the list. The probability is .12 that the subject will recall the first syllable on the list; the probability is .15 that the subject will recall the second syllable on the list; and the probability is .02 that the subject will recall them both. What is the probability that the subject will recall either syllable (or both)?

3.8 A criminologist says that the probability that an 18-year-old high school drop-out will be arrested for theft is 0.05, and that the probability that he or she will be arrested for either theft or homicide (or both) is 0.06.
(a) Is the probability that he or she will be arrested for homicide equal to 0.01? Why or why not?
(b) Is the probability that he or she will be arrested for homicide less than 0.01? Why or why not?

3.9 Coin *A* is loaded in such a way that heads are three times as likely to occur as tails. Coin *B* is loaded in such a way that tails are three times as likely to occur as heads. If both coins are tossed once, what is the probability that *at least* one comes up heads? (Note: The probability that *both* coins come up heads equals .1875.)

3.5 Multiplication Rule

The addition rule is very useful in cases where we are interested in the probability that *at least one* of several events will take place. Thus, in the previous section, we used it to determine the probability that *at least one* heads would come up in two flips of a true coin. In many situations, however, the statistician or decision maker is interested in the probability that *all* of several events will occur. We may want to determine the probability that *both* flips of a coin will be heads. Or we may want to determine the probability that *all four* of a firm's salesmen will be sick tomorrow. In cases of this sort, the multiplication rule, not the addition rule, is the one to apply. To understand the multiplication rule, it is essential that you be familiar with the concepts of joint probability, marginal probability, and conditional probability.

JOINT PROBABILITIES

In our discussion of the addition rule, we used *P*(*A* and *B*), the probability that *both* events *A* and *B* will occur. This is an example of a

Table 3.1
10,000 Persons
Classified by Attitude
toward Increased
Defense Spending and
by Political Party

Political party	Favors increased defense spending (A)	Does not favor increased defense spending (C)	Total
Democrat (B)	2,500	3,500	6,000
Republican (D)	2,500	1,500	4,000
Total	5,000	5,000	10,000

joint probability, which is defined as the *probability of the joint occurrence of two or more events.* As a concrete illustration of joint probabilities, consider Table 3.1, which shows the results of a hypothetical survey in which 10,000 individuals were asked whether or not they favored increased defense spending. As indicated in Table 3.1, 6,000 of those asked were Democrats and 4,000 were Republicans. Let A be the event that a person favored increased defense spending, let C be the event that he or she did not favor increased defense spending, let B be the event that the person was a Democrat, and let D be the event that he or she was a Republican.

Joint probabilities can be illustrated as follows. If a person is chosen at random from this group of 10,000, the joint probability that he or she is a Democrat and favors increased defense spending is

$$P(A \text{ and } B) = \frac{2,500}{10,000} = 0.25.$$

Similarly, the probability that a randomly selected person is a Democrat and does not favor increased defense spending is

$$P(B \text{ and } C) = \frac{3,500}{10,000} = 0.35.$$

The probability that a randomly selected person is a Republican and favors increased defense spending is

$$P(A \text{ and } D) = \frac{2,500}{10,000} = 0.25.$$

The probability that a randomly selected person is a Republican and does not favor increased defense spending is

$$P(C \text{ and } D) = \frac{1,500}{10,000} = 0.15.$$

Political party	Favors increased defense spending (A)	Does not favor increased defense spending (C)	Marginal probabilities
Democrat (B)	0.25	0.35	0.60
Republican (D)	0.25	0.15	0.40
Marginal probabilities	0.50	0.50	1.00

Table 3.2
Joint Probability Table for 10,000 Persons Classified by Attitude toward Increased Defense Spending and by Political Party

These joint probabilities can be shown in a *joint probability table* like Table 3.2. The probabilities in Table 3.2 are obtained by dividing each of the numbers in Table 3.1 by the total number of persons in the survey (10,000).

MARGINAL PROBABILITIES

In addition to showing the joint probabilities just mentioned, Table 3.2 shows the probability that a randomly chosen person is a Democrat, is a Republican, is in favor of increased defense spending, or is not in favor of increased defense spending. These probabilities, contained in the margins of the joint probability table, are called *marginal probabilities or unconditional probabilities*. (The name *marginal* stems from the position of these probabilities in the margins of the table.) For example, the marginal probability that a randomly chosen person favors increased defense spending is 0.50. The marginal probability that a randomly chosen person in this group is a Democrat is 0.60.

Unconditional Probability

Each marginal probability can be derived by summing the appropriate joint probabilities. For example, the marginal probability that a person favors increased defense spending is the sum of (1) the joint probability that a person is a Democrat and favors increased defense spending; and (2) the joint probability that a person is a Republican and favors increased defense spending. That this is true follows from the addition rule.[2] Consequently, if we did not know the marginal probability that a person favors increased defense spending, but knew the

[2] A person favors increased military spending if at least one of the following events occurs: (1) the person is a Democrat and in favor of increased military spending; (2) the person is a Republican and in favor of increased military spending. Since a person cannot be both a Democrat and a Republican, these two events are mutually exclusive. Thus, according to the addition rule, the probability that a person favors increased military spending equals the sum of the probability of the first of these events and the probability of the second of these events.

joint probabilities, we could find the marginal probability by adding .25 (the joint probability that a person is a Democrat and favors increased defense spending) and .25 (the joint probability that a person is a Republican and favors increased defense spending). The result of adding the two is of course .50.

CONDITIONAL PROBABILITIES

Statisticians frequently are interested in how the probability of one event is influenced by whether or not another event occurs. *The probability that one event will occur, given that another event is certain to occur, is a* **conditional probability.** In Table 3.2 we may be interested in calculating the probability that a person favors increased defense spending, given that he or she is a Democrat. This conditional probability is denoted by $P(A \mid B)$ and is read "the probability of A, given B." (Recall that event A is a person's favoring increased defense spending and event B is his or her being a Democrat.) The vertical line in $P(A \mid B)$ is read "given," and the event following the line (B in this case) is the one that is certain to occur.

Regardless of which events A and B stand for, *the conditional probability of* A, *given that* B *must occur, is*

Formula for P(A | B)

$$P(A \mid B) = \frac{P(A \text{ and } B)}{P(B)}.$$

(3.3)

In other words, the conditional probability of A, given the occurrence of B, is the joint probability of A and B divided by the marginal probability of B. (To rule out the possibility of dividing by zero it is assumed, of course, that the marginal probability of B is nonzero.)

To illustrate the use of the above definition, let's return to Table 3.2 and calculate the probability that a person favors increased defense spending, given that he or she is a Democrat. Since $P(A \text{ and } B) = .25$ and $P(B) = .6$,

$$P(A \mid B) = \frac{P(A \text{ and } B)}{P(B)} = \frac{.25}{.60} = .42.$$

This conditional probability is .42. To verify the correctness of this result we can go back to Table 3.1, which shows that the proportion of Democrats favoring increased defense spending is 2,500/6,000, or .42. Thus, the conditional probability of A given B is simply the proportion of times that A occurs out of the total number of times that B occurs.

STATEMENT OF RULE

Now that we have the definition of conditional probability in equation (3.3) at hand, the statement of the multiplication rule is quite simple. To obtain this rule we need only multiply both sides of equation (3.3) by $P(B)$, the result being as follows.

MULTIPLICATION RULE: *If* A *and* B *are two events, the joint probability that both* A *and* B *will occur equals the conditional probability of* A, *given* B, *times the probability of* B. *In other words,*

$$P(A \text{ and } B) = P(A \mid B)P(B).$$

(3.4)

*Formula for
P(A and B)*

Using the multiplication rule, we can determine the probability that a randomly selected person is both a Democrat and not in favor of increased defense spending. If we let C be the event that the person is not in favor of increased defense spending, and if we let B be the event that the person is a Democrat, then what we want to determine is $P(C$ and $B)$. We know that the probability that a person is a Democrat equals .6. In other words, $P(B) = .6$. We also know that the probability that a person is not in favor of increased defense spending, given that he or she is a Democrat, equals .58. That is, $P(C \mid B) = .58$. From these facts alone we can determine $P(C$ and $B)$, because the multiplication rule implies that

$$P(C \text{ and } B) = P(C \mid B)P(B)$$

$$= (.58)(.6) = .35.$$

Thus, the desired probability equals .35.

Going a step further, the multiplication rule can be extended to include more than two events, the result being as follows.

MULTIPLICATION RULE FOR n EVENTS: *If* E_1, E_2, $\cdots$, E_n *are n events, the joint probability that all these events will occur equals the probability of* E_1 *times the conditional probability of* E_2, *given* E_1, *times the conditional probability of* E_3, *given* E_1 *and* E_2, $\cdots$ *times the conditional probability of* E_n, *given* E_1, E_2, $\cdots$, E_{n-1}. *In other words,*

$$P(E_1 \text{ and } E_2 \text{ and } \cdots \text{ and } E_n) = P(E_1) \cdot P(E_2 \mid E_1) \cdot P(E_3 \mid E_1 \text{ and } E_2) \cdot \cdots \cdot P(E_n \mid E_1 \text{ and } E_2 \text{ and } \cdots \text{ and } E_{n-1}).$$

(3.5)

MISTAKES TO AVOID

THE CASE OF "CHUCK-A-LUCK"

"Chuck-a-luck" is a game that is often played at carnivals and gambling spots. The player pays a dollar to play. Three dice are thrown. If any 6's come up, the player gets back his or her dollar, plus one dollar for each 6 that comes up. Players often argue that this game is advantageous to them. They say that, since the probability of a 6 coming up on each die is $1/6$, the probability that the player will win is $1/6 + 1/6 + 1/6 = 1/2$. Thus, the player will win a dollar as often as he or she will lose one, and in addition will get an extra dollar when two 6's come up, and two extra dollars when three 6's come up. Are they right in believing that this game is advantageous to them? If not, what mistakes have they made?

SOLUTION

The probability that at least one 6 will come up is not $1/2$, because a 6 on the first die does not prevent a 6 on the second or third dies. Since they are not mutually exclusive, this probability is *not* $1/6 + 1/6 + 1/6 = 1/2$. (Recall the addition theorem.) Instead, the probability of exactly one 6 coming up is $75/216$; the probability of exactly two 6's coming up is $15/216$; and the probability of exactly three 6's coming up is $1/216$. (See footnote 3.) Suppose that a person plays this game repeatedly. In $125/216$ of the cases (over the long run), he or she will lose a dollar. In $75/216$ of the cases, he or she will win a dollar. In $15/216$ of the cases, he or she will win two dollars. In $1/216$ of the cases, he or she will win three dollars. Thus, on the average, the person will make

$$\frac{125}{216}(-\$1) + \frac{75}{216}(\$1) + \frac{15}{216}(\$2) + \frac{1}{216}(\$3) = -8 \text{ cents.}$$

[3] These probabilities can be derived as follows. There are three mutually exclusive ways in which exactly one 6 can come up. (Specifically, the first, second, or third die can show a 6.) The probability that each of these ways occurs is $(1/6)(5/6)(5/6)$. Thus, the probability of exactly one 6 is $(1/6)(5/6)(5/6) + (1/6)(5/6)(5/6) + (1/6)(5/6)(5/6) = 75/216$.

There are three mutually exclusive ways in which exactly two 6's can come up. (Specifically, the first, second, or third die can show other than a 6.) The probability that each of these ways occurs is $(1/6)(1/6)(5/6)$. Thus, the probability of exactly two 6's is $(1/6)(1/6)(5/6) + (1/6)(1/6)(5/6) + (1/6)(1/6)(5/6) = 15/216$.

There is only one way in which exactly three 6's can come up. (Specifically, all three dice must show a 6.) The probability that this occurs is $(1/6)(1/6)(1/6) = 1/216$.

In other words if this game is played repeatedly, the player will lose, on the average, 8 cents per play. Clearly, this game is not advantageous to the player.[4]

[4] This game is also discussed in W. A. Wallis and H. Roberts, *Statistics* (Glencoe: Free Press, 1956), pp. 332–33.

Consider the following example: (1) the probability that an undergraduate at the University of Michigan will go to graduate school is 1/2; (2) the probability that a Michigan undergraduate will get a master's degree, given that he or she goes to graduate school, is 2/3; and (3) the probability that a Michigan undergraduate will get a Ph.D., given that he or she goes to graduate school and gets a master's degree, is 1/5. What is the probability that a Michigan undergraduate will go to graduate school, will get a master's degree, and will get a Ph.D.? Using equation (3.5), this probability equals (1/2)(2/3)(1/5), or 1/15.

STATISTICAL INDEPENDENCE

The probability of the occurrence of an event is sometimes dependent on whether or not another event occurs. In Table 3.2, the probability that a person favors increased defense spending depends on whether the person is a Democrat or a Republican.[5] As we saw in a previous section, the probability that a person favors increased defense spending, given that he or she is a Democrat, equals .42. On the other hand, the probability that an individual favors increased defense spending, given that he or she is a Republican, equals .25/.40, or .62. Thus, these two events—a person's attitude toward increased defense spending and his or her political party—are *dependent* in the sense that the probability of one's occurring depends on whether or not the other occurs.

Not all events are dependent. For example, the probability that two physicians recommend surgery for Theresa, given that two recommend surgery for Jonathan, is equal to the marginal (or unconditional) probability that two recommend surgery for Theresa. (Both probabilities equal 1/4, as is evident from our discussion of Figure 3.3.) In other words, the probability that two recommend surgery for Jonathan is not influenced by whether or not two recommend surgery for Theresa. Similarly, if a fair coin is flipped a number of times, the probability of its coming up heads does not depend on whether the coin came up heads on the last flip. Why? Because the coin's behavior on one flip is

[5] Note once again that the figures in Table 3.2 are hypothetical. Neither Democrats, Republicans, nor independents should assume that we regard the figures as accurate. They are used only for illustrative purposes.

not influenced by its behavior on the last flip (or any other flip, for that matter). The definition of statistical independence is as follows.

> STATISTICAL INDEPENDENCE. *If events* A *and* B *are statistically independent, the probability of the occurrence of one event is not affected by the occurrence of the other. That is, each of the following equations is true:*

$$P(A \mid B) = P(A) \qquad (3.6a)$$

$$P(B \mid A) = P(B) \qquad (3.6b)$$

The example below illustrates how this definition can be used.

EXAMPLE 3.5 Two machines are drawn at random from a population of 10 machines, 3 of which are defective and 7 of which are not defective. The sampling is *not* carried out with replacement. (In other words, the first machine chosen is *not* put back into the population before the second machine is chosen.) Is whether or not the second machine is defective statistically independent of whether or not the first is defective?

SOLUTION: Given that the first machine selected is defective, what is the probability that the second machine selected will be defective? If the machine selected first is *not* put back into the population (and thus has no chance of being selected again), this probability equals 2/9, since only 2 of the 9 machines left in the population are defective. On the other hand, if the first selection were not defective, this probability would be 3/9. Thus, the probability of the second selection's being defective depends on whether the first selection is defective, which means that the results of each selection are *not* statistically independent. (However, *if the sampling were done with replacement, the results of each selection would be statistically independent, since the probability of getting a defective would be 3/10, regardless of the outcome of earlier selections.*)

THE MULTIPLICATION RULE WITH INDEPENDENT EVENTS

When two events are statistically independent, the multiplication rule can be simplified in the following way.

MULTIPLICATION RULE FOR TWO INDEPENDENT EVENTS. *If* A *and* B *are statistically independent events, the joint probability that both* A *and* B *will occur equals the unconditional probability of* A *times the unconditional probability of* B. *In other words,*

$$P(A \text{ and } B) = P(A) \cdot P(B) \qquad (3.7)$$

Moreover, the extension of the multiplication rule to situations where there are more than two events can be simplified in the following way when all of the events are statistically independent.

MULTIPLICATION RULE FOR n INDEPENDENT EVENTS. *If* E_1, E_2 ⋯ , E_n *are events, the joint probability that all* n *events will occur equals the product of their unconditional probabilities of occurrence if all the events are statistically independent. In other words,*

$$P(E_1 \text{ and } E_2 \text{ and } \cdots \text{ and } E_n) = P(E_1) \cdot P(E_2) \cdot \cdots \cdot P(E_n). \qquad (3.8)$$

APPLICATIONS

The multiplication rule is a powerful aid to the solution of a wide variety of practical problems. The following are three examples of how the multiplication rule and the concept of statistical independence can be used. Let's begin with a very simple example.

EXAMPLE 3.6 Two cards are chosen at random and without replacement from an ordinary deck of playing cards. What is the probability that both are hearts?

SOLUTION: Let *A* be the event that the first card is a heart and *B* be the event that the second card is a heart. Clearly, $P(A) = 13/52$, since 13 cards out of the 52 in the deck are hearts. And $P(B|A) = 12/51$, since the probability that the second card is a heart (*given that the first is a heart*) is $12/51$ because 12 cards out of the remaining 51 in the deck are hearts. Applying the multiplication rule,

$$P(A \text{ and } B) = P(B|A) \cdot P(A)$$

$$= 12/51(13/52) = 1/17.$$

Note that the multiplication rule implies that $P(A$ and $B)$ equals $P(B|A) \cdot P(A)$, as well as $P(A|B) \cdot P(B)$, which is the expression in equation (3.4).[6]

EXAMPLE 3.7 A salesman must call on all four of his customers in a certain area. The probability that he finds each of them in his or her office on a particular day is $1/2$, and whether or not one customer is in is statistically independent of whether any of the others is in. What is the probability that the salesman will find all of the customers in their offices on this day?

SOLUTION: Let E_1 be the event that the first customer is in his or her office, E_2 be the event that the second customer is in, E_3 be the event that the third customer is in, and E_4 be the event that the fourth customer is in. Since these events are statistically independent, the multiplication rule implies that

$$P(E_1 \text{ and } E_2 \text{ and } E_3 \text{ and } E_4) = P(E_1) \cdot P(E_2) \cdot P(E_3) \cdot P(E_4)$$

$$= (1/2)(1/2)(1/2)(1/2) = 1/16.$$

EXAMPLE 3.8 Johnstown's public school system has 4,000 employees. The number of employees in each income and age category is as follows:

Annual Income (dollars)	Employees under 45	Employees 45 and above	Total
Under 25,000	1,000	1,000	2,000
25,000 and over	500	1,500	2,000
Total	1,500	2,500	4,000

(a) What is the probability that a randomly chosen employee (1) will have an income of under $25,000; (2) will be under 45; (3) will have an income of under $25,000, given that he or she is under 45?

(b) Are age and income statistically independent?

SOLUTION: (a) The probability that an employee will have an income of under $25,000 is 2,000/4,000, or $1/2$. The probability that an employee will be under 45 is 1,500/4,000, or $3/8$. The probability that an employee will have an income of under $25,000, given that he or she is under 45, is 1,000/1,500, or $2/3$.

[6] Because the definition of a conditional probability in equation (3.3) implies that $P(B|A) = P(A$ and $B) \div P(A)$, it follows that $P(A$ and $B) = P(B|A)P(A)$.

(b) Age and income are not statistically independent because the probability that an employee has an income of under $25,000 is different if he or she is under 45 than if he or she is 45 or over. If an employee is under 45 this probability is 1,000/ 1,500, or 2/3. If an employee is 45 or over this probability is 1,000/2,500, or 2/5.

THE RELIABILITY OF THE APOLLO SPACE MISSION[7]

GETTING DOWN TO CASES

The Apollo mission with its objective of landing men on the moon involved severe problems of reliability. For the mission to be successful, all (or at least a great many) components of the Apollo system had to operate properly. A malfunction of any one of many components could have resulted in mission failure. Experts used statistical theory to develop sophisticated methods for estimating the probability that the undertaking would be successful. We are able here to present only simplified versions of what was done. A schematic version of the basic Apollo module is shown in Figure 3.8. We assume that the module works if, and only if, all five components function properly.

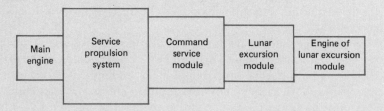

Figure 3.8
Simplified Representation of Apollo Module

(a) Suppose that the probability is 0.99 that each of the five Apollo components functions properly. If the components are independent, what is the probability that a mission will succeed?

One way of increasing the reliability of the Apollo system was through the use of a parallel configuration of components. For example, the second stage of the Saturn rocket used in the Apollo program had five rocket motors, and if any one of the motors failed, the others could be used for a satisfactory earth orbit. To see how the use of a parallel configuration of components increases reliability, suppose that the main engine component in Figure 3.8 is replaced by two engines, either of which is able to perform the tasks required by the mission. If either of these engines malfunctions, the other can take over.

[7] This case is based to some extent on a section from G. Lieberman, "Striving for Reliability," in J. Tanur et al. (eds.), *Statistics: A Guide to the Unknown.*

(b) If the probability is 0.99 that each of the engines functions properly, what is the probability that at least one of the engines functions properly?

(c) If the main engine component in Figure 3.8 is replaced by two such engines, what is the probability that a mission will succeed?

(d) Compare your result in (c) with that in (a) and interpret the difference between the two results.

(e) What are the disadvantages of a parallel configuration of components? Why not add as many redundant components as possible in order to increase reliability?

3.6 Subjective, or Personal Probability

Frequency Definition of Probability

In Section 3.3, the probability of an event was defined as the proportion of times that the event will occur in the long run if the relevant experiment is repeated over and over. This is the so-called *frequency definition of probability.* For example, as pointed out in Section 3.3, the probability that a particular die will come up a 1 can be viewed as the proportion of times that this will occur if the die is thrown innumerable times.

Some experiments are not easy to interpret in these terms because they cannot be repeated over and over. A new commercial product may have a different probability of succeeding if it is put on the market this month rather than next month. This is an "experiment" that cannot be performed over and over because market and other conditions vary from month to month. If the new product is *not* introduced this month but next, the experiment will be performed under different conditions and thus will be a different experiment.

Subjective Definition of Probability

In dealing with events and experiments of this sort statisticians and decision makers sometimes use a *subjective* or *personal definition of probability.* According to this definition, *the probability of an event is the degree of confidence or belief on the part of the statistician or decision maker that the event will occur.* For example, if the decision maker believes that event *A* is more likely to occur than event *B*, the probability of *A* is higher than the probability of *B*. If the decision maker believes that the odds are 50–50 that a particular event will occur, the probability attached to this event equals 0.50. The important factor in this concept of probability is what the decision maker believes.

To illustrate subjective probability, suppose that the governor of California must decide whether or not to announce that he is a candidate for President of the United States. The governor must estimate as accurately as possible the probability that he will get the nomination and be elected. The relevant experiment cannot be conducted over and over; that is, the governor cannot announce his candidacy again and again to determine the proportion of cases in which he gets the nomination and is elected president. (However, Harold Stassen, who unsuccessfully sought the Republican presidential nomination for decades, came close to doing this.) Instead, the governor must use his knowledge, experience, and intuition, together with whatever objective information can be obtained from public opinion polls and other sources, to estimate his chances of success. If the conclusion is that his probability of success is 0.20, then this is the governor's subjective probability of this event.

The same rules of probability apply, regardless of whether probabilities are given a frequency definition or a subjective definition. Therefore, all the principles presented in this chapter apply to both types of probabilities. Since subjective probabilities must conform to the same mathematical rules as probabilities based on the frequency definition, and since they can be manipulated in essentially the same way to solve problems, we seldom distinguish subjective probabilities from frequency-based probabilities in succeeding parts of this book.[8]

EXERCISES

3.10 An article in a New York newspaper by two well-known columnists stated that if the probability of downing an attacking airplane were 0.15 at each of five defense stations, and if a plane had to pass all five stations before arriving at the target, the probability that the plane would be downed before reaching the target was 0.75.[9]
(a) Do you agree with this reasoning?
(b) If not, what is the correct answer?

3.11 Suppose that $P(A) = 0.6$, $P(B) = 0.3$, and $P(A \text{ and } B) = 0.1$.
(a) Are A and B mutually exclusive events? Why, or why not?
(b) Are A and B statistically independent events? Why, or why not?

3.12 Grace Jones is selling her house. She believes there is a 0.2 chance that each person who inspects the house will purchase it. What is the probability that more than two people will have to inspect the house before Grace Jones finds a buyer? (Assume that the decisions of the people inspecting the house are independent.)

3.13 A true die is rolled twice. What is the probability of getting (a) a total of 6; (b) a total of less than 6; (c) a total of 7 or more?

[8] For two classic papers concerning alternative definitions of probability, see R. von Mises, "Probability: An Objective View," and L. J. Savage, "Probability: A Subjective View," in E. Mansfield, *Statistics for Business and Economics: Readings and Cases* (New York: Norton, 1980).

[9] W. A. Wallis and H. Roberts, *Statistics* (Glencoe: Free Press, 1956), p. 96.

3.14 Three cards are to be drawn (without replacement) from a deck with the five of hearts missing.
(a) What is the probability that all three cards will be clubs?
(b) What is the probability that all three cards will be of the same suit?

3.15 The faces of a true die showing a 5 and 6 are colored green; the other faces are colored red. The faces of another true die are colored the same way. If the two dice are tossed, what is the probability that both show a red face?

3.16 In a particular state legislature, 30 percent of the legislators are from rural areas and 10 percent are under 40 years of age. If a legislator is chosen (the probability of choosing each one being the same for all legislators), can we calculate the probability of choosing a legislator under 40 from a rural area by (a) multiplying 0.30 times 0.10; (b) adding 0.30 and 0.10? Why or why not?

3.17 The probability that a launch of a particular spacecraft will occur on time is 0.4. Whether or not a launch occurs on time is independent of whether previous launches occurred on time. If three such launches take place, compute the probability that
(a) the first two will occur on time, but the third will not.
(b) the number of launches occurring on time exceeds the number not occurring on time.

3.18 The probability that a lawyer passes the bar exam on his first try is 0.80. On the second try, it is 0.85, and on the third try it is 0.88. Calculate the probability that a lawyer:
(a) does not pass the bar exams after two tries.
(b) takes three tries to pass the bar exam.

3.19 At a particular hospital, the number of male employees equals the number of female employees. The probability is 1/2 that a male employee is a college graduate and 1/4 that a female employee is one. An employee who is a college graduate is chosen at random (that is, each such employee has the same probability of being chosen). What is the probability that such an employee is female?

3.20 The Jones family decides to buy a new car and narrows the choice down to a Ford, a Chevrolet, or a Toyota. The probability that they will buy a Ford is 0.3, and the probability that they will buy a Chevrolet is 0.4.
(a) What is the probability that they will purchase either a Ford or a Chevrolet?
(b) What is the probability that they will purchase a Toyota?
(c) If the Joneses were undecided as to whether or not they would purchase a car at all, would this influence your answer to (b)? If so, would you increase or decrease your answer to (b)?
(d) The probability that the Jones family will purchase a Ford is 0.3 and the probability that they will purchase a station wagon is 0.2. If the probability that they will purchase a Ford station wagon is 0.05, what is the probability that they will purchase either a Ford or a station wagon (or both)?

The Chapter in a Nutshell

1. Statisticians use the term *experiment* to describe any process by which data are obtained through the observation of uncontrolled events in nature or through controlled laboratory procedures. The *sample space* is defined as the set of all possible outcomes that may occur as a result of a particular experiment. In each occurrence of the experiment, one and only one outcome takes place. Sample spaces are often depicted as points on a graph or in tree diagrams.

2. An *event* is a subset of the sample space. The *probability* of an event is the proportion of times that this event occurs over the long run if the experiment is repeated many times under uniform conditions (according to the frequency definition). A probability must be greater than or equal to zero and less than or equal to one.

3. The *addition rule* states that, if $P(A)$ is the probability that event A occurs and if $P(B)$ is the probability that event B occurs, then $P(A \text{ or } B) = P(A) + P(B) - P(A \text{ and } B)$. Of course, if A and B are mutually exclusive, it follows that $P(A \text{ or } B) = P(A) + P(B)$. If $E_1, E_2, \cdots, E_n$ are mutually exclusive events, then $P(E_1 \text{ or } E_2 \text{ or } \cdots \text{ or } E_n) = P(E_1) + P(E_2) + \cdots + P(E_n)$.

4. The *conditional probability* of event A given the occurrence of event B—that is, $P(A \mid B)$—equals $P(A \text{ and } B) \div P(B)$. If $P(A \mid B) = P(A)$, event A is said to be statistically independent of event B. The *multiplication rule* states that if A and B are two events, the probability that both occur—$P(A \text{ and } B)$—equals $P(A \mid B)P(B)$. Of course, if A and B are independent, it follows that $P(A \text{ and } B)$ equals $P(A)P(B)$. If $E_1, E_2, \cdots, E_n$ are statistically independent events, $P(E_1 \text{ and } E_2 \text{ and } \cdots \text{ and } E_n) = P(E_1) \cdot P(E_2) \cdot \cdots \cdot P(E_n)$.

Chapter Review Exercises

3.21 If a gambler makes a single roll with a pair of true dice, what is the probability that the numbers on the two dice will differ?

3.22 Mary McCullough has been accepted at a number of law schools, but has not decided which one to attend. There are three states in which she can go to law school: Michigan, Illinois, and New York. In each state, she has been accepted at law schools located in a large city, a small city, and a small town.

(a) Draw a diagram similar to Figure 3.1 showing the nine different ways in which Mary McCullough can choose a law school to attend. Put the state in which the school is located along the horizontal axis, and the size of the community along the vertical axis. On the horizontal axis, let 1 equal Michigan, let 2 equal Illinois, and let 3 equal New York. On the vertical axis, let 1 equal large city, 2 equal small city, and 3 equal small town.

(b) Using the diagram you drew in (a), designate the subset of the sample space corresponding to each of the following events:
   (i) Mary McCullough chooses a Michigan law school.
  (ii) Mary McCullough chooses a law school in a small town.
 (iii) Mary McCullough chooses a law school in a large city in New York.
 (iv) Mary McCullough chooses a law school in a large city or a small town.
  (v) Mary McCullough chooses a law school in either a large city or New York (or both).

(c) Using the results you obtained in (a) and (b), indicate the set of outcomes, one of which must occur if
   (i) Mary McCullough chooses a Michigan law school.
  (ii) Mary McCullough chooses a law school in a small town.
 (iii) Mary McCullough chooses a law school in either a large city or a small town.

(d) Using your results from (a) and (b), indicate the set of events, all of which must occur if (i) Mary McCullough chooses a law school in a large city in New York, (ii) Mary McCullough chooses a law school in a small town in Michigan.

3.23 The probability that a student will get an A in statistics is 0.20, the probability that he will get an A in English is 0.25, and the probability that he will get an A in both is 0.05. What is the probability that he will get an A in neither subject?

3.24 A fair coin is flipped 6 times. What is the probability of 6 heads? What is the probability of 6 tails?

3.25 The Smith family is undecided about whether or not to buy a new car. If the probability is 0.9 that they will buy one, and if the probability is 0.3 that they will buy a Ford, and if the probability is 0.4 that they will purchase a car getting more than 20 miles per gallon, what is the probability

(a) that they will buy either a car getting more than 20 miles per gallon or a Ford, if there are no Ford cars that get more than 20 miles per gallon;

(b) that they will buy either a car getting more than 20 miles per gallon or a Ford (or both), if all Fords get more than 20 miles per gallon?

3.26 Distinguish between a subjective probability and one based on the frequency definition. Why are subjective probabilities ever used?

3.27 If a reporter feels that the odds are 2 to 1 that he can get an interview with the governor of his state, what is his personal probability that this will happen?

3.28 If you believe the odds are 3 to 1 that you will get at least a *B* in your statistics course, what is the probability (your subjective probability) that this is the case?

3.29 The Black Belt Sewing Machine Company has four plants. The number of motors shipped in 1986 by each supplier to each of Black Belt's plants is as follows:

		Supplier			
Plant	*I*	*II*	*III*	*IV*	*V*
1	200	100	300	100	400
2	300	200	400	200	300
3	200	300	300	300	100
4	100	300	200	400	200

(a) If the firm picks a motor from among those received by plant 2 (and if each motor has the same probability of being chosen) what is the probability that this motor came from supplier III?

(b) If the firm picks a motor from among those received at any of its plants, what is the probability that it came from supplier III?

(c) If event *X* is that a motor comes from supplier III, is this event independent of the plant receiving the motor?

(d) Calculate the probability
   (i) that a motor goes to plant 1, given that it comes from supplier II;
   (ii) that a motor comes from supplier II, given that it goes to plant 1;
   (iii) that a motor goes to plant 1, given that it comes from either supplier II or supplier III;
   (iv) that a motor comes from supplier II, given that it goes to either plant 1 or plant 2.

(e) The Black Belt Sewing Machine Company finds that the probability that a motor both comes from supplier I and is defective is 0.1. What is the probability that a motor purchased from supplier I is defective?

(f) If 20 percent of all the motors received by the Black Belt Sewing Machine Company are defective, what is the probability that a motor comes from supplier I, given that it is found to be defective?

3.30 A man has three coins in his pocket. Two are fair coins; the third has tails on both sides. He chooses a coin from his pocket (each coin having the same probability of being chosen), and tosses it twice. What is the probability that two heads will come up?

Appendix 3.1

COUNTING TECHNIQUES

In many practical situations, it is important to be able to calculate the number of ways in which an event can occur. Such calculations must often be carried out in order to compute the probability of an event.

SELECTION OF ITEMS

Frequently, a selection or a choice must be made in several steps, and at each step there are a number of alternatives. For example, suppose that a top student at a leading college is dressing for an interview with a potential employer, and that she has a choice of beige, blue, or green dress. Given her choice of dress color, she has a choice of beige, black, or navy shoes. In situations of this sort, one often must be able to calculate the total number of different outcomes that are possible. For example, in how many different ways can the applicant dress for the interview?

A tree diagram is a useful graphic device for solving such problems because it shows at each step the choices or alternatives that can be made and the possible outcomes at the end of the selection process. (Recall that at the beginning of this chapter tree diagrams were also used to represent sample spaces.) Figure 3.9 shows a tree diagram for the choices open to the applicant getting ready for the interview. As you can see, three branches are open in the first step, each branch corresponding to the color of dress. Once the young woman has picked one of these initial branches, she goes to another fork where there again are three branches, each corresponding to a different color of shoes. Once the second step has been completed, there clearly are nine different outcomes corresponding to the points on the right-hand side of the diagram where

Figure 3.9
Tree Diagram of
Student's Choices of
Dress Color and Shoe
Color

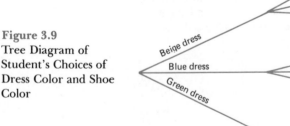

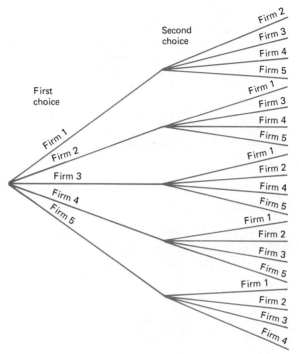

Figure 3.10
Tree Diagram of
Student's Choices with
Five Job Offers

she can wind up. That is, the young woman can choose a beige dress and beige shoes, a beige dress and black shoes, and so on.

In this case, there are three items of one kind (dress colors) and three items of a second kind (shoe colors), and we find that the number of ways we can pair an item of the first type with an item of the second type is (3)(3), or 9. Let's generalize this result. If there are *m* kinds of items, and *if there are* n_1 *items of the first kind,* n_2 *items of the second kind,* $\cdots$ *, and* n_m *items of the mth kind, the number of ways we can select one of each of the m kinds of items is* $n_1 \times n_2 \times \cdots \times n_m$. To illustrate, if subsequent to the interview the college student is offered a job by each of five companies, and if she is asked to list her first and second choices, how many different pairs could she list? There are five firms that she could list as her first choice. But once she makes her first choice, there are only four firms left that could be her second choice. Thus, in this case, $m = 2$, $n_1 = 5$, and $n_2 = 4$, so the answer is (5)(4), or 20. Figure 3.10 shows the relevant tree diagram.

The following example provides another illustration of how this result can be used.

EXAMPLE 3.9 Thirty firms bid on a particular job, and each firm submits a different bid. The names of the lowest bidder, the second lowest bidder, and the third lowest bidder are published. In how many ways can one select three firms ordered in this way?

SOLUTION: Clearly, any one of the 30 firms may be the lowest bidder. But once one selects any one of them as the lowest bidder, there are only 29 left that can be the second lowest bidder. And once one selects the lowest and second lowest bidders, there are only 28 firms that can be selected as the third lowest bidder. Consequently, one can pick the lowest bidder in 30 different ways, the pair of lowest and second lowest bidders in (30)(29) different ways, and the triad of lowest, second lowest, and third lowest bidders in (30)(29)(28) different ways. Thus, the answer is (30)(29)(28), or 24,360. In this case, $m = 3$, $n_1 = 30$, $n_2 = 29$, and $n_3 = 28$.

PERMUTATIONS

At this point, we can define a permutation. *If* x *items are selected (without replacement) from a set of* n *items, any particular sequence of these* x *items is called a* **permutation.** For example, Yale Harvard Brown is a permutation of three of the Ivy League colleges. That is, it is a possible sequence in which three of the eight names of the Ivy League colleges (Harvard, Yale, Princeton, Columbia, Pennsylvania, Dartmouth, Cornell, and Brown) can be listed. In this case, $x = 3$ and $n = 8$, since three items (college names) are being selected from a set of eight items. Turning to another example, *ab* is a permutation of two of the first three letters of the alphabet. In this case, $x = 2$ and $n = 3$ since two items (letters) are being selected from a set of three items. Another such permutation is *bc*. In other words, it is another pair of letters that can be made up out of *a*, *b*, and *c*. Note that the order of the items is of importance in a permutation. Thus, *ab* and *ba* are different permutations (from one another) even though they are composed of the same letters, *a* and *b*.

In general, *the number of permutations of* x *items that one can select from a set of* n *items is* $n(n-1)(n-2) \cdots (n-x+1)$. Based on our previous results, this is not difficult to prove. There are *x* items to be chosen. Suppose that the first

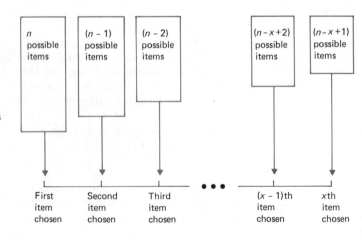

Figure 3.11
Number of Permutations of *x* Items from a Set of *n* Items

item that is chosen is put into the first slot along the horizontal line in Figure 3.11. Suppose the second item chosen is put into the second such slot, the third into the third slot, and so on. For our first choice, there are n possible items. Once the first selection is made, there are $(n - 1)$ items left for the second slot. Once the first and second selections have been made, there are $(n - 2)$ items left for the third slot and once the $(x - 1)$th selection is made, there are $(n - x + 1)$ items left for the last (that is, the xth) slot. So the total number of sequences of x items that can be chosen from n items must be $n(n - 1)(n - 2) \cdots (n - x + 1)$.

To check this result, let's see whether it gives the right answer in the case of the student with the job offers and in Example 3.9. With regard to the student offered a job by the five firms, we want to know the number of permutations of two items (firms) that one can select from a set of five items. According to the previous paragraph, this number should be $(5)(4)$, since $n = 5$ and $x = 2$. This result is just what we got before. In Example 3.9, the case of the 30 firms bidding on the job, we want the number of permutations of three items (firms) that one can select from a set of 30 items. Since $n = 30$ and $x = 3$, the answer according to the previous paragraph is $(30)(29)(28)$, which jibes with our previous result.

COMBINATIONS

In contrast to a permutation, where the order of the items matters, *a combination is a selection of items where the order does not matter.* For example, in the case of the Ivy League colleges, each permutation of three of the colleges shows a possible ranking of the top three schools in football. Clearly, the order here is important, since the first name in the sequence indicates first place and the last name indicates third place. But if instead we were interested in the number of three-way crew races that could occur in the Ivy League, the order wouldn't matter, since Harvard Cornell Brown and Cornell Brown Harvard would mean the same thing. Thus, in the latter case, one would want to calculate the number of combinations of three items that can be selected from a set of eight items.

Using the results we obtained concerning the number of permutations, it is easy to find out how many combinations there are. After all, the crucial point is that, from the point of view of a combination, many permutations are indistinguishable. For example, as we just observed, Harvard Cornell Brown and Cornell Brown Harvard are the same thing. Taking each particular combination, how many permutations correspond to this single combination? Clearly, there are $(3)(2)(1)$, since this is the number of permutations that can be created from three items. (This, of course, is an application of our formula for the number of permutations, both x and n being 3 in this case). To illustrate this, you can see that there are the following six permutations corresponding to the single combination consisting of Harvard, Cornell, and Brown:

Harvard Cornell Brown	Cornell Brown Harvard	Brown Cornell Harvard
Harvard Brown Cornell	Cornell Harvard Brown	Brown Harvard Cornell

Thus, since there are six permutations for each combination and since there are $(8)(7)(6) = 336$ permutations (because $n = 8$ and $x = 3$), it follows that there must be $336 \div 6 = 56$ combinations. In other words, there must be 56 different three-way crew races that can be arranged in the Ivy League.

In general, *the number of combinations of x items that one can select from a set of n items is*

$$\frac{n(n-1)\cdots(n-x+1)}{x(x-1)\cdots(2)(1)} \qquad (3.9)$$

Factorials

To prove this, recall that the numerator of this ratio is the number of permutations of x items that can be selected from a set of n items. Since there are $x(x-1)\cdots(2)(1)$ permutations that correspond to each combination of x items, the number of combinations must equal the numerator of (3.9) divided by this amount. Frequently, $x(x-1)\cdots(2)(1)$ is referred to as $x!$, which is read as "x factorial." For example, $4! = (4)(3)(2)(1) = 24$, and $6! = (6)(5)(4)(3)(2)(1) = 720$. (Note that by definition $0! = 1$.) Thus, the number of combinations of x items that can be chosen from a set of n items can also be written as

$$\frac{n(n-1)\cdots(n-x+1)}{x!} = \frac{n(n-1)\cdots(2)(1)}{(n-x)(n-x-1)\cdots(2)(1)x!} = \frac{n!}{(n-x)!x!}$$

To switch from crew to basketball, if we want to know the number of different basketball teams of five members each that can be formed from a group of eight players, this is the number of combinations of five items that can be selected from a set of eight items. Thus, the answer must be

$$\frac{(8)(7)(6)(5)(4)}{(5)(4)(3)(2)(1)} = \frac{6720}{120} = 56.$$

Expressed in terms of factorials, the answer is

$$\frac{8!}{3!5!} = \frac{40,320}{(6)(120)} = 56.$$

GALILEO ON GAMBLING: A CASE STUDY[10]

To illustrate the application of combinations and permutations, let's go back several hundred years to a case involving the famous mathematician and astronomer Galileo. In 1613, Galileo accepted a position as First and Extraordinary Mathematician of the University of Pisa and Mathematician to his Serene Highness Cosimo II of Tuscany. Besides receiving a title that betrayed no false modesty, Galileo received a large salary and no duties (a good combina-

---

[10] The discussion in this section is based on F. N. David, *Games, Gods, and Gambling* (New York: Hafner, 1962).

tion in any age), except that he was asked by Cosimo to work on certain problems in which his Serene Highness was interested. One problem that Cosimo apparently asked Galileo to solve was the following. Three dice are thrown. Although there are six ways of getting a 9—621 (that is, one die showing a 6, another a 2, and another a 1), 531, 522, 441, 432, and 333—and six ways of getting a 10—631, 622, 541, 532, 442, and 433—the probability of throwing a 9 seems in fact to be less than that of throwing a 10. Why?

Before considering Galileo's solution, it is important to recognize that this was not a purely academic exercise. At that time, there was a popular dice game in which three ordinary six-sided dice were thrown. Cosimo apparently had gambled sufficiently often at this game to find that the probability of getting a 9—that is, the probability that the sum of the numbers on the three dice would be 9—was lower than the probability of getting a 10. He wanted to know why this was the case.

Way of rolling the number	Total number of triads corresponding to this way of rolling the number	(Individual triads) First die	Second die	Third die
621	6	6	2	1
		6	1	2
		2	6	1
		2	1	6
		1	6	2
		1	2	6
531	6	5	3	1
		5	1	3
		3	5	1
		3	1	5
		1	5	3
		1	3	5
522	3	5	2	2
		2	5	2
		2	2	5
441	3	4	4	1
		4	1	4
		1	4	4
432	6	4	3	2
		4	2	3
		3	4	2
		3	2	4
		2	4	3
		2	3	4
333	1	3	3	3
Total	25			

Table 3.3
Number of Triads
Resulting in Dice
Showing a 9

**Table 3.4
Number of Triads
Resulting in Dice
Showing a 10**

Way of rolling the number	Total number of triads corresponding to this way of rolling the number	(Individual triads) First die	Second die	Third die
631	6	6	3	1
		6	1	3
		3	6	1
		3	1	6
		1	6	3
		1	3	6
622	3	6	2	2
		2	6	2
		2	2	6
541	6	5	4	1
		5	1	4
		4	5	1
		4	1	5
		1	5	4
		1	4	5
532	6	5	3	2
		5	2	3
		3	5	2
		3	2	5
		2	5	3
		2	3	5
442	3	4	4	2
		4	2	4
		2	4	4
433	3	4	3	3
		3	4	3
		3	3	4
Total	27			

Galileo began his reply by pointing out that if each side of each die is equally likely to come up, there are $(6)(6)(6) = 216$ configurations of the dice that can arise. To see this, note that the first die can have six outcomes (1 to 6). And since the second die can also have six outcomes, the number of pairs of outcomes is $(6)(6)$. Moreover, since the third die can also have six outcomes, the number of triads of outcomes is $(6)(6)(6)$. Further, since each side of each die is equally likely to come up, the probability of each of these triads is $1/216$. This is, of course, a simple illustration of what we discussed in previous sections of this chapter.

Next, Galileo noted that although it was true that both 9 and 10 could be rolled in six ways, the probability of each of these ways was not the same. For example, the probability of 333 was lower than that of 432. To see why, let's consider each of the three dice, and let's denote the outcome of a throw by three numbers in brackets, the first being the number on the first die, the second being the number on the second die, and the third being the number on the third die. Clearly, 432 can occur on the basis of six of these triads—[432], [423], [342], [324], [243], [234]—whereas 333 can occur on the basis of only one triad, [333]. Why six triads in the case of 432? Because this is the number of permutations of three items selected from a set of three items. Why 1 triad in the case of 333? Because there is only one triad in which all dice show a 3.

Galileo constructed a table showing the number of triads resulting in each of the six ways that 9 could be formed. As indicated in Table 3.3, there are 25 such triads. Similarly, he showed that there are 27 triads resulting in the six ways that 10 can be formed (as shown in Table 3.4). Since each triad has a probability of 1/216, it follows that the probability of rolling a 9 is 25/216, whereas the problem of rolling a 10 is 27/216. This was how Galileo solved the problem.

EXERCISES

3.31 The number of combinations of x items that one can select from a set of n items is often represented as $\binom{n}{x}$. Prove that $\binom{n}{x} = \binom{n}{n-x}$.

3.32 How many different poker hands can be drawn in which there is a straight (or straight flush)?

3.33 A panel of experts is asked to rank eight possible designs for a new type of shopping mall. How many different possible rankings could the panel give (assuming no ties)?

3.34 A shipment of goods contains 50 items, 5 of which are defective. To determine whether the shipment should be accepted, the buyer picks one item at random and then (without replacing the first item) picks another at random. If neither is defective, the shipment will be considered acceptable. What is the probability that the buyer will accept the shipment?

3.35 A publishing firm has six books in physics. Each year, it mails an advertisement to physics professors focusing on three of its physics books. The firm does not want to focus on the same three books more than once; it does not mind repeating one or two of them, but it does not want all three to be the same as in a previous mailing. How long can the firm go on before it will no longer be able to abide by this rule?

3.36 In the opinion of the chairman of a U.S. Senate committee, there are seven senators on the committee who are competent to serve on a subcommittee to investigate organized crime. In how many ways can the chairman choose a subcommittee of three senators, all of whom are on the committee and all of whom he regards as being competent to serve?

3.37 If there are five men and two women in the Senate who are competent to serve on a particular committee, in how many ways can the Senate leaders choose a committee of three if they want two men and one woman on the committee (and if all three are competent)?

3.38 The President of the United States wants to appoint six people as members of a commission on the problems of the aged. He does not want more than one of these people to be from the same state. A candidate for the commission is solicited from each state (including Alaska and Hawaii). The president's staff wants to make sure that each possible set of commissioners is considered and evaluated. If each such evaluation takes two hours of the staff's time, how many man-hours (of staff time) would it take to do this?

3.39 To prepare a patient for surgery, a nurse must carry out seven different tasks (A, B, C, D, E, F, and G). Tasks F and G must be carried out last, G following F. The other tasks can be performed in any order at all. The hospital administrator wants to evaluate each possible sequence of tasks to see which is most effective. If it takes one hour of an analyst's time to evaluate each sequence, how many man-hours (of analysts' time) would it take to do what the administrator wants?

3.40 The Black Belt Sewing Machine Company relies on five suppliers (I, II, III, IV, and V) to provide it with motors. Each of these suppliers can ship motors to each of Black Belt's four plants (1, 2, 3, and 4). Black Belt stamps a letter on each motor to indicate which supplier it came from and which plant used it. For example, it stamps an *A* on a motor from supplier I that is put into a sewing machine coming out of plant 1, a *B* on a motor from supplier II that is put into a sewing machine coming out of plant 2, a *C* on a motor from supplier II that is put into a sewing machine coming out of plant 1, and so on.
(a) How many letters will it take to represent all possible combinations of motors and plants?
(b) If the Black Belt Sewing Machine Company bought the same number of motors in 1986 from each supplier, and if each supplier shipped the same number of motors to each of Black Belt's plants, what proportion of all Black Belt's motors had an *A* stamped on them in 1986?
(c) Every week the Black Belt Sewing Machine Company tests one motor shipped by each of its five suppliers. The motors tested are ranked according to performance, from highest to lowest. The firm then sends this ranking to its plants. After repeating these tests hundreds of times, all possible orderings of the five suppliers have occurred. How many different rankings has the firm sent out to its plants?

3.41 John Monroe takes a test consisting of eight true-false questions. In how many different ways can he answer the exam? If he has no idea of the answer to any of the questions, what is the probability of his getting a perfect score?

3.42 There are four roads (*A, B, C,* and *D*) from Avon to Burgundy and three roads (*a, b,* and *c*) from Burgundy to Coventry. A criminal travels from Avon to Coventry via Burgundy. Sheriff Smith is asked by reporters what route the criminal took from Avon to Coventry. He doesn't know, and says that each possible route was equally likely. Eventually, he is forced to guess, and he guesses that the criminal went via road *A* to Burgundy and via road *b* to Coventry. What is the probability that he is correct?

4

OBJECTIVES

The concept of a probability distribution is central to much of statistical theory. The overall purpose of this chapter is to describe the nature, characteristics, and usefulness of probability distributions, as well as to discuss in detail the binomial distribution. Among the specific objectives are:

1 To define a random variable and a probability distribution.

2 To show how you can calculate a random variable's expected value, variance, and standard deviation.

3 To indicate the conditions under which the binomial distribution arises, and the ways in which you can calculate the probability that a binomial variable will equal a particular value (or values).

4 To show how the binomial distribution can be applied to solve a wide range of practical problems.

Probability Distributions and the Binomial Distribution

4.1 Introduction

In Chapter 3, we discussed basic concepts of probability. To understand statistical techniques, it is necessary to go more deeply into probability theory. In this chapter we will discuss the nature and characteristics of random variables and probability distributions. Then we will describe expected values and take up the most important discrete probability distribution, the binomial distribution. We shall also indicate some of the ways in which the binomial distribution is applied.

4.2. What Is a Random Variable?

One of the basic concepts in probability theory is that of the random variable, which is defined as follows.

> RANDOM VARIABLE: *A **random variable** is a numerical quantity the value of which is determined by an experiment. In other words, its value is determined by chance.*

In succeeding paragraphs we shall go into considerable detail to clarify what is meant by a random variable, but it is sufficient here to note two points concerning this definition. First, the definition assumes that

129

some sort of an experiment is conducted. (As stressed in Chapter 3, the term *experiment* should be interpreted very broadly and covers more than just laboratory procedures.) The outcomes of this experiment constitute a sample space, as pointed out in Chapter 3. Second, our definition implies that a random variable is one whose value is defined for each element of the sample space. This means that its value is defined for each outcome of the experiment, which is equivalent to saying that its value is determined by chance (because the outcome of the experiment is determined by chance).

To illustrate, if we take the sum of the numbers showing on two dice, this sum is clearly a random variable since its value is determined by the "experiment" of throwing the pair of dice. As pointed out in the previous chapter, the sample space generated by this experiment can be represented by the 36 points in Figure 3.1, which are reproduced in Figure 4.1. The value of this sum corresponding to each element of the sample space is shown above the point representing this element in Figure 4.1. Obviously, this sum fulfills the condition of the definition that a random variable must assume a value for each element of the sample space.

As another illustration, suppose that a gambler proposes the following bet on the outcome of the throw of a single die. He will give you $10 if you throw a 4, a 5, or a 6, and you will give *him* $10 if you throw a 1, a 2, or a 3. The amount you win (or lose)—+$10 (or −$10)—is a random variable since its value is determined by the outcome of an experiment. (That is, its value is determined by chance.) This experiment can result in six possible outcomes, corresponding to the die's coming up 1, 2, 3, 4, 5, or 6. For each of these outcomes (or points in the sample space), the amount you win is defined, which agrees with our definition of a random variable.

Figure 4.1
Sample Space for Roll of Two Dice

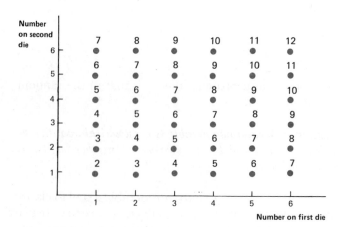

At this point, the various characteristics of a random variable should be coming into focus. First, *a random variable must take on numerical values.* In the experiment with the statistics course discussed in the previous chapter, the outcomes "pass" and "not pass" are not the values of a random variable because they are not in numerical form. However, if we arbitrarily let "pass" equal 0 and "not pass" equal 1, the result is a random variable. Second, *the value of a random variable must be defined for all possible outcomes of the experiment in question (that is, for all elements of the sample space).* For example, in the $10 bet based on the throw of the die, suppose that you win $10 if you throw a 4, a 5, or a 6, and that you lose $10 if you throw a 1 or a 2. But suppose also that *the amount you win or lose is undefined if you throw a 3.* Under these new rules the amount you win (or lose) is no longer a random variable. Why? Because how much you win (or lose) is no longer defined for *all* possible outcomes of the experiment. Specifically, if the outcome of the experiment is a 3, the value of the amount you win (or lose) is undefined.

It is important to recognize that the value of a random variable is unknown *before* the experiment in question is carried out. *After* the experiment is carried out, the value of the random variable is always known. For example, if you roll a pair of dice, the sum of the numbers on them is unknown before the roll; but after the roll, the value of the sum is known. Similarly, if you accept the gamble involving the roll of a single die, the amount you win or lose is unknown before you roll the die; after the roll, this amount is known.

DISCRETE AND CONTINUOUS RANDOM VARIABLES

Statisticians distinguish between two types of random variables: discrete and continuous. A *discrete random variable* can assume *only a finite or countable*[1] *number of distinct values.* For example, the sum of the numbers showing on two dice is a discrete random variable since the sum can only assume one of 11 possible values: 2, 3, 4, ⋯ , 11, or 12. The amount at stake in the gamble with the single die is also a discrete random variable since it can assume only one of two possible values: +$10 and −$10. Many—but by no means all—important random variables are discrete. Some random variables assume *any numerical value on a continuous scale.* Such random variables are called *continuous random variables.*

To illustrate the concept of a continuous random variable, consider the length of time elapsing between the arrival of successive

[1] Some discrete random variables, like the Poisson variable described in the appendix to Chapter 5, assume a countably infinite number of values.

voters at a polling booth. This length of time will vary considerably. Once a voter arrives, it may be a long time before the next voter arrives in some cases, a short time in others. The process by which voters arrive can be regarded as an experiment, and the outcome of this experiment is the time interval between successive voters' arrivals. This time interval is a random variable; but because it can vary continuously, it is not a discrete random variable. This means that the time interval between successive arrivals is not confined to certain values like 30 seconds, 1 minute, 2 hours, and so on, but can also assume *any* value in between them. Of course, in practice we might only measure the time interval to the nearest second, in which case the rounded time intervals are discrete. But the true time interval is continuous.

In this chapter we shall be concerned entirely with discrete random variables, and in the next we shall take up continuous random variables. Meanwhile, the following examples should be useful in illustrating the meaning and characteristics of a random variable.

EXAMPLE 4.1 A fair coin is flipped three times. If it comes up heads three times you receive $100; if it comes up heads twice, you receive $50; if it comes up heads once, you lose $75; and if it never comes up heads, you lose $100.

(a) Is the number of heads that comes up a random variable? If so, is it discrete or continuous?

(b) Is the amount you win (or lose) a random variable? If so, is it discrete or continuous?

SOLUTION: (a) The number of heads that comes up is a random variable, since there are eight possible outcomes (listed below) of this experiment and to each outcome there corresponds a certain number of heads:

heads/heads/heads	tails/tails/heads
heads/heads/tails	tails/heads/tails
heads/tails/heads	heads/tails/tails
tails/heads/heads	tails/tails/tails

This random variable is discrete because it can assume only four possible values: 0, 1, 2, and 3.

(b) The amount you win is a random variable, because for each of the outcomes of this experiment you win (or lose) a corresponding amount. The amount you win (or lose) is a discrete random variable because it can assume only four possible values: +$100, +$50, −$75, and −$100.

EXAMPLE 4.2 A shipment contains 20 machines, 4 of which are defective. The firm receiving the shipment chooses a random sample of 3 machines (without replacement). If any of the machines in the sample is defective, the firm rejects the shipment.

(a) Is the number of defective machines in the sample a random variable? If so, is it discrete or continuous?

(b) Is whether or not the shipment is rejected a random variable? If so, is it discrete or continuous?

SOLUTION: (a) The number of defective machines in the sample is a random variable since there are eight possible outcomes (listed below) of this experiment:

N N N	N N D	N D N	D N N
N D D	D N D	D D N	D D D

The first letter stands for the first machine in the sample and is D if it is defective or N if it is not defective. The second letter stands for the second machine in the sample and is D or N. The third letter, which is also D or N, stands for the third machine. To each of these outcomes there corresponds a number of defective machines in the sample. This random variable is discrete since it can assume only four possible values: 0, 1, 2, and 3.

(b) Whether or not the shipment is rejected is not a random variable because it is not in numerical form. However, it can be turned into a random variable by letting zero stand for rejection of the shipment and 1 stand for acceptance of it. (Obviously, any other pair of arbitrarily chosen numbers could also be used for this purpose.)

4.3 What is a Probability Distribution?

Based on the previous section, we know that the value of a random variable is determined by the outcome of a corresponding experiment. For example, the value of one particular sum of the numbers shown on a pair of dice is determined by one particular roll of the dice. And the amount you win or lose in the gamble involving the single die is determined by each roll of the die. Thus, since there is a certain probability of each outcome of the experiment, there must also be a certain probability of each value of the random variable. For example, since there is a 1/36 chance that each of the points in Figure 4.1 will occur (if the dice are true), it must be possible to deduce the probability that the sum of the numbers showing on the dice will equal 2, 3, 4, and so on. Similarly, since there is a 1/6 chance that a true die will show a 1, 2, 3, 4, 5, or 6, it must be possible to deduce the probability that the amount you will win in your gamble involving the single die will be +$10 or −$10. Such probabilities are provided by the probability distribution of the random variable, which is defined as follows.

> PROBABILITY DISTRIBUTION: *The **probability distribution** of a random variable* X *provides the probability of each possible value of the random variable. If* P(x) *is the probability that* x *is the value of the random variable, we can be sure that* ΣP(x) = 1, *where the summation is over all values that* X *takes on. This is because these values of* X *are mutually exclusive and one of them must occur.*

To illustrate a probability distribution, consider once again the sum of the numbers showing on two true dice. What is the probability distribution of this sum? Clearly, this sum can assume only the following values: 2, 3, 4, ⋯ , 11, and 12. *What is the probability of a 2?* This value of the sum can arise only if each die shows a 1, and the probability of this is 1/36. *What is the probability of a 3?* This value of the sum can arise if the first die shows a 1 and the second shows a 2, or if the first shows a 2 and the second shows a 1. Since the probability of each is 1/36, the probability of either one is 2/36. (Remember the addition rule.) *What is the probability of a 4?* This value of the sum can arise in three ways (a 1 on the first die and a 3 on the second, a 2 on the first die and a 2 on the second, or a 3 on the first die and a 1 on the second), each of which has a probability of 1/36. Thus, the probability of a 4 is 3/36. Based on similar reasoning, it is easy to figure out the probability of a 5, 6, ⋯ , 12. (One way to do this is to count the number of points in Figure 4.1 that result in a particular sum and multiply this number by 1/36, which is the probability of each point.) The results are shown in Table 4.1. As you can see, the sum of the probabilities in the probability distribution equals 1, which agrees with the definition above.[2]

Although it sometimes is convenient to represent a probability distribution by a *table* like Table 4.1, it is sometimes even more convenient to use a *graph*. We can present the probability distribution of the

Line Chart

sum of two dice in a *line chart*, as shown in Figure 4.2. Along the horizontal axis are ranged the various possible values of the random vari-

---

[2] Why must the sum of *P(x)* equal 1? Since the various values of *x* are mutually exclusive, the probability that *some* value of the random variable occurs is the sum of *P(x)*, as we know from the addition rule. Since some value of the random variable *must* occur, the probability that some value of the random variable occurs must equal 1. Thus, it follows that the sum of *P(x)* must equal 1.

Sum of numbers	Probability
2	1/36
3	2/36
4	3/36
5	4/36
6	5/36
7	6/36
8	5/36
9	4/36
10	3/36
11	2/36
12	1/36
Total	1

Table 4.1
Probability Distribution
of Sum of Numbers
Showing on Two True
Dice

able. At each point on the horizontal axis corresponding to such a value a vertical line is erected, the length of which is equal to the probability of this value's occurrence. Thus, in Figure 4.2 the vertical line at 3 is twice as long as that at 2 since the probability of a 3 is twice that of a 2. As you can see from Figure 4.2, the highest vertical line is at 7, and the line chart is symmetrical about this line in this case. (That is, the probability of a 6 equals the probability of an 8, the probability of a 5 equals the probability of a 9, and so on.)

Still another way to represent a probability distribution is by a mathematical function. As pointed out in our definition of a probability distribution, the probability that a random variable equals x is denoted by $P(x)$. In some cases we can express $P(x)$ as a relatively simple

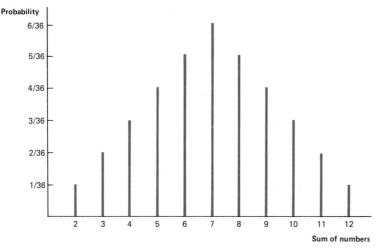

Figure 4.2
Probability Distribution
of Sum of Numbers
Showing on Two True
Dice

mathematical function of x. For example, if you flip a fair coin twice, the number of heads that comes up is a random variable whose probability distribution can be represented by the following equation:

$$P(x) = \frac{1/2}{x!(2-x)!}, \text{ for } x = 0, 1, 2, \tag{4.1}$$

where $x!$ is defined as $x(x-1)(x-2) \cdots (2)(1)$. Thus, $2! = (2)(1) = 2$, and $1! = 1$. (Also, note that $0! = 1$.)[3] The derivation of equation (4.1) is discussed in detail in Section 4.6. For now, it is essential only that we make sure that this equation really does represent the probability distribution of the number of heads. To see that this is true, note that the equation says that $P(0) = 1/2 \div [0!2!] = 1/4; P(1) = 1/2 \div [1!1!] = 1/2; P(2) = 1/2 \div [2!0!] = 1/4$. Clearly, each of these probabilities is right. Note, too, that $P(0) + P(1) + P(2) = 1$, as it should according to our definition of a probability distribution.

PROBABILITY DISTRIBUTIONS AND RELATIVE FREQUENCY
DISTRIBUTIONS

It is important to note the similarity between a probability distribution and a frequency distribution (discussed in Chapter 1). To see how similar they are, consider the probability distribution of the number coming up on a single die. If the die is true, this probability distribution is

$$P(x) = 1/6, \text{ for } x = 1, 2, 3, 4, 5, 6. \tag{4.2}$$

Suppose that this die is thrown 1,000 times and that the number of times each number comes up is as shown in Table 4.2. If we divide the number of times each number come up by the total number of throws (1,000), we get the proportion of times each number comes up (as shown in the third column of Table 4.2). Each of these proportions is the empirical counterpart of the corresponding probability. That is, the proportion of cases in which a 3 comes up is the empirical counterpart of the probability of a 3. If the frequency concept of probability applies, each of these proportions will tend to get closer and closer to the corresponding probabilities as the number of cases included in the frequency distribution becomes larger and larger.

---

[3] Readers of Appendix 3.1 will already have encountered $x!$, which is read as "x factorial."

Number on die	Number of times number comes up	Proportion of times number comes up
1	170	.170
2	159	.159
3	172	.172
4	158	.158
5	160	.160
6	181	.181
Total	1,000	1.000

Table 4.2
Number of Times
Each Number Shows
on a Die Cast 1,000
Times

Another way of stating this is to say that a probability distribution of a certain type is often regarded as the *relative frequency distribution* of the population. *A relative frequency distribution shows the proportion (not the number) of cases falling within each class interval.* For example, if we were to roll a true die again and again until finally data were collected concerning millions of rolls, what would be the relative frequency distribution of this population? After dividing the number of throws resulting in a 1, 2, 3, 4, 5, or a 6 by the total number of throws, we would obtain the proportion of throws coming up a 1, 2, 3, 4, 5, or a 6. This is the relative frequency distribution in this situation. In essence, this relative frequency distribution would be like the probability distribution in equation (4.2). Thus, *probability distributions are often used to represent or approximate population relative frequency distributions,* as we shall see in later chapters.

Relative Frequency Distribution

Since the concept of a probability distribution is so important in statistics, it is essential that it be well understood. The following two examples should help to clarify and illustrate its meaning.

EXAMPLE 4.3 As in Example 4.1, a fair coin is flipped three times. If it comes up heads three times you receive $100; if it comes up heads twice you receive $50; if it come up heads once you lose $75; if it never comes up heads you lose $100.

(a) Construct a line chart of the probability distribution of the number of heads.

(b) Construct a line chart of the probability distribution of the amount you win.

SOLUTION: (a) There are the following eight possible outcomes of this experiment, each with a probability of 1/8:

heads/heads/heads tails/tails/heads
heads/heads/tails   tails/heads/tails
heads/tails/heads   heads/tails/tails
tails/heads/heads   tails/tails/tails

Thus, the probability of no heads is 1/8; the probability of heads once is 3/8; the probability of heads twice is 3/8; and the probability of heads three times is 1/8. The line chart is as follows.

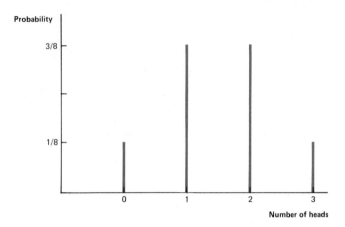

(b) Since the probability of no heads is 1/8, the probability that you will lose $100 is 1/8. Since the probability of heads once is 3/8, the probability that you will lose $75 is 3/8. Since the probability of heads twice is 3/8, the probability that you will win $50 is 3/8. Since the probability of heads three times is 1/8, the probability that you will win $100 is 1/8. Thus, the line chart is as follows.

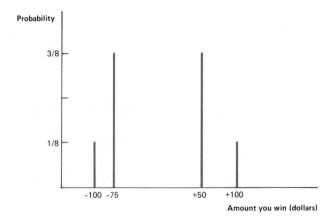

EXAMPLE 4.4 John Gillicuddy buys two tickets at $1 each on a car that is being raffled off. The organizers of the raffle sell a total of 5,000 tickets, and one ticket is picked at random to determine the winner. The car is worth $5,000.

(a) Construct a table showing the probability distribution of the outcome of the raffle, if 0 represents Gillicuddy's not winning the car and 1 represents his winning it.

(b) Construct a table showing the probability distribution of the amount that Gillicuddy wins or loses in the raffle.

SOLUTION: (a) Since John Gillicuddy has two tickets out of a total of 5,000, his probability of winning the car is 2/5,000 = 1/2,500 and his probability of not winning the car is 2,499/2,500. Thus, if zero stands for his not winning and 1 stands for his winning, the probability of each value of this random variable is

Value of random variable	Probability
0	2,499/2,500
1	1/2,500

(b) If Gillicuddy wins the car, he makes $4,998 (the value of the car less the amount spent on the tickets); if he does not win the car, he loses $2 (the amount spent on the tickets). Thus, the amount he wins or loses can assume two possible values, with the following probabilities:

Amount he wins or loses	Probability
−$2	$\dfrac{2,499}{2,500}$
+$4,998	$\dfrac{1}{2,500}$

EXERCISES

4.1 Three cards—the three, four, and five of spades—are in a box. One card is drawn at random. Then a second card is drawn at random. (The first card is *not* returned to the box before the second is drawn.) Let X represent the sum of the numbers on the two cards drawn from the box.
(a) What is the probability distribution of X?
(b) What is the probability that X is less than 8? Greater than 7?

4.2 Indicate whether the following random variables are continuous or discrete:
(a) the weight of a randomly chosen heavyweight boxer.
(b) the number of heavyweight boxers knocked out in a 7-day period.
(c) the height of a randomly chosen heavyweight boxer.

4.3 Which of the following cannot be a probability distribution?
(a) $P(x) = x/4$ $x = 1,2,3$
(b) $P(x) = x^2/8$ $x = 1,2,3$
(c) $P(x) = x/3$ $x = -1,+1,+3$

4.4 A box contains twice as many red marbles as green marbles. One marble is drawn at random from the box and replaced; then a second marble is drawn at random from the box. If both marbles are green, you win $5; if both are red, you lose $1; and if they are of different colors, you win or lose nothing.
(a) What is the probability distribution of the amount you win or lose?
(b) What is the probability that you at least break even?

4.5 The Alpha Corporation sells bicycles. Based on past experience, it feels that in the summer months it is equally likely that it will sell 0, 1, 2, 3, or 4 bicycles in a day. (The firm has never sold more than 4 bicycles per day.)
(a) Is the number of bicycles sold in a day a random variable?
(b) If so, what values can this random variable assume?
(c) If the number of bicycles sold in a day is a random variable, construct a table showing its probability distribution.
(d) Plot the probability distribution of the number of bicycles sold in a day in a line chart. What mathematical function can represent this probability distribution?
(e) What is the probability that the number of bicycles sold on a given day will be less than 1? Less than 3? Less than 4? Less than 6?
(f) The Alpha Corporation has only one salesman, whose income depends on the number of bicycles he sells per day. Specifically, he receives no commission on the first bicycle sold per day, a $20 commission on the second bicycle sold in a day, a $30 commission on the third, and a $40 commission on the fourth. (Thus, if he sells 3 bicycles in a given day, his commissions for this day total $50.) His income consists entirely of these commissions. Is his income on a particular day a random variable? If so, what values can this random variable assume? If his income is a random variable, construct a table showing its probability distribution.
(g) Plot the probability distribution of the salesman's income on a particular day in a line chart.
(h) What is the probability that his income in a particular day will exceed $20? $30? $40?
(i) Suppose that the number of bicycles sold one day is independent of the number sold the next day. Construct a table showing the probability distribution of the total number of bicycles sold in a two-day period. What is the probability that this number will exceed 2? 4? 6? 8?

(j) Unfortunately, the bicycle salesman has a weakness for gambling and is in debt to the Mob for $100. The Mob's local enforcer tells the salesman that he has two days to pay up, or he'll need a hospital bed. If the number of bicycles sold one day is independent of the number sold the next day, what is the probability that the salesman will earn enough during this two-day period to stay out of the hospital? (Assume that commissions are net of taxes and other deductions.)

(k) At the end of the two-day period described in (j) the salesman has earned only $70. The Mob's enforcer says that he will give the salesman a chance to win the extra $30. He will allow the salesman to flip a coin twice. If it comes up heads both times, he will give the salesman the $30 he needs; if it doesn't come up heads both times, he will take the salesman's $70 and begin the hospital-filling mayhem. Plot in a line chart the probability distribution for the salesman's monetary gains or losses in this game.

4.4 Expected Value of a Random Variable

As pointed out in the previous section, the probability distribution of a random variable is analogous to a frequency distribution. It is therefore not surprising that just as we found it useful to calculate the mean of a frequency distribution, we now find it useful to calculate the mean of a probability distribution. Another term often used to denote the mean of a probability distribution is the *expected value of the random variable,* which is defined as follows.

> EXPECTED VALUE OF A RANDOM VARIABLE: *The **expected value** of a discrete random variable* X, *denoted by* E(X), *is the weighted mean of the possible values that the random variable can assume, where the weight attached to each value is the probability that the random variable will assume this value. In other words,*

$$E(X) = \sum_{i=1}^{m} x_i P(x_i),$$ (4.3)   *Formula for E(X)*

where the random variable X *can assume* m *possible values,* x_1, x_2, $\cdots$, x_m *and the probability of its equaling* x_i *is* $P(x_i)$.

Clearly, the expected value of a random variable is analogous to the mean of a frequency distribution. If we compare the definition in equation (4.3) with our definition of the mean of a set of data calculated from a frequency distribution in equation (2.3), we see that the definitions are essentially the same.[4]

For the reason just stated, the expected value of any random variable X is often called the *mean* of X. If, as described in the previous section, the probability distribution in question is viewed as the population relative frequency distribution (which shows the proportion of cases with each value), the expected value of the random variable is really the population mean. To illustrate: We roll a true die a million times, and we regard the resulting relative frequency distribution of outcomes as the population. Clearly, the expected value of the number showing on a true die is

$$E(X) = \sum_{i=1}^{6} x_i P(x_i) = 1(1/6) + 2(1/6) + 3(1/6) + 4(1/6) +$$

$$5(1/6) + 6(1/6) = 3\frac{1}{2}.$$

Why? Because the random variable in question (the number coming up) can assume 6 possible values (thus, $m = 6$). These possible values are 1 to 6: thus $x_1 = 1$, $x_2 = 2$, $\cdots$, $x_6 = 6$. And the probability of each value is $1/6$: thus, $P(x_1) = P(x_2) = \cdots = P(x_6) = 1/6$. The mean of the population of 1 million outcomes of rolling a true die is 3 1/2, since the number of rolls in this population is so large that the proportion of cases when each number comes up is equal, for all practical purposes, to the probability of the number's coming up.

ROLE OF EXPECTED VALUES IN DECISION MAKING

The concept of a random variable's expected value (or mean, or *mathematical expectation,* as it is sometimes called) is extremely important in statistics. For example, consider the following gamble, which has already been described in Examples 4.1 and 4.3. A fair coin is flipped three times. If it comes up heads three times, you receive $100; if it comes up heads twice, you receive $50; if it comes up heads once, you lose $75; if it never comes up heads, you lose $100. Should you accept this gamble? Or put differently, if you were to repeat this gamble again and again, would you tend to come out ahead or would you tend to lose? The answer depends on the expected value of the amount

[4] To see how similar these definitions are, suppose that in each class interval the observations are all equal. (In other words, assume that every observation equals its class mark.) Then, since $f_j \div n$ in equation (2.3) is the relative frequency with which x_j occurs in the set of data, it can be regarded as the probability that x_j occurs in this set of data. And if $P(x_j)$ is substituted for $f_j \div n$ in equation (2.3), the result is equivalent to the right side of equation (4.3).

you win, an amount which (as we saw in Example 4.1) is a random variable. Why does it depend on this expected value? Because this expected value is the mean amount you will win (or lose) if you accept the gamble again and again for an indefinite number of times. In other words, the expected value is the mean of the infinite population of amounts you would win (or lose) if you accepted this gamble an infinite number of times.

What is the expected value of the amount you would win (or lose) from this gamble? Once again, let X stand for the relevant random variable, which in this case is the amount you would win (or lose). This random variable can assume only four values, $x_1 = 100$, $x_2 = 50$, $x_3 = -75$, and $x_4 = -100$, with the result that

$$E(X) = \sum_{i=1}^{4} x_i P(x_i) = 100(1/8) + 50(3/8) - 75(3/8) - 100(1/8)$$

$$= -9\frac{3}{8},$$

since (as we know from Example 4.3) $P(x_1) = P(x_4) = 1/8$ and $P(x_2) = P(x_3) = 3/8$. Because the expected value of the amount you would win from this gamble is negative, it is clear that you would lose money if you were to repeat the gamble over and over. More specifically, *if you averaged the results of repeating this gamble over and over, you would lose $9.375 per gamble.* Thus, the gamble is not a "fair bet."

Whether the expected value of one's winnings from a particular wager is positive or negative is obviously of importance in determining whether or not one should accept the wager. For example, if you have the choice of accepting a gamble where the expected value of your winnings is +$10 (as opposed to -$9.375 for the gamble described above), you may decide to accept the former rather than the latter. As you would expect, a decision maker often chooses the action or gamble that has the largest expected value of winnings or profits. But in general, although the expected value of one's winnings or profits is of importance, this alone is not the sole indicator of what action a decision maker should take.

The two examples below illustrate the calculation and interpretation of expected values.

EXAMPLE 4.5 At a certain gambling casino the following game is played: The dealer, who works for the house, allows you to pick two cards from a full deck. If both are hearts, you win $15; otherwise, you lose $1. What is the expected value of the amount you win (or lose) each time you play this game?

SOLUTION: The amount you would win is a random variable that can assume two possible values, +$15 or −$1. The probability that it equals $15 is the probability of your getting two hearts, which is 12/51(13/52) = 1/17, or .059. (Recall Example 3.6.) The probability that the amount will be −$1 must therefore be 1 − .059, or .941. Thus, the expected value of this random variable is

$$(\$15)(.059) + (-\$1)(.941) = -\$.056.$$

EXAMPLE 4.6 A marketing executive must decide whether or not to use a new label on a product. The firm will gain $800,000 if he adopts the new label and it turns out to be superior to the old label. The firm will lose $500,000 if the executive adopts the new label and it proves to be not superior to the old one. The firm will neither gain nor lose money if the executive sticks with the old label. The executive feels there is a 50-50 chance that the new label is superior to the old and a 50-50 chance that it is not. If he wants to take the action with the higher expected gain to the firm, should he decide to use the new label or not?

SOLUTION: If the executive decides to adopt the new label, the expected value of the firm's gain is

$$(\$800,000)(1/2) + (-\$500,000)(1/2) = \$150,000,$$

because the firm's gain is a random variable that can assume two possible values, +$800,000 and −$500,000, and the probability of each value is 1/2. If the executive decides not to adopt the new label, the expected value of the firm's gain is zero, since zero is the only possible value that it can assume. Since $150,000 exceeds zero, the expected gain if he adopts the new label is higher than if he does not. Thus, if he wants to take the action with the higher expected gain, he should adopt the new label.

4.5 Variance and Standard Deviation of a Random Variable

Just as the expected value of a random variable is analogous to the mean of a frequency distribution, the variance of a random variable is analogous to the variance of a frequency distribution. In particular, the variance of a random variable is defined as follows.

VARIANCE OF A RANDOM VARIABLE: *The **variance of a random variable** X, denoted by $\sigma^2(X)$, is the expected value of the squared deviation of the random variable from its expected value. In other words,*

$$\sigma^2(X) = E([X - E(X)]^2) = \sum_{i=1}^{m} [x_i - E(X)]^2 P(x_i),$$

(4.4)   *Formula for σ^2 (X)*

where the random variable can assume m *possible values,* $x_1, x_2, \cdots ,$ $x_m,$ *and* $P(x_i)$ *is the probability of its equaling* x_i.

To see the similarity between the variance of a random variable and the variance of a body of data, recall from Chapter 2 that the latter was defined as the mean of the squared deviations of the observations from their mean. Substitute "expected value" for "mean" in the previous sentence, and you have the definition of the variance of a random variable.

Similarly, the standard deviation of a random variable is analogous to the standard deviation of a body of data. Its definition is as follows.

STANDARD DEVIATION OF A RANDOM VARIABLE: *The **standard deviation of a random variable** is the positive square root of the random variable's variance. In other words, the standard deviation of a random variable* X, *denoted by* $\sigma(X)$, *is*

$$\sigma(X) = \sqrt{E([X - E(X)]^2)} = \sqrt{\sum_{i=1}^{m} [x_i - E(X)]^2 P(x_i)}.$$

(4.5)   *Formula for σ (X)*

Recall from Chapter 2 that the standard deviation of a set of data is the square root of the variance of the data. Thus, the definition here is analogous to the one given there.

Just as the standard deviation of a set of data indicates the extent of the dispersion or variability among the individual measurements within the set of data, *the standard deviation of a random variable indicates the extent of the dispersion or variability among the values that the random variable may assume.* For example, if it is *certain* that a random variable X will equal a certain number, then the difference between X and its expected value—that is, $X - E(X)$—will always be zero. Thus, the expected value of $[X - E(X)]^2$ will also be zero. Consequently, the variance and standard deviation of X will be zero. On the other hand, if it is likely that X will assume values *far removed* from its expected value, then the difference between X and its expected value—that is $X - E(X)$—will have a large probability of being big. Therefore, the expected value of $[X - E(X)]^2$ will also be big; and consequently, the variance and standard deviation of X will both be big.

To illustrate how one can calculate the variance and standard deviation of a random variable, let's calculate the variance and standard deviation of the number coming up on a true die. If this random variable is designated as X, it willl be recalled from the previous section of this chapter that the expected value of this random variable, $E(X)$, equals $3\frac{1}{2}$. Thus, the variance of this random variable is

$$\sigma^2(X) = \sum_{i=1}^{6} \left[x_i - 3\frac{1}{2} \right]^2 P(x_i)$$

$$= \left[1 - 3\frac{1}{2} \right]^2 \left(\frac{1}{6} \right) + \left[2 - 3\frac{1}{2} \right]^2 \left(\frac{1}{6} \right) + \left[3 - 3\frac{1}{2} \right]^2 \left(\frac{1}{6} \right)$$

$$+ \left[4 - 3\frac{1}{2} \right]^2 \left(\frac{1}{6} \right) + \left[5 - 3\frac{1}{2} \right]^2 \left(\frac{1}{6} \right)$$

$$+ \left[6 - 3\frac{1}{2} \right]^2 \left(\frac{1}{6} \right)$$

$$= \left(6\frac{1}{4} \right) \left(\frac{1}{6} \right) + \left(2\frac{1}{4} \right) \left(\frac{1}{6} \right) + \left(\frac{1}{4} \right) \left(\frac{1}{6} \right) + \left(\frac{1}{4} \right) \left(\frac{1}{6} \right)$$

$$+ \left(2\frac{1}{4} \right) \left(\frac{1}{6} \right) + \left(6\frac{1}{4} \right) \left(\frac{1}{6} \right)$$

$$= \frac{35}{12}.$$

And since the standard deviation of a random variable is the square roots of its variance,

$$\sigma(X) = \sqrt{35/12}.$$

The following example provides additional practice in calculating the variance and standard deviation of a random variable.

EXAMPLE 4.7 A fair coin is flipped three times. What is the variance and standard deviation of the number of times it comes up heads?

SOLUTION: If X denotes the number of heads that come up, we know from Example 4.3 that the probability that $X = 0$ is $1/8$, the probability that $X = 1$ is $3/8$, the probability that $X = 2$ is $3/8$, and the probability that $X = 3$ is $1/8$. The expected value of X is

$$E(X) = (0)(1/8) + (1)(3/8) + (2)(3/8) + (3)(1/8)$$

$$= 3/2.$$

Thus, the variance of X is

$$\sigma^2(X) = [0 - 3/2]^2(1/8) + [1 - 3/2]^2(3/8) + [2 - 3/2]^2(3/8)$$

$$+ [3 - 3/2]^2(1/8)$$

$$= (9/4)(1/8) + (1/4)(3/8) + (1/4)(3/8) + (9/4)(1/8)$$

$$= 3/4.$$

And the standard deviation of X is

$$\sigma(X) = \sqrt{3/4}.$$

EXERCISES

4.6 A pair of true dice are thrown. If the total number rolled is less than 6, you win \$1; if it is greater than 8, you lose \$1; otherwise you neither gain nor lose any amount. What is the expected value of the amount gained or lost?

4.7 If $Y = 2X$, what is the expected value of Y if the expected value of X is (a) 14; (b) −3? (For further discussion of problems of this kind, see Appendix 4.1 of this chapter.)

4.8 A fair coin is flipped 10 times. What is the expected value of the number of times it comes up heads? What is the expected value of the number of times it comes up tails?

4.9 Suppose that the probability distribution of X is as shown below:

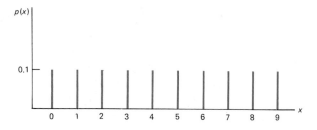

(a) Is X a continuous or discrete random variable?

(b) What real-life phenomena might the random variable X represent? (Hint: any number must end in 0, 1, $\cdots$, 9.)

(c) Calculate the expected value, variance, and standard deviation of X.

4.10 There is a 0.97 probability that no accident will occur at a particular race track during each day. The probability of one accident is 0.02, and the probability of two accidents is 0.01.

(a) What is the expected number of accidents in a day?

(b) What is the expected number of accidents in 10 days?

(c) What is the variance of the number of accidents in a day?

(d) What is the standard deviation of the number of accidents in a day?

4.11 The Alpha Corporation is equally likely to sell 0, 1, 2, 3, or 4 bicycles in a day. (a) What is the expected value of the number of bicycles sold by the Alpha Corporation in a particular day? (b) Is this a value that the random variable in question can assume? (c) What is the variance of the number of bicycles sold in a day? (d) What is the standard deviation of the number of bicycles sold in a day?

4.12 (a) Under the circumstances described in Exercise 4.5, what is the expected value of the income of the Alpha Corporation's salesman in a particular day? (b) Is this a value that the random variable in question can assume? (c) What is the standard deviation of the salesman's income in a day? (d) What is the expected value of the income of the Alpha Corporation's salesman in a particular two-day period? (e) Is this answer twice the answer to (a)? Why, or why not?

4.13 A bag contains 10 packages worth $20 each, 5 packages worth $50 each, and 1 package worth $100.

(a) If one package is chosen at random, what is the expected value of its worth?

(b) What is the standard deviation of its worth?

(c) In this case, is it possible for the random variable (the worth of the chosen package) to equal the expected value of the random variable?

(d) If you want to take the action that maximizes expected gain, should you pay $40 for the opportunity to pick a package at random from this bag?

4.6 The Binomial Distribution

Earlier in this chapter we discussed the nature and characteristics of random variables and probability distributions in general. We turn now to the description and analysis of the most important discrete probability distribution in statistics: the binomial distribution. Specifically, we indicate the conditions that generate the binomial distribution, the formula for the distribution (as well as its mean and standard deviation), and some of the ways in which this distribution has been used to help solve important practical problems.

BERNOULLI TRIALS

To understand the circumstances under which the binomial distribution arises, it is convenient to begin by describing a *Bernoulli process* or, what is the same thing, a series of *Bernoulli trials.* (James Bernoulli was a seventeenth-century Swiss mathematician who performed some of the early work on the binomial distribution.) Each Bernoulli trial takes place under the following circumstances. First, *each trial results in one of two possible outcomes, which is termed either "success" or "failure."* Second, *the probability of a success remains the same from one trial to the next.* Third, *the outcomes of the trials are independent of one another.*

An example of a Bernoulli trial is a game in which a true die is thrown, and if a 4, a 5, or a 6 comes up, you win. If a 1, a 2, or a 3 comes up, you lose. To make sure that this is a Bernoulli trial, let's see whether it meets the three conditions specified in the previous paragraph. Certainly, the first condition is met, since there are only two possible outcomes—win (success) or lose (failure). And the second condition is met, since the probability of winning (that is, of a success) remains constant (at $1/2$ from one trial to the next). Moreover, there is no reason to believe that your chances of winning are influenced in any way by the outcome of previous trials, since the die has no memory. Thus, all three conditions are met.

THE BINOMIAL DISTRIBUTION

Suppose that n Bernoulli trials occur and that the probability of a success on each trial equals Π. Under these circumstances[5] the number of successes occurring in these n trials has a binomial probability distribution, which means

$$P(x) = \frac{n!}{x!(n-x)!} \Pi^x (1 - \Pi)^{n-x}, \text{ for } x = 0, 1, 2, \cdots n \qquad (4.6)$$

Formula for Binomial Distribution

where $P(x)$ is the probability that the number of successes equals x. The number of successes, X, is a random variable, whereas n and Π are constants. It is customary to refer to X as a binomial random variable.

To illustrate the calculation of binomial probabilities, consider a case where a fair coin is flipped four times. What is the probability distribution of the number of heads? Since there are two mutually exclu-

[5] Of course, the probability of failure on each trial equals $1 - \Pi$, since success and failure on each trial are complementary events (as defined in Chapter 3).

sive outcomes each time the coin is flipped, since the probability of heads is 1/2 each time, and since the outcomes of the tosses are statistically independent, it follows that this is a situation where there are four Bernoulli trials and where the probability of a success (heads) equals 1/2. Substituting 4 for n and 1/2 for Π in equation (4.6), we can calculate the probability distribution of the number of heads, with the following result:

$$P(0) = \frac{4!}{0!4!} \left(\frac{1}{2}\right)^0 \left(\frac{1}{2}\right)^4 = \frac{1}{16},$$

$$P(1) = \frac{4!}{1!3!} \left(\frac{1}{2}\right)^1 \left(\frac{1}{2}\right)^3 = \frac{1}{4},$$

$$P(2) = \frac{4!}{2!2!} \left(\frac{1}{2}\right)^2 \left(\frac{1}{2}\right)^2 = \frac{3}{8},$$

$$P(3) = \frac{4!}{3!1!} \left(\frac{1}{2}\right)^3 \left(\frac{1}{2}\right)^1 = \frac{1}{4},$$

$$P(4) = \frac{4!}{4!0!} \left(\frac{1}{2}\right)^4 \left(\frac{1}{2}\right)^0 = \frac{1}{16}.$$

Thus, the probability of no heads is 1/16, of heads once is 1/4, of heads twice is 3/8, of heads three times is 1/4, and of heads four times is 1/16. This probability distribution is plotted in Figure 4.3.

Figure 4.3
Binomial Probability
Distribution, $n = 4$ and
$\Pi = 1/2$

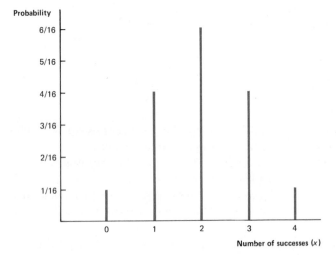

It is important to recognize that many random variables of considerable practical importance have a binomial distribution. For example, later in this chapter we show how this probability distribution has played a central role in psychological experiments concerning auditory perception and in quality control in industry and government.

TESTING THE FORMULA FOR $P(x)$

Before undertaking some practical applications of the binomial distribution, we want to verify the correctness of the formula for the binomial probability distribution in equation (4.6). To do this, let's show that the probabilities of 0, 1, or 2 heads (in 4 tosses) based on equation (4.6) are correct.

Probability of No Heads. Clearly, there is only one way for this outcome to occur: The coin must come up tails *four times in a row.* Recalling the multiplication rule, the probability of tails four successive times must be equal to $(1/2)^4$ or $1/16$, since the probability of tails each time is $1/2$ and the outcomes of various tosses are statistically independent of one another. This result agrees with our calculation of $P(0)$, as computed from equation (4.6).

Probability of Heads Once. There are four mutually exclusive ways that one heads can occur in four tosses; namely, heads can come up on the first, second, third, or fourth toss. The probability of each of these ways equals $1/16$. Why? Because the probability both of heads on a *particular* toss and tails on the *other three* tosses equals $(1/2)^4$, or $1/16$, since the probability of heads and the probability of tails equal $1/2$, and the outcomes of various tosses are statistically independent. (Again, recall the multiplication rule.) Since the probability of the occurrence of *each* of these ways is $1/16$ (and since they are mutually exclusive), the addition rule dictates that the probability that *any one* of these ways occurs must be $4(1/16)$, or $1/4$. This result is in accord with our calculation of $P(1)$ as computed from equation (4.6).

Probability of Heads Twice. If we designate heads by H and tails by T, there are the following six ways that two successes can be distributed among the four tosses of the coin: (1) *HHTT;* (2) *HTHT;* (3) *HTTH;* (4) *THHT;* (5) *THTH;* and (6) *TTHH.* The first way occurs when the first and second tosses come up heads and the third and fourth tosses come up tails; the second way occurs when the first and third tosses come up heads and the second and fourth tosses come up tails; and so on. Each of these six ways has a probability of occurrence of $(1/2)^4$ or $1/16$, because of the multiplication rule. Thus, because of the addition rule, the probability that any one of these (mutually exclusive) ways occurs must be $6(1/16)$, or $3/8$. This result is in accord with our calculation of $P(2)$, as computed from equation (4.6).

DERIVATION[6] OF THE FORMULA FOR $P(x)$

Although the foregoing discussion demonstrates that the formula for $P(x)$ results in the correct probabilities of 0, 1, and 2 heads out of four tosses of a coin, it does not demonstrate that this formula is always valid. To show that this is the case, it is necessary to derive the formula in equation (4.6). In the following four paragraphs, we provide such a derivation. Since the derivation is somewhat more technical than other parts of this chapter, some readers may want to skip these paragraphs. An understanding of subsequent sections does not depend on reading them.

To derive this formula, the first step is to note that there are a variety of ways that n Bernoulli trials can give rise to exactly x successes and $n - x$ failures. One way is that successes occur on each of the first x trials, after which all the failures occur. If S is a success and F is a failure, this sequence can be represented by

$$\overbrace{SSS \cdots S}^{x} \quad \overbrace{FFF \cdots F}^{n - x}$$

Another possible sequence is the following, where the first $x - 1$ trials result in successes, the next $n - x$ trials result in failures, and the last trial results in a success:

$$\overbrace{SS \cdots S}^{x - 1} \quad \overbrace{FFF \cdots FS}^{n - x}$$

Because the trials are independent, the probability that the first of these two sequences will occur is

$$\overbrace{\Pi\Pi\Pi \cdots \Pi}^{x} \quad \overbrace{(1 - \Pi)(1 - \Pi)(1 - \Pi) \cdots (1 - \Pi)}^{n - x} = \Pi^x(1 - \Pi)^{n - x}.$$

And the probability of occurrence of the second sequence is:

$$\overbrace{\Pi\Pi \cdots \Pi}^{x - 1} \quad \overbrace{(1 - \Pi)(1 - \Pi)(1 - \Pi) \cdots (1 - \Pi)\Pi}^{n - x} = \Pi^x(1 - \Pi)^{n - x}.$$

---

[6] Some instructors may prefer to omit this subsection, which can be skipped without loss of continuity.

which equals the probability of the first sequence. Clearly, the probability of obtaining *any* particular sequence of x successes and (n − x) failures must be the same as the probability of each of these two sequences.

How many ways can n Bernoulli trials give rise to exactly x successes and n − x failures? In other words, how many different sequences of this type can occur? There are n trials, which we can represent by n different cards (with a 1 on the first card, a 2 on the second card, . . . , and an n on the nth card). The problem is to determine how many different ways x of these cards can be chosen, since any set of x cards can be used to represent a sequence where successes occur on the trials corresponding to these cards (and where failures occur on the rest of the trials). Expressed in the language of Appendix 3.1, this problem amounts to asking how many combinations of x cards can be drawn from a set of n cards, since the order in which the cards are drawn does not matter. From Appendix 3.1 we know that the answer is $n! \div [(n - x)!x!]$. Consequently, this is the number of sequences in which exactly x successes occur.

Because each of these sequences constitutes one of the mutually exclusive ways in which x successes can occur in n trials, and because each such sequence has a probability of occurrence of $\Pi^x(1 - \Pi)^{n-x}$, the probability of x successes is obtained by adding $\Pi^x(1 - \Pi)^{n-x}$ as many times as there are sequences. Since there are $n! \div [(n - x)!x!]$ such sequences, it follows that this sum equals $\Pi^x(1 - \Pi)^{n-x}$ multiplied by $n! \div [(n - x)!x!]$. Thus, the probability that x successes will occur in n trials must equal

$$\frac{n!}{(n - x)!x!} \Pi^x(1 - \Pi)^{n-x},$$

which is the expression in equation (4.6). We have therefore proved that this formula is correct.

THE BINOMIAL DISTRIBUTION: A FAMILY OF DISTRIBUTIONS

It is important to recognize that the binomial probability distribution is not a single distribution, but a family of distributions. Depending on the values of n and Π, one can obtain a wide variety of probability distributions, all of which are binomial. For example, Figure 4.4 shows the binomial probability distribution when n = 6 and Π = 0.2, 0.3, 0.7, and 0.8. As you can see, it is skewed to the right when Π is less than 1/2 and skewed to the left when Π is greater than 1/2. Figure 4.5 shows the binomial probability distribution when Π = 0.4 and n = 5, 10, and 20. As you can see, the binomial probabil-

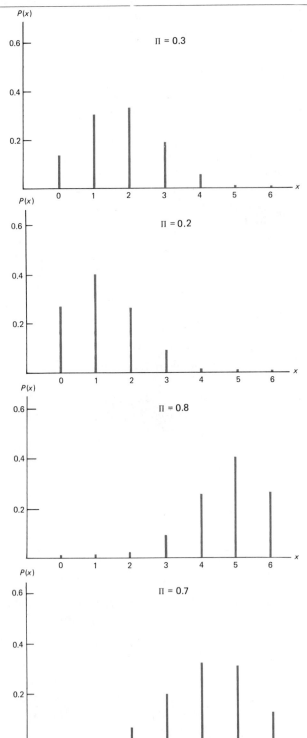

Figure 4.4
Four Binomial Distributions, All with $n = 6$

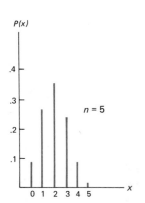

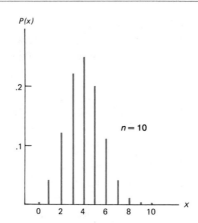

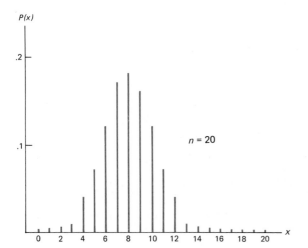

Figure 4.5
Three Binomial
Probability Distribu-
tions, All with $\Pi = 0.4$

ity distribution becomes increasingly bell-shaped as n increases in value. More will be said about this in the next chapter.

If n is fairly large, it is laborious to carry out the calculations involved in using equation (4.6) to compute $P(x)$. Fortunately, there are tables which give the value of $P(x)$ corresponding to various values of n and Π. Appendix Table 1 supplies the values of $P(x)$ for values of n from 1 to 20, and for values of Π from 0.05 to 0.50. For large values of n, approximations of the binomial distribution can be found, as we shall see in the following chapter.

If the value of Π exceeds 0.5, Appendix Table 1 can still be used. We need only switch the definitions of success and failure. For example, if the probability of a success is 0.8, what is the probability of four successes in 10 trials? To use Appendix Table 1, let's reverse the definitions of success and failure; that is, let's call a success what we formerly called a failure and vice versa. If we do this, what we want is the probability of six successes (formerly regarded as failures) in 10 trials. Since the probability of a success (formerly a failure) is 0.2, Appendix Table 1 shows that this probability equals .0055. (Look it up and see.)

MEAN AND STANDARD DEVIATION OF A BINOMIAL RANDOM VARIABLE

In Sections 4.4 and 4.5, we defined the expected value and standard deviation of a random variable. What are the expected value and standard deviation of a binomial random variable? First, let's consider the expected value, which (you will recall) is defined as the weighted mean of the possible values that a random variable can assume, each value being weighted by its probability of occurrence. Thus, the mean of a binomial random variable must equal

$$E(X) = \sum_{x=0}^{n} xP(x), \qquad (4.7)$$

since the possible values of X are from 0 to n.

To illustrate the calculation of the expected value of a binomial random variable, suppose once again that $n = 4$ and $\Pi = 1/2$. Then it follows from equation (4.7) that

$$E(X) = (0)P(0) + (1)P(1) + (2)P(2) + (3)P(3) + (4)P(4)$$

Substituting into this formula the values of $P(0), \cdots, P(4)$ obtained on page 150, we find that

$$E(X) = 0(1/16) + 1(1/4) + 2(3/8) + 3(1/4) + 4(1/16) = 2.$$

Thus, the expected value (or mean) of X, if $n = 4$ and $\Pi = 1/2$, is 2; or put somewhat differently, the mean number of successes under these conditions is 2.

Calculating the expected value of a binomial random variable in this way shows clearly what the basic computations are. In practical work, however, this is a highly inefficient procedure because it can be shown that for a binomial random variable

Expected Value of a Binomial Random Variable

$$E(X) = n\Pi. \qquad (4.8)$$

Thus, all one has to do is multiply the number of trials by the probability of success on each trial. Regardless of the values of n or Π, this simple computation will give the expected value of a binomial random variable. To make sure that this shortcut is correct, let's apply it to the case where $n = 4$ and $\Pi = 1/2$. Using equation (4.8), the expected value is $4(1/2)$, or 2, which is precisely the result we obtained by the more laborious procedure carried out in the previous paragraph. Of course, equation (4.8) certainly appeals to common sense. If n trials are carried out, and if the probability of a success in each trial is Π, it stands to reason that the mean number of successes will be $n\Pi$.

Next we turn to the standard deviation of a binomial random variable. From Section 4.5 we know that the standard deviation of any random variable is defined as

$$\sqrt{\Sigma(x_i - E(X))^2 P(x_i)},$$

where the summation is over all possible values of X. Since the possible values of a binomial random variable are $0, 1, \cdots, n$, its standard deviation must therefore equal

$$\sigma(X) = \sqrt{\sum_{x=0}^{n} (x - E(X))^2 P(x)}. \qquad (4.9)$$

To illustrate the calculation of the standard deviation of a binomial random variable, suppose once again that $n = 4$ and $\Pi = 1/2$. Then it follows from equation (4.9) that

$$\sigma(X) = [(0 - 2)^2(1/16) + (1 - 2)^2(1/4) + (2 - 2)^2(3/8)$$

$$+ (3 - 2)^2(1/4) + (4 - 2)^2(1/16)]^{1/2}$$

$$= \sqrt{1} = 1.$$

As in the case of the expected value, this rather laborious computation is useful because it shows clearly what the definition means. But in practical work there is no need to carry out these extensive calculations, since it can be shown that for a binomial random variable

$$\sigma(X) = \sqrt{n\Pi(1 - \Pi)}. \qquad (4.10)$$

Standard Deviation of a Binomial Random Variable

Thus, *all one has to do is multiply the number of trials by the product of the probability of success and probability of failure, and find the square root of the result.* Regardless of the values of n and Π, this will yield the standard deviation of a binomial random variable. For example, if $n = 4$ and $\Pi = 1/2$, $\sigma(X) = \sqrt{4(1/2)(1/2)} = 1$, according to equation (4.10), which is precisely the answer we got by the more laborious process earlier in this paragraph.

The following example, based on actual data, is designed to illustrate how binomial probabilities are calculated and interpreted, as well as to show how the expected value and standard deviation of a binomial random variable are computed.

EXAMPLE 4.8[7] In 1876, Sir Francis Galton, a famous British scientist and statistician, concluded that "young people hear shriller sounds than older people."[8] In recent years, psychologists like Bruce Schneider, Sandra Trehub, and Dale Bull of the University of Toronto have found that infants are even more sensitive than young adults to high-frequency sounds.[9] To study the auditory perception of infants, they have used the following experimental technique. The infant sits on a parent's lap in one corner of a sound-treated room. A sound is presented through one of two loudspeakers located in the adjacent corners. The loudspeaker presenting the sound is chosen at random. The experimenter waits for the infant to turn his or her head. The number of times that the infant turns his or her head in the direction of the sound is compared with what would be expected on the basis of chance.

(a) If there is a .7 probability that an infant will turn in the direction of the sound, and if the sound is presented 10 times, what is the probability that the infant will turn in the correct direction in less than 9 of the 10 times?

(b) If the sound is presented 10 times, what is the expected value and standard deviation of the number of times that the infant will turn in the correct direction?

(c) If the sound is presented 20 times, what is the probability that the infant will turn in the correct direction in less than 8 of the 20 times? Use Appendix Table 1 to find the answer.

SOLUTION: (a) Whether or not the infant turns in the direction of the sound can be viewed as a Bernoulli trial where the probability of success (that is, turning in the correct direction) is .7. Thus, the probability that the number of successes is 9 or 10 equals:

---

[7] I am indebted to Neil Macmillan of Massachusetts Institute of Technology for suggesting this example.

[8] B. Schneider, S. Trehub, and D. Bull, "High-Frequency Sensitivity in Infants," *Science*, February 29, 1980, p. 1003.

[9] Ibid.

$$P(9 \text{ or } 10) = P(9) + P(10) = \frac{10!}{9!1!} (.7)^9 (.3) + \frac{10!}{10!0!} (.7)^{10}$$

$$= .121 + .028 = .149.$$

Consequently, the probability that the infant will turn in the correct direction in less than 9 of the 10 times is $1 - .149 = .851$.

(b) The expected value of the number of times that the infant will turn in the correct direction is 10(.7), or 7, since $n = 10$ and $\Pi = .7$. The standard deviation of the number of times that the infant will turn in the correct direction is $\sqrt{10(.7)(.3)} = \sqrt{2.1}$, or 1.45, since $n = 10$, $\Pi = .7$, and $(1 - \Pi) = .3$.

(c) The desired probability equals the probability that the infant will *not* turn in the correct direction in *more* than 12 of the 20 times. Since the probability that the infant will *not* turn in the correct direction is 0.3, we look in Appendix Table 1 under $n = 20$ and $\Pi = .30$. The table shows that the probability that the infant will *not* turn in the correct direction in 13 of the 20 times is .0010, that the probability that the infant will *not* turn in the correct direction in 14 of the 20 times is .0002, and that the probability that the infant will *not* turn in the correct direction in more than 14 of the 20 times is essentially zero. Thus, the desired probability equals $.0010 + .0002 = .0012$.

4.7 Acceptance Sampling

We have already emphasized that the binomial distribution is of great practical importance. A major example of such application is acceptance sampling, which is one important field of quality control. As an illustration of acceptance sampling, suppose that a firm purchases a large number of metal fixtures from a particular supplier. To make sure that only a reasonably small percentage of the fixtures in any particular shipment is defective, *the firm takes a sample of 30 fixtures from the shipment and tests each to determine whether it is defective. If none is defective, the firm accepts the shipment; otherwise, it rejects it.* This is a very simple type of acceptance-sampling scheme, but it is a convenient starting point.

To see why the binomial distribution is important in acceptance sampling, we must note that if the sample is chosen at random,[10] the

---

[10] By *random sample* we mean one where each fixture in the sample has the same probability of being chosen. A more complete definition is given in Chapter 6.

probability that each fixture in the sample will be defective will equal the proportion of defective fixtures in the shipment. (This assumes that the sample is only a small proportion of the shipment, which is generally true, or that sampling is with replacement.[11]) For example, if 10 percent of the fixtures in a particular shipment are defective, the probability that a randomly chosen fixture in the shipment will be defective is .10. Thus, each choice of a fixture by the inspector can be regarded as a Bernoulli trial, where a success occurs if the fixture is found to be defective. On each such trial, the probability of a success equals Π, the proportion of defective fixtures in the shipment. Consequently, if the inspector chooses a sample of 30 fixtures from a shipment, the number of defective fixtures in the sample will be a binomial random variable. Specifically, the probability that x of these fixtures will be defective is

$$P(x) = \frac{30!}{(30 - x)!x!} \Pi^x(1 - \Pi)^{30 - x},$$

and thus the probability that none is defective is

$$P(0) = (1 - \Pi)^{30}.$$

This result is important because it enables the firm to calculate the probability that it will accept a shipment with various percentages of fixtures that are defective. For example, if 20 percent of the fixtures in a shipment are defective, what is the probability that the firm will accept such a shipment? Substituting .20 for Π, we find that this probability equals $(1 - .20)^{30}$, or .001. (Why? Because the firm will accept the shipment only if none are defective.) On the other hand, suppose that 10 percent of the fixtures in a shipment are defective. Under these circumstances, what is the probability that the firm will accept the shipment? Substituting .10 for Π, we find that this probability equals $(1 - .10)^{30}$, or .04. Finally, if 5 percent of the fixtures in a shipment are defective what is the probability that the firm will accept the shipment? Substituting .05 for Π, we find that this probability equals $(1 - .05)^{30}$, or .21. Calculations of this sort are used by firms and government agencies to determine the sample size and the type of sampling plan they should use. Thus, the above sampling plan may be quite appropriate if the firm wants to accept shipments with 5 percent defectives about 21 percent of the time, but the firm may not want to do this. It may, for example, want to accept such shipments only about 8 percent

[11] If the sampling is without replacement and if the sample is not a small proportion of the shipment, the hypergeometric distribution should be used. For a discussion of the hypergeometric distribution, see Appendix 4.2 of this chapter.

THE CASE OF RAINED-OUT TENNIS

The officials of tennis tournaments must make provisions for some matches being rained out. A tennis tournament will be held in Canada on August 4–6. There will be matches at 11 A.M. and 4 P.M. on each day. The morning matches are played on composition courts that dry quickly; the afternoon matches are played on clay courts that dry slowly. Since there are 6 groups of matches (3 days times 2 periods of the day) and since the officials of the tournament feel that the probability that each group will be rained out is 0.1, they calculate that the probability that more than one of these groups will be rained out is only .1143. To obtain this result, they use the binomial distribution where $n = 6$ and $\Pi = 0.1$. (See Appendix Table I.) Do you agree with their result? If not, what mistakes have they made?

SOLUTION

The binomial distribution does not seem to be appropriate in this situation for at least two reasons. First, the probability that a group of matches will be rained out does not seem to be the same from one group to another. A given amount of rain seems more likely to rain out the morning matches than the afternoon matches because of the difference in the type of court. Second, whether or not one group of matches is rained out does not seem to be independent of whether another group will be rained out. For example, if the morning matches on August 4 are rained out, the probability that the afternoon matches on that day will also be rained out seems higher than if the morning matches were not rained out (because rainy mornings often are followed by rainy afternoons). The moral is that one must be sure that the conditions underlying the binomial distribution are met; if they are not met, the binomial distribution should not be used.

of the time. If so, this objective can be achieved by taking a sample of 50 rather than 30 fixtures from each shipment, and by accepting the shipment only if none are defective. (As an exercise, prove that this is true. If you have difficulty, consult the note at the bottom of this page.[12])

[12] If $\Pi = .05$, $(1 - \Pi)^{50} = (1 - .05)^{50} = .078$. Thus, the probability of accepting a shipment with 5 percent defectives with this acceptance-sampling plan is about 8 percent.

**GETTING DOWN
TO CASES**

QUALITY CONTROL IN THE MANUFACTURE OF RAILWAY-CAR SIDE
FRAMES[13]

A producer of railway-car side frames wanted to establish a system to control the proportion of defective frames produced. Based on past data, the firm knew that 20 percent of the frames it produced were defective. It wanted to establish a procedure which would signal when the fraction of defectives jumped above 20 percent. After considerable discussion the firm decided to sample 10 frames from each day's output and find out the number of defectives. If this number exceeded 2 out of 10, the firm would stop its productive process to attempt to find out why such a relatively high percent was defective.

(a) Assuming that the 10 frames sampled were a very small proportion of the day's output, what was the probability that the firm's productive process would be stopped when in fact 20 percent of the day's output was defective?

(b) If defectives were to increase to 40 percent of a particular day's output, what is the probability that this inspection procedure would result in a stoppage of the firm's productive process?

(c) Based on your results in (a) and (b), write a one-paragraph report concerning the adequacy of the firm's sampling plan.

The firm sends a customer a shipment of five frames chosen at random from the day's output under the conditions described in (b). According to the terms of the firm's agreement with the customer, the firm must pay the customer $100 for every defective frame shipped as compensation for expenses incurred in receiving and handling defective materials.

(d) Graph the probability distribution of the amount the firm will have to pay the customer with regard to this shipment of five frames.

(e) What is the expected value of the amount the firm will have to pay the customer with regard to this shipment?

(f) What is the standard deviation of the amount the firm will have to pay the customer with regard to this shipment?

---

[13] The first half of this case is based in part on a section from A. J. Duncan, *Quality Control and Industrial Statistics* (Homewood, Ill.: Irwin, 1959). The numbers have been changed for pedagogical reasons. The reader may be surprised that so large a percentage (20 percent) of the frames was defective, but this was actually the case.

EXERCISES

4.14 The ratio of the variance of a binomial random variable to its expected value is 1/2. Can you determine n? Can you determine Π?

4.15 In how many ways can four successes be distributed among five trials? In how many ways can one success be distributed among five trials? Are your answers to these two questions the same?

4.16 A physician administers a new drug to 20 people with a certain disease. The probability is 0.15 that the drug will cure each person, and the result for one person is independent of that for another person.
(a) What is the expected value of the number of people in the sample cured by the drug? What is the standard deviation of this number?
(b) What is the expected value of the number of people *not* cured by the drug? What is the standard deviation of this number?

4.17 One fifth of a particular breed of rabbits are born with long hair. If 12 rabbits of this breed are born, calculate the probability that
(a) all have long hair.
(b) at least half have long hair.
(c) none has long hair.
(d) less than 2 have long hair.
(e) the number having long hair is less than the expected number with long hair.

4.18 Ten bills are sent to the Senate Finance Committee. The probability that each bill will be approved by this committee is 0.1.
(a) What is the expected value of the number of bills that will be approved?
(b) What is the standard deviation of the number of bills that will be approved? (Assume that the outcomes for various bills are independent.)
(c) Calculate the probability that the number of bills that will be approved is
  (i) less than the expected value.
  (ii) above the expected value, but less than one standard deviation above it.
  (iii) above the expected value, but less than two standard deviations above it.

4.19 The football team at the local state university has a 0.4 probability of winning each of the nine games that remain to be played this season. (The outcomes of the games are independent.) If the team wins at least eight of these games, the coach will receive a pay raise of $5,000; otherwise, he will receive a raise of $2,000.
(a) Is his pay raise a random variable? If so, does it have a binomial distribution?
(b) What is the expected value of his pay raise?
(c) What is the standard deviation of his pay raise?

4.20 John Martin, a plumber, installs ten hot water heaters in a particular housing development. The chance that each heater will last more than ten years is 0.3, and their lives are independent.

(a) Construct a line chart showing the probability distribution of the number of hot water heaters lasting more than ten years.

(b) Is this probability distribution skewed to the right or the left?

(c) Construct a line chart showing the probability distribution of the number of hot water heaters *not* lasting more than ten years.

(d) Is the probability distribution in (c) skewed to the right or the left?

4.21 A government agency receives a shipment of 500 hi-fi speakers. It chooses a sample of 9 speakers, and rejects the shipment if 2 or more of the speakers are defective. What is the probability that this agency will accept the shipment if the proportion defective is (a) 0.05; (b) 0.10; (c) 0.15?

The Chapter in a Nutshell

1. A *random variable* is a quantity the value of which is determined by an experiment; in other words, its value is determined by chance. A random variable must assume numerical values, and its value must be defined for all possible outcomes of the experiment in question; that is, for all elements of the sample space. Random variables are of two types: discrete and continuous. *Discrete random variables* can assume only a finite or countable number of numerical values, whereas *continuous random variables* can assume any numerical value on a continuous scale.

2. The *probability distribution of a discrete random variable* provides the probability of each possible value of the random variable. If $P(x)$ is the probability that x is the value of the random variable, the sum of $P(x)$ for all values of x must be 1. Probability distributions are represented by tables, graphs, or equations. A probability distribution can often be viewed as the relative frequency distribution (which shows the proportion of cases with each value) of a population.

3. If $x_1, x_2, \cdots, x_m$ are the possible values of a random variable X and if $P(x_1)$ is the probability that x_1 occurs, $P(x_2)$ is the probability that x_2 occurs, and so on, then the expected value of this random variable $E(X)$ is $\Sigma x_i P(x_i)$. In other words, the *expected value* is the weighted mean of the values that the random variable can assume, each value being weighted by the probability that it occurs. To shed light on decision problems, statisticians frequently calculate and compare the expected value of monetary gain if various courses of action are followed. However, such a comparison alone often cannot provide a complete solution to these problems.

4. The *variance of a random variable X,* denoted by $\sigma^2(X)$, is the expected value of the squared deviation of the random variable from its expected value. The *standard deviation of a random variable* is the positive square root of the random variable's variance. The standard deviation of a random variable is the most frequently used measure of the extent of dispersion or variability of the values assumed by the random variable. If it is certain that a random variable will equal a certain value, its standard deviation (and variance) is zero.

5. The most important discrete probability distribution is the *binomial distribution,* which represents the number of successes in n Bernoulli trials. That is, if each trial can result in a "success" or a "failure," and if the probability of a success equals Π on each trial (and if the outcomes of the trials are independent), the number of successes, x, that occur in n trials has a binomial distribution. Specifically,

$$P(x) = \frac{n!}{x!(n-x)!} \, \Pi^x (1-\Pi)^{n-x}, \text{ for } x = 0, 1, 2, \cdots n.$$

The expected value of x is $n\Pi$, and its standard deviation is $\sqrt{n\Pi(1-\Pi)}$.

Chapter Review Exercises

4.22 The Uphill Manufacturing Company, a maker of bicycle pedals, has seven suppliers that provide it with materials. The materials producers are faced with work stoppages because of labor problems, and Uphill's management feels that in the next six months there is a 10 percent probability that each supplier will be unable to provide Uphill with materials. Because the prospective labor problems are quite different from one supplier to another and the employees of the various suppliers do not act together, Uphill's management also thinks that whether any one supplier is unable to provide materials is independent of whether any other supplier can do so.

(a) What is the probability that none of the suppliers will be able to provide Uphill with materials?

(b) What is the probability that more than half of the suppliers will be unable to provide Uphill with materials?

(c) What is the expected number of suppliers that will be unable to provide Uphill with materials?

(d) What is the standard deviation of the number of suppliers that will be unable to provide Uphill with materials?

(e) If the employees of the various suppliers band together to negotiate with all their employers, what effect do you think this will have on your answers to (a), (b), (c), and (d)?

4.23 Wallingford College wants to know whether its students are satisfied with the quality of meals provided by the college cafeteria. Since it would be impossible for Wallingford's president to talk in depth with each of the college's 20,000 students, the president decides to pick 20 students and see how well each is satisfied with existing conditions. Suppose that the probability that each student in this sample will express dissatisfaction is 0.20, and that whether or not one student expresses dissatisfaction is statistically independent of whether of not another student does so.

(a) What is the probability that a majority of the sample will express dissatisfaction with existing conditions?

(b) What is the probability that 2 or less of the sample will express dissatisfaction with existing conditions?

(c) Wallingford's director of dining services proposes the following gamble to the president: if the number of students expressing dissatisfaction is more than 20 percent of the sample, he will pay the president $100. If this is not the case, the president will pay him $100. Is this a fair bet? Why or why not?

(d) What is the expected number of students who will express dissatisfaction? What is the standard deviation?

4.24 It is important that the Durham Manufacturing Company maintain careful control over the quality of the bolts it uses. Each shipment (which contains 10,000 bolts) is subjected to the following acceptance-sampling procedure: A sample of 15 bolts is taken from the shipment and each is tested for defects. If more than one of the bolts are found to be defective, the shipment is rejected.

(a) If a shipment contains 20 percent defectives, what is the probability that it will be accepted?

(b) If a shipment contains 10 percent defectives, what is the probability that it will be accepted?

(c) If a shipment contains 5 percent defectives, what is the probability that it will be accepted?

(d) Durham's suppliers protest that this inspection procedure is inappropriate because they guarantee only that no more than 10 percent of a shipment is defective. Do you agree that it is inappropriate? Why, or why not?

4.25 Which of the following are random variables?

(a) the birthday of President Reagan;

(b) the number of years in a century;

(c) the number of presidents in the United States in the nineteenth century.

4.26 The probability distribution of X is

$$P(x) = \tfrac{1}{10} \qquad x = 1, 2, \cdots, 10.$$

Draw a line chart representing this probability distribution.

* **4.27** Use Minitab (or some other computer package) to calculate the binomial distribution for $n = 21$ and $\pi = 0.4$.

4.28 An insurance company offers a 50-year-old woman a $1,000 one-year term insurance policy for an annual premium of $18. If the annual number of deaths per 1,000 is six for women in this age group, what is the expected gain for the insurance company from a policy of this kind?

4.29 An insurance company offers a 52-year-old man a $1,000 one-year term insurance policy for an annual premium of $25. If the annual number of deaths per 1,000 is seven for men in this age group, what is the standard deviation of the gain for the insurance company from a policy of this kind?

4.30 (a) Considering only the monetary gains and losses of the gamble in part (k) of Exercise 4.5, is this a fair gamble? (b) What is the expected value of the salesman's winnings (or losses) in this game? (c) Omitting the non-monetary considerations (namely, the threat of injury), would the salesman agree to this gamble if he is interested in maximizing the expected value of his monetary gains?

4.31 A candidate for mayor is considering whether or not to debate her opponent on television. If she does so, and wins the debate, she would gain an extra 100,000 votes. If she does so, and loses the debate, she would lose 150,000 votes. She believes that the probability is 0.55 that she will win the debate (and that the probability is 0.45 that she will lose it). If she is interested in maximizing the expected value of the number of votes she receives, should she debate her opponent? Why, or why not?

* Exercise requiring computer package, if available.

Appendix 4.1

EXPECTED VALUE OF A LINEAR FUNCTION OF A RANDOM VARIABLE

Frequently, one is interested in the expected value of a linear function of a random variable. (If X is a random variable and $Y = a + bX$, where a and b are constants, then Y is a *linear function* of X.) For example, suppose that a firm's total annual costs are a linear function of the number of units of output it sells per year, and that the price per unit of output is constant. Then the firm's annual profit is a linear function of the number of units of output it sells per year, as shown in the illustration below. If the number of units of output sold per year is a random variable with a known expected value, what is the expected value of the firm's annual profit? To solve this sort of problem, we use the following result:

EXPECTED VALUE OF A LINEAR FUNCTION OF A RANDOM VARIABLE. *If* Y *is a linear function of a random variable* X—*that is, if* Y = a + bX—*the expected value of* Y, E(Y), *equals* a + bE(X), *where* E(X) *is the expected value of* X.

This result is not difficult to prove. Given the definition of an expected value in equation (4.3), it follows that

$$E(Y) = \sum_{i=1}^{m} y_i P(y_i),$$

(4.11)

where the summation is over all possible values of Y, and $P(y_i)$ is the probability that Y equals y_i. Since y_i equals $a + bx_i$, the probability that $Y = y_i$ must equal the probability that $X = x_i$. (Why? Because Y can equal y_i if and only if X equals x_i.) Thus, $P(y_i)$ must equal $P(x_i)$. Substituting $P(x_i)$ for $P(y_i)$ and $(a + bx_i)$ for y_i in equation (4.11), we have

$$E(Y) = \sum_{i=1}^{m} (a + bx_i) P(x_i),$$

and since

$$(a + bx_i) P(x_i) = aP(x_i) + bx_i P(x_i),$$

$$E(Y) = \sum_{i=1}^{m} [aP(x_i) + bx_i P(x_i)] = \sum_{i=1}^{m} aP(x_i) + \sum_{i=1}^{m} bx_i P(x_i).$$

Since a and b are constants, they can be moved in front of the summation signs, so

$$E(Y) = a \sum_{i=1}^{m} P(x_i) + b \sum_{i=1}^{m} x_i P(x_i).$$

Finally, since

$$\sum_{i=1}^{m} P(x_i) = 1 \text{ and } \sum_{i=1}^{m} x_i P(x_i) = E(X),$$

$$E(Y) = a + bE(X),$$

which is the result we set out to prove.

To illustrate the application of this result, consider the following. A certain plant manufacturing TV sets has a fixed cost of \$1 million per year. The gross profit from each TV set sold—that is the price less the unit variable cost —is \$20. The number of sets the plant sells per year is a random variable with an expected value of 100,000. What is the expected value of this plant's annual profit? Let Y equal the plant's annual profit. Since this profit equals its gross profit less its fixed costs,

$$Y = -1,000,000 + 20X,$$

where X is the number of TV sets sold per year. Thus,

$$E(Y) = -1,000,000 + 20E(X) = -1,000,000 + 20(100,000)$$

$$= 1,000,000.$$

Consequently, the expected value of the plant's annual profit is $1 million.

In some cases, one also is interested in the standard deviation of a linear function of a random variable. To solve such a problem, we use the following result.

STANDARD DEVIATION OF A LINEAR FUNCTION OF A RANDOM VARIABLE. *If* Y *is a linear function of a random variable* X—*that is, if* Y = a + bX—*the standard deviation of* Y, σ(Y), *equals* bσ(X), *where* σ(X) *is the standard deviation of* X.

To illustrate the above, in the case of the TV manufacturer, what is the standard deviation of the annual profit if the standard deviation of the number of TV sets sold per year is 10,000? Since $Y = -1,000,000 + 20X$, the standard deviation of Y equals $20\sigma(X)$, or $200,000, since $\sigma(X)$ equals 10,000.

Appendix 4.2

THE HYPERGEOMETRIC DISTRIBUTION[14]

In acceptance sampling and in many other areas, the hypergeometric distribution is an important probability distribution. To understand the conditions under which this distribution arises, suppose that a finite population consists of two types of items which we call successes and failures. If there are A successes and $(N - A)$ failures in the population, and if we draw a random sample (without replacement) of n items from the population, what is the probability distribution of the number of successes in the sample? Letting x equal the number of successes in the sample, the answer is

$$P(x) = \frac{\dfrac{A!}{x!(A-x)!}\dfrac{(N-A)!}{(n-x)!(N-A-n+x)!}}{\dfrac{N!}{(N-n)!n!}}, \text{ for } x = 0, 1, 2, \cdots, n. \qquad (4.12)$$

This is called the hypergeometric distribution.[15]

[14] We assume here that the reader has read Appendix 3.1.
[15] Note that x cannot exceed A, and that $n - x$ cannot exceed $N - A$.

It is not too difficult to derive the result in equation (4.12). How many different combinations of x successes can be made from the A successes in the population? As we know from Appendix 3.1, the answer is $A! \div [x!(A-x)!]$. How many different combinations of $(n-x)$ failures can be made from the $(N-A)$ failures in the population? As we know from Appendix 3.1, the answer is $(N-A)! \div [(n-x)!(N-A-n+x)!]$. Thus, the total number of different ways that we can combine the x successes with $(n-x)$ failures is

$$\left(\frac{A!}{x!(A-x)!} \right) \left(\frac{(N-A)!}{(n-x)!(N-A-n+x)!} \right)$$

As we also know from Appendix 3.1, the total number of different combinations of n items that can be made from the N items in the population is $N! \div [(N-n)!n!]$. Thus, since the probability of x successes in a sample of n is the number of equally likely samples with x successes and $(n-x)$ failures divided by the total number of equally likely samples of size n, this probability is

$$\frac{\dfrac{A!}{x!(A-x)!} \dfrac{(N-A)!}{(n-x)!(N-A-n+x)!}}{\dfrac{N!}{(N-n)!n!}},$$

which is what we set out to prove.

To illustrate the use of the hypergeometric distribution, suppose that a shipment of motors contains 10 motors, 2 of which are defective and 8 of which are not. What is the probability that a sample of three motors chosen at random from this shipment *without replacement* will contain one defective? Since $A = 2, N = 10, n = 3$, and $x = 1$, equation (4.12) implies that this probability equals

$$\frac{\left(\dfrac{2!}{1!1!} \right) \left(\dfrac{8!}{2!6!} \right)}{\dfrac{10!}{7!3!}} = \frac{2 \left[\dfrac{8(7)}{2} \right]}{\dfrac{10(9)(8)}{3(2)(1)}} = \frac{56}{120} = \frac{7}{15}.$$

If one is sampling without replacement from a finite population, the hypergeometric distribution, not the binomial distribution, is the correct one to use. (If the sampling is with replacement, the binomial is the correct one.) However, if the sample is a small percentage of the population, the binomial distribution provides a very good approximation to the hypergeometric distribution. For example, suppose that a shipment of goods contains 1,000 items, 20 of which are defective and 980 of which are not defective. A random sample of 20 items is drawn from the shipment without replacement, and we want to know the probability of x defectives in this sample. Even though the sample is without replacement, the sample is so small a proportion of the shipment that for an adequate approximation we should use the binomial probability distribution based on $n = 20$ and $\Pi = .02$ (that is, 20/1000). In general, if the sample is 5 percent or less of the population, this approximation is adequate.

If sampling is without replacement, it can be shown that the mean and standard deviation of the number of successes is

$E(X) = n\Pi$

$$\sigma(X) = \sqrt{n\Pi(1 - \Pi)\left(\frac{N - n}{N - 1}\right)},$$

where $\Pi = A/N$, the proportion of successes in the population. In other words, these are the mean and standard deviation of the hypergeometric distribution. If the sample is a small percent of the population, $(N - n)/(N - 1)$ will be approximately equal to 1, since n/N is very small. Thus, under these conditions, the mean and standard deviation of the number of successes will be approximately equal to those given for the binomial distribution in equations (4.8) and (4.10).

Appendix 4.3
Joint Probability Distributions and Sums of Random Variables

JOINT PROBABILITY DISTRIBUTIONS

In Chapter 3, we discussed experiments involving two events, and characterized the probability of each outcome in terms of a joint probability table. (Recall the joint probability table for persons classified by attitude toward increased defense spending and by political party, shown in Table 3.2.) Let's now consider cases where each event is represented by a numerical value. Experiments of this sort involve two random variables. For example, suppose that two coins (coin A and coin B) are each flipped once. Let X be the random variable corresponding to the number of times that coin A comes up tails, and let Y be the random variable corresponding to the number of times that coin B comes up tails. Table 4.3 shows the *joint probability distribution* for X and Y. The joint probabilities in Table 4.3 show the probabilities that X and Y assume particular values. For example, the probability that $X = 1$ and $Y = 0$, denoted by $P(X = 1$ and $Y = 0)$, equals $1/4$.

Table 4.3
Joint Probability
Distribution for X
and Y

	Value of X		
Value of Y	*0 (no tails)*	*1 (one tail)*	*Total (marginal probabilities)*
0 (no tails)	1/4	1/4	1/2
1 (one tail)	1/4	1/4	1/2
Total (marginal probabilities)	1/2	1/2	1

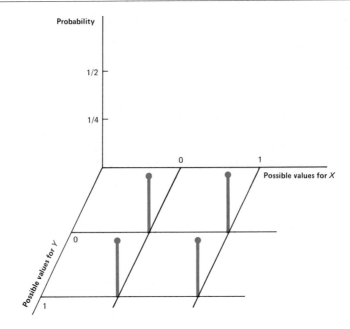

Figure 4.6
Joint Probability
Distribution for *X*
and *Y*

A joint probability distribution can be represented graphically, as illustrated in Figure 4.6. The value of *X* is measured along one axis, the value of *Y* is measured along another axis, and the joint probability is measured along the third (vertical) axis. Each possible outcome is represented by a spike from the floor of the graph (at the point corresponding to this outcome's value of *X* and *Y*); the height of the spike measures the probability of this outcome. (Of course, the sum of these heights for all possible outcomes equals 1.) In the particular case in Figure 4.6, the height of each of the four spikes equals 1/4, in accord with Table 4.3.

In Chapter 3 we referred to the row and column totals in a joint probability table as marginal probabilities. The column totals in Table 4.3 correspond to *P(x)* and constitute the *marginal probability distribution of* X. The row totals in Table 4.3 correspond to *P(y)* and constitute the *marginal probability distribution of* Y. In other words, the marginal probability distributions are as shown in Table 4.4.

Table 4.4
Marginal Probability
Distributions of *X*
and *Y*

Marginal probability distribution of X		Marginal probability distribution of Y	
Value of X	*Probability*	*Value of* Y	*Probability*
0	1/2	0	1/2
1	1/2	1	1/2
Total	1	Total	1

It is important to distinguish between a marginal probability distribution and a conditional probability distribution. To obtain the conditional probability that Y assumes a particular value (say, 1) given that X assumes a particular value (say, 0), we divide the joint probability that $Y = 1$ and $X = 0$ by the marginal probability that $X = 0$. In other words,

$$P(Y = 1 \mid X = 0) = \frac{P(Y = 1 \text{ and } X = 0)}{P(X = 0)} = \frac{.25}{.50} = .50.$$

To obtain the conditional probability that $Y = 0$ given that $X = 0$, we divide the joint probability that $Y = 0$ and $X = 0$ by the marginal probability that $X = 0$:

$$P(Y = 0 \mid X = 0) = \frac{P(Y = 0 \text{ and } X = 0)}{P(X = 0)} = \frac{.25}{.50} = .50.$$

Thus, the *conditional probability distribution of* Y *given that* X $= 0$ is as shown in Table 4.5. (Prove, as an exercise, that in this case the conditional probability distribution of Y, given that $X = 1$, is the same as the conditional probability distribution of Y, given that $X = 0$.)

Conditional probability distribution of Y, *given that* X $= 0$		*Conditional probability distribution of* Y, *given that* X $= 1$	
Value of Y	*Probability*	*Value of* Y	*Probability*
0	1/2	0	1/2
1	1/2	1	1/2
Total	1	Total	1

Table 4.5
Conditional Probability Distributions of Y

The conditional probability distribution of X can be derived in a similar way. To obtain the conditional probability that X assumes a particular value (say, 1), given that Y assumes a particular value (say, 1), we divide the joint probability that $X = 1$ and $Y = 1$ by the marginal probability that $Y = 1$. In other words,

$$P(X = 1 \mid Y = 1) = \frac{P(X = 1 \text{ and } Y = 1)}{P(Y = 1)} = \frac{.25}{.50} = .50.$$

To obtain the conditional probability that $X = 0$, given that $Y = 1$, we divide the joint probability that $X = 0$ and $Y = 1$ by the marginal probability that $Y = 1$:

$$P(X = 0 \mid Y = 1) = \frac{P(X = 0 \text{ and } Y = 1)}{P(Y = 1)} = \frac{.25}{.50} = .50.$$

Thus the *conditional probability distribution of* X *given that* Y = 1 is as shown in Table 4.6. (Prove, as an exercise, that in this case the conditional probability distribution of *X*, given that *Y* = 0, is the same as the conditional probability distribution of *X*, given that *Y* = 1.)

In Chapter 3 we took up the definition of statistical independence of events. Now we define statistical independence of random variables.

> STATISTICAL INDEPENDENCE OF RANDOM VARIABLES: *Two random variables* X *and* Y *are statistically independent if the conditional probability distribution of* X, *given any value of* Y, *is identical to the marginal probability distribution of* X, *and if the conditional probability distribution of* Y, *given any value of* X, *is identical to the marginal probability distribution of* Y.

In the case of the two coins, a comparison of Table 4.4 with Tables 4.5 and 4.6 shows that *X* and *Y* are independent. Why? Because both conditional probability distributions of *Y* in Table 4.5 are identical to the marginal probability distribution of *Y* in Table 4.4, and both conditional probability distributions of *X* in Table 4.6 are identical to the marginal probability distribution of *X* in Table 4.4.

Table 4.6
Conditional Probability
Distributions of *X*

Conditional probability distribution of X, given that Y = 0		Conditional probability distribution of X, given that Y = 1	
Value of X	*Probability*	*Value of* X	*Probability*
0	1/2	0	1/2
1	1/2	1	1/2
Total	1	Total	1

EXAMPLE 4.9 Let *X* equal 0 if a particular flight from Seattle to Chicago is on time, and 1 if it is not. Let *Y* equal 1 if the flight encounters severe turbulence, and 0 if it does not. The joint probability distribution for *X* and *Y* is as follows:

	Value of Y	
Value of X	*0 (No turbulence)*	*1 (Turbulence)*
0 (on time)	.75	.05
1 (not on time)	.15	.05

What is the marginal probability distribution of X? What is the conditional probability distribution of X, given that $Y = 0$? Are X and Y statistically independent?

SOLUTION: The marginal probability distribution of X is given by the horizontal row totals; that is, the marginal probability that $X = 0$ is $.75 + .05 = .80$, and the marginal probability that $X = 1$ is $.15 + .05 = .20$. The conditional probability that $X = 0$, given that $Y = 0$, is

$$P(X = 0 \mid Y = 0) = \frac{P(X = 0 \text{ and } Y = 0)}{P(Y = 0)} = \frac{.75}{.90} = .83,$$

and the conditional probability that $X = 1$, given that $Y = 0$, is

$$P(X = 1 \mid Y = 0) = \frac{P(X = 1 \text{ and } Y = 0)}{P(Y = 0)} = \frac{.15}{.90} = .17.$$

Since the marginal probability distribution of X is not identical to the conditional probability distribution of X, given that $Y = 0$, X and Y are not statistically independent.

SUMS OF RANDOM VARIABLES

To solve problems of many kinds, statisticians often find it necessary to determine the expected value and variance of the sum of a number of random variables. The following propositions are very helpful in solving such problems.

EXPECTED VALUE OF A SUM OF RANDOM VARIABLES: *If* $X_1, X_2, \cdots, X_m$ *are* m random variables,

$$E(X_1 + X_2 + \cdots + X_m) = E(X_1) + E(X_2) + \cdots + E(X_m).$$

That is, the expected value of the sum of random variables is equal to the sum of the expected values of the random variables. This proposition is true whether or not the random variables are statistically independent.

VARIANCE OF A SUM OF RANDOM VARIABLES: *If* $X_1, X_2, \cdots, X_m$ *are* m *statistically independent random variables,*

$$\sigma^2(X_1 + X_2 + \cdots + X_m) = \sigma^2(X_1) + \sigma^2(X_2) + \cdots + \sigma^2(X_m).$$

That is, the variance of the sum of statistically independent random variables is equal to the sum of the variances of the random variables.

The example below illustrates how these propositions can be applied.

EXAMPLE 4.10 In order to commercialize a new product, the Ozone Chemical Company must carry out research, do pilot-plant work, and build a production facility. The cost of each of these three steps is a random variable. The expected value and standard deviation of each of these random variables are as follows:

Random variable	Expected value (dollars)	Standard deviation (dollars)
Cost of carrying out research	25,000	20,000
Cost of pilot-plant work	50,000	20,000
Cost of production facility	200,000	30,000

What is the expected value of the total cost incurred in these three steps? If the costs incurred in various steps are statistically independent, what is the standard deviation of the total cost incurred in these three steps?

SOLUTION: Let X_1 be the cost of carrying out research, X_2 be the cost of pilot-plant work, and X_3 be the cost of building a production facility. Based on the above proposition,

$$E(X_1 + X_2 + X_3) = E(X_1) + E(X_2) + E(X_3).$$

Since $E(X_1) = \$25,000$, $E(X_2) = \$50,000$, and $E(X_3) = \$200,000$,

$$E(X_1 + X_2 + X_3) = \$25,000 + \$50,000 + \$200,000 = \$275,000.$$

That is, the expected value of the total cost incurred in these three steps is \$275,000. If X_1, X_2, and X_3 are statistically independent,

$$\sigma^2(X_1 + X_2 + X_3) = \sigma^2(X_1) + \sigma^2(X_2) + \sigma^2(X_3).$$

Since $\sigma^2(X_1) = (\$20,000)^2$, $\sigma^2(X_2) = (\$20,000)^2$, and $\sigma^2(X_3) = (\$30,000)^2$,

$$\sigma^2(X_1 + X_2 + X_3) = (\$20,000)^2 + (\$20,000)^2 + (\$30,000)^2,$$

and

$$\sigma(X_1 + X_2 + X_3) = \sqrt{(\$20{,}000)^2 + (\$20{,}000)^2 + (\$30{,}000)^2}$$

$$= \$41{,}231.$$

That is, the standard deviation of the total cost incurred in these three steps is $41,231.

5

OBJECTIVES

The normal distribution is of key importance in statistics. The overall purpose of this chapter is to describe the nature, characteristics, and uses of the normal distribution. Among the specific objectives are:

1 To describe how observations can be expressed in standard units, that is, as Z values.

2 To show how the table of the standard normal distribution can be used to find the probability that a normal variable will fall in a specified range.

3 To indicate how the normal distribution can be used to approximate the binomial distribution.

4 To show how the normal distribution can be applied to solve a wide range of practical problems.

The Normal Distribution

5.1 Introduction

In Chapter 4, we studied the nature and characteristics of the binomial distribution, the most important discrete probability distribution. We now turn to the normal distribution, a continuous probability distribution that plays a central role in statistics. We will take up the nature and characteristics of the normal distribution and some of the ways in which this distribution is used in actual applications. In the chapter appendix, we describe the Poisson distribution, another important probability distribution.

5.2 Continuous Distributions

First we will describe how the probability distribution of a continuous random variable can be characterized graphically. To see how this is done, consider the actual case of a manufacturing plant that produces pieces of metal in which two holes are stamped. The distance between the hole centers varies from one piece of metal to another and is a continuous random variable since it can assume any numerical value on a continuous scale. If we were to examine 1,000 pieces of metal produced by this plant, we could find the proportion of these pieces where the distance between the hole centers is 2.994 inches and under 2.995 inches, the proportion where the distance is 2.995 inches and under 2.996 inches, and so on. Then, as shown in panel A of Figure 5.1, we could construct a histogram in which *the area of the bar representing each class interval is equal to the proportion of all pieces within this class interval.* (In other words, the height of the bar equals the proportion of all pieces within this class interval divided by the width of the class interval.) The area under this histogram must equal 1 because the sum of the proportions in all class intervals must equal 1. Such a histogram is said to use a *density* scale.

To obtain more complete information about the relative frequency distribution of these distances, we could examine 1,000,000 rather than 1,000 pieces of metal and construct another histogram like the one in panel A of Figure 5.1. Because of the greatly increased number of observations, we would now be able to divide the class intervals *more finely.* As a result, we would have *more* bars, each of which is *narrower* than in the histogram in panel A. This second histogram is shown in panel B of Figure 5.1. The proportion of pieces of metal in which the distance between hole centers lies in a particular range can be read from this histogram by measuring the area of the bars lying within this range. For example, the proportion of pieces of metal where the distance between the hole centers is 3.003 inches and under 3.005 inches is measured by the shaded area of the histogram in panel B of Figure 5.1.

Finally, in order to obtain even more complete information, we could examine countless numbers of these pieces of metal, with the result that the class intervals could (and would) be made ever finer and more numerous. In the limit, the histogram would become a *smooth curve,* as shown in panel C of Figure 5.1. As in panels A and B, the total area under this smooth curve would equal 1, since the proportion of observations in all class intervals must total 1. Also, in panel C, as in the other panels of Figure 5.1, the proportion of pieces of metal in which the distance between the hole centers lies in a particular range can be found by measuring the area under the smooth curve in this range. Thus, if we wanted to know the proportion of pieces of metal in which the distance between the hole centers is between 3.004 inches and 3.005 inches, we would measure the shaded area under the smooth curve in panel C.

The smooth curve in panel C of Figure 5.1 is important because we can use it to determine the probability that the distance between hole centers lies within a particular range, such as between 3.004 inches and 3.005 inches. As we have just seen, the proportion of cases in the long run where the distance lies in this range is equal in value to the area under the smooth curve in this range. Thus, *the probability that the distance between hole centers lies within a particular range is equal in*

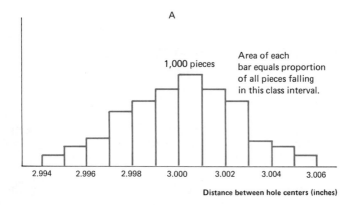

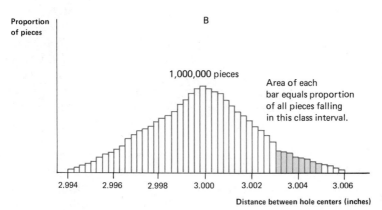

Figure 5.1
Histograms of Distances between Hole Centers of Pieces of Metal, Based on Data for 1,000 Pieces and 1,000,000 Pieces; and Probability Density Function

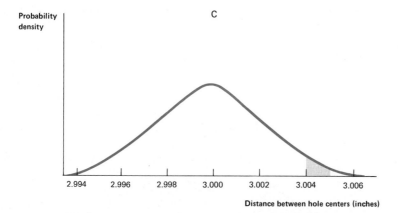

value to the area under the smooth curve in this range. In Figure 5.1 the probability that the distance between hole centers is between 3.004 inches and 3.005 inches equals the shaded area under the smooth curve in panel c.

5.3 Probability Density Function of a Continuous Random Variable

The smooth curve in panel c of Figure 5.1 is called a probability density function. We turn now from the specific case in Figure 5.1 to some generalizations concerning the probability density function of a continuous random variable.

> PROBABILITY DENSITY FUNCTION OF A CONTINUOUS RANDOM VARIABLE. *In the limit, as more and more observations are gathered concerning a continuous variable, and as class intervals become narrower and more numerous, the histogram (using a density scale) of the variable becomes a smooth curve (as in Figure 5.1) called a* **probability density function.** *The total area under any probability density function must equal 1. The probability that a random variable will assume a value between any two points,* a *and* b, *equals the area under the random variable's probability density function over the interval from* a *to* b.[1]

To illustrate the use and interpretation of a random variable's probability density function, consider the length of time it takes individuals to solve an algebra problem. The length of time it takes a randomly chosen individual to solve this particular problem can be viewed as a random variable. Suppose the probability density function of this random variable is as shown in Figure 5.2. This curve is of basic im-

Figure 5.2
Probability Density
Function of Length of
Time to Solve Problem

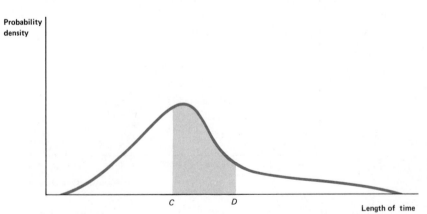

[1] The probability that a continuous random variable is precisely equal to a particular value is zero since the area under the probability density function at this particular value is a line of zero width.

portance because it enables us to calculate the probability that this length of time will be in any particular range in which we are interested. If we want to know the probability that this length of time will be between C and D in Figure 5.2, all we have to do is determine the area between C and D under the probability density function. This is the shaded area in Figure 5.2.

5.4 The Normal Distribution

The most important continuous probability distribution is the normal distribution. The formula for the probability density function of a normal random variable (that is, a random variable with a normal distribution) is

$$f(x) = \frac{1}{\sqrt{2\pi}\,\sigma}\, e^{-\frac{1}{2}[(x-\mu)/\sigma]^2} \,, \qquad (5.1)$$

The Normal Curve

where μ is the random variable's expected value (or mean), σ is its standard deviation, e is approximately 2.718 and is the base of the natural logarithms, and π is approximately 3.1416. Like the binomial distribution, the normal distribution is really a family of distributions. Depending on its mean and standard deviation, the location and shape of the normal probability density function—or ***normal curve,*** as we shall call it for short—can vary considerably.

To show how much the normal curve can vary, Figure 5.3 presents three normal curves, one with a mean of 15 and a standard

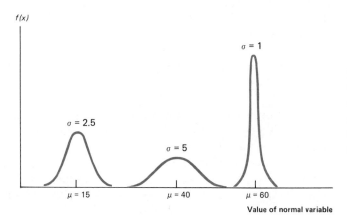

Figure 5.3
Three Normal Curves, with $\mu = 15$ and $\sigma = 2.5$; $\mu = 40$ and $\sigma = 5$; and $\mu = 60$ and $\sigma = 1$

deviation of 2.50, one with a mean of 40 and a standard deviation of 5, and one with a mean of 60 and a standard deviation of 1. As you can see, all three are bell-shaped and symmetrical, but the curves are located at quite different points along the horizontal axis because they have different means, and they exhibit quite different amounts of spread or dispersion because they have different standard deviations. Because of the differences in their means and standard deviations, some normal curves (like the middle one in Figure 5.3) are short and squat, whereas others (like the one on the right in Figure 5.3) are tall and skinny. But in accord with the definition of a probability density function in the previous section, the total area under any normal curve must equal 1.

Although normal curves vary in shape because of differences in mean and standard deviation, all normal curves have the following characteristics in common.

1. SYMMETRICAL AND BELL-SHAPED. All normal curves are symmetrical about the mean. In other words, the height of the normal curve at a value that is a certain amount *below* the mean is equal to the height of the normal curve at a value that is the same amount *above* the mean. Because of this symmetry, the mean of a normal random variable equals both its median and its mode. (Recall the discussion of the relative position of the mean, median, and mode in Chapter 2.) Besides being symmetrical, the normal curve is bell-shaped, as in Figure 5.3. And a normal random variable can assume values ranging from $-\infty$ to $+\infty$.

2. PROBABILITY THAT A VALUE WILL LIE WITHIN k STANDARD DEVIATIONS OF THE MEAN. Regardless of its mean or standard deviation, the probability that the value of a normal random variable will lie within *one* standard deviation of its mean is 68.3 percent, the probability that it will lie within *two* standard deviations of its mean is 95.4 percent, and the probability that it will lie within *three* standard deviations of its mean is 99.7 percent. Panel B of Figure 5.4 shows the distance from the mean, μ, in units of the standard deviation, σ. Clearly, almost all the area under a normal curve lies within three standard deviations of the mean.

3. LOCATION AND SHAPE DETERMINED ENTIRELY BY μ AND σ. The location of a normal curve along the horizontal axis is determined *entirely* by its mean μ. For example, if the mean of a normal curve equals 4, it is centered at 4; if its mean equals 400, it is centered at 400. The amount of spread in a normal curve is determined *entirely* by its standard deviation σ. If σ increases, the curve's spread widens; if σ decreases, the curve's spread narrows.

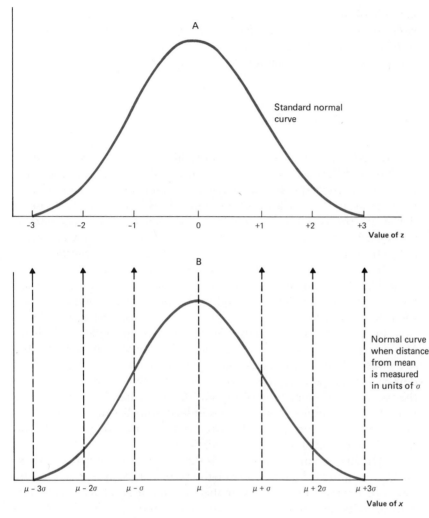

Figure 5.4
Comparison between
Standard Normal
Curve and Any
Normal Curve when
Distance from Mean Is
Measured in Units of σ

Why is the normal distribution so important in statistics? Basically, for three reasons. First, *the normal distribution is a reasonably good approximation to many populations.* Experience has shown that many (but by no means all) population histograms (using a density scale) are approximated quite well by a normal curve. For example, the histogram (using a density scale) of heights, weights, or IQs is likely to be reasonably close to a normal curve. Second, it can be shown that under circumstances described in the following chapter, *the probability distribution of the sample mean should be close to the normal distribution.* This is one of the most fundamental results in statistics. We must postpone discussing it until the next chapter, but it is an important reason for the key role played by the normal distribution in statistics.[2] Third, and related to the previous point, *the normal distribution can be used in many instances to approximate the binomial distribution.* In Section 5.8, we shall show how this approximation can be employed.

5.5 The Standard Normal Curve

As stressed in the previous section, normal curves vary greatly in shape because of differences in the mean μ, and in the standard deviation σ. However, if one expresses any normal random variable as a deviation from its mean, and measures these deviations in units of its standard deviation, the resulting random variable, called a **standard normal variable,** has the probability distribution shown in panel A of Figure 5.4. This probability distribution is called the **standard normal curve.**

Standard Units

If the weights of adult males are normally distributed, with a mean of 170 pounds and a standard deviation of 20 pounds, it is possible to express the weight of each adult male in *standard units* by finding the deviation of his weight from the mean and expressing this deviation in units of the standard deviation. For example, if William Morris's weight is 190 pounds, it is +1.0 in standard units. Why? Because his weight is 20 pounds above the mean, and since the standard deviation is 20 pounds, this amounts to a positive (+) deviation from the mean of 1 standard deviation. On the other hand, if John Jarvis's weight is 160 pounds, it is −0.5 in standard units because his weight is 10 pounds below the mean, which amounts to a negative (−) deviation from the mean of 0.5 standard deviations.

The important point to note is that if any normal variable is expressed in standard units, its probability distribution is given by the standard normal curve. Thus, if the weights of adult males are normally distributed and if we express them in standard units, their probability distribution is given by the standard normal curve. Put more formally, if X is a normally distributed random variable, then

The Standard Normal Variable

$$Z = \frac{X - \mu}{\sigma}$$

(5.2)

has the standard normal distribution regardless of the values of μ and

[2] Under some circumstances, statistics other than the sample mean also have a normal distribution. We single out the sample mean only because of its great importance in statistical applications.

σ. Thus, if X is the weight of an adult male $(X - 170) \div 20$ has the standard normal distribution.

As a further illustration, given the fact that the heights of adult females are normally distributed with a mean of 66 inches and a standard deviation of 2 inches, if X is the height of an adult female, what is Z? Z is X expressed in standard units. That is,

$$Z = \frac{X - 66}{2}.$$

What is the value of Z corresponding to a height of 67 inches? It is $(67 - 66) \div 2$, or 0.5. What is the height corresponding to a Z value of -2.0? Since $(X - 66) \div 2 = -2.0$, X must equal 62 inches.

Figure 5.4 shows what happens when we express a normal variable, X, in standard units. Panel B shows the probability distribution of X. Note that in this panel the value of X is measured in units of the standard deviation (σ) from the mean (μ). When we express X in standard units, the value of Z corresponding to each value of X is shown by the arrows. Thus, if X equals $\mu - 3\sigma$, the corresponding Z value is -3; if X equals $\mu - 2\sigma$, the correponding Z value is -2; and so on. Clearly, *the mean of the standard normal distribution is zero,* since zero in panel A corresponds to μ in panel B. Also, *the standard deviation of the standard normal distribution is* 1, since a distance of σ along the horizontal axis in panel B corresponds to a distance of 1 in panel A.[3]

5.6 How to Calculate Normal Probabilities

It frequently is necessary to calculate the probability that the value of a normal random variable lies between two points. To calculate this probability, two steps must be carried out.

---

[3] Using the results of Appendix 4.1, it is easy to prove that the mean of the standard normal variable is zero. Since $Z = (X - \mu) \div \sigma$, it follows that

$$Z = -\frac{\mu}{\sigma} + \frac{1}{\sigma} X.$$

Thus, in accord with Appendix 4.1,

$$E(Z) = -\frac{\mu}{\sigma} + \frac{1}{\sigma} E(X) = -\frac{\mu}{\sigma} + \frac{\mu}{\sigma} = 0.$$

Also, the results of Appendix 4.1 imply that the standard deviation of Z must equal the standard deviation of X (that is, σ) multiplied by $1/\sigma$, or 1.

1. FIND THE POINTS ON THE STANDARD NORMAL DISTRIBUTION COR-
REPONDING TO THESE TWO POINTS. For example, if the heights of adult
women are normally distributed with a mean of 66 inches and a standard deviation of 2 inches and if we want to know the probability that
the height of an adult woman lies between 65 and 68 inches, our first
step is to find the points on the standard normal distribution corresponding to 65 and 68 inches. Since $\mu = 66$ and $\sigma = 2$, these points are
$(65 - 66) \div 2$ and $(68 - 66) \div 2$, respectively. Simplifying terms, they
are -0.5 and $+1.0$.

2. DETERMINE THE AREA UNDER THE STANDARD NORMAL CURVE BE-
TWEEN THE TWO POINTS WE HAVE FOUND. If we want to know the proba-

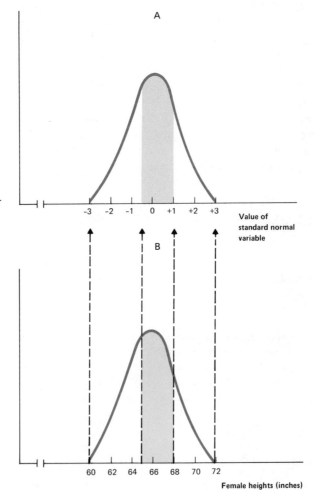

Figure 5.5
Normal Distribution of
Female Heights and
Standard Normal
Curve

bility that the height of an adult woman lies between 65 and 68 inches, we determine the area under the standard normal curve between the points on the curve corresponding to 65 and 68 inches. Since these points are −0.5 and 1.0 (as we know from the preceding paragraph), we must determine the area under the standard normal curve between −0.5 and 1.0.

Why does this procedure give the correct answer? *Because the area under any normal curve between two points is equal to the area under the standard normal curve between the corresponding two points.* A comparison between panels A and B in Figure 5.5 demonstrates that this is true. In panel B the normal curve shows the distribution of adult female heights. The probability that the height of an adult woman lies between 65 and 68 inches equals the area under the curve between 65 and 68 inches (that is, the shaded area in panel B). In panel A we show the standard normal distribution. The points on the standard normal distribution corresponding to 65 and 68 inches are −0.5 and 1.0, as we know from previous paragraphs. The shaded area under the standard normal curve equals the probability that the standard normal variable lies between −0.5 and 1.0. As you can see, the two shaded areas are equal. Thus, the probability that a height lies between 65 and 68 inches equals the area under the standard normal curve between −0.5 and 1.0.

5.7 How to Use the Table of the Standard Normal Distribution

The area under the standard normal distribution between various points is tabled. To carry out the second step in the procedure described above, one must be able to use this table, which is contained in Appendix Table 2. In the following paragraphs, we indicate how this table is used in various situations.

1. AREA BETWEEN ZERO AND SOME POSITIVE VALUE. Each number in the body of Appendix Table 2 shows the area between zero (the mean of the standard normal distribution) and the positive number (z) given in the left-hand column (and top) of the table. For example, to determine the area between zero and 1.10, look at the row labeled 1.1 and the column labeled .00; the area is .3643. This is the shaded area in panel A of Figure 5.6. Similarly, to determine the area between zero and 1.63, look at the row labeled 1.6 and the column labeled .03; the area is .4484.

Figure 5.6
Areas under the
Standard Normal
Curve

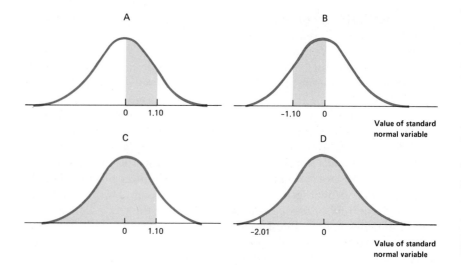

2. AREA BETWEEN ZERO AND SOME NEGATIVE VALUE. Because the standard normal curve is symmetrical, the area between zero and any negative value is equal to the area between zero and the same positive value. Hence Appendix Table 2 can readily be used to evaluate the desired area. For example, the area between zero and −1.10 equals that between zero and +1.10; thus, this area (shaded in panel B of Figure 5.6) must be .3643. Similarly, the area between zero and −1.63 equals that between zero and +1.63, which we know to be .4484.

3. AREA TO THE LEFT OF SOME POSITIVE VALUE. Suppose that we want to determine the area to the left of 1.10. This area (shaded in panel C of Figure 5.6) is composed of two parts: the area to the left of zero, and the area between zero and 1.10. The area to the left of zero is .5 because the standard normal curve is symmetrical about zero and because the area under the entire curve equals 1. The area between zero and 1.10 can be determined from Appendix Table 2, as we already know. Since it is .3643, the area we want is .5000 + .3643 = .8643.

4. AREA TO THE RIGHT OF SOME NEGATIVE NUMBER. What is the area to the right of −2.01? This area (shaded in panel D of Figure 5.6) is composed of two parts: the area to the right of zero, and the area between zero and −2.01. The area to the right of zero is .5 because the standard normal curve is symmetrical about zero and because the area under the entire curve equals 1. The area between zero and −2.01 can

be determined from Appendix Table 2, since it equals the area between zero and 2.01, which is .4778. Thus, the area we want equals .5000 + .4778 = .9778.

5. AREA TO THE RIGHT OF SOME POSITIVE VALUE. What is the area to the right of 1.65? This area (shaded in panel A of Figure 5.7) plus the area between zero and 1.65 must equal .5000, because the total area to the right of zero equals .5000. Thus, the area we want equals .5000 minus the area between zero and 1.65. Since Appendix Table 2 shows that the area between zero and 1.65 is .4505, the area we want equals .5000 − .4505 = .0495.

6. AREA TO THE LEFT OF SOME NEGATIVE VALUE. What is the area to the left of −1.05? This area (shaded in panel A of Figure 5.7) plus the area between zero and −1.05 must equal .5000, because the total area to the left of zero is .5000. Thus, the area we want equals .5000 minus the area between zero and −1.05. The area between zero and −1.05 equals the area between zero and 1.05, which is .3531, according to Appendix Table 2. Hence the area we want equals .5000 − .3531 = .1469.

7. AREA BETWEEN TWO POSITIVE VALUES. What is the area between 1.10 and 1.82? This area (shaded in panel B of Figure 5.7) equals the difference between (a) the area between zero and 1.82, and (b) the

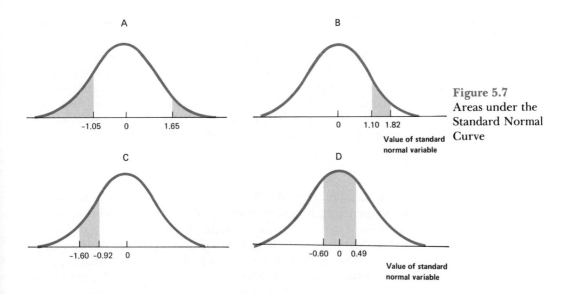

Figure 5.7
Areas under the
Standard Normal
Curve

area between zero and 1.10. Appendix Table 2 shows that the former area is .4656 and that the latter area is .3643. Thus, the area we want is .4656 − .3643 = .1013.

8. AREA BETWEEN TWO NEGATIVE VALUES. What is the area between −1.60 and −0.92? This area (shaded in panel c of Figure 5.7) equals the difference between (a) the area between zero and −1.60, and (b) the area between zero and −0.92. The area between zero and −1.60 equals the area between zero and 1.60, which is .4452, according to Appendix Table 2. The area between zero and −0.92 equals the area between zero and 0.92, which is .3212, according to Appendix Table 2. Thus, the area we want is .4452 − .3212 = .1240.

9. AREA BETWEEN A NEGATIVE AND A POSITIVE VALUE. Finally, suppose we want to determine the area between −0.60 and 0.49. This area (shaded in panel d of Figure 5.7) equals the sum of (a) the area between zero and −0.60, and (b) the area between zero and 0.49. The area between zero and −0.60 equals the area between zero and 0.60, which is .2257, according to Appendix Table 2. The area between zero and 0.49 is .1879, according to Appendix Table 2. Thus, the area we want is .2257 + .1879 = .4136.

Because of the central importance of the normal distribution in statistics, it is essential that you be able to calculate the probability that a normal random variable lies in a given range. The following three examples are designed to illustrate how this is done.

EXAMPLE 5.1 Find the probability that the value of the standard normal variable will lie between −1.23 and +1.14.

SOLUTION: Appendix Table 2 shows that the area under the standard normal curve between 0 and 1.23 is .3907, so the area between 0 and −1.23 must also be .3907. Appendix Table 2 shows that the area between 0 and 1.14 is .3729. Thus, the area between −1.23 and +1.14 equals .3907 + .3729 = .7636, which means that the probability we want equals .7636.

EXAMPLE 5.2 Although yardsticks are supposed to be exactly 36 inches long, their length varies somewhat because of errors in construction. For a particular type of yardstick, the lengths are normally distributed, the mean length being 36 inches and the standard deviation being .001 inches. What is the probability that a yardstick of this type has a length that is (a) between 35.9990 and 36.0005 inches; (b) less than 35.9985 inches; (c) greater than 36.0004 inches?

SOLUTION: (a) The first step is to find the points on the standard normal distribution corresponding to 35.9990 and 36.0005. These points are $(35.9990 - 36) \div .001$ and $(36.0005 - 36) \div .001$, or -1.0 and $+0.5$, respectively. According to Appendix Table 2, the area under the standard normal curve between zero and 1.0 is .3413, which means that the area between zero and -1.0 also is .3413. According to Appendix Table 2, the area between zero and 0.5 is .1915. Thus, the area between -1.0 and 0.5 is $.3413 + .1915 = .5328$. This is the probability that a yardstick's length is between 35.9990 and 36.0005 inches.

(b) The point on the standard normal distribution corresponding to 35.9985 inches is $(35.9985 - 36) \div .001$, or -1.5. According to Appendix Table 2, the area under the standard normal curve between 0 and 1.5 is .4332, so the area between 0 and -1.5 also is .4332. Thus, the area to the left of -1.5 equals $.5000 - .4332 = .0668$. This is the probability that a yardstick's length is less than 35.9985 inches.

(c) The point on the standard normal distribution corresponding to 36.0004 inches is $(36.0004 - 36) \div .001$, or $+0.4$. According to Appendix Table 2, the area under the standard normal curve between zero and 0.4 is .1554, so the area to the right of 0.4 must equal $.5000 - .1554 = .3446$. This is the probability that a yardstick's length is more than 36.0004 inches.

EXAMPLE 5.3 A newspaper reporter writes an article saying that 90 percent of the yardsticks (of the type described in Example 5.2) have lengths of 36.0020 inches or less, while 10 percent have lengths of more than 36.0020 inches. He is incorrect. What figure should be substituted for 36.0020?

SOLUTION: The first step is to find the number which the value of the standard normal variable will exceed with a probability of .10. Since the probability must be $.50 - .10$ (or .40) that the value of the standard normal variable lies between zero and this number, we must look in Appendix Table 2 for the value of Z corresponding to a probability of .40. This value is 1.28. The next step is to find the length that corresponds to this value of the standard normal variable. In other words, we must find the value of X in equation (5.2) that corresponds to $Z = 1.28$. Clearly, this value is $\mu + 1.28\sigma$, which equals $36 + 1.28(.001)$, or 36.00128 inches. Thus, the probability is 0.10 that a yardstick's length will exceed 36.00128 inches. The figure of 36.00128 inches, not the reporter's, is correct.

MISTAKES TO AVOID

THE CASE OF WYOMING CITIES

According to the 1980 Census of Population, Wyoming contains 17 cities (urban places, in census terminology) with population of 5,000 or more. The mean population of these cities was 14,900. The standard deviation of these cities' populations was 13,500. Based on these data, an urban sociologist wanted to estimate the number of Wyoming cities with 1980 population exceeding 25,000. Assuming that the populations of these cities were normally distributed, he calculated the proportion of these cities with populations exceeding 25,000. Since the point on the standard normal distribution corresponding to 25,000 is $(25,000 - 14,900) \div 13,500$, or .75, he found the area under the standard normal curve to the right of .75, which is $.5000 - .2734$, or .2266. (See Appendix Table 2.) Then he multiplied .2266 times 17 (the number of Wyoming cities), the result being about 4. Thus, he estimated that 4 Wyoming cities had populations exceeding 25,000 in 1980. Do you agree with this result? If not, what mistakes did he make?

SOLUTION

The crucial mistake is his assumption that the populations of these 17 cities were normally distributed. In fact, their distribution was far from normal, as shown below.

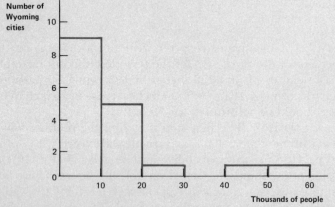

Consequently, his calculation (based on the assumption that the distribution was normal) is very much in error. Rather than four Wyoming cities with populations exceeding 25,000, there really were only two (Casper and Cheyenne). The moral here is that one cannot assume without any good reason that a variable is normally distributed. A great many variables are *not* normally distributed, as we shall see in this and subsequent chapters.

5.8 Using the Normal Distribution as an Approximation to the Binomial Distribution

As pointed out in an earlier section of this chapter, one reason why the normal distribution is so important is that it can be used as an approximation to the binomial distribution under certain circumstances. These are described below.

> NORMAL APPROXIMATION TO THE BINOMIAL DISTRIBUTION. *If n (the number of trials) is large and Π (the probability of success) is not too close to zero or 1, the probability distribution of the number of successes occurring in n Bernoulli trials can be approximated by a normal distribution. Experience indicates that the approximation is fairly accurate as long as nΠ > 5 when Π ⩽ 1/2 and n (1 − Π) > 5 when Π > 1/2.*

The fact that the normal distribution can approximate the binomial distribution under the circumstances described above is useful because, as noted in Chapter 4, it is tedious to calculate the binomial probabilities when *n* is large.

The following illustration shows how the normal distribution is used to estimate binomial probabilities. If a true coin is flipped 1,600 times, what is the probability distribution of the number of times that the coin comes up heads? Since there are 1,600 Bernoulli trials, the number of times heads comes up is clearly a binomial random variable. Moreover, since $n = 1,600$ and $\Pi = 1/2$, its mean is 800 (=$n\Pi$) and its standard deviation is 20 (= $\sqrt{n\Pi(1 - \Pi)}$). The probability distribution of the number of times heads comes up is shown (as a histogram) in panel A of Figure 5.8. Since it is difficult to evaluate each of the binomial probabilities (in equation 4.6) when *n* is as large as 1,600, we would like to approximate this probability distribution with another that is easier to calculate. Fortunately, as noted above, the normal distribution—with the same mean (800) and standard deviation (20) as the binomial distribution—is a good approximation.

A visual comparison of the normal distribution in panel B of Figure 5.8 with the binomial distribution in panel A (each of which has a mean of 800 and a standard deviation of 20) certainly indicates that the former is shaped much like the latter. But to make sure that this is a good approximation, we must investigate in greater detail. Let's look carefully at the segment of both probability distributions between $x = 788$ and $x = 790$ (where *x* is the number of heads that comes up).

A

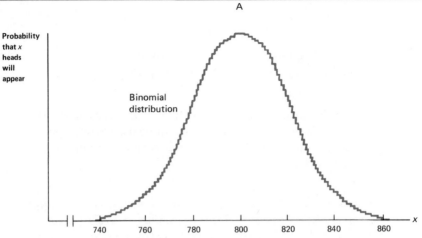

Figure 5.8
Comparison of
Binomial Distribution
and Normal Distribu-
tion, Both with Mean of
800 and Standard
Deviation of 20

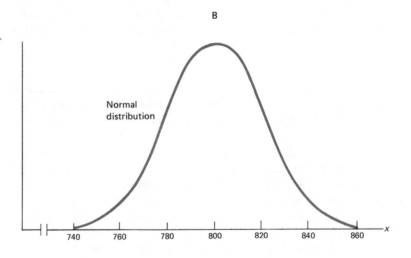

Figure 5.9 shows a "blow up" of each of the probability distributions
over the relevant range. If the approximation is accurate, the area
under the normal curve between 787.50 and 790.50 must be approximately equal to the area under the binomial distribution between
787.50 and 790.50. In other words, the shaded area under the continuous curve should be approximately equal to the sum of the areas of
the three rectangles (A, B, and C) shown in Figure 5.9. The sum of the
areas of these three rectangles equals the true probability that the
number of heads is 788, 789, or 790, whereas the shaded area is the

approximation to this probability. Based on Figure 5.9, it certainly appears that the approximation is good.

To find the probability that the number of heads is 788, 789, or 790, we do *not* find the area under the normal curve between 788 and 790; instead, we find the area *between 787.50 and 790.50*. As shown in Figure 5.9, in order to approximate the three rectangles (*A, B,* and *C*) corresponding to the probabilities of 788, 789, and 790 heads, we must include the area under the continuous curve from 787.50 to 790.50. This is often called a *continuity correction,* a correction due to the fact that a discrete probability distribution is being approximated by a continuous one. In general, to find the probability that a binomial variable equals at least *c* but no more than *d* (where $c < d$), we find the probability that a normal variable (with mean $n\Pi$ and standard deviation $\sqrt{n\Pi(1 - \Pi)}$) lies between $(c - 1/2)$ and $(d + 1/2)$.

Continuity Correction

Based on the normal approximation, what is the probability that the number of heads is 788, 789, or 790? As we have seen, this probability is approximately equal to the probability that the value of a normal random variable with mean equal to 800 and standard deviation equal to 20 lies between 787.50 and 790.50. The value of the standard normal variable corresponding to 787.50 is −.625, and that corresponding to 790.50 is −.475. Using Appendix Table 2, we find that the area under the standard normal curve between zero and 0.625 is approximately .234, which means that the area between zero and

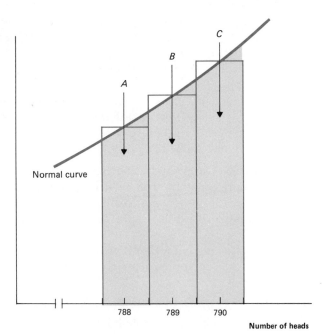

Figure 5.9
Normal Approximation to the Binomial Distribution in the Range between 787.50 and 790.50

**GETTING DOWN
TO CASES**

How Ma Bell Solved an Engineering Problem

The Bell Telephone System has been a pioneer in using probability theory to solve many kinds of engineering problems. The following case (with some simplifications) is derived from actual practice.

A telephone exchange at *A* was to serve 2,000 telephones in a nearby exchange at *B*. Since it would have been too expensive to install 2,000 trunklines from *A* to *B*, it was decided to install enough trunklines so that only 1 out of every 100 calls would fail to find an unutilized trunkline immediately at its disposal.

During the busiest hour of the day, each of these 2,000 telephone subscribers requires a trunkline to *B* for an average of two minutes. Thus, at a fixed moment during the busiest hour, there are 2,000 telephone subscribers each of which has a probability of 1/30 that it will require a trunkline to *B*. Under normal conditions, whether or not one subscriber requires a trunkline to *B* is independent of whether another subscriber does so. (Under abnormal conditions, such as a flood, or an earthquake, this assumption of independence is unlikely to hold, since many people are likely to want to make calls; however, the telephone company was interested in solving the problem under typical conditions.)

As stated above, the telephone company wanted to determine how many trunklines it should install so that when 1 out of the 2,000 subscribers puts through a call requiring a trunkline to *B* during the busiest hour of the day, he or she would find an unutilized trunkline to *B* immediately at his or her disposal in 99 out of 100 cases.

(a) Solve this problem using the normal distribution.

(b) Write a one-paragraph report to the telephone company describing the results you have obtained.

—0.625 also is approximately .234. Similarly, the area between zero and 0.475 is approximately .183, which means that the area between zero and −0.475 also is approximately .183. Thus, the probability that the number of heads is 788, 789, or 790 equals (approximately) .234 − .183 = .051.

The following is another illustration of how the normal distribution can be used to approximate the binomial distribution.

EXAMPLE 5.4 According to a study carried out at Cornell University's Medical College, the probability is 1/2 that a physician's recommendation for voluntary knee surgery will not be confirmed if the patient goes to another physician for an opinion.[4] If a hospital schedules 100 patients (who each received only one medical opinion) for such knee surgery, what is the probability that at least 60 of them will cancel the operation if each one of them obtains a second opinion, and refuses surgery if the second physician does not confirm the recommendation of the first? (Assume that whether the recommendation for one patient is confirmed is independent of whether the recommendation for another patient is confirmed.)

SOLUTION: The number of patients who will cancel the operation has a binomial distribution with mean equal to 100 (1/2), or 50, and standard deviation equal to $\sqrt{100(1/2)(1/2)}$, or 5. Because of the continuity correction, the probability that the number who will cancel the operation is 60 or more can be approximated by the probability that the value of a normal variable with mean equal to 50 and standard deviation equal to 5 exceeds 59.50. The value of the standard normal variable corresponding to 59.50 is $(59.50 - 50) \div 5$, or 1.9. Appendix Table 2 shows that the area under the standard normal curve between zero and 1.9 is .4713, so the area to the right of 1.9 must equal .5000 − .4713 = .0287. This is the (approximate) probability that at least 60 patients will cancel the operation.

[4] M. Hinkel, H. Ruchlin, and S. Parsons, *Eight Years' Experience with a Second Opinion Elective Surgery Program* (Washington, D.C.: Health Care Financing Administration, 1981). Of course, this figure of 1/2 pertains only to the population of patients included in the study, and is rough in various respects. But it is accurate enough for present purposes.

5.9 Calculating the Probabilities of Wind Changes for Hurricanes: A Case Study[5]

Since hurricanes annually cause property damage totaling billions of dollars, government agencies and private organizations have a strong interest in studying hurricanes in order to reduce their impact. In 1970, statisticians at the Stanford Research Institute (now SRI Interna-

[5] The case study is based on R. Howard, J. Matheson, and D. North, "The Decision to Seed Hurricanes," *Science,* June 1972. The present discussion is simplified, and some numbers have been changed. For a much more complete and accurate (but also more technical) account, see the above article in *Science.*

It is assumed that *if no change occurs in maximum sustained wind speed,* a hurricane of the sort considered here will cause $100 million in property damage.

tional) began a study for the U.S. Department of Commerce of the behavior and impact of hurricanes. The normal distribution was used to compute the probability that wind changes of various magnitudes would occur for an unseeded hurricane during a 12-hour period before landfall. Based on various meteorological studies, it appears that *the percentage change in maximum sustained wind speed during this period is normally distributed with a mean of zero and a standard deviation of 15.6 percent.* In other words, the distribution of the percentage change in an unseeded hurricane's wind speed during such a period is as shown in Figure 5.10.

The amount of property damage caused by a hurricane is related to the extent of the change in its maximum sustained wind speed. Thus, if a change of +32 percent or more occurs, the Stanford statisticians estimate that property damage of well over $300 million might be expected. On the other hand, if a change of −34 percent or less occurs, the estimated property damage is less than $20 million. From this we see that it is important to know the probability that the change in maximum sustained wind speed will lie in various ranges. What is the probability that the change in wind speed will exceed +32 percent, thus resulting in well over $300 million in property damage? Obviously, this is an important question for many firms and government agencies.

If X is the percentage change in maximum sustained wind speed, we want to evaluate the probability that $X > 32$. This probability is denoted by

$$Pr \{X > 32\}.$$

Using the procedures described in earlier sections of this chapter, it is relatively simple to evaluate this probability. To do so, we must find the

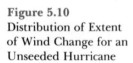

Figure 5.10
Distribution of Extent
of Wind Change for an
Unseeded Hurricane

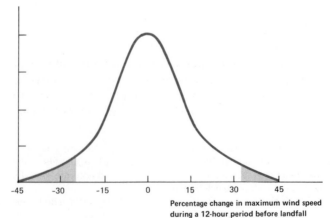

Percentage change in maximum wind speed
during a 12-hour period before landfall

value of the standard normal variable corresponding to 32 percent. This value is $(32 - 0) \div 15.6$, or 2.05. (Why? Because the mean of X is zero and its standard deviation is 15.6, as pointed out above.) Appendix Table 2 shows that the area under the standard normal curve between zero and 2.05 is .4798, so the area to the right of 2.05 equals $.5000 - .4798 = .0202$. Thus, the probability is about .02 that an increase of more than 32 percent in wind speed will occur. (The shaded area to the right of zero in Figure 5.10 is equal to this probability.)

It is also useful to be able to answer a somewhat different type of question: What is the percentage change in wind speed that will be exceeded with a specified probability? For example, what is the percentage change in wind speed that will be exceeded with a probability of .95? To answer this question, we must find the value of x_0 such that

$$Pr \{X > x_0\} = .95.$$

As a first step, note that Appendix Table 2 shows that the area under the standard normal curve between zero and 1.64 is .45. Thus, the area between zero and -1.64 also is .45, which means that the area to the right of -1.64 is .95. Having determined that the probability is .95 that a standard normal variable will exceed -1.64, we must now find the value of the normal variable corresponding to this value of the standard normal variable. In other words, we must find the value of X in equation (5.2) when $Z = -1.64$. Clearly, the desired value of the normal variable is $\mu - 1.64\sigma$, which here equals $0 - 1.64(15.6) = -25.6$ percentage points. Thus, a -25.6 percentage point change in wind speed is the value that will be exceeded with a probability of .95. (The shaded area to the left of zero in Figure 5.10 equals .05.)

EXERCISES

5.1 Find the probability that the standard normal variable lies (a) above 2.3; (b) below -3.0; (c) above 0.7; (d) between 1 and 2; (e) between -1 and 2.

5.2 If X is a normal variable with $\mu = 2$ and $\sigma = 3$, show how it can be converted into the standard normal variable.

5.3 "If you know that the probability that a normal variable exceeds a certain number, Q, is .10, you can be sure that the probability that this variable is less than $-Q$ is also .10." Do you agree? Why, or why not?

5.4 Find the area under the standard normal curve which lies: (a) between 0 and 1.82; (b) between -1.32 and 0; (c) between -1.08 and 1.08; (d) between 1.32 and 1.46; (e) between -1.08 and -0.23; (f) between -0.48 and 2.01.

5.5 Find the value of z if
  (a) the area under the standard normal curve between 0 and z is 0.1985.
  (b) the area under the standard normal curve to the right of z is 0.2776.
  (c) the area under the standard normal curve between $-z$ and 0 is 0.0910.
  (d) the area under the standard normal curve to the left of z is 0.8051.
  (e) the area under the standard normal curve between $-z$ and z is 0.1820.

5.6 A random variable is normally distributed with mean equal to 300 and standard deviation equal to 60. Calculate the probability that the value of this random variable
  (a) is less than 280.
  (b) exceeds 350.
  (c) lies between 185 and 265.
  (d) lies between 305 and 375.

5.7 The IQs of students at a particular college are normally distributed with mean equal to 125 and standard deviation equal to 10. Determine the percentage of the students with IQs
  (a) below 115.
  (b) above 140.
  (c) between 120 and 130.
  (d) between 100 and 115.

5.8 Suppose X is a normal random variable with mean μ and standard deviation σ.
  (a) Under what circumstances is X/σ the standard normal variable?
  (b) Under what circumstances is $(X - \mu)$ the standard normal variable?

5.9 If the weights of adult males are normally distributed with mean equal to 170 pounds and standard deviation equal to 20 pounds, the probability that a certain weight will be exceeded is .05. What is this weight?

5.10 In the previous exercise, the probability that an adult male's weight will be less than a certain amount is .10. What is this amount?

The Chapter in a Nutshell

1. In the limit, as more and more observations are gathered concerning a continuous random variable, and as class intervals become narrower and more numerous, the histogram (using a density scale) of the variable becomes a smooth curve called a *probability density function*. The total area under any probability density function must equal 1. The probability that a random variable will assume a value between any two points is equal in value to the area under the random variable's probability density function between these two points.

2. The most important continuous probability distribution is the

normal distribution, whose probability density function is called the *normal curve.* The location and spread of a normal curve depend on its mean and standard deviation, but all normal curves are symmetrical and bell-shaped. If one expresses any normal variable as a deviation from its mean and measures these deviations in units of its standard deviation, the result is called the *standard normal variable.* The standard normal variable is a normal variable with a mean of zero and a standard deviation of 1. The areas under the standard normal curve are given in Appendix Table 2.

3. To calculate the probability that the value of any normal variable (with mean μ and standard deviation σ) lies between two points, *a* and *b*, find the points on the standard normal distribution corresponding to *a* and *b*. These points are $(a - \mu) \div \sigma$ and $(b - \mu) \div \sigma$. Then use Appendix Table 2 to determine the area under the standard normal curve between $(a - \mu) \div \sigma$ and $(b - \mu) \div \sigma$. This area equals the probability that the value of the normal variable is between *a* and *b*.

4. If *n* (the number of trials) is large and Π (the probability of success) is not too close to zero or 1, the probability distribution of the number of successes in *n* Bernoulli trials can be approximated by a normal distribution. Specifically, to approximate the probability that the number of successes is from *c* to *d*, find the probability that the value of a normal variable (with mean $n\Pi$ and standard deviation $\sqrt{n\Pi(1 - \Pi)}$) lies between $(c - 1/2)$ and $(d + 1/2)$. (Of course, *c* is presumed to be less than *d*.)

Chapter Review Exercises

5.11 The probability that a marksman will hit the target is $1/3$. If he takes 50 shots, what is the probability that he will hit the target less than 10 times? (Use the normal distribution as an approximation to the binomial distribution.)

5.12 The Martin Company announces that it will give $500 bonuses to its sales people who are among the top 10 percent in sales in 1987. It believes that its salespeople will be normally distributed with respect to their 1987 sales, the mean being $400,000 and the standard deviation being $100,000. If this is correct, how big must a salesperson's sales be to get him or her the bonus?

5.13. A manufacturing process turns out 20 percent defective items. A sample of 50 items is taken from the 3,000 produced on a particular day. Determine the probability that
(a) 7 items or less in the sample are defective.
(b) 21 items or more in the sample are defective.
(c) more than 10 and less than 19 items in the sample are defective.
(d) more than 8 and less than 12 items in the sample are defective.

5.14 Two hundred students enroll in a course in experimental psychology. The probability that each will pass the course is 0.9. Calculate the probability that
(a) 8 or more will not pass.
(b) 20 or more will not pass.
(c) more than 7 but less than 20 will not pass.
(d) the number that pass will differ by one or less from its expected value.

5.15 In a particular Congressional district, one-third of the registered voters are Republicans. A political polling organization chooses a random sample of 300 of the registered voters in this district. What is the probability that less than 85 or more than 115 of voters in the sample are Republicans? (Use the normal approximation to the binomial distribution.)

5.16 The probability that a person will complete a particular part of the Wechsler Intelligence Test in less than 90 seconds is 0.10. If 1,000 people take this part of the test, the probability is 0.05 that more than _____ of these people will complete it in less than 90 seconds.

5.17 Ann McCutcheon takes a multiple choice test composed of 20 questions, each of which has four possible answers. If she has no idea of the correct answers, and if she chooses answers at random, what is the probability that more than 3 but less than 8 of these answers are correct?
(a) Use the normal approximation to the binomial distribution to answer this question.
(b) Use the binomial distribution to obtain an exact answer to this question.
(c) How close is the answer based on the normal approximation to the exact answer?

5.18 The length of time it takes a rat to get through a particular maze is normally distributed. Based on past experience, it is known that 30 percent of all rats get through it in less than 1.30 minutes and that 40 percent of all rats get through it in more than 1.71 minutes. What is the mean length of time it takes to get through it? What is the standard deviation of the lengths of time it takes to get through it?

5.19 From past experience, the Uphill Manufacturing Corporation knows that the deviations of the width of its bicycle pedals from their mean width are normally distributed with a standard deviation of .02 inches.
(a) What is the probability that a pedal's width is more than .03 inches above the mean?
(b) What is the probability that a pedal's width is more than .05 inches below the mean?
(c) What is the probability that a pedal's width differs (either positively or negatively) from the mean by less than .015 inches?
(d) During 1985, the mean width of all bicycle pedals produced by Uphill was exactly equal to what the design called for. However, during 1986, the mean width of all bicycle pedals produced by the firm was .01 inches greater than the design called for. In both years, the standard deviation of the pedal widths was .02 inches, and the pedal

widths were normally distributed. In 1985, what was the probability that a pedal would be wider than called for by the design?

(e) In 1986, what was the probability that a pedal would be wider than called for by the design?

(f) In 1985, what was the probability that a pedal would be more than .04 inches wider than called for by the design?

(g) In 1986, what was the probability that a pedal would be more than .04 inches wider than called for by the design?

Appendix 5.1

THE POISSON DISTRIBUTION

Another important probability distribution is the Poisson distribution, which is named after a nineteenth-century Swiss mathematician. The *Poisson distribution* is a discrete probability distribution which has the following formula:

$$P(x) = \frac{\mu^x e^{-\mu}}{x!} , \text{ for } x = 0, 1, 2, \cdots \tag{5.3}$$

where $P(x)$ is the probability that a variable with a Poisson distribution equals x, μ is the mean or expected value of the Poisson distribution, and e is approximately 2.718 and is the base of the natural logarithms.[6] Like the binomial and normal distributions, the Poisson distribution is really a family of distributions. Depending on the value of μ, the shape of the probability distribution will vary considerably.

One reason why the Poisson distribution is so important in statistics is that it can be used as an approximation to the bionomial distribution under the circumstances described below.

POISSON APPROXIMATION TO THE BINOMIAL DISTRIBUTION. *If* n *(the number of trials) is large and* Π *(the probability of success) is small, the probability of* x *successes occurring in* n *Bernoulli trials can be approximated by the Poisson distribution where* $n\Pi = \mu$. *Experience indicates that this approximation is adequate for most practical purposes if* n *is at least 20 and* Π *is no greater than .05.*

Whereas the normal distribution approximates the binomial distribution when Π is *not* very small, the Poisson distribution approximates it when Π *is* very small; thus, the two approximations complement one another.

To illustrate how the Poisson distribution can be used in this way, let's consider the following situation. You drive to work 15,000 times in a 30-year

[6] A Poisson random variable, like a binomial random variable, is discrete. Unlike the binomial, it does not assume a finite number of possible values; instead, it assumes a countably infinite number of possible values.

period, and the probability of your having an accident each time you drive to work is .0001. In this case, each trip can be considered a "trial" and each accident can be considered a "success" (although only for your garage mechanic and/or mortician). Thus, $n = 15{,}000$ and $\Pi = .0001$. Since n is very large and Π is very small, the Poisson distribution should be a good approximation to the binomial distribution. Since $\mu = n\Pi = 15{,}000(.0001)$, or 1.5, equation (5.3) can be used to obtain the probability of 0, 1, 2, $\cdots$ accidents during this thirty-year period, the results being:

$$P(0) = \frac{1.5^0 e^{-1.5}}{0!} = e^{-1.5} = 0.22,$$

$$P(1) = \frac{1.5^1 e^{-1.5}}{1!} = 1.5 e^{-1.5} = 0.33,$$

$$P(2) = \frac{(1.5)^2 e^{-1.5}}{2!} = \frac{2.25 e^{-1.5}}{2} = 0.25,$$

$$P(3) = \frac{(1.5)^3 e^{-1.5}}{3!} = \frac{3.375 e^{-1.5}}{6} = 0.13,$$

$$P(4) = \frac{(1.5)^4 e^{-1.5}}{4!} = \frac{5.062 e^{-1.5}}{24} = 0.05,$$

$$P(5) = \frac{(1.5)^5 e^{-1.5}}{5!} = \frac{7.594 e^{-1.5}}{120} = 0.01.$$

We could, of course, compute the probability of 6, 7, 8, $\cdots$ accidents, but these probabilities are less than .005. Figure 5.11 shows this Poisson distribution graphically. As you can see, the distribution is not symmetrical, but skewed to the right. (Recall our discussion of skewness in Chapter 2.)

Figure 5.11
Poisson Distribution,
with $\mu = 1.5$

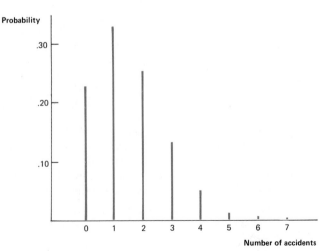

As pointed out earlier in this section, *the expected value of any Poisson random variable equals* μ. To demonstrate this, in the present illustration the expected number of accidents equals

$$E(X) = (0)(0.22) + (1)(0.33) + (2)(0.25) + (3)(0.13) + (4)(.05)$$

$$+ (5)(.01) + (6)(.004) + (7)(.001) + \cdots$$

$$= 1.50.$$

Also, *the standard deviation of any Poisson random variable equals* $\sqrt{\mu}$. (As an exercise, prove that this is true in this illustration; that is, prove that the standard deviation of the number of accidents equals $\sqrt{1.50}$. If you have difficulty, consult the footnote on this page).[7]

The computations involved in evaluating $P(x)$ can be onerous. To reduce the computational burden, Appendix Table 3 can be used. This table shows $P(x)$ for selected values of μ. The following example will provide additional practice in using the Poisson distribution.

EXAMPLE 5.5. A small private school has 100 students. On a particular day, the probability is .02 that each student will not attend school because of illness or some other reason.

   (a) What is the probability that 0, 1, 2, or 3 students will not attend school on this day?

   (b) What is the expected number of students who will not attend school?

   (c) What is the standard deviation of the number of students who will not attend school?

---

[7] To determine the standard deviation, first determine the variance by inserting the values of $P(x)$ into the following expression:

$$\Sigma(x - \mu)^2 P(x).$$

The result is

$$\sigma^2 = (0 - 1.5)^2(0.22) + (1 - 1.5)^2(0.33) + (2 - 1.5)^2(0.25) + (3 - 1.5)^2(0.13)$$

$$+ (4 - 1.5)^2(.05) + (5 - 1.5)^2(.01) + (6 - 1.5)^2(.004)$$

$$+ (7 - 1.5)^2(.001) + \cdots$$

Thus,

$$\sigma^2 = 2.25(0.22) + 0.25(0.33) + 0.25(0.25) + 2.25(0.13) +$$

$$6.25(.05) + 12.25(.01) + 20.25(.004) + 30.25(.001) + \cdots$$

$$= 0.495 + .082 + .063 + .292 + .313 + .122 + .081 + .030 + \cdots$$

$$= 1.50.$$

The standard deviation is the square root of the variance, or $\sqrt{1.50}$.

SOLUTION: (a) Since $\mu = n\Pi = 100(.02)$ or 2.0, equation (5.3) yields the following probabilities:

$$P(0) = 2^0 e^{-2} \div 0! = .1353$$

$$P(1) = 2^1 e^{-2} \div 1! = .2707$$

$$P(2) = 2^2 e^{-2} \div 2! = .2707$$

$$P(3) = 2^3 e^{-2} \div 3! = .1804$$

These probabilities can be found in the column of Appendix Table 3 where $\mu = 2.0$.

(b) The expected number of students who will not attend school equals $\mu = n\Pi = 100(.02)$ or 2.

(c) The standard deviation of the number of students who will not attend school equals $\sqrt{\mu} = \sqrt{n\Pi} = \sqrt{2}$.

ADDITIONAL USES FOR THE POISSON DISTRIBUTION

Besides being a useful approximation to the binomial distribution, the Poisson distribution is very important in its own right. Assume that events of a particular kind occur at random during a particular time span. To make things more concrete, suppose that the events in question are demands by a firm's customers for a particular type of spare part. If the following four conditions are met, the probability distribution of the number of such events (that is, the number of demands for this type of spare part) in a fixed period of time will be a Poisson distribution:

1. THE PROBABILITY THAT EACH EVENT OCCURS IN A VERY SHORT TIME IN-TERVAL MUST BE PROPORTIONAL TO THE LENGTH OF THIS TIME INTERVAL. Thus, the probability that a spare part of this type is demanded in a two-minute interval must be double the probability that a spare part of this type is demanded in a one-minute interval. Why? Because the former time interval is twice as long as the latter time interval.

2. THE PROBABILITY THAT TWO OR MORE EVENTS OF THE RELEVANT KIND OCCUR IN A VERY SHORT TIME INTERVAL MUST BE SO SMALL THAT IT CAN BE RE-GARDED AS ZERO. Thus, the probability that more than one order will occur for this type of spare part in a one-second time interval must be essentially zero. This assumption seems reasonable in this case. If the time interval is only one second long, it would be difficult indeed for two different orders for this type of spare part to be received by the company.

3. THE PROBABILITY THAT A PARTICULAR NUMBER OF THESE EVENTS OCCURS IN A PARTICULAR TIME INTERVAL MUST NOT DEPEND ON WHEN THIS TIME INTERVAL BEGINS. Thus, the probability that an order is received by the company in a

one-minute time interval beginning at noon tomorrow must be the same as the probability that an order will be received in a one-minute time interval beginning at 2 P.M. today. This is because it is assumed that this probability depends only on the length of the time interval, not on when the time interval begins. The fact that the one time interval begins at noon tomorrow and the other time interval begins at 2 P.M. today must not influence this probability at all.

4. THE PROBABILITY THAT A PARTICULAR NUMBER OF THESE EVENTS (DE-MANDS) OCCURS IN A PARTICULAR TIME INTERVAL MUST NOT DEPEND ON THE NUMBER OF THESE EVENTS THAT OCCURRED PRIOR TO THE BEGINNING OF THIS TIME INTERVAL (or in some shorter time interval prior to the beginning of this time interval). For example, suppose that five orders for this type of spare part were received prior to 2 P.M. today. The fact that five orders (rather than four, six, or some other number) were received prior to that time should not influence the probability of receiving an order in the one-minute time interval beginning at 2 P.M. today. Of course, this assumption may be violated if there is a tendency for these events to bunch together in time. For example, if orders tend to bunch together, the probability of receiving an order in the one-minute time interval beginning at 2 P.M. today may be dependent on whether an order was received just before 2 P.M. If so, the Poisson distribution is not appropriate.

If these four conditions are met, it can be shown that *the probability that* x *such events will occur in a time interval of length* Δ *(delta) is*

$$P(x) = \frac{(\lambda\Delta)^x e^{-\lambda\Delta}}{x!}, \qquad (5.4)$$

where λ *(lambda) is the mean number of such events per unit of time.* This, of course, is the same probability distribution as in equation (5.3), the only difference being that $\lambda\Delta$ is used here in place of μ. However, since $\lambda\Delta$ is the expected value of x, it is the same as what we formerly called μ.[8]

REPLACEMENT OF PARTS ON POLARIS SUBMARINES: A CASE STUDY[9]

The United States Department of Defense faced a problem in the operation of Polaris submarines that illustrates the practical utility of the Poisson distribution. Each submarine goes out on a mission of relatively fixed length (about 60 days), after which it is resupplied by a tender. During each mission a submarine must rely upon its own supply of spare parts. At the end of the mission, the tender replenishes the items that have been taken out of the subma-

---

[8] Sometimes the Poisson distribution is used to characterize events distributed at random in space rather than in time. For example, the Poisson might be used to find the probability of a submarine's being located in a particular area.

[9] This section is based in part on S. Haber and R. Sitgreaves, "An Optimal Inventory Model for the Intermediate Echelon when Repair Is Possible," *Management Science*, February 1975.

rine's supplies. How many spare parts should each tender carry in order to replace the spare parts that are used up in the preceding mission? Analysts have made substantial use of the Poisson distribution in solving this problem.

Obviously, the number of spare parts of a particular type that a tender must replenish for a particular submarine equals the number of such parts that failed during the preceding mission. Thus, the answer to the Defense Department's problem depends in considerable part (but not wholly)[10] on the probability distribution of the number of parts of a particular type which will fail during a mission. To estimate this probability distribution, analysts have found it useful to assume that failures meet the four conditions that were discussed in the previous section. In other words, the probability that a particular kind of part will fail during a short time interval is proportional to the length of time interval and is independent of when the interval occurs; it is also independent of how many such parts have failed prior to that time interval. Also, the probability that more than one part of a particular type will fail in a very short interval is so small that it can be regarded as zero.

Given that failures of a particular type of part can be represented in this way, we know (from the previous section) that the probability that x failures of a particular type of part will occur during a mission is

$$P(x) = \frac{(\lambda\Delta)^x e^{-\lambda\Delta}}{x!},$$

where Δ is the length of the mission and λ is the average number of failures per unit of time for the particular type of part. According to studies based on the first 61 patrols of Polaris submarines, the value of $\lambda\Delta$ varies from very close to zero to as high as 5.0, depending on the particular part.

If the Defense Department is specifically interested in a part where the value of $\lambda\Delta$ is estimated to be 1.0, what is the probability distribution of the number of such parts that will fail during a mission? Based on the column of Appendix Table 3 where $\mu = 1.0$, the answer is

$$P(0) = \frac{1^0 e^{-1}}{0!} = .3679,$$

$$P(1) = \frac{1^1 e^{-1}}{1!} = .3679,$$

$$P(2) = \frac{1^2 e^{-1}}{2!} = .1839,$$

$$P(3) = \frac{1^3 e^{-1}}{3!} = .0613,$$

$$P(4) = \frac{1^4 e^{-1}}{4!} = .0153.$$

[10] Ibid.

We can ignore $P(5)$, $P(6)$, $\cdots$ since each of these probabilities is less than .005.

Obviously, decision makers in the Defense Department have found it very helpful to know that the chances are about 37 out of 100 that *no* spare parts of this type will need to be replaced after a mission, that the chances are about 37 out of 100 that *one* will have to be replaced, that the chances are about 18 out of 10 that *two* will have to be replaced, and so on. Information of this sort, if properly applied, can promote a much more effective inventory policy for submarine tenders.

EXERCISES

5.20 If $\mu = 2$, what is the probability that a Poisson random variable, X, equals (a) 1; (b) 2; (c) 3?

5.21 If $\Pi = 1/3$ and $n = 100$, should you use the normal distribution or the Poisson distribution as an approximation to the binomial distribution?

5.22 Given $\Pi = .01$ and $n = 300$, should you use the normal distribution or the Poisson distribution as an approximation to the binomial distribution?

5.23 If a Poisson random variable has an expected value of 3.0. what is its variance?

5.24 If a Poisson random variable's coefficient of variation—that is, its standard deviation divided by its mean—equals 2, what is its mean?

5.25 Given $n = 20$ and $\Pi = .05$, what is the probability that $X = 0$, based on (a) the binomial distribution; and (b) the Poisson approximation to the binomial distribution?

5.26 R. D. Clarke reported the following data concerning the number of hits by buzz bombs during World War II in south London. (Each area covers 1/4 square kilometer.)

Number of hits	Number of areas
0	229
1	211
2	93
3	35
4	7
5 or more	1
Total	576

Are the results what might be expected on the basis of the Poisson distribution, where the mean number of hits per area was 1.0?

5.27 The number of accidents occurring in a given month at one of the Morris Company's plants is known to conform to the Poisson distribution. The standard deviation of this distribution is 1.732 accidents per month. What is the probability that no accidents will occur in this plant during this month?

5.28 One hundred packages are mailed to customers in New York City. Each is mailed on a different day, and the probability that each will arrive within 48 hours of mailing is 0.05. How long it takes for one package to arrive is independent of how long it takes for another package to arrive. Determine the probability that

(a) 4 of the packages arrive within 48 hours of mailing.

(b) 5 of the packages arrive within 48 hours of mailing.

(c) 6 of the packages arrive within 48 hours of mailing.

5.29 The office of a particular U.S. Senator has on the average five incoming calls per minute. Using the Poisson distribution, find the probability that there will be

(a) exactly two incoming calls during any given minute.

(b) exactly three incoming calls during any given minute.

(c) no incoming calls during any given minute.

5.30 According to the records of Wallingford College, the number of students who fail the course in social psychology is a random variable with a Poisson distribution, its mean being 4. Determine the probability that the number of failures

(a) equals 5.

(b) is less than 3.

(c) is greater than 6.

5.31 The number of automobile accidents on a particular turnpike between 1 P.M. and 2 P.M. has a mean of 3, and is distributed according to the Poisson distribution. What is the probability that it will exceed or fall short of its mean by more than one standard deviation?

5.32 The mean number of defects in ten yards of cloth produced by the ABC Textile Company is two. The number of defects is distributed according to the Poisson distribution. If more than four defects are present in ten yards of cloth, it is substandard, and the company makes no profit on it. What is the probability that ten yards of cloth are substandard?

5.33 A psychiatrist receives an average of six telephone calls while he is away during his lunch hour (from noon to 1 P.M.). The number of calls he receives then is distributed according to the Poisson distribution. During his lunch hour, another psychiatrist agrees to answer this psychiatrist's telephone twice, but not more. No one else is available to answer his phone, since his secretary is sick.

(a) What is the probability that this psychiatrist will receive at least one call during his lunch hour that is not answered?

(b) What is the probability that this psychiatrist will receive exactly one call during his lunch hour that is not answered?

5.34 The sociology department of Wallingford College contains 20 faculty members. The probability that each of them publishes a book this year is 0.05. Whether one of them publishes a book is independent of whether another does so.

(a) Use the binomial distribution to determine the exact probability that two of them publish a book this year.

(b) Use the Poisson approximation to the binomial distribution to approximate this probability.

(c) Based on your answers to (a) and (b), how close is the Poisson approximation?

5.35 The Howe Company wants no more than 1/10 of 1 percent of the motors it produces to be defective. Production quality is checked by examining a certain number of motors chosen at random from each day's output, and the manufacturing process is stopped if any motor is defective. If the firm wants the probability to be about .05 that the process will be stopped when it is producing 1/10 of 1 percent defectives, how many motors must be examined from each day's output?

5.36 Metropolitan Hospital's switchboard receives an average of four incoming calls per minute.

(a) What conditions must be satisfied if the number of incoming calls in any given minute is to be represented by a Poisson distribution?

(b) If these conditions are met, what is the probability that there will be exactly five incoming calls in a given minute? Exactly six incoming calls in a given minute? Exactly seven incoming calls in a given minute?

(c) Describe how the probability distribution of the number of incoming calls in a given minute might be useful in deciding how much capacity the hospital should have to handle calls.

(d) What is the standard deviation of the number of incoming calls in a given minute?

* **5.37** Use Minitab (or some other computer package) to calculate the Poisson distribution when $\mu = 1.45$.

* Exercise requiring computer package, if available.

Sampling, Estimation, and Hypothesis Testing

OBJECTIVES

Sampling plays a central role in statistics. The overall purpose of this chapter is to indicate the ways in which statisticians draw samples, as well as to describe the important concept of a sampling distribution. Among the specific objectives are:

1   To describe the nature of simple random sampling, systematic sampling, stratified random sampling, and cluster sampling.

2   To show how a table of random numbers can be used to pick a random sample..

3   To describe the sampling distribution of the sample mean and sample proportion.

Sample Designs and Sampling Distributions

6.1 Introduction

As stressed throughout the previous chapters, the field of statistics is concerned with the nature and effectiveness of sampling techniques. To comprehend how statistical methods are used, it is essential that you be familiar with the major kinds of sample designs and that you understand the concept of a sampling distribution. In this chapter, we begin by describing the various kinds of commonly used sample designs. Then we discuss the concept of a sampling distribution and present some fundamental results concerning the sampling distributions of the sample mean and proportion. Our treatment of these topics makes extensive use of the probability theory contained in Chapters 3 to 5.

6.2 Probability Samples and Judgment Samples

At the outset of any sampling investigation, one must determine whether the sample is to be a probability sample or a judgment sample. These are the two broad classes of sample designs that can be used, and each is defined as follows.

A **probability sample** *is one where the probability that each element (that is, each member) of the population is included in the sample is known. In a* **judgment sample,** *personal judgment plays a major role in determining which elements of the population are selected, and this probability is not known.*

For example, in constructing a sample of cities in Illinois, if we pick randomly 5 of Illinois's biggest cities and if we pick randomly 10 of the other Illinois cities, the resulting sample of 15 Illinois cities is a probability sample because we know the probability that each Illinois city in the population is included. On the other hand, if we go down the list of Illinois cities and choose 15 cities that we consider "typical" or "average," this is a judgment sample because it is not based on random methods of selection and we do not know the probability that each city in the population will be included.

The most important disadvantage of a judgment sample is that there is no way to tell how "far off" a sample result is likely to be. That is, one cannot estimate the difference between the sample result and the population parameter one is trying to measure. In contrast, *if a probability sample is taken, we can estimate how large the sampling error is likely to be,* as we shall see in subsequent sections of this chapter. Nonetheless, there are situations where judgment samples are used. In some cases, a probability sample is too expensive or impractical. For example, if our sample of Illinois cities must be confined to a single city because of budget limitations, we might well decide to choose the city on the basis of expert judgment rather than on the basis of chance. In a case of this sort, the coverage of the sample will be so narrow that what can be inferred concerning the population as a whole will be largely a matter of judgment in any event, regardless of whether a probability or a judgment sample is used.

Quota Sampling

One type of judgment sample encountered frequently is a quota sample. *In a quota sample, the population as a whole is split into a number of groups or strata, and whoever is charged with drawing the sample is instructed to include a certain number of members of each group.* For example, if it is known that 20 percent of the individuals in a particular town are black men, 25 percent are black women, 30 percent are white men, and 25 percent are white women, a quota sample of 100 people in the town might specify that the interviewers pick 20 black men, 25 black women, 30 white men, and 25 white women. In this way an attempt is made to construct a sample that seems representative of the population as a whole. *The reason why quota sampling is a form of judgment sampling is that the choice of members from each group or stratum to be included in the sample is not determined by random selection, but by the interviewers.* Since interviewers tend to choose members of each group that can be contacted most readily, a variety of unknown biases may result. And like any form of judgment sample, there is no way to determine how large the sampling errors are likely to be.[1] Nonetheless, quota sampling is often used because it is convenient and inexpensive.

6.3 Types of Probability Samples

Four types of probability samples are frequently used: simple random samples, systematic samples, stratified random samples, and cluster samples. Following are brief descriptions of each type.

SIMPLE RANDOM SAMPLE

Simple random sampling is the method that serves as the best introduction to probability sampling.

If the population contains N elements, a simple random sample of n elements is a sample chosen so that every combination of n elements has an equal chance of selection. Assuming that the sampling is without replacement, this means that each element in the population has a probability of $1/N$ of being the first chosen, that each of the $(N - 1)$ elements not chosen on the first draw has a probability of $1/(N - 1)$ of being the second chosen, $\cdots$, and that each of the $(N - n + 1)$ elements not chosen on the $(n - 1)th$ draw has a probability of $1/(N - n + 1)$ of being the last chosen.

An alternative and equally useful definition is as follows.

*A **simple random sample** is a sample chosen so that the probability of selecting each element in the population is the same for each and every element, and the chance of selecting one element is independent of whether some other element is chosen.*

To illustrate what we mean by a simple random sample, suppose that four subjects (*A, B, C,* and *D*) have agreed to participate in a psychological experiment, and that the psychologist wants to pick a simple random sample of two of these subjects. There are six different samples of size 2 that can be drawn from this population of four subjects.[2]

---

[1] For further comparison of probability and judgment samples, see M. Hansen and W. Hurwitz, "Dependable Samples for Market Surveys," in E. Mansfield, *Statistics for Business and Economics: Readings and Cases* (New York: Norton, 1980).

[2] More accurately, in keeping with definitions in Chapter 1, the population consists of *measurements* concerning these four subjects, and the sample consists of *measurements* concerning two of them. Also, we assume that sampling is without replacement.

(As shown in Appendix 3.1, the number of different samples of size n that can be drawn without replacement from a population of N elements equals $N! \div [(N - n)!n!]$.) Specifically, these samples are (AB); (AC); (AD); (BC); (BD); and (CD). For this sample to be a simple random sample, each of these six samples must have a probability of 1/6 of being selected. If sampling is without replacement, this can be achieved by choosing the first subject in such a way that each of the four subjects has a probability of 1/4 of being chosen, and then by choosing the second subject in such a way that each of the remaining three subjects has a probability of 1/3 of being chosen.

If the population is infinite, the number of possible samples of size n is also infinite. Thus, the alternative definition (given above) of a simple random sample must be used: A simple random sample is one where the probability of selecting each element in the population is the same, and the chance of selecting one element is independent of whether some other element is chosen. For example, if a coin were tossed repeatedly for an indefinite period of time, an infinite population would result and each element contained in this population would be heads or tails. If the coin were tossed six times, this would be a simple random sample of size 6 if the probability of heads (or tails) remained the same from one toss to the next and if the result of one toss was independent of the result of another toss.

How do statisticians actually pick a simple random sample? In cases where the number of elements in the population is small, we could (1) number each of the elements in the population; (2) record each number on a slip of paper; (3) place all the slips of paper in a hat or bowl, where they are mixed well; and (4) draw n slips of paper from the hat or bowl. The numbers on the slips of paper would indicate which elements in the population are included in the sample. Although this procedure is straightforward, we could run into problems if the number of elements in the population is large or if the slips of paper are not well mixed. To avoid such problems statisticians frequently use a table of random numbers in choosing a simple random sample. We shall describe how such a table is used in Section 6.4.

SYSTEMATIC SAMPLE

To illustrate how a systematic sample is chosen, suppose that a statistician has a list of 1,000 people from which to pick a sample of 50. Since there are 1,000 people, this can be accomplished by taking every 20th person on the list. Once a choice has been made among the first 20 persons on the list, the entire sample has been chosen since there are only 20 samples that can be drawn. The first possible sample consists of the 1st, 21st, 41st, ⋯ persons on the list. The second possible sample consists of the 2nd, 22nd, 42nd, ⋯ persons on the list. And so on. To determine which of these possible samples will be drawn, the statistician draws at random a number between 1 and 20. The chosen number identifies which of the first 20 persons he or she will begin with, which in turn identifies which sample will be drawn. In general, a systematic sample is defined as follows.

> A **systematic sample** *is obtained by taking every* k*th element on a list of all elements in the population. To determine which of the first* k *elements is chosen, a number from* 1 *to* k *is chosen at random.*

A systematic sample is often viewed as being essentially the same as a simple random sample. It is important to recognize that this is true only *if the elements of the population are in random order on the list.* One cannot always be sure that the phenomenon being measured does not have a periodicity or other type of pattern on the list. For example, W. A. Wallis and H. Roberts have pointed out that on census record sheets "the first names on the sheets tend to be predominantly male, gainfully employed, and above average in income. The reason is that the enumerators are instructed to start in a certain block at the corner house (which tends to have a higher rental value than houses in the middle of the block), and in the household to start with the head (usually male and the breadwinner)."[3] If there is a periodicity of this sort, a systematic sample may be far from random and may be less precise than a simple random sample.

On the other hand, circumstances exist where because the elements of the population are not listed in random order, a systematic sample may be *more* precise than a simple random sample. Basically, the reason is that a systematic sample insures that the elements in the sample are distributed evenly throughout the list. If there is a strong tendency for the characteristic being measured to increase or decrease steadily as one progresses from the beginning to the end of the list, this "even distribution" increases the precision of the sample estimate.

STRATIFIED RANDOM SAMPLE

In designing a sample, it frequently is useful to recognize that the population can be divided into various groups or *strata.* For example, if a sample of students is to be drawn in order to estimate the mean height of freshmen at your college or university, it would probably be

[3] W. Allen Wallis and Harry V. Roberts, *Statistics: A New Approach,* pp. 488–489.

MISTAKES TO AVOID

THE CASE OF THE LITERARY DIGEST

In 1936, the *Literary Digest* sampled voters to predict whether the Democratic candidate, Franklin D. Roosevelt, or Alfred M. Landon, his Republican opponent, would win the 1936 presidential election. The *Digest* sent questionnaires to about 10 million people. The frame consisted of lists of voters owning a telephone or a car. Over 2 million questionnaires were filled out and returned to the *Digest*. The results, shown below, were disastrously inaccurate.

| | *Percentage of vote* | |
Candidate	Literary Digest's estimate	Actual
Roosevelt	43	62
Landon	57	38

What fundamental mistakes did the *Digest* make?

SOLUTION: Until 1976, it was widely held that the basic reason for this statistical debacle was that the *Digest* sampled the wrong population. It sampled only those voters with a telephone or a car. Since poorer voters often did not have telephones or cars in 1936, and since the poor tended to be much more heavily for Roosevelt than for Landon, this meant that the population sampled by the *Digest* was much more heavily Republican than the total population of voters. However, in 1976, Maurice Bryson of Colorado State University[4] argued persuasively that the more important problem was the *Digest*'s reliance on voluntary response. The Landon supporters felt more strongly about the outcome than the Roosevelt supporters—and thus were more likely to fill out and return the questionnaire.

What was the result? Although the *Literary Digest*'s sample survey was less lethal, it was to statistics what the *Titanic* (the famous British ocean liner that sank in the North Atlantic in 1912, with over 1,500 lives lost) was to ocean navigation. As for the *Literary Digest* itself, it subsequently folded.

[4] M. Bryson, "The *Literary Digest* Poll: Making of a Statistical Myth," *American Statistician*, November 1976.

wise to stratify the population (in this case, the freshman class) into two groups: males and females.

> *In general, **stratified random sampling** is sampling in which the population is divided into strata and a random sample is taken from the elements in each stratum.*

Why is stratified random sampling employed? Because a more precise estimate can often be obtained from a sample of a given size if stratified random sampling is used rather than simple random sampling.

To illustrate the application of stratified random sampling, suppose that a sociologist is about to carry out a sample survey to estimate the proportion of all adults in a particular area who are unemployed. If the sociologist believes that the unemployment rate tends to be higher among nonwhites than among whites, he can divide the population into two *strata*, one containing whites, and one containing nonwhites. Then, if 80 percent of all adults in this area are white, and 20 percent are nonwhite, the proportion of all adults in this area who are unemployed is

$$\pi = 0.8\pi_1 + 0.2\pi_2$$

where π_1 is the proportion of *all* white adults in this area who are unemployed, and π_2 is the proportion of *all* nonwhite adults in this area who are unemployed. Consequently, *if a simple random sample is taken from each stratum,* an estimate of the proportion of all adults in this area who are unemployed can be obtained by computing

$$p = 0.8\,p_1 + 0.2\,p_2, \tag{6.1}$$

where p_1 is the proportion of the *sample* of white adults who are unemployed, and p_2 is the proportion of the *sample* of nonwhite adults who are unemployed.

How to Form Strata

The basic idea in formulating strata is to subdivide the population so that these subdivisions differ greatly with regard to the characteristic being measured, and so that there is as little variation as possible within each stratum (or subdivision) with regard to the characteristic under measurement. For example, in the problem discussed above, the sociologist should stratify adults so that the differences *among* the strata in the proportion unemployed are *large*, while the variation *within* each stratum in this regard is *small*.

If properly constructed, a stratified sample can generally result in more precise results than those obtained by using a simple random sample. This is why stratified sampling is used, as we stressed above.[5]

Once the strata have been defined, there remains the problem of determining how the total sample is to be divided among the strata. In other words, how many elements are to be chosen from each stratum? Two possible answers are proportional allocation and optimum allocation.

Proportional Allocation. This method of allocating the sample *makes the sample size in each stratum proportional to the total number of elements in the stratum.* For example, in the case of the sociologist cited above, this would mean that 80 percent of the sample would be chosen from the white population, while 20 percent would be chosen from the nonwhite population. Why? Because 80 percent of *all* adults in the relevant area are white and 20 percent are nonwhite. At first glance, it certainly seems reasonable to sample the same proportion of elements in each stratum, and this allocation method is frequently used. But if one has some knowledge of the population standard deviation in each stratum, it may be preferable to use optimum allocation, described below.

Optimum Allocation. This method prescribes that *the sample size in each stratum be proportional to the product of the number of elements (in the population) in the stratum and the standard deviation of the characteristic being measured in the stratum.* For example, suppose that we want to estimate the average assets of banks in a given state, and that we subdivide banks into two strata; national banks and state banks. If there are 300 national banks and 500 state banks, and if the standard deviation of the assets of the national banks is \$100 million and that of the assets of the state banks is \$20 million, then the number of national banks in the sample should be proportional to $300 \times \$100$ million, while the number of state banks in the sample should be proportional to $500 \times \$20$ million. In other words, the number of national banks in the sample should be proportional to \$30 billion, while the number of state

---

[5] Stratified sampling results in greater precision because the error in the estimate for the population as a whole is due only to errors in the estimates for the various strata, whereas in simple random sampling there are also errors due to weighting the strata incorrectly. For example, in the case described above, the proportion of adults who are unemployed in a *simple random sample* would be

$$\frac{n_1}{n} p_1 + \frac{n_2}{n} p_2,$$

where n_1 is the number of whites in the *sample*, p_1 is the proportion of these whites who are unemployed, n_2 is the number of nonwhites in the *sample*, p_2 is the proportion of these nonwhites who are unemployed, and $n = n_1 + n_2$. This expression is the same as equation (6.1) except that n_1/n will differ (due to chance variation) from the true proportion of adults in the area who are white, whereas the true proportion, 0.8, is used in equation (6.1). Similarly, n_2/n in this expression will differ (due to chance variation) from the true proportion of adults in the area who are nonwhite, whereas the true proportion, 0.2, is used in equation (6.1).

THE DECENNIAL CENSUS OF POPULATION

Since 1790, a census has been taken every 10 years in the United States, the basic constitutional purpose being to apportion the membership of the House of Representatives among the states. These censuses have provided an enormous amount of valuable data concerning the age, location, sex, race, marital status, education, occupation, industry, and income of our nation's citizens. The job of organizing and taking the census is a huge undertaking which involves the recruiting and training of about 150,000 people, most of them for only a few weeks of work.

Sampling was first employed in collecting census data in the 1940 census. For some questions, such as those concerning wage and salary income, a 5 percent sample of the population was used. In later censuses, the use of sampling was extended, and in 1970 only the basic listing of the population, with data on age, sex, race, marital status, and family relationship, was collected on a 100 percent basis.

(a) Suppose that a question (for example, about a person's health or income) requires an average of 20 seconds of work for each person counted in the census. With about 200 million people in the population, about how many hours of work would be eliminated if only 5 percent of the people were asked this question? If people's time is valued at $4 per hour, how much would the saving be in dollar terms?

(b) What are the advantages of a probability sample rather than a judgment sample in the case of the census?

(c) It has long been recognized that the enumerators (the people who ask the questions) can influence the answers they obtain, but it was not until the 1950 census that the magnitude of these influences was investigated. The results indicated that these influences can be surprisingly large. Consequently, the Bureau of the Census changed its procedures in several ways. Can you guess what the changes were?

(d) If sampling is so effective, why not take the entire census with a sample? Isn't it wasteful to take a complete census?

banks in the sample should be proportional to $10 billion, which means that 3/4 of the banks in the sample should be national banks and 1/4 should be state banks. The reason why optimum allocation is "optimum" is that it minimizes the expected sampling errors in the estimate of the population mean (or proportion). Despite its advantages, it may be impractical to use optimum allocation if little or nothing is known about the population standard deviation in each stratum.

CLUSTER SAMPLE

Still another important sampling technique is cluster sampling.

*In a **cluster sample,** one divides the elements in the population into a number of clusters or groups. One then begins by choosing at random a sample of these clusters, after which a simple random sample of the elements in each chosen cluster is selected.*

The sociologist's problem of estimating the proportion of adults in a particular area that are unemployed can be used to illustrate the application of a cluster sample. In this case, clusters can be formed by geographical location. All adults in the relevant area can be classified by the block in which they live, and a simple random sample of these blocks can be chosen. Then, within each of the chosen blocks, a simple random sample of the adults themselves can be picked.

The major advantage of cluster sampling is that *it is cheaper to sample elements that are physically or geographically close to one another.* Thus, it is cheaper to sample 50 adults, all of whom are concentrated in five blocks, than to sample 50 adults who are scattered all over a large area. In general, the results of a cluster sample are less precise than those of a simple random sample (assuming that sample size is constant) because the elements in a particular cluster tend to be relatively similar. However, *per dollar spent on the survey* a cluster sample may be more effective than simple random sampling. Why? Because, for the same total cost, cluster sampling provides a much larger sample than simple random sampling.

6.4 How to Use a Table of Random Numbers

Here and in subsequent chapters we shall concentrate on simple random sampling because it provides the simplest and best introduction to probability sampling. One way to select a simple random sample is to use a table of random numbers.

A *table of random numbers* is a table of numbers generated by a random process. For example, suppose that we want to construct a table of five-digit random numbers. To do so, we could write 0 on one slip of paper, a 1 on a second slip, a 2 on a third, ⋯ , and a 9 on a tenth slip. We could then put the 10 slips of paper into a hat and draw out one at random. The number on this slip of paper would be the first digit of a random number. After replacing the slip of paper in the hat, we could make another random draw. The number on the new slip of paper would be the second digit of the random number. This procedure could then be repeated three more times, yielding the third, fourth, and fifth digits of the random number. The result is the first five-digit random number. Given enough time and persistence, we could formulate as many such random numbers as needed. Fortunately, it is not necessary to go through such a time-consuming process because tables of random numbers have already been formulated. Although these tables generally have been calculated on a computer rather than by drawing slips of paper from a hat, the principles of their construction are essentially the same as if slips of paper had been used.

A table of random numbers is given in Appendix Table 5. As you can see, it contains column after column of five-digit random numbers. (The left-hand column shows the line number and should not be confused with the random numbers.) To illustrate how the table can be used to pick a simple random sample, suppose that we want to draw a simple random sample of 50 people from a population of 1,000 persons. Our first step would be to number each of the persons from 0 to 999. We would then turn to any page of the table and proceed in any systematic way to pick our sample. For example, we might begin at the top, left-hand column of page A14 and work down. Since we need three-digit, not five-digit random numbers, we would pay attention to only the first three digits of each five-digit random number. The topmost five-digit number in the left-hand column of page A14 is 53535; its first three digits are 535. Thus, the first person to be picked is number 535. Reading down, the next five-digit number in the left-hand column of page A14 is 41292; its first three digits are 412. Thus, the second person to be picked is number 412. This procedure should be repeated until the entire sample of 50 has been selected.

Several points should be noted concerning this procedure. (1) If any previously chosen number comes up again, it should be ignored. For example, if 412 were to come up again before the 50 persons were chosen, it should be ignored. (2) If a number comes up which is not admissible because it does not correspond to any item in the population, it also should be ignored. For example, if there were 500, not 1,000, persons in the population, and if 680 came up, it should be ignored. (3) It makes no difference whether you read the table down, up, to the left, to the right, or diagonally. As long as you proceed from number to number in a systematic fashion the table will work properly.

THE 1970 DRAFT LOTTERY

MISTAKES TO AVOID

In the late 1960s, there were charges of unfairness in the procedures by which some local draft boards decided who should be drafted into the American armed forces. Thus, for the 1970 draft, it was decided that the eligible candidates would be randomly ordered for induction. Capsules representing each of the 366 days of the year[6] were put in a cage and were randomly withdrawn by an individual. Since September 14 was chosen first, men born on that date were the first ones inducted in 1970. Since April 24 was chosen second, men born on that date were the second ones inducted; and so on.

Each day of the year was given a number from 1 to 366. For example, September 14 received the number 1 and April 24 received the number 2, as we have just seen. The average number for the days in each month are shown in the figure below. Is there any indication that the numbers were really not chosen at random? What sort of departure from randomness, if any, seems to exist?

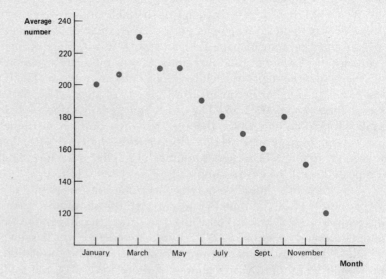

SOLUTION: Later months (like November and December) tend to have lower numbers than earlier months (like January and February). Thus, men with birthdays in later months tended to be drafted before men with birthdays in earlier months. Had the drawing been completely random, one would have expected the averages to be about the same, on the average, in the later months as in the early ones. Some statisticians believe that the capsules were put into the cage in month order and that they were not thoroughly mixed, the result being that the capsules for the later months tended to be withdrawn before those for the ear-

---

[6] This includes February 29, which occurs every four years.

lier months. The moral is that, even if a sample is drawn by picking capsules out of a cage or by picking numbers out of a hat, unsuspected biases may occur. Thus, it is wise to use a table of random numbers.[7]

[7] For further discussion, see S. Fienberg, "Randomization and Social Affairs: The 1970 Draft Lottery," *Science*, January 22, 1971; and B. Williams, *A Sampler on Sampling* (New York: Wiley, 1978), p. 7.

EXERCISES

6.1 (a) If a finite population consists of the elements *A, B, C, D,* and *E,* list all possible samples of size 2 that can be drawn (without replacement) from this population.

(b) List all possible samples of size 3 that can be drawn (without replacement) from this population.

(c) If each of these possible samples of size 3 has the same probability of being selected, what is the probability that *A* will be included in the sample that is selected? What is the probability that both *A* and *B* will be included in this sample?

6.2 How many different samples of size 4 can be selected (without replacement) from a finite population consisting of the five boroughs of New York City (Manhattan; the Bronx; Brooklyn; Queens; Staten Island)? How many different samples of size 3 can be selected (without replacement) from this population?

6.3 Use a table of random numbers to select a sample of two drugstores from among those listed in the yellow pages of your telephone directory.

6.4 The tickets issued by a movie theater are numbered serially. On October 28, 1985 the theater sold tickets with numbers beginning at 20860 and ending at 23102. Use a table of random numbers to select a sample of 10 of these tickets.

6.5 Hotels often leave questionnaires concerning the quality of their service in their rooms, and ask their guests to fill them out. Can the results be regarded as a random sample? Why or why not?

6.6 The Educational Testing Service, in one of its studies, wanted to obtain a representative sample of U.S. college students. To do so, they stratified all colleges and universities into a number of groups (e.g., public universities with 25,000 or more students, and private four-year colleges with 1,000 or fewer students). Then they picked one college or university in each stratum that they regarded as representative. Each of the chosen colleges or universities was asked to select a sample of its students. Write a one-paragraph evaluation of this procedure.[8]

[8] For further discussion of this case, see D. Freedman, R. Pisani, and R. Purves, *Statistics* (New York: Norton, 1978).

6.7 In a television appearance, a congressman asks voters in his district to send him a postcard indicating whether he should vote for or against a particular proposed piece of legislation. He receives 4,100 replies. Since voters from the western part of his district tend to be Democrats, while those from the eastern part tend to be Republicans, he divides the replies on the basis of their geographical origin. Based on the results, he finds that 68 percent of the replies from the western part of his district are against the legislation, while 36 percent of the replies from the eastern part are against it.
(a) Did the congressman carry out a stratified sample?
(b) Did he carry out a stratified random sample?
(c) What criticisms would you make concerning his sample design?
(d) What improvements would you suggest?

6.8 A firm wants to determine the mean income of the families in a particular suburb of Washington, D.C. The firm divides the suburb into seven parts, each of which contains the same number of families. After picking two of these parts at random, it chooses a random sample of 50 families in each of these two parts.
(a) What are the advantages of this sample design over a simple random sample of 100 families drawn from the entire suburb?
(b) Is the result of this sample design likely to be as precise as the result of a random sample of 100 families drawn from the entire suburb?
(c) Regardless of the sample design, can you identify some problems in obtaining the desired information?

6.9 As pointed out in Chapter 1, Philadelphia's school teachers went on strike for a long period in 1981. Suppose that Philadelphia's school board had carried out a sample of public opinion concerning its behavior regarding the strike by picking a random sample of Philadelphia school children and sending a questionnaire to their parents.
(a) Would such a sample be a random sample of the opinions of Philadelphia residents?
(b) From what population would such a sample be drawn?
(c) What biases might result from such a sample?

6.10 A lawyer must estimate the mean value of the pieces of furniture contained in a 15-room house after the house's owner dies. Such information is needed to fill out the estate tax return. In all, there are 135 pieces of furniture, including sofas, chairs, tables, and so on. The lawyer decides to estimate the mean value on the basis of a sample.
(a) Would you advise him to stratify the relevant population?
(b) If so, what strata would you advise him to use?

6.11 A population consists of two strata, *A* and *B*. There are 100 items in *A*, and the standard deviation is 20. There are 50 items in *B*, and the standard deviation is 100.
(a) If you want to carry out a stratified random sampling with optimum allocation, what proportion of the sample will you take from stratum *A*? From stratum *B*?

(b) If you want to use proportional allocation, what proportion of the sample will you take from stratum *A*? From stratum *B*?

(c) Which of the allocation schemes in (a) and (b) will result in smaller expected sampling errors?

6.5 Concept of a Sampling Distribution

Sampling distributions are of central importance in statistics. The idea of a sampling distribution stems directly from the fact that there generally are a large number of possible samples that can be selected from a population. Thus, the value of any statistic computed from a sample will vary from sample to sample. For example, suppose that a population consists of the eight numbers in Table 6.1, that we are going to take a simple random sample of two of these numbers, and that the sample statistic in which we are interested is the sample mean. Since the sample is randomly chosen, the probability that each possible sample will be selected is known and it is possible to deduce the probability distribution of the value of the sample statistic. This is an example of a sampling distribution, which is defined as follows.

> A ***sampling distribution*** *is the probability distribution of the value of a statistic.*

What is the sampling distribution of the mean of a simple random sample of size 2 from the population in Table 6.1? There are 28 different samples of size 2 that can be drawn without replacement from this population (since there are only eight numbers in the population). Table 6.2 lists each of these samples, as well as its mean. Since the probability of selecting each of these samples is $1/28$, the probability that the sample mean assumes each of the 28 values (not all of which are

Measurements included in the population
2
4
5
7
8
0
1
3

Table 6.1
Population of Eight Numbers

Table 6.2
Mean and Range of 28 Possible Samples of Size 2 from the Population in Table 6.1

Sample	Numbers included in sample	Sample mean	Sample range
1	2,4	3.0	2
2	2,5	3.5	3
3	2,7	4.5	5
4	2,8	5.0	6
5	2,0	1.0	2
6	2,1	1.5	1
7	2,3	2.5	1
8	4,5	4.5	1
9	4,7	5.5	3
10	4,8	6.0	4
11	4,0	2.0	4
12	4,1	2.5	3
13	4,3	3.5	1
14	5,7	6.0	2
15	5,8	6.5	3
16	5,0	2.5	5
17	5,1	3.0	4
18	5,3	4.0	2
19	7,8	7.5	1
20	7,0	3.5	7
21	7,1	4.0	6
22	7,3	5.0	4
23	8,0	4.0	8
24	8,1	4.5	7
25	8,3	5.5	5
26	0,1	0.5	1
27	0,3	1.5	3
28	1,3	2.0	2

different from one another) is 1/28. Thus, the probability distribution of the value of the sample mean in this case is as shown in Figure 6.1. This probability distribution is a sampling distribution.

Sampling distributions can be constructed for any sample statistic, not just the sample mean. For example, Table 6.2 also shows the sample range for each of the 28 samples of size 2 from the population in Table 6.1. Since the probability of each of the 28 values (not all of which are different from one another) is 1/28, the probability distribution of the value of the sample range in this case is as shown in Figure 6.2 This is another example of a sampling distribution.

Any sampling distribution is based on the assumption that the sample size and the sample design remain fixed. Thus, the sampling

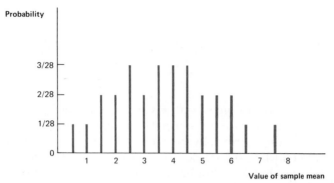

Figure 6.1
Probability Distribution of the Mean of a Simple Random Sample of Size 2 from Population in Table 6.1

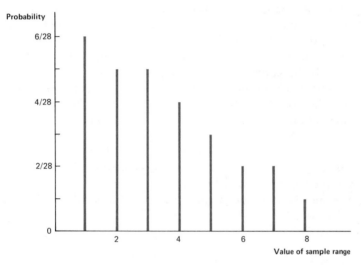

Figure 6.2
Probability Distribution of the Range of a Simple Random Sample of Size 2 from Population in Table 6.1

distributions in Figures 6.1 and 6.2 are based on the assumption that samples of size 2 are taken, and that the samples are simple random samples. In general, it is not possible to derive sampling distributions in the way employed in Table 6.2 because most populations contain so many elements that it is essentially impossible to list all possible samples. Even if there are only 1,000 elements in a population, there are 1,000!/(998! 2!) or 499,500 different samples of size 2 that can be drawn from this population. Obviously, the listing of all such samples would be a formidable task; and fortunately (as we shall see in later sections) there are other ways of finding the sampling distribution of a sample statistic. The reason why we presented this laborious procedure is not that it is used to obtain sampling distributions in actual cases, but that it is useful in explaining the concept of a sampling distribution.

6.6 The Sampling Distribution of the Sample Proportion

Many statistical investigations are aimed at estimating the proportion of the members of some population who have a specified characteristic. For example, we may be interested in the proportion of St. Louis women who prefer Brand A beer to Brand B, or in the proportion of fifth graders in Baltimore who have learning disabilities. In investigations of this sort, the sample proportion is generally used as an estimate of the population proportion. If the sample is a small percentage of the whole population (say, 5 percent or less), we can determine the sampling distribution of the sample proportion without much difficulty, based on our results in Chapter 4 concerning the binomial distribution.

If the sample is a simple random sample, and if it is a small percentage of the population, each observation has a probability Π of having the specified characteristic, where Π is the proportion of the population with this characteristic. If the sample contains n observations, each observation selected can be viewed as a Bernoulli trial where there is a probability Π of success, where success is defined as the observation's having the specified characteristic. Thus, as we know from Chapter 4, the number of observations in the sample that have the specified characteristic must be a binomial random variable. In other words,

$$P(x) = \frac{n!}{(n-x)!x!} \, \Pi^x (1 - \Pi)^{n-x}, \text{ for } x = 0,1,2, \cdots n$$

where $P(x)$ is the probability that x observations in the sample will have the specified characteristic.

The sample statistic whose probability distribution we want to derive is the sample proportion. It is important to note that the value of the sample proportion is determined entirely by the number of observations in the sample with the specified characteristic. For example, if the sample size is 5 the sample proportion can be 1/5 if and only if exactly one observation in the sample has the desired characteristic; it can be 2/5 if and only if exactly two observations in the sample have the desired characteristic; and so on. Thus, *the probability that the value of the sample proportion is x/n must equal the probability that the number of observations in the sample with the specified characteristic is x.* Consequently, the sampling distribution of the sample proportion must be as follows:

$$Pr\left(p = \frac{x}{n}\right) = \frac{n!}{(n-x)!x!} \Pi^x(1-\Pi)^{n-x}, \qquad (6.2)$$

Formula for Sample Proportion's Probability Distribution

where $Pr(p = x/n)$ denotes the probability that the sample proportion p equals x/n.

To illustrate how this formula can be used, suppose that a market-research firm is about to ask a sample of beer drinkers whether they prefer Budweiser over Miller (presenting both in unmarked cans). If the firm selects a simple random sample of 20 beer drinkers, and if the proportion of the population preferring Budweiser over Miller is 0.50, what is the sampling distribution of the sample proportion? The formula in equation (6.2) tells us that the probability that the sample proportion equals zero can be determined by finding (in Appendix Table 1) the probability that a binomial variable (with $n = 20$ and $\Pi = 0.5$) equals zero, that the probability that the sample proportion equals $1/20$ can be determined by finding the probability that this binomial variable equals 1, that the probability that the sample proportion equals $2/20$ can be determined by finding the probability that this binomial variable equals 2, and so on. The resulting sampling distribution is shown in Figure 6.3

The sampling distribution of a sample statistic is useful in indicating the amount of error the statistic is likely to contain. The following example shows how the sampling distribution of the sample proportion can be used in this way.

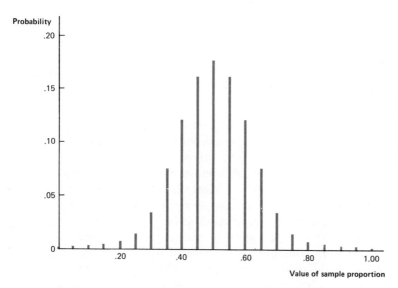

Figure 6.3
Sampling Distribution of the Sample Proportion, $n = 20$ and $\Pi = 0.50$

EXAMPLE 6.1 A high school contains 1,000 students. The school's principal draws a random sample of 20 students and determines the proportion of these students who would like to participate in a trip to the United Nations. If the proportion of students in the entire school who would like to participate equals .15, what is the probability that the sample proportion will be in error by more than .10? (That is, what is the probability that the sample proportion will differ by more than .10 from the population proportion?)

SOLUTION: Since the population proportion is .15, the sample proportion will *not* differ by more than .10 from the population proportion if the sample proportion equals .05, .10, .15, .20, or .25. Using Appendix Table 1, we can obtain the probability of the sample proportion's equaling each of these values. The probability that it equals .05 is .1368, since this is the probability of one success (in this case, one student wanting to participate) in 20 trials, given that $\Pi = .15$. The probability that it equals .10 is .2293, since this is the probability of two successes in 20 trials, given that $\Pi = .15$. The probability that it equals .15 is .2428, since this is the probability of three successes in 20 trials, given that $\Pi = .15$. The probability that it equals .20 is .1821, since this is the probability of four successes in 20 trials, given that $\Pi = .15$. The probability that it equals .25 is .1028, since this is the probability of five successes in 20 trials, given that $\Pi = .15$. Thus, the probability that the sample proportion differs from the population proportion by more than .10 is $1 - (.1368 + .2293 + .2428 + .1821 + .1028)$, or .1062.

6.7 The Sampling Distribution of the Sample Mean

Perhaps the most frequent objective of statistical investigations is to estimate the mean of some population. For example, a psychologist may want to estimate the mean score of college freshmen on a particular type of test, or a political scientist may want to estimate the mean amount of money spent by mayoral candidates on their 1986 campaigns. In investigations of this sort, the sample mean is generally used as an estimate of the population mean. To measure the extent of the sampling errors that may be present in the sample mean, it is essential to know the sampling distribution of the sample mean, which is the subject of this section. We will begin with the case where the population is known to be normal, then take up the case where it is not normal, and conclude with the case where the population contains relatively few observations.

WHERE THE POPULATION IS NORMAL

Suppose that simple random samples of size 10 are selected repeatedly from a normal population and that the mean of each such sample is calculated. If a very large number of such sample means are calculated, what does the probability distribution of their values look like? As shown in Figure 6.4, this probability distribution has the same mean as the population, namely μ. Thus, if a large number of sample means were calculated, *on the average* the sample mean would equal the population mean. Also, the probability distribution of the sample mean is bell-shaped and symmetrical about the population mean, which means that it is equally likely that a sample mean will fall below or above the population mean. Further, because of the averaging, the dispersion of the sampling distribution of the sample mean is less than the dispersion of the population. In other words, the standard deviation of the sampling distribution of the sample mean, denoted by $\sigma_{\bar{x}}$, is less than the standard deviation of the population, denoted by σ.

Now suppose that simple random samples of size 40 are selected repeatedly from the same normal population, and that the mean of each such sample is calculated. If a very large number of such sample means are calculated, what does the probability distribution of their values look like? As shown in Figure 6.4, this probability distribution,

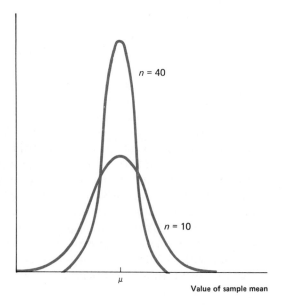

Figure 6.4
Sampling Distribution of Sample Mean, $n = 10$ and $n = 40$

like the one for samples of size 10, has the same mean as the population and is bell-shaped and symmetrical. The most obvious difference between this sampling distribution and the one for samples of size 10 is this distribution's much smaller standard deviation. Because the sample size is larger, it is likely that the mean of a sample of size 40 will be closer to the population mean than the mean of a sample of size 10. This, of course, accounts for the smaller dispersion in the sampling distribution of the sample mean for samples of size 40.

A close inspection of both sampling distributions in Figure 6.4 suggests that they are normal distributions. At least this is how they look, and mathematicians have proved that this is indeed the case. A remarkable feature of a normal population is that the sampling distribution of means of simple random samples from such a population is also normal. This is true for a sample of any size. Given this result as well as the others noted above, it is possible to provide the following complete description of the sampling distribution of the sample mean from a normal population.

SAMPLING DISTRIBUTION OF THE SAMPLE MEAN (NORMAL POPULATION): *The sample mean is normally distributed, the mean of its sampling distribution equals the mean of the population (μ), and the standard deviation of its sampling distribution ($\sigma_{\bar{x}}$) equals the standard deviation of the population divided by the square root of the sample size. That is,*

Formula for $\sigma_{\bar{x}}$

$$\sigma_{\bar{x}} = \frac{\sigma}{\sqrt{n}},$$

(6.3)

where σ is the population standard deviation and n *is the sample size.*

Equation (6.3) is a very important result. *The standard deviation of the sampling distribution of the sample mean, $\sigma_{\bar{x}}$, is a measure of "how far off" the sample mean is likely to be from the population mean. If $\sigma_{\bar{x}}$ is large,* the distribution of the sample mean contains *a great deal of dispersion,* which means there is a relatively *high* probability that the sample mean will depart considerably from the population mean. If $\sigma_{\bar{x}}$ is *small,* the distribution of the sample mean contains *relatively little dispersion,* which means there is a relatively *low* probability that the sample mean will depart considerably from the population mean. For example, panel A of Figure 6.5 shows the distribution of the sample mean in a *Standard Error of the Sample Mean* case where its standard deviation—often called the **standard error of the sample mean**—is relatively large. Panel B shows the distribution of the sample mean in a case where its standard deviation is relatively

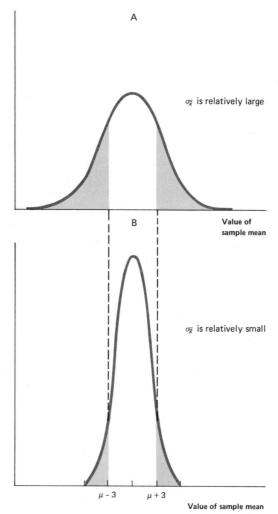

Figure 6.5
Sampling Distribution of the Sample Mean, when $\sigma_{\bar{x}}$ Is Relatively Large and Relatively Small

small. As shown in Figure 6.5, the probability that the sample mean differs from the population mean by more than an arbitrary amount (say 3, which is shown in Figure 6.5) is much higher in the case shown in panel A than in that shown in panel B. (In each case this probability is equal in value to the shaded area in Figure 6.5.)

Given that $\sigma_{\bar{x}}$ is a measure of how "far off" the sample mean is likely to be, it is obviously important to specify what determines $\sigma_{\bar{x}}$. According to equation (6.3), $\sigma_{\bar{x}}$ is determined by two things: the standard deviation of the population, σ, and the sample size, n. *Holding sample size constant, $\sigma_{\bar{x}}$ is proportional to the standard deviation of the population.* This is reasonable because one would expect that sample means drawn from more variable populations would themselves tend to be more variable. *Holding constant the standard deviation of the population, $\sigma_{\bar{x}}$ is inversely proportional to the square root of the sample size.* This means that, in a sense, diminishing returns set in as the sample size is increased. If the sample size is quadrupled, $\sigma_{\bar{x}}$ is cut only in half; if the sample size is multiplied by 25, $\sigma_{\bar{x}}$ is cut only to one-fifth of its previous amount.

To prevent confusion, it is extremely important that you understand the distinction between $\sigma_{\bar{x}}$ (the standard deviation of the sampling distribution of the sample mean) and σ (the standard deviation of the population). The former is the standard deviation of the distribution of the sample mean. (Such distributions have been presented in Figure 6.4 and Figure 6.5) The latter is the standard deviation of the population from which the samples are drawn. Since the distribution of the sample mean is obviously quite different from the population being sampled, the standard deviations of the two are quite different, although not unrelated. (As pointed out in the previous paragraph, $\sigma_{\bar{x}}$ is proportional to σ, if sample size is held constant.)

Our results concerning the sampling distribution of the sample mean have important applications, some of which are illustrated by the following example.

EXAMPLE 6.2 A psychological experiment is carried out to measure how long it takes subjects to carry out a particular task. The investigator picks a random sample of 100 subjects, measures how long it takes each one to carry out this task, and calculates the sample mean. If, in the population as a whole, the length of time it takes to carry out this task is normally distributed with a standard deviation of .01 hours, what is the probability that the sample mean will exceed the population mean by more than .0015 hours?

SOLUTION: The distribution of the sample mean is normal with mean equal to the population mean μ and standard deviation equal to $\sigma/\sqrt{n}$, which in this case is $.01 \div \sqrt{100}$, or .001 hours. Figure 6.6 shows the distribution of the sample mean. The probability we want equals the area under this curve to the right of $\mu +$ .0015. The value of the standard normal variable corresponding to $\mu + .0015$ is $[(\mu + .0015) - \mu] \div \sigma_{\bar{x}} = .0015 \div .001 = 1.5$. Thus, we must evaluate the area under the standard normal curve to the right of 1.5. Appendix Table 2 shows that the area between zero and 1.5 is .4332, so the area to the right of 1.5 must equal $.5000 - .4332 = .0668$. This is the probability that the sample mean will exceed the population mean by more than .0015 hours.

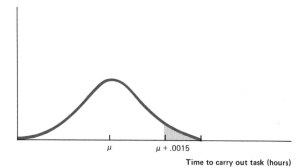

Figure 6.6
Probability that the
Sample Mean Exceeds
the Population Mean by
More than .0015
Hours

WHERE THE POPULATION IS NOT NORMAL

As we know from the previous paragraphs, if the population is normal the sampling distribution of the sample mean is normal. But what if, as frequently occurs, the population is *not* normal? A remarkable mathematical theorem, known as the **central limit theorem**, has shown that if the sample size is moderately large, the distribution of the sample mean can be approximated by the normal distribution, even if the population is not normal. This very important theorem can be stated as follows.

> CENTRAL LIMIT THEOREM. *As the sample size* n *becomes large, the sampling distribution of the sample mean can be approximated by a normal distribution with a mean of μ and a standard deviation of $\sigma/\sqrt{n}$, where μ is the mean of the population and σ is its standard deviation.*

Central Limit Theorem

What is particularly impressive about this result is that the normal approximation seems to be quite good so long as the sample size is larger than about 30, regardless of the nature of the population. And if the population is reasonably close to normal, sample sizes of much less than 25 are likely to result in the normal approximation's being serviceable.[9]

To illustrate the working of the central limit theorem, suppose

---

[9] Using the results of Appendix 4.3 and 4.1, we can prove that the expected value of $\bar{x}$ equals μ. From Appendix 4.3 we know that $E(\Sigma x_i) = E(x_1) + E(x_2) + \cdots + E(x_n)$. Since the expected value of each observation in a random sample equals μ, it follows that $E(\Sigma x_i) = n\mu$. And using the results of Appendix 4.1, $E(\Sigma x_i/n) = \mu$. Note that this proof does not depend on an assumption that the population is normal.

Based on Appendix 4.3 and 4.1, it is also possible to prove that $\sigma_{\bar{x}} = \sigma/\sqrt{n}$. Since the observations in a random sample are statistically independent, we know from Appendix 4.3 that $\sigma^2(\Sigma x_i) = \sigma^2(x_1) + \sigma^2(x_2) + \cdots + \sigma^2(x_n)$. Since the variance of each observation equals σ^2, it follows that $\sigma^2(\Sigma x_i) = n\sigma^2$. And using the results in Appendix 4.1, $\sigma^2(\Sigma x_i/n) = n\sigma^2/n^2 = \sigma^2/n$. This proof does not assume that the population is normal.

that the population frequency distribution is as shown in panel A of Figure 6.7. This population is such that it is equally likely than an observation in the population falls anywhere between the lower limit *A* and the upper limit *B*. (A population of this sort is said to have a *uniform distribution*.) Clearly, this population is far from normally distributed. Panels B, C, and D of Figure 6.7 show the sampling distribution of the sample mean for samples of size 2, 4, and 25 from this population. Even for samples of size 4, this sampling distribution is bell-shaped and close to normal. For samples of size 25, it is very close to normal.

The central limit theorem applies to discrete populations as well as to continuous ones, a fact which has important implications for the sampling distribution of the sample proportion. *The sample proportion is really a sample mean from a population where a success is denoted by a* 1 *and a failure is denoted by a zero.* The mean of such a population is Π, the proportion of successes in the population; and the mean of the sample is *p*, the proportion of successes in the sample. Since the sample proportion is really a sample mean from a population consisting of zeros and 1's, its sampling distribution must tend toward normality as the sample size increases, according to the central limit theorem. This fact will be used frequently in subsequent chapters.

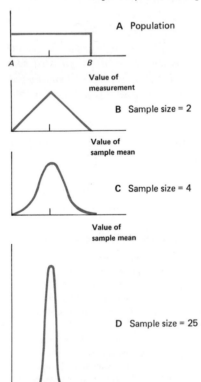

Figure 6.7
Sampling Distribution of Means of Samples of Sizes 2, 4, and 25 from a Population with a Uniform Distribution

The following example illustrates some of the practical implications of the central limit theorem. In subsequent chapters this theorem will have repeated application.

EXAMPLE 6.3 An engineer tests a random sample of 64 of the light bulbs received in a large shipment to determine the longevity of each. From past experience, the engineer knows that the standard deviation of the longevity of such light bulbs is 160 hours. What is the probability that the sample mean will differ from the population mean by more than 40 hours?

SOLUTION: Because the sample size is well above 30, we can be reasonably certain that the sampling distribution of the sample mean is very close to normal, regardless of the nature of the population. More specifically, the distribution of the sample mean should be essentially normal with a mean equal to the population mean μ and a standard deviation equal to $\sigma/\sqrt{n}$, where σ is the population standard deviation. Since $\sigma = 160$ and $n = 64$, $\sigma_{\bar{x}} = 160 \div \sqrt{64} = 20$ hours. What we want is the probability that the sample mean lies above $\mu + 40$ or below $\mu - 40$. Figure 6.8 shows the distribution of the sample mean. The probability we want equals the sum of the values of the two shaded areas. The value of the standard normal variable corresponding to $\mu + 40$ is $[(\mu + 40) - \mu] \div \sigma_{\bar{x}} = 40 \div 20 = 2$; the value corresponding to $\mu - 40$ is $[(\mu - 40) - \mu] \div \sigma_{\bar{x}} = -40 \div 20 = -2$. Thus, we must evaluate the areas under the standard normal curve to the right of 2 and to the left of -2. Since Appendix Table 2 shows that the area between zero and 2 is .4772, the area to the right of 2 must be .5000 − .4772 = .0228. Because of the symmetry of the standard normal curve, the area to the left of -2 must also be .0228. Thus, the total area to the right of 2 and to the left of -2 equals .0228 + .0228 = .0456. This is the probability that the sample mean will differ from the population mean by more than 40 hours.

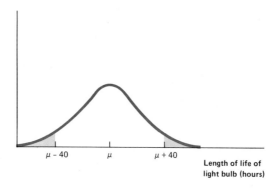

Figure 6.8
Probability that the Sample Mean Differs from the Population Mean by More than 40 Hours

WHERE THE POPULATION IS SMALL

In previous parts of this section we have assumed that the population is infinite, while in fact many populations in the real world are not. If the population is large relative to the size of the sample, our previous results can be applied without modification even though the population is finite. However, if the population is less than 20 times the sample size, the following simple correction should be applied to the standard deviation of the sample mean.

STANDARD DEVIATION OF THE SAMPLE MEAN (FINITE POPULATIONS). *Whereas $\sigma_{\bar{x}}$ equals $\sigma/\sqrt{n}$ for infinite populations, in the case of finite populations the following formula applies:*

Formula for $\sigma_{\bar{x}}$ when the Population is Small

$$\sigma_{\bar{x}} = \frac{\sigma}{\sqrt{n}} \sqrt{\frac{N-n}{N-1}}, \qquad (6.4)$$

where σ is the standard deviation of the population, N is the number of observations in the population, and n is the sample size.

Finite Population Correction Factor

In other words, the value of $\sigma_{\bar{x}}$ for an infinite population must be multiplied by $\sqrt{(N-n)/(N-1)}$ to make it appropriate for a finite population. The multiplier $\sqrt{(N-n)/(N-1)}$ is often called the *finite population correction factor*. When n is small relative to N, this correction factor is so close to 1 that it can be ignored.[10]

If sampling is with replacement (that is, if each element that is included in the sample is put back in the population so that it can be chosen more than once), there is no need to use the finite population correction factor, even if the number of elements in the population is small relative to the sample size. In other words, equation (6.3) is valid under these circumstances. However, in most practical situations, sampling is not carried out with replacement.

The following illustrates the application of equation (6.4). A taxi company has 17 cabs and chooses a simple random sample of 13 of them. On the basis of the sample, the firm computes the mean number of miles that a taxi has been driven since its tires were inspected. If the population standard deviation is 5,000 miles, the standard deviation of the sample mean equals

---

[10] In Appendix 4.2 it was stated that the standard deviation of the hypergeometric distribution equals the standard deviation of the binomial distribution times the finite population correction factor.

$$\sigma_{\bar{x}} = \frac{\sigma}{\sqrt{n}} \sqrt{\frac{N-n}{N-1}} = \frac{5,000}{\sqrt{13}} \sqrt{\frac{17-13}{17-1}}$$

$$= \left(\frac{5,000}{3.606}\right)\left(\frac{1}{2}\right) = 693.$$

Thus, the standard deviation of the sample mean is 693 miles, which is one-half of what it would have been had the population been infinite.

Since the standard deviation of the sample mean is a measure of how far off the sample mean is likely to be, an important implication of equation (6.4) is that *if the sample is small relative to the population, the accuracy of the sample mean depends entirely on the sample size and not on the fraction of the population included in the sample.* To test this, we can draw a simple random sample of size 100 from two populations, both of which have a standard deviation of 1,000. One population contains 10,000 observations, so the sample is 1 percent of the population; the other contains 100,000 observations, so the sample is 1/10 of 1 percent of the population. In the first population, the standard deviation of the sample mean is

$$\sigma_{\bar{x}} = \frac{1,000}{\sqrt{100}} \sqrt{\frac{10,000-100}{10,000-1}} = 100 \sqrt{\frac{9,900}{9,999}} \doteq 100.$$

In the second population, it is

$$\sigma_{\bar{x}} = \frac{1,000}{\sqrt{100}} \sqrt{\frac{100,000-100}{100,000-1}} = 100 \sqrt{\frac{99,900}{99,999}} \doteq 100.$$

Thus, the standard deviation of the sample mean is essentially the same, although the fraction of the population included in the sample is 10 times bigger in the first population than in the second. This is an important point, and one some people find hard to believe. To repeat, the sample size, not the fraction of the population included in the sample, determines the accuracy of the sample mean if the sample is small relative to the population.

Once the modification of $\sigma_{\bar{x}}$ in equation (6.4) has been made, the results for infinite populations can be applied to finite populations where the sample size is large relative to the population. The following example shows how this modification can be made and applied.

EXAMPLE 6.4 The Los Angeles Police Department receives a shipment of 100 car radios. A simple random sample of 36 radios is chosen, and each is tested for its closeness to specifications. Based on this test, each radio is given a rating from zero to 10 points. If the Los

Angeles Police Department knows from past experience that the standard deviation of the ratings in a shipment is 2 points, what is the probability that the mean rating in the sample will be more than one point below the mean rating for all radios in the shipment?

SOLUTION. Since the sample is considerably more than 5 percent of the population, equation (6.4) must be used to calculate $\sigma_{\bar{x}}$, which is $(2/\sqrt{36}) \sqrt{(100 - 36)/99} = (2/6) \sqrt{64/99} = .27$ points. Because the sample size is well above 25, the sampling distribution of the sample mean should be close to normal. The probability that the sample mean is more than one point below the population mean equals the probability that the sample mean is more than $1/.27 = 3.70$ standard deviations below the population mean. To evaluate this probability, we find the area under the standard normal curve to the left of -3.70. Appendix Table 2 shows that the area between zero and 3.70 exceeds .499, so the area to the right of 3.70 must be less than .001. Because of the symmetry of the standard normal curve, the area to the left of -3.70 must also be less than .001. Thus, the probability that the mean rating for the sample will be more than one point below the mean rating for all radios in the shipment is less than .001.

EXERCISES

6.12 If we are sampling from an infinite population, how much reduction occurs in the standard deviation of the sample mean if (a) the sample size is increased from 2 to 4; (b) the sample size is increased from 4 to 6; (c) the sample size is increased from 100 to 102?

6.13 A team of school psychologists administers a test to a very large number of subjects. It is known that the standard deviation of the scores is about 100 points. A random sample of 100 of the scores is drawn and the mean of this sample is determined. What is the probability that this sample mean will differ from the population mean by more than 15 points?

6.14 A simple random sample of size 5 is taken (without replacement) from a population of 10 items. If the standard deviation of the population is 10, what is the standard error of the sample mean?

6.15 (a) Show that for a finite population the standard error of the sample mean equals approximately

$$\frac{\sigma}{\sqrt{n}} \sqrt{1 - \frac{n}{N}}$$

if N is of reasonable size. The quantity, n/N, is often called the *sampling fraction*.

(b) Using the above formula, find the standard error of the sample mean if $\sigma = 2$, $n = 16$, and (i) $N = 100$; (ii) $N = 10,000$.

6.16 A furniture factory produces tables, the mean length of its tables being 28.0 inches, the standard deviation being 0.02 inches.

(a) The factory's manager says that 99.7 percent of the tables produced are between 27.94 and 28.06 inches long. Do you agree? Why or why not?

(b) The factory's foreman says that, if simple random samples of 100 tables are drawn from the factory's output, the mean length of the sample will be between 27.994 and 28.006 inches in about 99.7 percent of the samples. Do you agree? Why or why not? (Assume that each such sample is a very small percentage of the factory's output.)

6.17 There are 3,000 gas stations in a particular region of the United States. Their mean annual sales in 1986 equals $950,000, and the standard deviation of their 1986 sales is $310,000. Random samples of 16 gas stations are taken repeatedly, and the mean 1986 sales of the gas stations in each sample is computed.

(a) What is the mean of the sampling distribution of the sample mean?

(b) What is the standard deviation of the sampling distribution of the sample mean?

(c) Can we be sure that the sampling distribution of the sample mean is normal?

6.18 (a) If the variance of an infinite population is 100, and if a simple random sample of size 64 is drawn from this population, what is the probability that the sample mean will be 40 or more, if the population mean is 38?

(b) If the variance of the population is 10,000 rather than 100, what is the above probability?

(c) Why does the answer to (b) exceed that of (a)?

6.19 (a) A random sample of size 100 is drawn from an infinite population with mean equal to 80 and standard deviation equal to 9. What is the probability that the sample mean is less than 78?

(b) If the sample size is 49 rather than 100, what is the above probability?

(c) Why does the answer to (b) exceed that of (a)?

6.20 Wallingford College's garage has 50 spark plugs. The average life of these spark plugs is 40,000 miles, the standard deviation being 3,000 miles. Eight of these spark plugs will be chosen at random and installed on one of the college's cars.

(a) What is the standard deviation of the sampling distribution of the average life of these eight spark plugs?

(b) If the lives of the 50 spark plugs are (approximately) normally distributed, what is the probability that the average life of these eight spark plugs is less than 38,000 miles?

6.21 A class contains 85 students. The mean and standard deviation of their IQ's are 115 and 10, respectively. A random sample of 50 students is drawn from this class, and the mean IQ of the students in the sample is calculated.

(a) What is the probability that this mean IQ exceeds 120?

(c) If the class had contained twice as many students, how would the answer to (a) have changed? Why?

6.22 A shipment of 2,000 textbooks is delivered to the Baltimore school system, which selects a sample of 20 of the textbooks to inspect. The shipment contains 200 defective textbooks. What is the probability that the proportion defective in the sample will depart from the true proportion defective by more than 0.05?

6.23 A ship with 3,000 refugees arrives in Miami, Florida. In fact, 1,000 of these refugees have relatives in the United States. A random sample of 100 refugees is questioned to determine whether they have relatives in the United States. What is the probability that the proportion of the sample having relatives in the U.S. differs from the population proportion by more than 0.02?

The Chapter in a Nutshell

1. Sample designs can be divided into two broad classes: probability samples and judgment samples. A *probability sample* is one where the probability that each element of the population will be chosen in the sample is known. A *judgment sample* is chosen in such a way that this probability is not known. One of the most important disadvantages of a judgment sample is that one cannot calculate the sampling distribution of a sample statistic and thus there is no way of knowing how big the sampling errors in the sample results are likely to be.

2. If a population contains N elements, a *simple random sample* is one chosen so that each of the different samples of size n that could be chosen has an equal chance of selection. Alternatively, a simple random sample can be defined as one in which the probability of selecting each element in the population is the same and in which the chance that any one element is chosen is independent of the choice of any other element. A table of random numbers can be used in choosing a simple random sample. A *systematic sample* is obtained by taking every kth element of the population on a list. If the elements of the population are in random order, this is tantamount to a simple random sample.

3. One of the most important ways that expert judgment can be used to design a sample is in the construction of strata, or subdivisions of the population. In *stratified random sampling* (where the population is stratified first, and a simple random sample is chosen from each stratum), the resulting estimate is often more precise for a sample of given size than if simple random sampling is used. In formulating strata, one should subdivide the population so that the strata differ greatly with regard to the characteristic being measured, and so that there is as little variation as possible *within* each stratum with regard to this characteristic. To maximize the precision of the sample results, the sample size in each stratum should be proportional to the product of the number of elements in the stratum and the standard deviation of the characteristic being measured in the stratum. In *cluster sampling,* the population is divided into a number of clusters and a sample of these clusters is chosen, after which a simple random sample of the elements in each chosen cluster is selected. The major advantage of cluster sampling is that it is relatively cheap to sample elements that are physically or geographically close together.

4. A *sampling distribution* is the probability distribution of the value of a particular sample statistic. If the sample is a small proportion of the population (or if the sampling is with replacement), the binomial distribution can be used to derive the sampling distribution of the sam-

ple proportion. If the population is normal, the sampling distribution of the sample mean is normal with a mean equal to the population mean and with a standard deviation equal to the population standard deviation divided by the square root of the sample size. Even if the population is not normal, the sampling distribution of the sample mean will have approximately these properties as long as the sample size is larger than about 30. (If the population is at all close to normal, a sample size of much less than 30 often will suffice.)

5. The *standard deviation of the sampling distribution of the sample mean* $\sigma_{\bar{x}}$ is a measure of the extent of the sampling error that is likely to be contained by the sample mean. This is often called the *standard error of the sample mean*. If the population is finite, it equals the value of $\sigma_{\bar{x}}$ for an infinite population multiplied by $\sqrt{(N-n)/(N-1)}$; this multiplier is called the *finite population correction factor*. If the sample is a small percentage of the population, this multiplier is so close to 1 that it can be ignored. Thus, if the sample is small relative to the population, the accuracy of the sample mean depends entirely on the sample size, not on the fraction of the population included in the sample.

Chapter Review Exercises

6.24 In 1948, the Gallup Poll predicted that Harry Truman would get 44 percent of the vote and that Thomas Dewey, his Republican opponent for president, would get 50 percent of the vote. These results were based on a quota sample of 50,000 potential voters.
(a) How close were these results? (Look up the answer in the *World Almanac* or some similar source.)
(b) In the presidential elections of 1936, 1940, and 1944, the Gallup Poll always overestimated the Republican percentage of the vote. Can you give some reasons for this?
(c) Since 1948, the Gallup Poll has used probability samples in its estimates of the outcome of presidential elections. Its estimate of the percent of votes going to the winner has been in error, on the average, by about 2 percentage points since 1948. In the 1970's, its sample size was under 4,000. How can you explain its better performance in recent years than in 1948, even though its sample size has decreased greatly?

6.25 A polling organization asks a sample of voters which of two presidential candidates they favor. Do you think that the order in which the names of the candidates are presented will affect the results? If so, do you think that the candidate named first would be at an advantage or disadvantage?

6.26 The Uphill Manufacturing Company is interested in determining how consumers in a particular town rate the performance of various makes of bicycles. It hires a polling organization to call every hundredth number listed in the telephone directory and ask whoever answers to rate the comparative performance of various makes.

(a) Is this a random sample? Why, or why not?

(b) What pitfalls and disadvantages do you see in the design of this sample?

(c) The polling organization carries out this survey. The report of its findings indicates that the telephone calls were made between 10 A.M. and 2 P.M. on Mondays, Tuesdays, and Wednesdays; if there was no answer at a particular telephone number, that number was dropped from the sample. Give several reasons why this survey might yield distorted results.

6.27 The Department of Agriculture wants to estimate the total acres of corn planted in a certain county of 3,000 farms where corn is the major crop.

(a) If it can obtain an alphabetical list of all the names of farm owners in the county, describe how the Department of Agriculture can draw a random sample of 100 farms. (Assume each farm has a single owner.)

(b) The Department of Agriculture decides to pick a number from 1 to 30 at random, and the number turns out to be 16. It then picks the 16th, 46th, 76th, 106th, ⋯ farms on the list. Is this a random sample? If the number of acres of a farm planted with corn is not related in any way to the name of its owner, will a sample of this sort be essentially equivalent to a simple random sample?

(c) A statistician working for a local university criticizes the Department of Agriculture's survey design. The statistician says that more precise results could be obtained by stratifying the sample according to last year's farm size. Do you agree? Why, or why not?

(d) Available data indicate that the number of farms, classified by last year's size, is as shown in the middle column below:

Farm size last year (acres)	Number of farms	Standard deviation of number of acres of corn
0–50	1,000	10
51–100	500	10
101–150	500	10
151–200	400	15
201–250	400	15
Over 250	200	50
Total	3,000	

If proportional allocation is used, and if the total sample size is 100, what will be the number of farms chosen at random from each of the six strata?

(e) The right-hand column in the table in (d) shows the estimated standard deviation of the number of acres of corn among the farms in each stratum. (For example, among farms with a total acreage last year of 50 or less, the standard deviation of the number of acres

planted with corn this year is estimated to be 10.) If optimum allocation is used, and if the total sample size is 100, what will be the
number of farms chosen at random in each of the six strata?

(f) Still another statistician takes issue with the stratified sample proposed by the statistician in (c). The second statistician suggests that
the list of farms in alphabetical order by owner's names be used to
construct a cluster sample. All farms owned by people with names beginning with *A* would make up one cluster, all farms owned by people
with names beginning with *B* would constitute a second cluster, and
so on. The statistician suggests that a random sample of five clusters
be chosen and that 20 farms be chosen at random in each cluster. If
the principal cost of the survey is the expense of driving from one
farm to another to obtain information concerning the number of
acres planted with corn, is this suggestion likely to reduce the survey's
costs? Why, or why not?

6.28 Martingale University draws a simple random sample of its bills to students to determine what proportion of bills contain numerical errors. In
fact, 10 percent of all such bills contain numerical errors.

(a) If there are 10,000 such bills, and if the sample contains 100 bills,
what is the probability that the proportion of bills in the sample containing numerical errors exceeds 16 percent?

(b) How large must the sample be if the standard deviation of the sample
proportion equals .06?

6.29 If the population is roughly normal, the *standard deviation of the sample
median* equals $\sqrt{\pi/2} \cdot \sigma/\sqrt{n}$. If a sample median is to have a standard deviation equal to that of a sample mean (from the same population), how
much bigger must the sample size be?

6.30 A statistician calculates the standard error of a sample mean (from an infinite population) on the basis of incorrect information concerning the
sample size. His result is double what it should be. He originally thought
that the sample size was 100. What was the true sample size?

6.31 The Energetic Corporation is a producer of electric light bulbs. The
company's production process does not result in precisely similar bulbs.
Instead, bulb length is normally distributed, with a mean of 3.00 inches
and a standard deviation of .10 inches.

(a) What is the sampling distribution of the mean length of a simple random sample of four bulbs? That is, what is the mean of this sampling
distribution? What is its standard deviation? What is its shape?

(b) If the population is normal, the distribution of the sample mean is
normal, regardless of how small the sample may be. Using this fact,
determine the probability that the sample mean in (a) will exceed 3.01
inches.

(c) In (a), how large a sample must the firm take to make the standard
deviation of the sample mean equal .01 inches?

6.32 A librarian wants to estimate the mean number of books on anthropology
in the nation's 100 largest libraries in 1986. A random sample of 25 of

these libraries is drawn (without replacement). If the actual mean number of anthropology books in these 100 libraries was 50,000 and the standard deviation was 5,000, what is the expected value of the sample mean? What is the standard deviation of the sample mean?

7

OBJECTIVES

One of the principal aims of statistical investigations is to make estimates. The overall purpose of this chapter is to describe how statisticians go about doing this. Among the specific objectives are:

1   To indicate how point estimates can be calculated for the population mean, proportion, and standard deviation.

2   To indicate how confidence intervals can be calculated for the population mean and proportion (as well as for the difference between two means or proportions).

3   To show how you can determine how large a sample must be in order for an estimate (of a mean or proportion) to have a desired level of accuracy.

Statistical Estimation

7.1 Introduction

Having taken up the necessary aspects of probability theory and sampling techniques in previous chapters, we can now go on to statistical inference, the branch of statistics that shows how rational decisions can be made on the basis of sample information. Statistical inference deals with two types of problems: *estimation* and *hypothesis testing*. This chapter covers estimation, and the next two chapters deal with hypothesis testing. Since statistical estimation is of enormous importance, we will provide a rather detailed discussion of various estimation techniques and how they can be used.

7.2 Point Estimates and Interval Estimates

Many statistical investigations are carried out in order to estimate a parameter of some population. Why? Because a decision maker's proper course of action often depends on the value of a particular parameter. For example, studies have been conducted in which ninth graders have been taught algebra using a programmed learning text with no formal lectures, the purpose being to determine the mean grade of the students on standardized tests designed to measure their comprehension of algebra. School administrators (and others) are interested in the mean grade of these students because it helps to indicate how effective this sort of educational procedure is.

In estimating a particular parameter, the decision maker uses a statistic calculated from a sample. In the studies of the effects of using a programmed learning text with no formal lectures, the sample mean was used as an estimate of the population mean. A statistic which is used to estimate a parameter is an **estimator.** An **estimate** is the *numerical value* of the estimator that is used. It is important to distinguish between an *estimator* and an *estimate.* For example, if a sample mean is used to estimate a population mean, and if the sample mean equals 30, the estimator used is the sample mean, whereas the estimate is 30.

Statisticians differentiate between two broad classes of estimates: point estimates and interval estimates. A *point estimate is a single number.* For example, if a psychologist estimates that 39 percent of the 7-year-old children in a particular area spend over 4 hours per day watching television, 39 percent is a point estimate. An *interval estimate, on the other hand, is a range of values within which the parameter is thought to lie.* Thus, if the above psychologist estimates that between 31 and 47 percent of the 7-year-old children in a particular area spend over 4 hours per day watching television, 31 to 47 percent is an interval estimate.

In some circumstances, only a point estimate is required. The estimate may be used in a complex series of computations, and the users may want only a single number. More often, however, an interval estimate is preferable to a point estimate because the former indicates how much error is likely to be in the estimate. For example, consider the point estimate that 39 percent of the 7-year-old children in a particular area spend over 4 hours per day watching television. Such an estimate provides no idea of how much error it is likely to contain. That is, it is impossible to tell whether the data indicate that the proportion is likely to be very close to 39 percent, or whether it may well depart considerably from 39 percent. The advantage of interval estimates is that they provide such information. The decision maker can construct an interval estimate so that he or she has a specified amount of confidence that it will include the desired parameter. For example, in constructing his interval estimate of the percent of children spending over 4 hours per day watching TV, the psychologist in question can establish a .95 probability that such an interval will include the population percentage.

7.3 Point Estimation

There are a variety of estimators that can be used to make a point estimate of a particular parameter of a given population. For example, if we want to estimate the mean of a certain population, we can use either the mean or the median of a sample drawn from this population. Which should we use? Obviously, we want to choose the one that will be closer to the population mean, but there is no way of knowing which is closer because we do not know the value of the population mean. All we can do is compare the sampling distribution of the sample mean with that of the sample median. Such a comparison will show which of these estimators is more likely to depart considerably from the population mean. In choosing among estimators, statisticians consider the following three criteria: unbiasedness, efficiency, and consistency. Each of these is described below.

UNBIASEDNESS

As indicated in previous chapters, the expected value of a statistic's sampling distribution is a measure of central tendency which shows the statistic's long-term mean value. If a statistic's sampling distribution is as shown in Figure 7.1, and if this statistic is used as an estimator of θ, this statistic clearly is not a very reliable estimator. Why? Because the expected value of this estimator is 3θ, which means that if such an estimator were used repeatedly, the average estimate would be about three times the parameter we wish to estimate. To avoid estimators of this sort, statisticians use unbiasedness as one criterion.

UNBIASEDNESS. *An **unbiased** estimator is a statistic the expected value of which equals the parameter being estimated.*

To illustrate how this definition can be applied, the sample mean is an unbiased estimator of the population mean because, as we saw in the

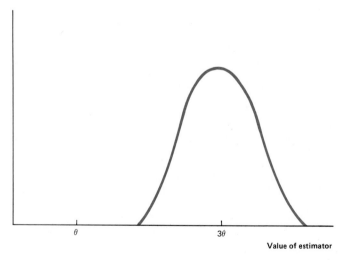

Figure 7.1
Sampling Distribution of a Biased Estimator of θ

θ 3θ

Value of estimator

previous chapter, the mean of its sampling distribution equals the population mean.

The concept of bias used here is somewhat different from that discussed in Chapter 1 where bias was described as a systematic, persistent sort of error due to faulty selection of a sample. Even if a sample is a properly chosen random sample, a bias can result if the estimator does not, on the average, equal the parameter being estimated. For example, had we defined the sample variance as

$$\sum_{i=1}^{n} \frac{(x_i - \bar{x})^2}{n},$$

it would have been a biased estimator of the population variance, σ^2. As shown in panel A of Figure 7.2, its expected value would have been $[(n-1)/n]\sigma^2$, not σ^2. It is for this reason (as we pointed out in Chapter 2) that we define the sample variance with $(n-1)$, not n, in the denominator. If $(n-1)$ is the denominator, the sample variance is an unbiased estimate of σ^2, as shown in panel B of Figure 7.2.

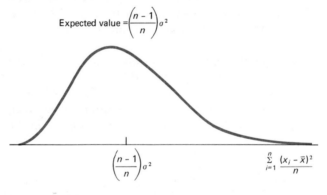

A: Sampling distribution of $\sum_{i=1}^{n} \frac{(x_i - \bar{x})^2}{n}$

Expected value $= \left(\dfrac{n-1}{n}\right)\sigma^2$

$\left(\dfrac{n-1}{n}\right)\sigma^2$ $\sum_{i=1}^{n} \frac{(x_i - \bar{x})^2}{n}$

Figure 7.2
Sampling Distributions
of $\displaystyle\sum_{i=1}^{n} \frac{(x_i - \bar{x})^2}{n}$ and

$\displaystyle\sum_{i=1}^{n} \frac{(x_i - \bar{x})^2}{n-1}$

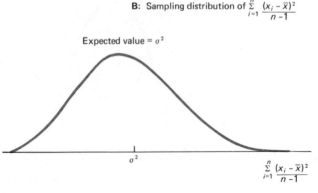

B: Sampling distribution of $\sum_{i=1}^{n} \frac{(x_i - \bar{x})^2}{n-1}$

Expected value $= \sigma^2$

σ^2 $\sum_{i=1}^{n} \frac{(x_i - \bar{x})^2}{n-1}$

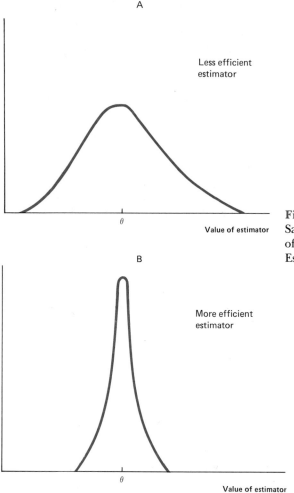

Figure 7.3
Sampling Distributions
of Two Unbiased
Estimators of θ

EFFICIENCY

The unbiasedness of an estimator does not necessarily mean that the estimator is likely to be close to the parameter we want to estimate. For example, the estimator whose sampling distribution is shown in panel A of Figure 7.3 has a high probability of differing considerably from its mean, which (since it is unbiased) is the parameter we want to estimate. This is because its sampling distribution contains a great deal of dispersion or variability. On the other hand, the estimator whose sampling distribution is shown in panel B of Figure 7.3 has a low probability of differing considerably from its mean because its sampling distribution exhibits little dispersion or variability. Statisticians would say that the estimator in panel B is "more efficient" than the one in panel A, since its sampling distribution is concentrated more tightly about the parameter we want to estimate. Efficiency, defined as follows, is an important criterion used by statisticians to choose among estimators.

> EFFICIENCY. *If two estimators are unbiased, one is more **efficient** than the other if its variance is less than the variance of the other.*[1]

To illustrate, let's go back to the question of whether the sample mean or the sample median should be used to estimate the population mean. As we know from Chapter 6, the variance of the sample mean is $\sigma^2 \div n$; and it can be shown that if the population is normal, the variance of the sample median is approximately $1.57\sigma^2 \div n$ if the sample is large. (See Exercise 6.29.) Thus, the sample mean is *more efficient* than the sample median because the sample median's variance is 57 percent greater than that of the sample mean. This implies that the sample median is more likely than the sample mean to differ considerably from the population mean. (In other words, holding constant the sample size, the sample mean is more likely than the sample median to be close to the population mean.) Because it is more efficient, the sample mean is generally preferred over the sample median as an estimator of the population mean.

CONSISTENCY

Still another criterion used for choosing among estimators is consistency. Some estimators are consistent, while others are not. The statistical definition of consistency is as follows.

> CONSISTENCY. *A statistic is a **consistent** estimator of a parameter if the probability that the statistic's value is very near the parameter's value increasingly approaches 1 as the sample size increases.*

In other words, if a statistic is a consistent estimator of a particular parameter, the statistic's probability distribution becomes increasingly concentrated on this parameter as the sample size increases.

It is desirable for an estimator to be consistent because this means that the estimator becomes more reliable as the sample size increases. For example, consider the sample mean, which is a consistent estimator

---

[1] For biased estimators, one estimator is more efficient than another if its mean-square error is less than that of the other. If X is an estimator of θ, its mean-square error is the expected value of its squared deviation from θ. That is, it equals

$$E[(X - \theta)^2].$$

of the population mean. Since the standard deviation of the sample mean's sampling distribution equals $\sigma/\sqrt{n}$, it clearly tends to zero as n increases in value. This means that the probability distribution of the sample mean becomes concentrated ever more tightly about the population mean. If the population is normal, the sample median is also a consistent estimator; thus, both the sample mean and the sample median satisfy this criterion.

7.4 Point Estimates for μ, σ, and Π

We have just described the criteria that statisticians frequently use to choose among estimators. Based on these criteria, certain statistics are generally preferred over others as estimators of the population mean (μ), population variance (σ^2), population standard deviation (σ), or population proportion (Π). This does not mean that other estimators are not preferred under special circumstances or that it is incorrect to use other estimators. It does mean that based on the criteria given in the previous section, the following estimators are the standard ones used by statisticians to estimate these parameters.

SAMPLE MEAN. This is the most common estimator of the population mean. As we know, it is unbiased and consistent. Moreover, it can be shown that if the population is normal the sample mean is the most efficient unbiased estimator available. For these reasons the sample mean is generally the preferred estimator of the population mean.

SAMPLE VARIANCE AND STANDARD DEVIATION. The sample variance is an unbiased and consistent estimator of the population variance. It is relatively efficient as compared with other estimators. Its square root, the sample standard deviation, is generally used as an estimator of the population standard deviation even though it is not unbiased. The sample standard deviation is also relatively efficient.

SAMPLE PROPORTION. This is an unbiased, consistent, and relatively efficient estimator of the population proportion. For these reasons, it is generally the preferred estimator of the population proportion.

EXERCISES

7.1 Does each of the following statements pertain to an estimate or an estimator? If the statement pertains to an estimate, is it a point estimate or an interval estimate?

(a) The fire chief in a particular community says that 80 percent of the houses in the community are within 3 miles of the nearest fire department.

(b) The U.S. Bureau of the Census prefers the median to the mean as a measure of central tendency for the distribution of persons by number of years of schooling completed.

(c) A physician who delivers an 8 lb. baby boy estimates that, when he is 1 year old, the child will weigh between 20 and 25 lbs.

7.2 "The sample standard deviation is biased; thus, it should not be used, except under exceptional circumstances." Comment and evaluate.

7.3 A test is administered to a random sample of 25 students at a local medical school, the purpose of the test being to determine each student's proficiency in biochemistry. The lowest possible score is 0; the highest is a 6.0. The results are as follows:

4.1	4.6	4.6	4.6	5.1
4.3	4.7	4.6	4.8	4.8
4.5	4.2	5.0	4.4	4.7
4.7	4.1	3.8	4.2	4.6
3.9	4.0	4.4	4.0	4.5

The medical school is interested in estimating the mean test score for all of the students. Provide a point estimate of this population parameter.

7.4 Is the sample proportion a consistent estimator of the population proportion? Why, or why not?

7.5 A public-interest law firm picks a random sample of 60 hi-fi stores in a particular area, and asks each of them to repair a hi-fi set. In each case the law firm determines whether the store makes unnecessary repairs in order to inflate its bill. The law firm finds that eight of the stores are guilty of this practice.

(a) Provide a point estimate of the proportion of all such stores in the area that inflate bills in this way.

(b) After the law firm has presented the results of its study, an attorney representing the hi-fi repair stores objects that a sample of this size is quite unreliable. The attorney also maintains that the sample percentage of hi-fi repair stores engaging in such shady practices is a biased estimate of the percentage of all such stores engaging in such practices. Evaluate the attorney's objections.

7.6 A statistic's mean-square error can be used as a measure of its reliability as an estimator. If X is a statistic that is used as an estimator of θ, X's mean-square error is

$$E[(X - \theta)^2].$$

(a) If X is an unbiased estimator of θ, what is another name for its mean-square error?

(b) It can be shown that an estimator's mean-square error equals

$$(\mu - \theta)^2 + \sigma^2,$$

where μ is its mean, θ is the parameter to be estimated, and σ is its standard deviation. Explain why $(\mu - \theta)$ is often called its bias.

(c) Based on the formula in (b), explain why a biased estimator may be preferred to an unbiased one if the former has a much smaller variance than the latter.

7.5 Confidence Intervals for the Population Mean

Interval estimates are generally preferred over point estimates because the latter provide no information concerning how much error they are likely to contain. Interval estimates, on the other hand, do provide such information. To illustrate an interval estimate and how it is constructed, in this section we show how such an estimate is made of the population mean. We begin with the case where the population standard deviation is known and the sample size is large ($n > 30$). Then we take up the more realistic case where the standard deviation is unknown, both when the sample size is large and when it is small.

WHERE σ IS KNOWN: LARGE SAMPLE

In Chapter 5 we defined $Pr\{X > 32\}$ as the probability that X is greater than 32. Now we define $Pr\{a < X < b\}$ as the probability that X lies between a and b. Thus, the probability that the value of the sample mean lies between $\mu - 1.96\sigma/\sqrt{n}$ and $\mu + 1.96\sigma/\sqrt{n}$ is denoted by $Pr\{\mu - 1.96\sigma/\sqrt{n} < \overline{X} < \mu + 1.96\sigma/\sqrt{n}\}$. To construct an interval estimate of the population mean, we begin by noting that our results from the previous chapter concernng the sampling distribution of the sample mean imply that

$$Pr\left\{\mu - 1.96\frac{\sigma}{\sqrt{n}} < \overline{X} < \mu + 1.96\frac{\sigma}{\sqrt{n}}\right\} = 0.95, \tag{7.1}$$

where μ is the population mean, σ is the population standard deviation, n is the sample size, and $\overline{X}$ is the sample mean. What equation (7.1) says is that the probability that the sample mean will lie within 1.96 standard errors of the population mean equals 0.95. (Recall that the standard error of the sample mean equals $\sigma/\sqrt{n}$.) Since we know from the previous chapter that if $n > 30$ (and if the population is large relative to the sample size), the sample mean is normally distributed with a mean of μ and a standard deviation of $\sigma/\sqrt{n}$, and since we know (from Appendix Table 2) that the probability that any normal random variable will lie within 1.96 standard deviations of its mean is 0.95, it follows that equation (7.1) is true.[2]

To construct an interval estimate for the population mean, we rearrange the terms inside the brackets on the left side of equation (7.1). If we subtract μ from $\mu - 1.96\sigma/\sqrt{n}$, $\overline{X}$, and $\mu + 1.96\sigma/\sqrt{n}$, we get

$$Pr\left\{-1.96\,\frac{\sigma}{\sqrt{n}} < \overline{X} - \mu < 1.96\,\frac{\sigma}{\sqrt{n}}\right\} = .95$$

And if we subtract $\overline{X}$ from all three terms of the inequality in brackets and multiply each of the resulting terms by -1, we get

$$Pr\left\{\overline{X} - 1.96\,\frac{\sigma}{\sqrt{n}} < \mu < \overline{X} + 1.96\,\frac{\sigma}{\sqrt{n}}\right\} = .95. \qquad (7.2)$$

(See footnote 3 if you have difficulty deriving equation (7.2).)[3]

Since equation (7.2) is of basic importance in the construction of an interval estimate for a population mean, it is essential that you know

---

[2] How do we know (from Appendix Table 2) that the probability that any normal variable will be within 1.96 standard deviations of its mean is 0.95? The reasoning is as follows. We want to determine the probability that a normal variable lies between $\mu - 1.96\sigma$ and $\mu + 1.96\sigma$. The points on the standard normal curve corresponding to these points are $[(\mu - 1.96\sigma) - \mu] \div \sigma$ and $[(\mu + 1.96\sigma) - \mu] \div \sigma$. That is, they are -1.96 and $+1.96$. Appendix Table 2 shows that the area under the standard normal curve between zero and 1.96 equals .4750. Thus, the area between -1.96 and $+1.96$ equals $2(.4750) = .95$.

[3] If we subtract $\overline{X}$ from all three terms, we get

$$Pr\left\{-\overline{X} - 1.96\,\frac{\sigma}{\sqrt{n}} < -\mu < -\overline{X} + 1.96\,\frac{\sigma}{\sqrt{n}}\right\} = .95.$$

The important fact to note at this point is that if all three terms inside the brackets are multiplied by -1 the inequalities must be reversed. (To see that this is true, consider the simple inequality $y \geqslant 1$. If this is true, it follows that $-y \leqslant -1$. As we indicated, the inequality is reversed after multiplication by -1.) Applying this fact, it follows that

$$Pr\left\{\overline{X} - 1.96\,\frac{\sigma}{\sqrt{n}} < \mu < \overline{X} + 1.96\,\frac{\sigma}{\sqrt{n}}\right\} = .95,$$

which is what we set out to derive.

exactly what it means. Does it mean that the probability is .95 that the population mean lies between $\overline{X} - 1.96\sigma/\sqrt{n}$ and $\overline{X} + 1.96\sigma/\sqrt{n}$? In one sense yes, in another no. Before the sample has been drawn and while the value of the sample mean is still unknown, there is a .95 probability that the interval between $\overline{X} - 1.96\sigma/\sqrt{n}$ and $\overline{X} + 1.96\sigma/\sqrt{n}$ will include the population mean. Thus, if we were to draw one sample after another, in 95 percent of the samples (over the long run) this interval—which varies, of course, from sample to sample because $\overline{X}$ varies—would include the population mean.

To see that this is true, look at Figure 7.4, which shows what would occur if we were to construct a large number of interval estimates for a certain population mean μ. Panel A of Figure 7.4 shows the sampling distribution of the sample mean, $\overline{X}$. If the sample mean lies between $\mu - 1.96\sigma/\sqrt{n}$ and $\mu + 1.96\sigma/\sqrt{n}$, the interval estimate will

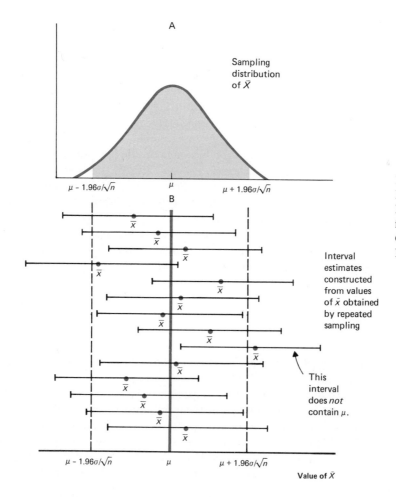

Figure 7.4
Display of Interval Estimates Constructed from Values of $\overline{X}$ Obtained from Repeated Sampling

contain μ. Since the area under the sampling distribution between $\mu - 1.96\sigma/\sqrt{n}$ and $\mu + 1.96\sigma/\sqrt{n}$ equals .95, the probability that $\overline{X}$ will lie in this range equals .95. Panel B of Figure 7.4 shows the interval estimates that would result from repeated sampling. Over the long run, 95 percent of these intervals would contain μ.

Note, however, that the previous discussion has pertained only to the situation before the sample has been drawn. Once a particular sample has been drawn and the sample mean has been calculated, it no longer is correct to say that the probability is 0.95 that the interval between $\overline{x} - 1.96\sigma/\sqrt{n}$ and $\overline{x} + 1.96\sigma/\sqrt{n}$ includes the population mean. To see why, suppose that $\sigma = 10$, $n = 100$, and $\overline{x} = 2.00$. Under these circumstances, equation (7.2) may be incorrectly interpreted to say that

$$Pr\left\{2.00 - 1.96\,\frac{10}{\sqrt{100}} < \mu < 2.00 + 1.96\,\frac{10}{\sqrt{100}}\right\} = .95,$$

or

$$Pr\{.04 < \mu < 3.96\} = .95.$$

But since the population mean μ is a constant, it makes no sense to say that the probability that it is between .04 and 3.96 is .95. Either its value lies in this interval or it doesn't. *All that one can say is that if intervals of this sort are calculated repeatedly, they will include the population mean in about 95 percent of the cases.*

The interval within the brackets on the left side of equation (7.2) is called a **confidence interval.** As you can see, this is an interval which has a certain probability of including the population mean, this probability being called the **confidence coefficient.** Thus, the confidence coefficient in equation (7.2) is .95. Although .95 is frequently used as a confidence coefficient, there is no reason why other confidence coefficients should not be chosen. In general, *if the confidence coefficient is set equal to* $(1 - \alpha)$, *the confidence interval for the population mean is*

Confidence Interval for μ (σ Known)

$$\overline{x} - z_{\alpha/2}\,\frac{\sigma}{\sqrt{n}} < \mu < \overline{x} + z_{\alpha/2}\,\frac{\sigma}{\sqrt{n}}, \qquad (7.3)$$

where $z_{\alpha/2}$ is the value of the standard normal variable that is exceeded with a probability of $\alpha/2$.

To illustrate the use of expression (7.3), suppose that we want to construct a 90 percent confidence interval rather than a 95 percent confidence interval in the case presented in the paragraph before last. (Note that confidence coefficients are often expressed as percentages.) Since $1 - \alpha = .90$, it follows that $\alpha/2 = .05$. From Appendix Table 2, we find that $z_{.05} = 1.64$. Thus, inserting $z_{.05}$, $\overline{x}$, σ, and n into expression

(7.3), we get

$$2.00 - 1.64 \frac{10}{\sqrt{100}} < \mu < 2.00 + 1.64 \frac{10}{\sqrt{100}},$$

which implies that the 90 percent confidence interval for the population mean is .36 to 3.64.

Two important points should be noted here. First, *holding constant the size of the sample, the width of the confidence interval tends to increase as the confidence coefficient increases.* For example, if the confidence coefficient is 90 percent, the width of the confidence interval is $2(1.64)\sigma/\sqrt{n}$, whereas if the confidence interval is 95 percent, the width[4] of the confidence interval is $2(1.96)\sigma/\sqrt{n}$. This makes sense. If you want to be more and more confident that the interval estimate includes the population mean, you must widen the interval if the sample size is fixed. Second, *for a fixed confidence coefficient (and a fixed population standard deviation), the only way to reduce the width of the confidence interval is to increase the sample size.* For example, if the confidence coefficient is 95 percent, the width of the confidence interval is $2(1.96)\sigma/\sqrt{n}$, as we saw above. If σ is fixed, the only way to reduce this width is to increase n.

WHERE σ IS UNKNOWN: LARGE SAMPLE

In most actual cases, the standard deviation of the population is unknown. If the sample size exceeds 30 it is a relatively simple matter to adapt the results presented previously to the situation where σ is unknown. As indicated earlier in this chapter, the sample standard deviation s is generally used as an estimator of the population standard deviation. Mathematicians have shown that if the sample is large, we can simply substitute the sample standard deviation for the population standard deviation in the results obtained in the previous part of this section. Thus, if we want to construct a 95 percent confidence interval —that is, a confidence interval with a confidence coefficient of 95 percent—we can substitute s for σ in equation (7.2), the result being

$$Pr\left\{\overline{X} - 1.96 \frac{s}{\sqrt{n}} < \mu < \overline{X} + 1.96 \frac{s}{\sqrt{n}}\right\} = .95.$$

Consequently, the interval estimate is from $\overline{x} - 1.96s/\sqrt{n}$ to $\overline{x} + 1.96s/\sqrt{n}$.

---

[4] The 95 percent confidence interval is $\overline{x} - 1.96\sigma/\sqrt{n}$ to $\overline{x} + 1.96\sigma/\sqrt{n}$; thus, the difference between its upper and lower limits (that is, its width) is $2[1.96\sigma/\sqrt{n}]$. Similarly, since the 90 percent confidence interval is $\overline{x} - 1.64\sigma/\sqrt{n}$ to $\overline{x} + 1.64\sigma/\sqrt{n}$, the difference between its upper and lower limits (that is, its width) is $2[1.64\sigma/\sqrt{n}]$.

In general, *if the confidence coefficient is set equal to* $(1 - \alpha)$, *the confidence interval for the population mean is*

Confidence Interval for μ (σ Unknown and n > 30)

$$\bar{x} - z_{\alpha/2} \frac{s}{\sqrt{n}} < \mu < \bar{x} + z_{\alpha/2} \frac{s}{\sqrt{n}}, \qquad (7.4)$$

where s *is the sample standard deviation, and* $z_{\alpha/2}$ *is the value of the standard normal variable that is exceeded with a probability of* $\alpha/2$. Like expression (7.3), expression (7.4) is applicable only if the population is large relative to the sample or if sampling is with replacement.[5] As we saw in the previous chapter, if these two conditions are both not true the finite population correction factor must be included. Thus, *if sampling is without replacement, and if the population is not large,* the confidence interval for the population mean is

Confidence Interval for μ (σ Unknown and n > 30) when Population is Small Relative to Sample and Sampling Is Without Replacement

$$\bar{x} - z_{\alpha/2} \frac{s}{\sqrt{n}} \sqrt{\frac{N-n}{N-1}} < \mu < \bar{x} + z_{\alpha/2} \frac{s}{\sqrt{n}} \sqrt{\frac{N-n}{N-1}}, \qquad (7.5)$$

where N is the number of items in the population.

These results concerning the confidence interval for a population mean are of enormous importance. The following example based on a real case[6] (although the numbers have been changed) should help illustrate how these results are used and interpreted.

---

[5] If the population is not large relative to the sample and if sampling is without replacement, $\sigma/\sqrt{n}$ in (7.3) should be multiplied by the finite population correction factor $\sqrt{(N-n)/(N-1)}$.

[6] Owen L. Davies, *The Design and Analysis of Industrial Experiments* (London: Oliver and Boyd, 1956), p. 72. The actual analysis was more extensive and complicated than our discussion in this section, which is simplified to emphasize the basic points considered here.

EXAMPLE 7.1 A chemist wants to estimate the mean strength of a new synthetic fiber. To measure this fiber's strength, she determines the number of pounds that can be supported by one strand before breaking. A random sample of 36 strands of the fiber is taken, with the following results:

Strand	Breaking load (pounds)	Strand	Breaking load (pounds)	Strand	Breaking load (pounds)
1	2.2	13	2.2	25	2.3
2	2.2	14	2.3	26	2.4
3	2.2	15	2.3	27	2.3
4	2.3	16	2.3	28	2.4
5	2.3	17	2.2	29	2.4
6	2.3	18	2.2	30	2.3
7	2.3	19	2.2	31	2.3
8	2.3	20	2.2	32	2.3
9	2.4	21	2.2	33	2.4
10	2.4	22	2.4	34	2.3
11	2.4	23	2.3	35	2.4
12	2.2	24	2.4	36	2.3

Construct a 95 percent confidence interval for the mean breaking load of a strand of this new fiber.

SOLUTION: If x_i is the breaking load (in pounds) of the ith strand in the sample, we find that

$$\sum_{i=1}^{36} x_i = 82.8$$

$$\bar{x} = 82.8/36 = 2.3$$

$$\sum_{i=1}^{36} (x_i - \bar{x})^2 = \sum_{i=1}^{36} x_i^2 - \frac{1}{36} \left(\sum_{i=1}^{36} x_i \right)^2 = 0.20$$

$$\sum_{i=1}^{36} (x_i - \bar{x})^2/(n-1) = 0.20/35 = .00571$$

$$s = \sqrt{.00571} = .0756.$$

Since the population is very large (because a very large number of such strands of fiber can be produced), expression (7.4) is appropriate. Because a 95 percent confidence interval is wanted, $z_{\alpha/2} = z_{.025} = 1.96$. Thus, the desired confidence interval is

$$2.30 - 1.96 \left(\frac{.0756}{\sqrt{36}} \right) < \mu < 2.30 + 1.96 \left(\frac{.0756}{\sqrt{36}} \right).$$

Simplifying terms, this confidence interval is 2.275 to 2.325 pounds.

As stressed before, the proper interpretation of this result is *not* that the probability is 95 percent that the population mean lies between 2.275 and 2.325 pounds. Instead, this result means that if confidence intervals of this sort were constructed in a great number of cases they would include the population mean 95 percent of the time.

**GETTING DOWN
TO CASES**

THE EFFECT OF A NEW ENZYME ON A PHARMACEUTICAL PROCESS[7]

An engineer was attempting to estimate the extent to which a new enzyme altered the yield of a certain pharmaceutical process. For a given batch of product, the yield was measured by the ratio of the actual output to the theoretical output as calculated from formulas based on past experience. Thus a yield of 1.03 meant that 3 percent more output was gotten from the batch than the formula indicated; 0.98 meant that 2 percent less output was gotten from the batch than the formula indicated.

The engineer tested the new enzyme on 36 batches and obtained the following yields:

1.28	1.31	1.48	1.10	0.99	1.22
1.65	1.40	0.95	1.25	1.32	1.23
1.43	1.24	1.73	1.35	1.31	0.92
1.10	1.05	1.39	1.16	1.19	1.41
0.98	0.82	1.22	0.91	1.26	1.32
1.71	1.29	1.17	1.74	1.51	1.25

(a) Calculate a 90 percent confidence interval for the true mean yield gotten from the enzyme.

(b) Calculate a 95 percent confidence interval for the true mean yield gotten from the enzyme.

(c) Calculate a 99 percent confidence interval for the true mean yield gotten from the enzyme.

(d) Specify the assumptions underlying your results in (a), (b), and (c).

(e) Provide an unbiased point estimate of the true mean yield from the enzyme.

(f) If the yields are normally distributed is the estimator you used in (e) at least as efficient as any other unbiased estimator?

(g) If the assumptions specified in (d) are correct, would you conclude that the true mean yield gotten from the enzyme almost certainly exceeds 1.00? Explain.

(h) What is the mistake (if any) in the following argument, put forth by another analyst: "There is no good theoretical reason for believing that the mean yield should be higher than 1.00 with the enzyme. Moreover, in the 36 batches studied, the stan-

---

[7] This case is based on a section from W. Allen Wallis and Harry V. Roberts, *Statistics: A New Approach.* Some of the numbers have been changed for expository purposes.

dard deviation of individual yields was .228. In my opinion, .228 represents a large fraction of the difference between [the sample mean] 1.268 and 1.00. There is no real evidence that the enzyme increases yield ⋯ ."[8]

[8] Ibid. Some of the numbers have been changed.

WHERE σ IS UNKNOWN: SMALL SAMPLE

In many cases, a confidence interval must be constructed on the basis of a sample where $n \leq 30$. In such cases, the expressions given in the previous part of this section are not appropriate. No longer can we simply substitute the sample standard deviation for the population standard deviation as we did in expression (7.4). However, if the population is normal it is possible to construct a confidence interval for the population mean even if the sample size is 30 or less. Such a confidence interval is based on the t distribution described below.

THE T DISTRIBUTION. *If the population sampled is normally distributed,* $(\overline{X} - \mu) \div s/\sqrt{n}$ *has the* **t distribution.** *The* t *distribution is symmetrical, bell-shaped and has zero as its mean.*

The t distribution is a sampling distribution: Specifically, it is the distribution of the statistic $(\overline{X} - \mu) \div s/\sqrt{n}$. Suppose that simple random samples of size n are taken repeatedly from a normal population with expected value of μ and standard deviation of σ. If we calculate the value of $(\overline{X} - \mu) \div s/\sqrt{n}$ for each sample, we can construct a sampling distribution for this statistic. If a sufficient number of samples are chosen this sampling distribution will conform to the t distribution. The t distribution is really a family of distributions, each of which corresponds to a particular *number of degrees of freedom*. In this context the number of degrees of freedom equals $(n - 1)$; in other contexts it will equal other amounts, as we shall see in later chapters.

Degrees of Freedom

It is not easy to give an adequate intuitive, nonmathematical interpretation of the number of degrees of freedom. From a mathematical point of view, the number of degrees of freedom is simply a parameter in the formula for the t distribution. However, one way to interpret the number of degrees of freedom (that is, $[n - 1]$) in the present context is to say that it equals the number of independent deviations from the sample mean (that is, $[x_i - \overline{x}]$) in the computation of the sample standard deviation s. Since the sum of these deviations—that is, $\Sigma(x_i - \overline{x})$—equals zero, it follows that if we know $(n - 1)$ of these

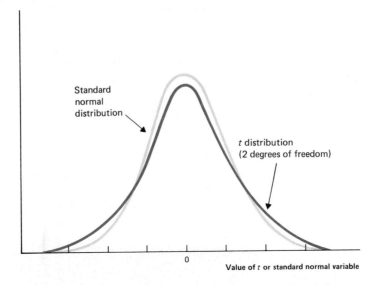

Figure 7.5
The *t* Distribution
(with 2 Degrees of
Freedom) Compared
with the Standard
Normal Distribution

deviations we can determine the value of the remaining deviation by using the fact that their sum equals zero. Thus, only $(n-1)$ of the deviations are independent. In other words, as the statistician might put it, there are $(n-1)$ degrees of freedom.

The shape of the *t* distribution is rather like that of the standard normal distribution. Figure 7.5 compares the *t* distribution (with 2 degrees of freedom) to the standard normal distribution. As you can see, both are symmetrical, bell-shaped, and have a mean of zero. The *t* distribution is somewhat flatter at the mean and somewhat higher in the tails than the standard normal distribution. As the number of degrees of freedom becomes larger and larger, the *t* distribution tends to become exactly the same as the standard normal distribution. The *t* distribution is often called Student's *t* distribution because the statistician W. S. Gosset, who first derived this distribution, published his findings under the pseudonym Student.[9]

To find the probability that the value of *t* exceeds a certain number, we can use Appendix Table 6. As you can see, each row of this table corresponds to a particular number of degrees of freedom. The numbers in each row are the numbers that are exceeded with the indicated probability by a *t* variable. For example, the first row indicates that if a *t* variable has one degree of freedom, there is a .40 probability that its value will exceed .325, a .25 probability that its value will exceed 1.000, a .05 probability that its value will exceed 6.314, a .01

---

[9] Gosset used this pseudonym because his employer, Guinness Brewery, forbade publication of scientific research of this sort by employees under their own names.

probability that its value will exceed 31.821, and so on. Since the t distribution is symmetrical, it follows that if a t variable has one degree of freedom there is a .40 probability that its value will lie below $-.325$, a .25 probability that its value will lie below -1.000, and so on.

If the sample size is 30 or less and the population is normal (and large relative to the sample), a confidence interval for the population mean can be constructed by using the t distribution in place of the standard normal distribution in expression (7.4). In other words, *if the confidence coefficient is set equal to $(1 - \alpha)$, the confidence interval for the population mean is:*

$$\bar{x} - t_{\alpha/2} \frac{s}{\sqrt{n}} < \mu < \bar{x} + t_{\alpha/2} \frac{s}{\sqrt{n}},$$

(7.6)

Confidence Interval for μ (σ Unknown and n $\leq$ 30)

where $t_{\alpha/2}$ *is the value of a* t *variable (with* n $- 1$ *degrees of freedom) that is exceeded with a probability of* $\alpha/2$. Thus, if a sample of 16 observations is chosen, and if the sample mean is 20 and the sample standard deviation is 4, the 95 percent confidence interval for the population mean is

$$20 - 2.131 \left(\frac{4}{\sqrt{16}}\right) < \mu < 20 + 2.131 \left(\frac{4}{\sqrt{16}}\right),$$

since Appendix Table 6 shows that for 15 degrees of freedom the value of t that will be exceeded with a probability of .025 is 2.131. Simplifying terms, it follows that the confidence interval in this case is 17.869 to 22.131. Statisticians frequently construct confidence intervals in this way. The following example illustrates how it is done.

EXAMPLE 7.2 An engineer wants to estimate the mean length of life of a new type of bulb which is designed to be extremely durable. He tests nine of these bulbs and finds that the length of life (in hours) of each is as follows:

5000	5100	5400
5200	5400	5000
5300	5200	5200

Previous experience indicates that the lengths of life of individual bulbs of a particular type are normally distributed. Construct a 90 percent confidence interval for the mean length of life of all bulbs of this new type.

SOLUTION: If x_i is the length of life of the ith light bulb in the sample, we find that

$$\sum_{i=1}^{9} x_i = 46,800$$

$$\bar{x} = 5,200$$

$$\sum_{i=1}^{9} (x_i - \bar{x})^2 = 180,000$$

$$\sum_{i=1}^{9} (x_i - \bar{x})^2/(n - 1) = 22,500$$

$$s = \sqrt{22,500} = 150.$$

Since $n = 9$, expression (7.6) is appropriate. Because a 90 percent confidence interval is wanted, $t_{\alpha/2} = t_{.05}$; and the number of degrees of freedom $(n - 1)$ is 8. Appendix Table 6 shows that if there are 8 degrees of freedom, $t_{.05} = 1.86$. Thus, the desired confidence interval is

$$5200 - 1.86 \left(\frac{150}{\sqrt{9}}\right) < \mu < 5200 + 1.86 \left(\frac{150}{\sqrt{9}}\right).$$

Simplifying terms, the confidence interval is 5107 to 5293 hours.

WHAT COMPUTERS CAN DO

OBTAINING A CONFIDENCE INTERVAL

As pointed out in Chapter 1, computer packages like Minitab are available to carry out many statistical calculations. To illustrate how Minitab can be used to calculate a confidence interval, suppose that one wants to compute a 95 percent confidence interval for a mean (the standard deviation of the population being unknown). The sample is 12, 18, 15, 16, 20, 21.

To calculate such a confidence interval, all that one has to do is to type the following three lines:

```
-SET INTO C2
12 18 15 16 20 21
-TINTERVAL WITH 95 PERCENT CONFIDENCE, DATA IN C2
```

The first line says that the numbers should be put in column C2 of the worksheet that Minitab maintains in the computer. The next line contains the numbers themselves. The third line commands the computer to calculate the desired confidence interval.

After the above three lines are typed, the computer automatically prints out the following:

```
C2    N =    6     MEAN =     17.000     ST.DEV. =     3.35
A 95.00 PERCENT C.I. FOR MU IS (    13.4869,     20.5131)
```

The first line contains the number of observations and the sample mean and standard deviation, while the second line contains the confidence interval. Thus, the 95 percent confidence interval for the mean is 13.4869 to 20.5131. This is the answer we want.

EXERCISES

7.7 What is the probability that the value of a random variable with the t distribution with 4 degrees of freedom will lie
(a) above 3.747;
(b) below −4.604;
(c) between 2.132 and 2.776?

7.8 Compare the t distribution with an infinite number of degrees of freedom in Appendix Table 6 to the standard normal distribution in Appendix Table 2. In particular, show that the probability is the same that each will exceed (a) 1.645; (b) 1.960; (c) 0.674. Do you find this surprising? Why, or why not?

7.9 A hospital's records indicate that, in a simple random sample of 90 cases, the mean fee of the physician for a particular kind of operation was $810, the sample standard deviation being $85. These 90 cases are a very small percentage of the operations of this type carried out in the hospital.
(a) Construct a 90 percent confidence interval for the mean fee of the physician for this type of operation in all such cases in this hospital.
(b) Construct a 95 percent confidence interval of this sort.
(c) Construct a 99 percent confidence interval of this sort.

7.10 It is known that the difference between a person's true weight and his or her weight as indicated by a particular scale is normally distributed with a standard deviation of 0.8 ounces. Thirty-one people weigh themselves on this scale. If we subtract each person's true weight from his or her weight according to this scale, the results are 0.6, 0.4, −0.1, 0.2, 0.5, 0.7, 0.4, −0.2, −0.1, 0.6, 0.8, 0.9,1.7, 1.2, 1.5, 2.0, 1.5, −1.4, −1.3, −1.2, −1.1, −1.8, −1.2, 1.4, 0.9, −0.7, 0.8, 0.5, 1.0, 1.2, and 0.2 ounces.
(a) Construct a 95 percent confidence interval for the mean difference between a person's true weight and his or her weight according to this scale.
(b) Construct a 90 percent confidence interval of this sort.

7.11 A social worker wants to estimate the mean number of years of school completed by the residents of a particular neighborhood. A simple random sample of 90 residents is taken, the mean years of school completed being 8.4 and the sample standard deviation being 1.8. The neighborhood contains about 2,500 residents.

(a) Calculate a 90 percent confidence interval for the mean years of school completed by the residents of the neighborhood.

(b) Calculate a 98 percent confidence interval of this sort.

7.12 The city of Atlantis draws a random sample of ten employees from its labor force of 5,000 people. The number of years that each of these employees has worked for the city is 8.2, 5.6, 4.7, 9.6, 7.8, 9.1, 6.4, 4.2, 9.1, and 5.6. (The number of years an employee has worked for the city is normally distributed.)

(a) Calculate an 80 percent confidence interval for the mean number of years that all the city's employees have worked for the city.

(b) Calculate a 90 percent confidence interval for this mean.

7.13 An oil company wants to determine the mean weight of a can of its oil. It takes a random sample of 80 such cans (from several thousand cans in its warehouse), and finds the mean weight is 31.15 ounces and the standard deviation is 0.08 ounces.

(a) Compute a 95 percent interval for the mean weight of the cans in the firm's warehouse.

(b) Is your answer to (a) based on the assumption that the weights of the cans of oil in the warehouse are normally distributed? Why, or why not?

7.14 A telephone company wants to determine the mean height of its telephone poles. It takes a random sample of 12 poles, and finds that the heights (in feet) are: 35.9, 35.8, 36.2, 36.4, 36.6, 36.2, 35.8, 35.9, 36.0, 36.1, 35.9, 36.0.

(a) Compute a 90 percent confidence interval for the mean height of the firm's telephone poles.

(b) Is your answer to (a) based on the assumption that the heights of the company's telephone poles are normally distributed? Why, or why not?

7.15 There are 60 apartments in a San Diego apartment building. The owner of the building wants to estimate the mean number of people living in an apartment. He draws a random sample (without replacement) of 36 apartments in the building. The number of people living in each apartment is as follows:

1	2	1	1	2	1	2	3	1
2	2	1	2	2	1	1	2	3
2	3	1	2	3	2	2	3	2
3	2	2	3	1	1	2	2	1

(a) Compute the desired 95 percent confidence interval.

(b) Is your answer to (a) based on the assumption that the numbers of people in the apartments are normally distributed? Why, or why not?

7.6 Confidence Intervals for the Population Proportion

Statisticians find it important to estimate the population proportion as well as the population mean. We have seen that the purpose of a statistical investigation may be to estimate the proportion of children in a particular area that spend over 4 hours per day watching television. Or a political pollster may want to estimate the percentage of voters who say they will vote for a particular candidate in the next election. In this section we describe how confidence intervals are constructed for population proportions. We will begin with the case where the normal distribution can be used and then take up the case where special graphs are required.

USE OF THE NORMAL DISTRIBUTION

The sample proportion equals X/n, where X is the number of successes in the sample and n is the sample size. A success occurs when a child spends over 4 hours per day watching television, or when a voter says he or she will vote for the candidate in the next election. As we know from Chapter 4, the expected value of X, the number of successes, is $n\Pi$, where Π is the population proportion. Since the sample proportion is X divided by n, its expected value must equal the expected value of X divided by n. (For a proof, see Appendix 4.1.)[10] Consequently,

$$E\left(\frac{X}{n}\right) = \Pi.$$

In other words, the expected value of the sample proportion equals the population proportion. For example, if 70 percent of the population say that they will vote for the candidate in the next election, the expected value of the sample proportion equals 0.70.

The standard deviation of the sampling distribution of the sample proportion equals

$$\sigma\left(\frac{X}{n}\right) = \sqrt{\frac{\Pi(1-\Pi)}{n}}.$$

---

[10] In Appendix 4.1 we showed that if a random variable is a linear function of another random variable, its expected value is a linear function of the expected value of the other random variable. The sample proportion is a linear function of X; specifically, it equals $(1/n)X$. Thus, its expected value equals $1/nE(X) = 1/n(n\Pi) = \Pi$.

To see why, recall from Chapter 4 that the standard deviation of X equals $\sqrt{n\Pi(1-\Pi)}$. The standard deviation of X/n equals the standard deviation of X divided by n—that is, $\sqrt{n\Pi(1-\Pi)}$ divided by n or $\sqrt{\Pi(1-\Pi)/n}$. (This follows from Appendix 4.1). For example, if a random sample of 10 people is taken (with replacement) and if $\Pi = .70$, the standard deviaton of the sample proportion saying that they will vote for the candidate in the next election equals $\sqrt{.7(.3)/10} = .145$.

Thus, the sampling distribution of the sample proportion has a mean equal to Π and a standard deviation equal to $\sqrt{\Pi(1-\Pi)/n}$. We know from previous chapters that if the sample size is sufficiently large (and if Π is not very close to zero or 1) the sampling distribution can be approximated by the normal distribution. Thus, under these conditions, the sample proportion is approximately normally distributed with a mean of Π and a standard deviation of $\sqrt{\Pi(1-\Pi)/n}$, which means that

$$Pr\left\{\Pi - z_{\alpha/2}\sqrt{\frac{\Pi(1-\Pi)}{n}} < \frac{X}{n} < \Pi + z_{\alpha/2}\sqrt{\frac{\Pi(1-\Pi)}{n}}\right\} = 1 - \alpha,$$

where $z_{\alpha/2}$ is the value of the standard normal variable that is exceeded with the probability of $\alpha/2$. If we rearrange terms within the brackets on the left side of this equation, it follows that[11]

$$Pr\left\{\frac{X}{n} - z_{\alpha/2}\sqrt{\frac{\Pi(1-\Pi)}{n}} < \Pi < \frac{X}{n} + z_{\alpha/2}\sqrt{\frac{\Pi(1-\Pi)}{n}}\right\} = 1 - \alpha.$$

[11] The reasoning here is like that leading up to equation (7.2). Specifically, if we subtract Π from all three terms inside the brackets,

$$Pr\left\{-z_{\alpha/2}\sqrt{\frac{\Pi(1-\Pi)}{n}} < \frac{X}{n} - \Pi < z_{\alpha/2}\sqrt{\frac{\Pi(1-\Pi)}{n}}\right\} = 1 - \alpha.$$

Subtracting X/n from all three terms inside the brackets, we get

$$Pr\left\{-\frac{X}{n} - z_{\alpha/2}\sqrt{\frac{\Pi(1-\Pi)}{n}} < -\Pi < -\frac{X}{n} + z_{\alpha/2}\sqrt{\frac{\Pi(1-\Pi)}{n}}\right\} = 1 - \alpha.$$

Multiplying all three terms inside the brackets by -1, and reversing the inequalities, we get

$$Pr\left\{\frac{X}{n} - z_{\alpha/2}\sqrt{\frac{\Pi(1-\Pi)}{n}} < \Pi < \frac{X}{n} + z_{\alpha/2}\sqrt{\frac{\Pi(1-\Pi)}{n}}\right\} = 1 - \alpha,$$

which is what we set out to derive.

As it stands, this equation cannot be used to construct a confidence interval for Π because without a knowledge of Π we cannot compute $\sqrt{\Pi(1-\Pi)/n}$. However, if the sample is sufficiently large it is permissible to substitute the sample proportion for Π in this expression, the result being

$$
Pr\left\{ \frac{X}{n} - z_{\alpha/2} \sqrt{\frac{\left(\frac{X}{n}\right)\left[1-\frac{X}{n}\right]}{n}} < \Pi < \frac{X}{n} + z_{\alpha/2} \sqrt{\frac{\left(\frac{X}{n}\right)\left[1-\frac{X}{n}\right]}{n}} \right\}
$$

$$
= 1 - \alpha.
$$

(7.7)

Based on equation (7.7), it is clear that *if the sample is sufficiently large and if the confidence coefficient is set equal to $(1-\alpha)$ the confidence interval for the population proportion is*

$$
p - z_{\alpha/2} \sqrt{\frac{p(1-p)}{n}} < \Pi < p + z_{\alpha/2} \sqrt{\frac{p(1-p)}{n}},
$$

(7.8)   *Confidence Interval for Π (if n is Large)*

where p is x/n, the sample proportion. Of course, this expression assumes that the population is large relative to the sample or that sampling is carried out with replacement; otherwise $\sqrt{p(1-p)/n}$ must be multiplied by $\sqrt{(N-n)/(N-1)}$. This result is of widespread usefulness. The following example illustrates how it is applied.

EXAMPLE 7.3 A polling organization selects a random sample of 400 residents of Houston, Texas and asks each person whether the United States should increase its nonmilitary aid to countries in Central America. Sixty percent of the people in the sample favor increased aid. Calculate a 95 percent confidence interval for the percentage of Houston residents in favor of increased aid.

SOLUTION: Since $(1-\alpha) = .95$, $z_{\alpha/2} = z_{.025} = 1.96$. Inserting .60 for p and 400 for n in expression (7.8), we have

$$
.60 - 1.96 \sqrt{\frac{.60(.40)}{400}} < \Pi < .60 + 1.96 \sqrt{\frac{.60(.40)}{400}}.
$$

Simplifying terms, this confidence interval is

$$
.60 - .048 < \Pi < .60 + .048,
$$

which means that our interval estimate of the percentage of Houston residents favoring increased aid is 55.2 percent to 64.8 percent.

USE OF SPECIAL GRAPHS

Expression (7.8) provides a confidence interval for the population proportion when the sample size is sufficiently large. But how large is "sufficiently large"? According to William Cochran,[12] the minimum sample size needed to insure the validity of this expression is about 30 if the population proportion is about .50; about 50 if the population proportion is about .40 or .60; about 80 if it is about .30 or .70; about 200 if it is about .20 or .80; and about 600 if it is about .10 or .90. If the sample size does not meet these standards one must then use special graphs that have been computed for this purpose. Appendix Tables 7a and 7b respectively contain 95 percent confidence intervals and 99 percent confidence intervals for the population proportion. These graphs are used in the following way.

If the sample proportion is less than 50 percent, one finds the point on the bottom horizontal axis that equals the sample proportion. For example, if the sample proportion is 45 percent, the appropriate point on the horizontal axis is .45. Then, using the two curves that pertain to the relevant sample size, one finds the two points on the vertical axis that correspond to this point on the horizontal axis. For example, if $n = 100$, the two curves are as shown in Figure 7.6 for a 95 percent confidence interval. If the sample proportion is 0.45, the two points on the vertical axis corresponding to 0.45 on the horizontal axis are .35 and .55, as shown in Figure 7.6. Thus, the 95 percent confidence interval is 35 percent to 55 percent.

If the sample proportion is greater than 50 percent, one uses the top (rather than the bottom) horizontal scale and the right-hand (rather than the left-hand) vertical scale in Appendix Tables 7a and 7b. For example, suppose that the sample proportion is 0.65 and $n = 100$. As shown in Figure 7.7, we begin by finding .65 on the *top* horizontal scale, after which we find the two points on the *right-hand* vertical scale corresponding to this location on the two curves. Since these two

[12] See W. G. Cochran, *Sampling Techniques* (New York: Wiley, 1953), p. 41. As Cochran points out, smaller sample sizes than those given above may be acceptable if the statistician is willing to accept somewhat larger risks of error than is Cochran. As pointed out in Chapter 5, many statisticians regard the normal approximation as acceptable so long as $n\Pi > 5$ when $\Pi \le 1/2$ and $n(1 - \Pi) > 5$ when $\Pi > 1/2$.

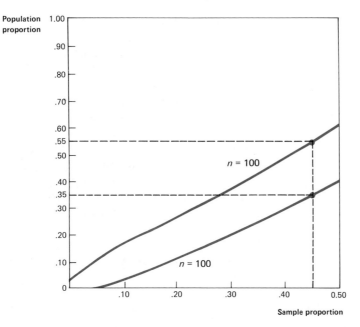

Figure 7.6
Derivation of a Confidence Interval for the Population Proportion, when the Sample Proportion Is 0.45

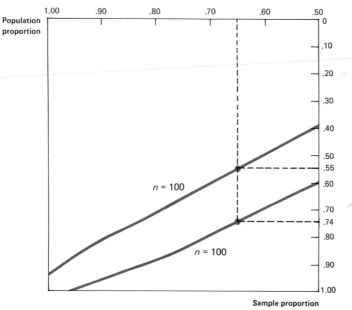

Figure 7.7
Derivation of a Confidence Interval for the Population Proportion, when the Sample Proportion Is 0.65

points are 55 and 74 percent, the 95 percent confidence interval is 55 percent to 74 percent.

These graphs are of considerable use to statisticians. The following is a further illustration of how they are employed.

EXAMPLE 7.4 In a random sample, 42 of 60 students interviewed in a very large college psychology class reported that they had a dental examination during the previous year. Obtain a 95 percent confidence interval for the proportion of the students in the class who had such an examination then.[13]

SOLUTION: Since the sample proportion exceeds 50 percent, we must use the top horizontal and right-hand vertical scales in Appendix Table 7a. First, we find .70 (the sample proportion) on the top horizontal scale. Then we must find the points on the two curves (for $n = 60$) that correspond to this horizontal location. Since these two points correspond to 57 percent and 81 percent on the right-hand vertical axis, the desired confidence interval is 57 percent to 81 percent.

[13] For relevant data and discussion, see R. Broyles and C. Lay, *Statistics in Health Administration* (Germantown, Md.: Aspen Systems, 1979).

7.7 Confidence Intervals for the Difference Between Two Means or Two Proportions[14]

Frequently, the purpose of a statistical investigation is to estimate the difference between two population means or two population proportions. For example, suppose that a psychologist wants to estimate the difference between two types of children in the mean number of hours per day spent watching television. In this section we will show how a confidence interval can be constructed for the difference between two population means. We will also show how a confidence interval can be constructed for the difference between two population proportions.

DIFFERENCE BETWEEN TWO MEANS: INDEPENDENT SAMPLES

Let's take two populations, one with a mean of μ_1 and a standard deviation of σ_1, the other with a mean of μ_2 and a standard deviation of σ_2. A simple random sample of n_1 observations is chosen from the first population and a simple random sample of n_2 observations is chosen from the second. These two random samples are entirely *independent;*

[14] Some instructors may want to skip this section. If so, this can be done without loss of continuity. Note that the methods discussed in this selection are by no means the only ones that can be applied to obtain a confidence interval for the difference between two means or two proportions.

in particular, the observations in one sample are not paired in any way with those in the other sample.[15] *If both samples are large, and if the confidence coefficient is set equal to* $(1 - \alpha)$, *the confidence interval for the difference between the population means is*

$$\bar{x}_1 - \bar{x}_2 - z_{\alpha/2}\sqrt{\frac{s_1^2}{n_1} + \frac{s_2^2}{n_2}} < \mu_1 - \mu_2 < \bar{x}_1 - \bar{x}_2 + z_{\alpha/2}\sqrt{\frac{s_1^2}{n_1} + \frac{s_2^2}{n_2}},$$ (7.9)

Confidence Interval for $\mu_1 - \mu_2$

where s_1^2 *is the variance of the sample taken from the first population and* s_2^2 *is the variance of the sample taken from the second population.*

The following example shows how this result can be used under practical circumstances.

[15] If they are paired, the results of this paragraph are not applicable. For a description of how a confidence interval can be obtained from paired comparisons, see Section 7.9. Also, it is assumed here and in expression (7.10) that the population is large relative to the sample or that sampling is with replacement.

EXAMPLE 7.5 Yale's Dorothy and Jerome Singer have found that children who report a belief in an unfriendly or dangerous world are more likely to be in families emphasizing heavy television watching.[16] A random sample of 100 children is drawn from among those who report such a belief. The mean number of hours per day spent watching television is 5.02, and the sample variance is 1.44 (hours)². A random sample of 100 children is drawn from among those who do not report such a belief. The mean number of hours per day spent watching television is 4.28, and the sample variance is 1.21 (hours)². Constuct a 90 percent confidence interval for the difference between children with and without this belief in the mean number of hours per day spent watching television.

SOLUTION: Since $(1 - \alpha) = .90$, $z_{\alpha/2} = z_{.05} = 1.64$. Inserting 5.02 for $\bar{x}_1$, 4.28 for $\bar{x}_2$, 1.44 for s_1^2, 1.21 for s_2^2, and 100 for n_1 and n_2, equation (7.9) becomes the following:

$$.74 - 1.64(.16) < \mu_1 - \mu_2 < .74 + 1.64(.16)$$

or

$$.48 < \mu_1 - \mu_2 < 1.00.$$

[16] J. and D. Singer, "Psychologists Look at Television: Cognitive, Developmental, Personality, and Social Policy Implication," *American Psychologist,* July 1983.

Thus, the desired confidence interval for the difference between children with and without this belief in the mean number of hours per day watching television is 0.48 to 1.00 hours.

DIFFERENCE BETWEEN TWO PROPORTIONS: INDEPENDENT SAMPLES

Suppose that there are two populations, one where the proportion having a certain characteristic is Π_1, the other where the proportion with this characteristic is Π_2. A simple random sample of n_1 observations is chosen from the first population and a simple random sample of n_2 observations is chosen from the second. These two random samples are entirely *independent;* in particular, the observations in one sample are not paired in any way with those in the other. *If both samples are sufficiently large, and if the confidence coefficient is set at $(1 - \alpha)$, the confidence interval for the difference between the population proportions is*

Confidence Interval for $\Pi_1 - \Pi_2$

$$p_1 - p_2 - z_{\alpha/2} s_{p_1 - p_2} < \Pi_1 - \Pi_2 < p_1 - p_2 + z_{\alpha/2} s_{p_1 - p_2}, \qquad (7.10)$$

where p_1 is the sample proportion in the first population, p_2 is the sample proportion in the second population, and $s_{p_1 - p_2} = \sqrt{\dfrac{p_1(1 - p_1)}{n_1} + \dfrac{p_2(1 - p_2)}{n_2}}$.

The following example shows how this result can be used.

EXAMPLE 7.6 Sandra Hofferth and Kristin Moore published a study of women who had children at a relatively early age in the *American Sociological Review.*[17] They found that, among a sample of 256 women who had their first child before the age of 19, 83 percent were currently married and living with their husbands. Among a sample of 1,012 women who had their first child at the age of 19 or more, 91 percent were currently married and living with their husbands. Construct a 95 percent confidence interval for the difference between the proportion of women in these two groups who were currently married and living with their husbands.

SOLUTION: Since $(1 - \alpha) = .95$, $z_{\alpha/2} = z_{.025} = 1.96$. If we sub-

[17] S. Hofferth and K. Moore, "Early Childbearing and Later Economic Well-being," *American Sociological Review,* October 1979.

stitute .91 for p_1, .83 for p_2, 1,012 for n_1, and 256 for n_2 in expression (7.10), we get

$$.91 - .83 - 1.96(.025) < \Pi_1 - \Pi_2 < .91 - .83 + 1.96(.025),$$

or

$$.03 < \Pi_1 - \Pi_2 < .13.$$

Thus, the desired confidence interval for the difference between the percentage of women in these two groups that were currently married and living with their husands is 3 to 13 percentage points.

EXERCISES

7.16 The admissions office of a large university wants to estimate the proportion of the applicants for admission that are in the top 10 percent of their high school classes. The admissions director draws a random sample of 24 of the applicants and finds that 25 percent of them are in the top 10 percent. Construct a 95 percent confidence interval for the proportion in the top 10 percent.

7.17 A county government plans to estimate the differences between two municipalities in the average distance of houses from a fire station. In the first municipality, 100 houses are chosen at random; the sample mean is 5.3 miles and the sample variance is 1.6. In the second community, 400 houses are chosen at random; the sample mean is 5.8 miles and the sample variance is 1.8. Calculate a 90 percent confidence interval for the difference between the two communities in the mean distance of a house from a fire station.

7.18 An airline wants to determine the proportion of passengers on its New York-Chicago flights who carry only hand luggage. The airline picks a random sample of 40 passengers traveling on these flights and finds that 14 percent carry only hand luggage. Calculate a 95 percent confidence interval for the proportion of hand-luggage passengers on these flights.

7.19 A botanist wants to determine the proportion of tulip bulbs of a particular type that will bloom. He selects a random sample of 100 of such bulbs, and finds that 36 percent of them bloom.
(a) Calculate a 90 percent confidence interval for this proportion.
(b) Calculate a 95 percent confidence interval for this proportion.

7.20 A St. Louis newspaper wants to estimate the proportion of its subscribers who believe that the government should be allowed to tap telephones without a court order. It picks a random sample of 250 of its subscribers, and finds that 19 percent of them believe that the government should have this power.

(a) Calculate a 98 percent confidence interval for this proportion.

(b) Calculate a 90 percent confidence interval for this proportion.

7.21 A San Francisco medical researcher selected a random sample of 40 drug users (all of whom had injected drugs), and asked each one the kind of drug he or she first injected. The researcher finds that 16 of them first injected heroin.

(a) Determine a 95 percent confidence interval for the proportion of drug users in San Francisco for whom heroin was the first drug injected.

(b) Determine a 99 percent confidence interval for this proportion.

7.22 In a study of the impact of race on beliefs about the American opportunity structure, a random sample of 12 blacks are asked whether they believe that rich and poor have equal opportunity to get ahead. Five of the people in the sample believe that this is true.

(a) Determine a 95 percent confidence interval for the proportion of blacks who believe that this is true.

(b) Determine a 99 percent confidence interval for this proportion.

7.23 A New York market research firm wants to compare the average price of a pair of shoes (of a particular type) in Chicago with that in New York. It picks a random sample of 50 shoe stores in Chicago, and finds that the mean price is $56.35, the standard deviation being $3.42. It picks a random sample of 50 shoe stores in New York, and finds that the mean price is $58.15, the standard deviation being $4.13.

(a) Compute a 95 percent confidence interval for the difference between the mean price of such a pair of shoes in New York and the mean price of such a pair of shoes in Chicago.

(b) Is your answer to (a) based on the assumption that the prices in each city are normally distributed? Why or why not?

7.24 A school board is responsible for two elementary schools. It wants to determine how the mean IQ of the students at school A compares with the mean IQ of those at school B. It chooses a random sample of 90 students from each school. At school A, the sample mean IQ is 109 and the sample standard deviation is 11. At school B, the sample mean IQ is 98 and the standard deviation is 9.

(a) Compute a 90 percent confidence interval for the difference between the mean IQ at school A and mean IQ at school B.

(b) Compute a 99 percent confidence interval for this difference.

7.25 A U.S. senator from Michigan believes that the percentage of voters favoring a particular proposal is higher in Detroit than in other parts of the state. He picks a random sample of 200 Detroit voters and finds that 59 percent favor the proposal. He picks a random sample of 200 Michigan voters from outside Detroit and finds that 52 percent favor the proposal.

(a) Construct a 90 percent confidence interval for the difference between the proportion in Detroit favoring the proposal and the proportion outside Detroit favoring it.

(b) Construct a 99 percent confidence interval for this difference.

7.8 Determining the Size of the Sample

In setting out to estimate a particular parameter a statistician must choose the kind of estimator to be used and the sample size. We have already discussed the various kinds of estimators, and now we take up the determination of the size of the sample. First, we will indicate how this decision can be made in estimating a population mean; then we will indicate how to pick a sample size when estimating a population proportion.

WHERE THE POPULATION MEAN IS ESTIMATED

In many cases, the decision maker authorizing the experiment or survey wants the resulting estimate to have a specified degree of precision.[18] For example, consider the engineer (Example 7.2) who needed to estimate the mean longevity of a new type of bulb. He might want the probability to be .90 that the sample mean will differ from the population mean by no more than 30 hours. Given this specified degree of precision, we can determine how large the sample must be if we know the standard deviation of the life of bulbs of the new type.

To see how the sample size can be determined, recall once again that for reasonably large samples, the sample mean $\overline{X}$ is distributed normally with mean equal to the population mean μ and standard deviation equal to $\sigma/\sqrt{n}$ (where σ is the population standard deviation and n is the sample size). Thus,

$$Pr\left\{\mu - 1.64\frac{\sigma}{\sqrt{n}} < \overline{X} < \mu + 1.64\frac{\sigma}{\sqrt{n}}\right\} = .90, \qquad (7.11)$$

which says that the probability that the sample mean will lie within 1.64 standard deviations of the population mean equals .90. As we know from previous discussions, this is true of any normal variable. Another way of stating the same thing is

$$Pr\left\{-1.64\frac{\sigma}{\sqrt{n}} < \overline{X} - \mu < 1.64\frac{\sigma}{\sqrt{n}}\right\} = .90.$$

[18] Ideally, the degree of precision specified should reflect the costliness of errors in the estimate and the costliness of sampling. However, it frequently is difficult to quantify the cost of errors of a particular size; and in practice decision makers often specify degrees of precision based on intuitive and informal judgments concerning these more basic factors.

In other words, the probability that the sample mean will differ from the population mean by less than $1.64\sigma/\sqrt{n}$ is .90.

If the desired precision is to be obtained, the probability that the sample mean will differ from the population mean by less than 30 hours must equal .90. This means that

$$1.64\sigma/\sqrt{n} = 30,$$

since we know from the previous paragraph that the probability that the sample mean will differ from the population mean by less than $1.64\sigma/\sqrt{n}$ equals .90. Suppose (contrary to Example 7.2) that the engineer knows that σ equals 160 hours. Then

$$\frac{1.64(160)}{\sqrt{n}} = 30$$

or

$$n = \left[\frac{1.64(160)}{30}\right]^2,$$

which means that n must equal 77.

In general, *if it is desired that the probability be* $(1 - \alpha)$ *that the sample mean differ from the population mean by no more than some number δ (delta), the sample size must equal*

Formula for n in Estimating μ

$$n = \left(\frac{z_{\alpha/2}\sigma}{\delta}\right)^2,$$

$$(7.12)$$

where σ *is the population standard deviation and* $z_{\alpha/2}$ *is the value of the standard normal variable which has a probability $\alpha/2$ of being exceeded.*[19]

Equation (7.12) is of great practical importance. Even if you do not know the value of σ, rough estimates of its value can be inserted into (7.12) to get some idea of how large the sample must be. The following example shows how one can determine the sample size using this formula.

---

[19] This assumes that the sample is large and that the population is large relative to the sample (or that sampling is with replacement).

EXAMPLE 7.7 A bank wants to estimate the mean balance in the checking accounts of its depositors 65 years old or over. There are a very large number of such accounts, and the bank manager believes that the standard deviation of the balances held by such individuals is about \$160. If the bank wants the probability to be .95 that the sample mean will differ from the population mean by no more than \$20, how big a sample must be taken?

SOLUTION: Since $(1 - \alpha) = .95$, $z_{\alpha/2} = z_{.025} = 1.96$. Substituting 160 for σ and 20 for δ in (7.12), we have

$$n = \left[\frac{(1.96)(160)}{20} \right]^2 = 246.$$

Thus, the sample size should be 246.

WHERE THE POPULATION PROPORTION IS ESTIMATED

In investigations aimed at estimating a population proportion, a degree of precision is generally specified. For example, consider the polling organization (Example 7.3) that needs to estimate the proportion of Houston residents who favor increased nonmilitary aid to Central America. The sponsors of this survey may decide that for the results to be useful to them the probability must be .90 that the sample percentage differs from the population percentage by no more than five percentage points. To see how this statement of desired precision can be used to determine the sample size, recall that for sufficiently large samples the sample proportion p is approximately normally distributed with mean equal to the population proportion Π and standard deviation equal to $\sqrt{\Pi(1 - \Pi)/n}$. Thus,

$$Pr\left\{ - 1.64 \sqrt{\frac{\Pi(1 - \Pi)}{n}} < p - \Pi < 1.64 \sqrt{\frac{\Pi(1 - \Pi)}{n}} \right\} = .90,$$

which says that the probability that the sample proportion will differ from the population proportion by no more than 1.64 standard deviations equals .90. This, of course, is true of all normal variables.

If the desired precision is to be obtained, the probability that the sample proportion will differ from the population proportion by .05 (that is, five percentage points) must be .90. This means that

$$1.64 \sqrt{\frac{\Pi(1 - \Pi)}{n}} = .05,$$

since we know from the previous paragraph that the probability that the sample proportion will differ from the population proportion by less than $1.64\sqrt{\Pi(1-\Pi)/n}$ equals .90. Although the polling organization does not know the value of Π, it is likely to have some idea of its approximate value. For example, suppose that Π is believed to be in the neighborhood of 0.5. Then

$$1.64\sqrt{\frac{(.5)(.5)}{n}} = .05,$$

or

$$n = \frac{(1.64)^2}{(.05)^2}(.5)(.5),$$

which means that n must equal about 269.

In general, *if it is desired that the probability be* $(1-\alpha)$ *that the sample proportion differs from the population proportion by no more than some number* δ*, and if the population proportion is believed to be approximately equal to* $\hat{\Pi}$*, the sample size must equal*

Formula for n *in Estimating* Π

$$n = \left(\frac{z_{\alpha/2}}{\delta}\right)^2 \hat{\Pi}(1-\hat{\Pi}).$$
(7.13)

This result assumes that the sample is large enough for the normal approximation to be used and that the population is large relative to the sample (or that sampling is with replacement). It is worth noting that if one wants a conservative estimate of n (that is, an estimate that tends to err on the high side), it is best to shade one's estimate of Π in the direction of 0.5. Why? Because the right side of equation (7.13) gets larger as Π approaches 0.5. Thus, if Π is shaded toward 0.5, the estimate of n will tend to err on the high side.

Equation (7.13), like equation (7.12), is of great practical importance. The following example illustrates how this formula is used.

EXAMPLE 7.8 A government agency plans to estimate the percentage of welfare recipients in a particular area who are over 60 years of age. A reasonable estimate is that this percentage is about 30. The agency wants the probability to be .99 that the sample percentage differs from the population percentage by less than 5 percentage points. How large a sample should the agency take?

SOLUTION: Since $(1 - \alpha) = .99$, $z_{\alpha/2} = z_{.005} = 2.58$. Substituting .30 for $\hat{\Pi}$ and .05 for δ in equation (7.13), we have

$$n = \left(\frac{2.58}{.05}\right)^2 (.30)(.70)$$

$$= 2662.56(.21)$$

$$= 559.$$

Thus, the sample size should be about 559.

THE CASE OF THE YOUNG PINE TREES

Firms that produce and sell young trees often estimate, in early spring or late winter, the number of healthy young trees that are growing on their land, in order to establish policy toward the solicitation and acceptance of orders. The manager of Marietta Tree Company divides his firm's lands into 1,000 plots (of equal size) and asks that a random sample of 100 plots be examined to determine the number of healthy young pine trees growing on each plot. To estimate the total number of healthy young pine trees on the firm's lands, the manager says he will compute the mean number of such trees per plot in the sample, and multiply the result by 1,000. He wants the probability to be 0.90 that his estimate (of the total number of healthy young pine trees growing on the firm's lands) will be in error by no more than 500 trees. From past experience, he knows that the standard deviation of the number of healthy young pine trees in a plot is about 2. The cost of examining a plot and determining the number of healthy young pine trees growing on it is $100. When the manager's boss hears of what he is doing, he calls the manager a fathead. Was the manager wrong? If so, what mistake did he make?

MISTAKES TO AVOID

SOLUTION

The manager wants the probability to be 0.90 that the mean number of healthy young pine trees per plot in the sample will differ from the true mean by 500/1000, or 0.5. For this to be true, the sample size must be

$$\left(\frac{1.64(2)}{0.5}\right)^2 = 43,$$

according to equation (7.12). Thus, the sample size of 100 is much too large. By reducing the sample size from 100 to 43 plots, the manager could have saved 57 times $100, or $5,700, without having the precision fall below the desired level.

Going a step further, it is possible that the Marietta Tree Company really does not need the probability to be 0.90 that this estimate will be in error by no more than 500 trees. Perhaps a lower probability (such as 0.75 or 0.60) will do. There are costs as well as benefits in obtaining greater accuracy. If a lower probability will do, the sample size can be reduced further.[20]

[20] For further discussion of the use of statistical methods to estimate the number of young trees and seedlings, see W. Cochran, *Sampling Techniques*.

7.9 Statistical Estimation in Chemical Research: A Case Study

In Chapter 1 we pointed out that British chemists carried out the following experiment to estimate the effect of a chlorinating agent on the abrasion resistance of a certain type of rubber.[21] Ten pieces of this type of rubber were cut in half, and one half-piece was treated with the chlorinating agent, while the other half-piece was untreated. Then the abrasion resistance of each half was evaluated on a machine, and the difference between the abrasion resistance of the treated half-piece and the untreated half-piece was computed. Table 7.1 shows the 10 differences (one corresponding to each of the pieces of rubber in the sample). Based on this experiment, the chemists were interested in estimating the mean difference between the abrasion resistance of a treated and untreated half-piece of this type of rubber. In other words, if this experiment were performed again and again, an infinite population of such differences would result. The chemists were interested in estimating the mean of this population, since the mean is a good measure of the effect of the chlorinating agent on this type of rubber's abrasion resistance.

If you were a statistical consultant for the chemists, how would you analyze these data? Recalling the material in Section 7.4, you would recognize that a good point estimate of the mean of this population is the sample mean, which Table 7.1 shows to be 1.27. Thus, your first step would be to advise them that if they want a single number as

[21] Owen L. Davies, *The Design and Analysis of Industrial Experiments*, p. 13.

Piece	Difference
1	2.6
2	3.1
3	−0.2
4	1.7
5	0.6
6	1.2
7	2.2
8	1.1
9	−0.2
10	0.6

$$\sum_{i=1}^{10} x_i = 12.7$$

$$\bar{x} = 1.27$$

$$s = 1.1265$$

Table 7.1
Differences in Abrasion Resistance (Treated Material Minus Untreated Material), 10 Pieces of Rubber

Source: Owen L. Davies, *The Design and Analysis of Industrial Experiments*, p. 13.

an estimate, 1.27 is a good number to use. Next, mindful of one of the central points of this chapter, you would point out that such a point estimate contains no indication of how much error it may contain, whereas a confidence interval does contain such information. Since the population standard deviation is unknown and the sample is small, expression (7.6) should be used in this case to calculate a confidence interval. Assuming that the chemists want a confidence coefficient of 95 percent, the confidence interval is 0.464 to 2.076, because $t_{.025} = 2.262$, $s = 1.1265$, and $n = 10$. The chances are 95 out of 100 that such a confidence interval would include the population mean.[22]

The above analysis is, in fact, exactly how the statisticians working with the chemists proceeded. Despite the fact that the sample consisted of only 10 observations, the evidence was very strong that the chlorinating agent had a positive effect on abrasion resistance. After all, the 95 percent confidence interval was that the mean difference between abrasion resistance of rubber with and without treatment was an increase of between 0.464 and 2.076. (For that matter, the statisticians found that the 98 percent confidence interval was that the mean difference was an increase of between 0.265 and 2.275.) The best esti-

[22] Note that this analysis assumes that the population is normally or approximately normally distributed.

mate was that the chlorinating agent resulted in an increase of about 1.27 in abrasion resistance.

In conclusion, note that it would have been incorrect to have viewed the abrasion resistance of the 10 treated half-pieces as one sample and the abrasion resistance of the 10 untreated half-pieces as another, and to have used expression (7.9) to obtain a confidence interval for the difference between the mean abrasion resistance of treated half-pieces and the mean abrasion resistance of untreated half-pieces. This would have been incorrect because the two samples are *paired* or *matched;* that is, each half-piece in one sample has a mate (the other half of the piece it comes from) in the other sample. Thus, the observations in one sample are not independent of those in the other sample, because if one half-piece is relatively resistant to abrasion due to chance variation in its production or other factors, its mate in the other sample is likely to be resistant as well. Because the samples are not independent, expression (7.9) is not appropriate, since it assumes independence.[23] Instead, the proper technique is to consider the difference between each observation in one sample and its mate in the other sample as a single observation, as we did in Table 7.1.

Matched Pairs

EXERCISES

7.26 A local government plans to estimate the percentage of vacant buildings in a particular area of several square miles. A reasonable guess is that about 20 percent of the buildings in this area are vacant. The government wants the probability to be .90 that the sample percentage differs from the population percentage by no more than 2 percentage points. How large should the sample of buildings be?

7.27 A municipal government wants to determine the mean market value of the 700 houses in the eastern part of the municipality. For the results to be useful, it is believed that the probability that the sample mean will differ by less than $5,000 from the population mean should equal .95. A rough estimate of the standard deviation of the market values of the houses is $15,000. How large a sample should be drawn?

7.28 A spokesman for a hi-fi repair shop claims that in about 40 percent of the cases where a hi-fi set is repaired the customer is undercharged due to clerical errors. A public-interest law firm decides to estimate the proportion of cases where undercharging occurs. The law firm plans to construct a random sample so that the probability of the sample proportion's being in error by more than .01 is .05. How large should the sample be?

7.29 Suppose that in the experiment described in Section 7.9 the chlorinated half-pieces of rubber were evaluated on a different machine than the un-

---

[23] Also, the formula in expression (7.9) is not appropriate because it assumes that the sample sizes are large.

treated half-pieces. Would this make the results of the experiment more difficult to interpret? If so, how?

7.30 A clothing store wants to estimate the percentage of its customers that have seen a certain magazine ad. It intends to take a simple random sample of its customers to estimate this percentage. It wants the probability to be 0.95 that this estimate differs by no more than 1 percentage point from the true percentage. The store thinks that about 30 percent of its customers may have seen the ad. How big a sample should be taken by the store?

7.31 An airline wants to estimate the percentage of people traveling on its flights between Boston and Miami whose tickets are bought more than a week before their flight. It believes that this percentage is about 40. It is about to select a random sample of people traveling on its flights between Boston and Miami. If it wants the probability to be 0.01 that the sample percentage will be in error by more than 2 percentage points, how big a sample should it select?

The Chapter in a Nutshell

1. Estimates are of two types: *point estimates* and *interval estimates.* A statistic used to estimate a population parameter is called an *estimator.* In choosing among estimators, statisticians consider the following three criteria: lack of bias, efficiency, and consistency. An *unbiased estimator* is one whose expected value equals the parameter being estimated. If two estimators are unbiased, one is more efficient than the other if its variance is less than the variance of the other. A statistic is a consistent estimator of a parameter if the probability that the statistic's value is very near that of the parameter approaches 1 as the sample size increases. Based on these criteria, the sample mean, sample proportion, and sample standard deviation are judged to be very good estimators of the population mean, population proportion, and population standard deviation, respectively.

2. If the population standard deviation σ is known, the large-sample confidence interval for the population mean μ is

$$\bar{x} - z_{\alpha/2} \frac{\sigma}{\sqrt{n}} < \mu < \bar{x} + z_{\alpha/2} \frac{\sigma}{\sqrt{n}},$$

where $z_{\alpha/2}$ is the value of the standard normal variable that is exceeded with a probability of $\alpha/2$. Before the sample is drawn, there is a probability of $(1 - \alpha)$ that this interval will include μ. Thus, the confidence coefficient is said to be $(1 - \alpha)$. If σ is unknown, the sample standard deviation s can be substituted for σ in this expression if the sample size n is at least 30.

3. When the population standard deviation is unknown and the sample size is less than 30, the t distribution can be used to formulate a confidence interval for the population mean, providing the population is at least approximately normal. The *t distribution* is a family of distributions, each of which corresponds to a certain number of degrees of freedom. As the number of degrees of freedom increases, the t distribution moves increasingly closer to the standard normal distribution. In this context, the number of degrees of freedom equals $(n - 1)$. If the confidence coefficient is set equal to $(1 - \alpha)$, the confidence interval for the population mean is

$$\bar{x} - t_{\alpha/2} \frac{s}{\sqrt{n}} < \mu < \bar{x} + t_{\alpha/2} \frac{s}{\sqrt{n}},$$

where $t_{\alpha/2}$ is the value of a t variable (with $[n - 1]$ degrees of freedom) that is exceeded with a probability of $\alpha/2$.

4. If the sample size is sufficiently large, and if the confidence coefficient is set equal to $(1 - \alpha)$, the confidence interval for the population proportion Π is

$$p - z_{\alpha/2} \sqrt{\frac{p(1 - p)}{n}} < \Pi < p + z_{\alpha/2} \sqrt{\frac{p(1 - p)}{n}},$$

where p is the sample proportion. For small samples, one must use special graphs (in Appendix Tables 7a and 7b) to obtain a confidence interval for the population proportion.

5. If independent samples of sizes n_1 and n_2 are chosen from two populations, we can calculate a confidence interval for the difference between the population means $\mu_1 - \mu_2$. The following formula is applicable if both n_1 and n_2 are large:

$$\bar{x}_1 - \bar{x}_2 - z_{\alpha/2} \sqrt{\frac{s_1^2}{n_1} + \frac{s_2^2}{n_2}} < \mu_1 - \mu_2 < \bar{x}_1 - \bar{x}_2 + z_{\alpha/2} \sqrt{\frac{s_1^2}{n_1} + \frac{s_2^2}{n_2}},$$

where the confidence coefficient equals $(1 - \alpha)$. Also, we can calculate a confidence interval for the difference between the population proportions, $\Pi_1 - \Pi_2$. The following formula is applicable if both n_1 and n_2 are sufficiently large:

$$p_1 - p_2 - z_{\alpha/2} s_{p_1 - p_2} < \Pi_1 - \Pi_2 < p_1 - p_2 + z_{\alpha/2} s_{p_1 - p_2},$$

where $s_{p_1 - p_2} = \sqrt{\dfrac{p_1(1 - p_1)}{n_1} + \dfrac{p_2(1 - p_2)}{n_2}}.$

Neither of these formulas is correct if the two samples are matched, since this violates the assumption that the samples are independent.

6. If it is desired that the probability be $(1 - \alpha)$ that the sample mean differs from the population mean by no more than a given number δ, the sample size must equal

$$n = \left(\frac{z_{\alpha/2} \sigma}{\delta} \right)^2.$$

If it is desired that the probability be $(1 - \alpha)$ that the sample proportion differs from the population proportion by no more than some number δ, the sample size must equal

$$n = \left(\frac{z_{\alpha/2}}{\delta} \right)^2 \hat{\Pi}(1 - \hat{\Pi}),$$

where $\hat{\Pi}$ is an estimate of Π. These results assume that simple random sampling is used, that the sample is large enough so that the normal distribution can be used, and that the population is large relative to the sample (or that sampling is with replacement). The last assumption is made throughout this section (The Chapter in a Nutshell).

Chapter Review Exercises

7.32 A firm with 50 overseas plants chooses (without replacement) a random sample of 40 plants. For each plant in this sample the firm determines the number of days the plant was shut down in 1985 by labor disputes. The sample mean turns out to be 9.8 days.

(a) If the standard deviation of the number of days the firm's overseas plants were shut down by labor disputes in 1985 was 2, calculate a 90 percent confidence interval for the mean number of days that all the firm's overseas plants were shut down for this reason in 1985.

(b) If the firm had 100 rather than 50 overseas plants, calculate a 90 percent confidence interval for the mean number of days that all the firm's overseas plants were shut down by labor disputes in 1985. Explain the difference between your answer here and in (a).

7.33 (a) The Educational Testing Service knows that the standard deviation of the scores on a particular achievement test is 400 points. Calculate the 95 percent confidence interval for the mean score on this test, based on a sample of 40 people where the mean score is 4,500.

(b) A government statistician says that the 90 percent confidence interval for the mean score on another achievement test is 4,500 to 4,800, based on a sample of 36 people. The statistician also says that the standard deviation of the scores on this test equals 500. Is there any contradiction between these statements? If so, what is the contradiction?

(c) The statistician in (b) says that, if the standard deviation of the scores is known, the width of the 95 percent confidence interval for the mean is always about 20 percent greater than the width of the 90 percent confidence interval (if the sample size is held constant). Is this true? Why, or why not?

7.34 The Mercer Company does not know the standard deviation of the lengths of life of a particular component. It therefore chooses a random sample of 36 of these components and obtains the following lengths of life (in thousands of hours):

4.2	5.0	4.6	4.9	5.0	5.1
4.3	4.9	4.5	4.8	4.9	4.6
4.4	5.1	4.7	4.4	4.8	4.6
4.8	4.7	4.4	4.5	4.8	4.8
4.9	4.8	4.3	4.6	4.7	4.5
5.1	4.8	4.6	4.6	4.7	5.0

(a) Compute a 90 percent confidence interval for the mean length of life of these components.

(b) Compute a 95 percent confidence interval for the mean length of life of these components.

(c) Since Mercer also does not know the standard deviation of the lengths of life of motors received from a particular supplier, it chooses a random sample of 9 motors from this supplier and determines the life of each. The results (in thousands of hours) are as follows:

4.3	4.6	3.8
4.2	4.3	3.9
4.1	3.9	4.0

Compute a 90 percent confidence interval for the mean length of life of motors received from this supplier.

(d) Compute a 95 percent confidence interval for the mean length of life of motors received from this supplier.

(e) What major assumption underlies your calculations in (c) and (d)?

7.35 For sufficiently large samples it can be shown that the sample standard deviation s is approximately normally distributed, with a mean equal to the population standard deviation σ, and with a standard deviation equal to $\sigma \div \sqrt{2n}$. Use these results to show that if n is sufficiently large *a confidence interval for σ is*

$$\frac{s}{1 + \dfrac{z_{\alpha/2}}{\sqrt{2n}}} < \sigma < \frac{s}{1 - \dfrac{z_{\alpha/2}}{\sqrt{2n}}}$$

where the confidence coefficient equals $(1 - \alpha)$. (In Chapter 9 we take up the χ^2 distribution, which can be used to construct a confidence interval for σ when n is small.)

7.36 There are two methods of re-roofing a house, method A and method B. A research organization chooses a random sample of 200 houses re-roofed according to method A and finds that 18 percent experienced leaks within four years. Another random sample of 200 houses re-roofed according to method B is chosen. It is found that 29 percent of these houses experienced leaks within four years. Compute a 95 percent confidence interval for the difference between method B and method A in the percentage of houses experiencing leaks within four years.

7.37 A public-interest law firm picks a random sample of 100 hi-fi repair stores and asks them to fix a hi-fi set. In 34 of the cases the customer is undercharged. (That is, the bill is lower than the law firm considers appropriate.)

(a) Use the normal approximation to obtain a 90 percent confidence interval for the proportion of cases where such undercharging occurs.

(b) If a sample is large, the sample standard deviation is approximately normally distributed, with a mean equal to the population standard deviation σ and with a standard deviation equal to $\sigma \div \sqrt{2n}$. In the circumstances described above, the sample standard deviation of the amounts charged by the 100 hi-fi repair stores to fix the hi-fi set was $8.70. Provide a 90 percent confidence interval for the population standard deviation. (See Exercise 7.35.)

7.38 There are 10,000 people in a particular neighborhood in Brooklyn. A cable television station wants to estimate the average number of hours on a particular day that a person in this neighborhood spent watching its programs. Its executives think that the standard deviation of the number of hours a person spent watching its programs was about 3.2 hours. The station's president proposes that a random sample be taken of the people in the neighborhood. If it is desired that the probability be 0.98 that the sample mean differs by no more than 0.5 hours from the true mean, how big must such a sample be?

7.39 A state university wants to estimate the average amount that its students paid in state income taxes in 1986. The standard deviation of state income tax paid by the students is estimated to be about $60. The university wants the probability to be 0.95 that the sample mean will differ by no more than $10 from the true mean. If a random sample of the students is selected, how big should it be?

7.40 A bank wants to determine the proportion of its depositors who also have deposits in any of the local savings and loan associations. A random sample is constructed of 100 depositors, and it is determined that 46 percent have deposits in a local savings and loan association.
   (a) Use the normal approximation to determine a 95 percent confidence interval for this proportion.
   (b) Use Appendix Table 7a to determine this confidence interval, and compare the result with your finding in (a).

7.41 The Educational Testing Service wants to estimate the mean score on the achievement test in Exercise 7.33(a), and it wants the probability to be .95 that the error is no more than 20 points. Based on the information in Exercise 7.33(a), how large a sample must the Educational Testing Service take to be certain this is true?

*7.42 Minitab was used to calculate a 99 percent confidence interval for the mean difference in abrasion resistance between treated and untreated half-pieces of rubber. The data are in Table 7.1. Interpret the computer printout, which is as follows:

```
-SET FOLLOWING DATA INTO C3
 2.6 3.1 -0.2 1.7 0.6 1.2 2.2 1.1 -0.2 0.6
-TINTERVAL WITH 99 PERCENT CONFIDENCE, DATA IN C3
 C3  N = 10  MEAN = 1.2700  ST.DEV. = 1.13
-A 99.00 PERCENT C.I. FOR MU IS (0.1121, 2.4279)
```

*7.43 A random sample of ten women who had legal abortions in 1986 indicates that their ages (in years at the time of the abortion) were 21, 18, 28,

---

* Exercise requiring some familiarity with Minitab.

16, 22, 26, 31, 21, 19, 30. Minitab is used to calculate a 95 percent confidence interval for the mean age of women having legal abortions in 1986. Interpret the computer printout, which is as follows:

```
—SET FOLLOWING DATA INTO COL C2
 21 18 28 16 22 26 31 21 19 30
—TINTERVAL WITH 95 PERCENT CONFIDENCE FOR DATA IN C2
 C2  N = 10  MEAN = 23.200  ST.DEV. = 5.22
—A 95.00 PERCENT C.I. FOR MU IS (19.4621, 26.9379)
```

$\mathscr{8}$

OBJECTIVES

A principal aim of many statistical investigations is to test hypotheses. The overall purpose of this chapter is to describe the testing procedures that statisticians use. Among the specific objectives are:

1   To describe how you define the null and alternative hypotheses, and how you calculate the probability of a Type I and a Type II error.

2   To indicate how you test whether a population mean equals a specified value (and whether the difference between two means equals zero).

3   To indicate how you test whether a population proportion equals a specified value (and whether the difference between two proportions equals zero).

Hypothesis Testing

8.1 Introduction

As indicated in the previous chapter, statistical inference, the part of statistics that shows how probability theory can be used to help make decisions based on sample data, deals with two types of problems: estimation and hypothesis testing. Having discussed estimation problems in the previous chapter, we turn now to hypothesis testing. Since statisticians are continually engaged in testing hypotheses, and since the methods they use and the results they obtain are of great significance, it is important to understand the concepts involved in hypothesis testing and how these concepts can be applied to practical problems.

8.2 Hypothesis Testing: An Illustration

Hypothesis testing deals with decision making—that is, with the rules for choosing among alternatives. Since statistical decisions must be made under conditions of uncertainty, there is a nonzero probability of error; and the object of the statistician's theory of hypothesis testing is to develop decision rules that will control and minimize the probability of error.

Null and Alternative Hypotheses (H_0 and H_1)

To introduce some of the essential features of the statistical theory of hypothesis testing, let's look at an actual problem that confronted engineers engaged in manufacturing a metal piece whose height was of central importance in determining whether it met the relevant specifications. Based on previous experience, the engineers knew that the mean height of the metal pieces produced was .8312 inches. A test was needed for detecting any change in this mean height so that whatever factors were responsible for such a change could be corrected.[1]

The engineers picked a random sample from each day's output. Based on this sample, they wanted to test the hypothesis that a change had *not* occurred in the mean height of all the metal pieces produced in a day. In other words, the engineers wanted to test the hypothesis that the mean height still was .8312 inches. Let's denote this hypothesis by H_0 and call it the null hypothesis. In statistics the *null hypothesis* is the basic hypothesis that is being tested for possible rejection. The engineers had to choose between this hypothesis and the *alternative hypothesis,* denoted by H_1, which maintained that a change *had* occurred in the mean height of all the metal pieces produced in a day. In other words, the alternative hypothesis maintained that the mean height no longer was .8312 inches. (Much more will be said in the following section about the ways in which the null and alternative hypotheses are defined.)

To understand the nature of the decision that faced the engineers, it is essential to recognize that, on any given day, four possible situations could arise:

1. No change occurs in the mean height of the metal pieces produced (H_0 is true), and the engineers conclude that this is the case. Thus, the engineers accept H_0, which is the correct decision under these circumstances.

2. No change occurs in the mean height of the metal pieces produced (H_0 is true), but the engineers conclude that such a change has occurred. Thus, the engineers reject H_0 (and accept H_1), which is the wrong decision under the circumstances. (One cost of this wrong decision is the output that is lost while the firm's productive process is shut down needlessly so that the nonexistent deterioration in quality can be corrected.)

3. A change occurs in the mean height of the metal pieces produced (H_0 is not true), and the engineers conclude that this is the case. Thus, the engineers reject H_0 (and accept H_1), which is the correct decision under the circumstances.

[1] Lester Kauffman, "Statistical Quality Control at the St. Louis Division of American Stove Company," as quoted and described in A. J. Duncan, *Quality Control and Industrial Statistics.*

4. A change occurs in the mean height of the metal pieces produced (H_0 is not true), but the engineers conclude that no change has occurred. Thus, the firm accepts H_0 (and rejects H_1), which is the wrong decision under the circumstances. (One cost of this wrong decision is that the deterioration in quality is allowed to go uncorrected, with the result that an unusually high percentage of the firm's output must be scrapped because it fails to meet specifications.)

Based on the listing of the possibilities above, it is clear that two kinds of error can be committed. First, one can reject the null hypothesis when it is true—which is possibility 2 above. Second, one can accept the null hypothesis when it is false—which is possibility 4 above. These two kinds of error are called Type I and Type II errors and are defined as follows:

> A ***Type I error*** *occurs if the null hypothesis is rejected when it is true. A* ***Type II error*** *occurs if the null hypothesis is accepted when it is not true.*

Type I and Type II Errors

Thus, if the engineers conclude that a change occurs in the mean height of the metal pieces produced when in fact no such change takes place, they commit a Type I error. If they conclude that no change occurs when there is a change, they commit a Type II error.[2]

Table 8.1 provides a condensed description of the situation facing the engineers. The two possible *states of nature* shown in the table are that hypothesis H_0 is true or that it is not true. It is not known which of these states of nature—or states of the world—is valid, but one of the two alternative *courses of action* (the acceptance or rejection of the null hypothesis) must be chosen. The *outcome*—that is, the result of each course of action when each state of nature is true—is shown in Table 8.1. As you can see, a correct decision is the result of the two cases corresponding to possibilities 1 and 3 above; a Type I error is the result of the case corresponding to possibility 2 above; and a Type II error is the result of the case corresponding to possibility 4 above.

Alternative courses of action	State of nature	
	H_0 *is true*	H_0 *is not true**
Accept H_0	Correct decision	Type II error
Reject H_0	Type I error	Correct decision

Table 8.1
Possible Outcomes of the Engineers' Decision Problem

*In other words, H_1 is true.

[2] Whether the engineers should be more concerned about avoiding a Type I or a Type II error depends on the relative costs of such errors. Some of these costs have been cited above.

8.3 Basic Concepts of Hypothesis Testing

To carry out the desired test, the engineers need some criterion or decision rule for choosing between the null hypothesis (that the mean height is still .8312 inches) and the alternative hypothesis (that the mean height is no longer .8312 inches). The classical statistical theory of hypothesis testing presented in this chapter provides such decision rules based on the results of a random sample. These rules are constructed so that the probability of a Type I error and that of a Type II error can each be measured (if possible) and, to some extent at least, be reduced. Before describing these rules, we must discuss how one specifies the null hypothesis and the alternative hypothesis, and we must distinguish between one-tailed and two-tailed tests.

NULL AND ALTERNATIVE HYPOTHESES

As mentioned above, the null hypothesis is the basic hypothesis that is being tested for possible rejection. Typically, the null hypothesis corresponds to the *absence* of the effect that is being investigated. For example, the engineers want to detect a change in the mean height of the metal pieces produced. If such a change is *absent,* the mean height is still .8312 inches. Thus, the null hypothesis is that the mean height is .8312 inches. In testing any null hypothesis, there is an alternative hypothesis which the decision maker accepts if the null hypothesis is rejected. For example, the alternative hypothesis in the previous section is that the mean height differs from .8312 inches. There are two quite different kinds of alternative hypotheses, which correspond to one-tailed tests and two-tailed tests.

TWO-TAILED TESTS

In many statistical investigations, the purpose is to see whether a certain population parameter has changed or whether it differs from a particular value. This is true in the case of the engineers, who want to detect changes *in either direction* in the mean height of the metal pieces produced. Because the proportion of the firm's output not meeting specifications will increase if the mean height is *either* too large *or* too small, the engineers want to detect changes in either direction. Thus, the null hypothesis and alternative hypothesis in this case are

$$H_0: \mu = .8312 \qquad H_1: \mu \neq .8312,$$

where μ is the population mean.

How can the engineers carry out such a test? They can calculate the sample mean of the heights, $\bar{x}$. If the null hypothesis is true (that is, if the population mean is .8312 inches), the sampling distribution of the sample mean is as shown in Figure 8.1. Specifically, as we know from Chapter 6, the sample mean is approximately normally distributed and has a mean of .8312 inches and a standard deviation of $\sigma/\sqrt{n}$ (where σ is the population standard deviation and n is the sample size). Thus, if σ is known, the engineers can establish certain values of the sample mean that are very unlikely to occur, given that the null hypothesis is true. In particular, as pointed out in Figure 8.1, the probability is only .05 that the value of the sample mean will differ from .8312 inches by more than $1.96\sigma/\sqrt{n}$ under these circumstances. Consequently, if the value of the sample mean does in fact differ by that much from .8312 inches, it is evidence that the null hypothesis is not true.

Several things should be noted about this hypothesis-testing procedure. First, *the decision maker or statistician bases his or her decision on the value of a **test statistic,** which is a statistic computed from the sample.* In this case, the test statistic is the sample mean since this is the statistic used by the engineers to test the null hypothesis. Second, *the set of all possible values of the test statistic is referred to as the **sample space.*** For example, in this case the set of all possible values of the sample mean is the sample space. Third, *the testing procedure divides the sample space into two mutually exclusive parts: the **acceptance region** and the **rejection region.** If the test statistic's value falls in the acceptance region, the null hypothesis is accepted; if its value falls in the rejection region, the null hypothesis is rejected.* For example, in this case the acceptance region consists of values of the sample mean that differ from .8312 inches by less than $1.96\sigma/\sqrt{n}$.

Test Statistic

Acceptance Region and Rejection Region

Why is this hypothesis-testing procedure called a two-tailed test? The answer is indicated in Figure 8.1, which shows that the rejection region contains values of the test statistic (that is, the sample mean) that lie under *both* tails of its (that is, the sample mean's) sampling distribution. Basically, the reason why the rejection region is of this sort is that the engineers want to reject the null hypothesis (that is, $\mu = .8312$ inches) *either if μ is less than .8312 inches or if μ is greater than .8312 inches.* Regardless of whether the mean height is too small or too large, the engineers want to detect departures from the specified value of .8312 inches.

ONE-TAILED TESTS

In many statistical investigations, the decision maker or statistician is concerned about detecting departures from the null hypothesis *in only one direction.* For example, suppose that the Federal Trade

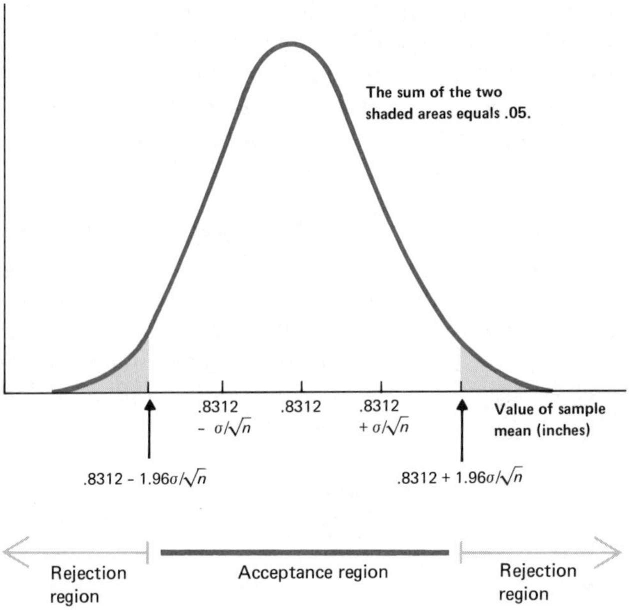

Figure 8.1
Sampling Distribution
of the Mean Height of
Metal Pieces in Sample
(if Null Hypothesis Is
True)

The sum of the two
shaded areas equals .05.

$.8312$ $.8312$ $.8312$ Value of sample
$-\sigma/\sqrt{n}$ $+\sigma/\sqrt{n}$ mean (inches)

$.8312 - 1.96\sigma/\sqrt{n}$ $.8312 + 1.96\sigma/\sqrt{n}$

Rejection   Acceptance region   Rejection
region   region

Commission wants to test whether the proportion of defective tires in a particular shipment is .06, as claimed by the tire manufacturer. The commission does not want to reject the null hypothesis (that the proportion defective is .06) if the proportion defective is in fact less than .06 since, regardless of whether this proportion is .06 or less than .06, the manufacturer's claim does not exceed its performance. *Only if the proportion defective exceeds .06 does the commission want to admonish the manufacturer for making a fake claim.* Thus, in this case the null and alternative hypotheses are

$$H_0: \Pi = .06 \qquad H_1: \Pi > .06,$$

where Π is the proportion of the shipment that is defective.[3]

Whereas the Federal Trade Commission wants to reject the null hypothesis only if the parameter in question is too high, other situations call for a rejection of the null hypothesis only if the parameter in question is too low. For example, a large public school system wants to test whether the mean length of life of the light bulbs in a particular incoming shipment is less than 2,000 hours. In this case, the school sys-

---

[3] In this case, it is possible to define the null hypothesis as $\Pi \leq .06$ rather than $\Pi = .06$. (And in the case in the next paragraph, one can define the null hypothesis as $\mu \geq 2,000$ rather than $\mu = 2,000$.) However, it simplifies the exposition to define the null hypothesis in the manner shown above. More is said about this in footnotes 4 and 9.

tem does not want to reject the null hypothesis (that the mean length of life is 2,000 hours) if the population mean is in fact greater than 2,000 hours, since, regardless of whether the mean is 2,000 hours or more than 2,000 hours, the school system wants to keep the shipment. *Only if the population mean is less than 2,000 hours does the school system want to reject the shipment.* Thus, in this case the null and alternative hypotheses are

$$H_0: \mu = 2,000 \qquad H_1: \mu < 2,000$$

where μ is the population mean.

When the decision maker or statistician is interested only in detecting departures from the null hypothesis in one direction, the hypothesis-testing procedure is a one-tailed test. To see why, consider the case of the school system that wants to test whether the mean length of life of the light bulbs in the shipment equals 2,000 hours. If the population mean equals 2,000 hours, the sampling distribution of the sample mean is as shown in Figure 8.2. That is, if a random sample of n light bulbs is selected from the shipment, the mean length of life of the bulbs in the sample will be distributed approximately normally with a mean of 2,000 hours and a standard deviation of $\sigma/\sqrt{n}$ (where σ is the population standard deviation). Thus, if the null hypothesis is true (that is, if $\mu = 2,000$), the probability is only .05 that the value of the sample mean will fall below 2,000 hours by an amount exceeding $1.64\sigma/\sqrt{n}$. Conse-

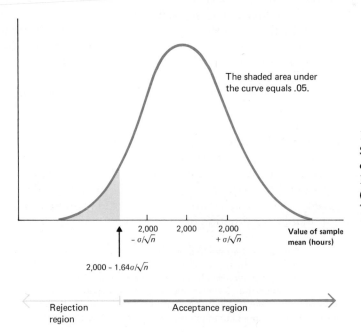

The shaded area under the curve equals .05.

2,000 $- \sigma/\sqrt{n}$ 2,000   2,000 $+ \sigma/\sqrt{n}$ Value of sample mean (hours)

2,000 $- 1.64\sigma/\sqrt{n}$

Rejection region   Acceptance region

Figure 8.2
Sampling Distribution of Mean Length of Life of Bulbs in Sample (if Null Hypothesis Is True)

quently, if the value of the sample mean does in fact fall below 2,000 hours by an amount exceeding $1.64\sigma/\sqrt{n}$, this is evidence that the null hypothesis is not true. In other words, the sample mean is the test statistic, and the rejection region contains all values of the sample mean falling below 2,000 hours by an amount exceeding $1.64\sigma/\sqrt{n}$. This rejection region (shown in Figure 8.2) is under only *one* tail of the test statistic's (that is, the sample mean's) sampling distribution, which accounts for the term *one-tailed test.*

DECISION RULES AND THE NULL HYPOTHESIS

It is important to recognize that *a test procedure (whether a one-tailed or two-tailed test) specifies a **decision rule** indicating whether the null hypothesis should be accepted or rejected.* For example, in the case of the engineers, the decision rule is (1) reject the null hypothesis if the sample mean differs by more than $1.96\sigma/\sqrt{n}$ from .8312 inches; (2) otherwise, accept the null hypothesis. In the case of the school system receiving the shipment of light bulbs, the decision rule is (1) reject the null hypothesis if the sample mean falls below 2,000 hours by more than $1.64\sigma/\sqrt{n}$; (2) otherwise, accept the null hypothesis.

If the null hypothesis is accepted, this doesn't mean that we are sure that it is true. All it means is that, on the basis of the available evidence, we have no good reason to reject the null hypothesis. Thus, acceptance of the null hypothesis should be interpreted as saying that we do not reject it, not that we are convinced beyond any doubt that it is completely accurate. Each decision rule can be expected to result in a wrong decision for a certain percentage of the time. In other words, each decision rule contains a certain probability of Type I error and a certain probability of a Type II error. In later sections, we shall indicate how these probabilities can be calculated.

It is also important to note that *the null hypothesis always states that a population parameter is equal to some value, not that it is unequal to some value.*[4] For example, a beer manufacturer needs to determine whether beer drinkers prefer a new brand of its beer over the old brand. To find out, a sample of beer drinkers is given some of the new brand, and these beer drinkers are asked to rate the new brand on a scale from 1 (very poor) to 90 (perfect). Then the mean rating in the sample is compared with 60, which is known to be the population mean rating for the old brand. In a situation of this sort the preferred procedure, as noted

---

[4] In footnote 3 we pointed out that in the one-tailed tests discussed there, the null hypothesis may be defined as $\Pi \leq .06$ or $\mu \geq 2,000$. If such a definition is used, the point in the text still stands: The null hypothesis cannot be that a population parameter is unequal to some value. For example, neither the hypothesis that $\Pi \leq .06$ nor that $\mu \geq 2,000$ is a statement that a population parameter is *unequal* to some value; they are both statements that a population parameter is *at most* or *at least* some value.

earlier, is to regard the existence of *no difference* between the mean rating of the new brand and that of the old brand as the null hypothesis. One reason why statisticians prefer this procedure is that, if we regard the null hypothesis as being that the population mean for the new brand *equals* 60, we can calculate the sampling distribution of the sample mean. Thus, we can control the probability of a Type I error (as will be indicated in greater detail in the next section). But if we regard the null hypothesis as being that the mean rating for the new brand *exceeds* that of the old brand, we cannot calculate the probability of a Type I error unless we know how large the difference between the means is. Since statisticians find it desirable to specify and control the probability of a Type I error, they have a preference for making the null hypothesis specific in this way.[5]

EXERCISES

8.1 In the 1960s the National Academy of Sciences carried out an extensive study to determine whether a new anesthetic called holothane results in higher death rates than other anesthetics like pentothal or cyclopropane.
   (a) What is the null hypothesis? What is the alternative hypothesis?
   (b) What is the consequence of a Type I error and a Type II error?

8.2 A firm's engineers test the hypothesis that 2 percent of the items coming off its assembly line are defective. They pick a random sample of 5 items each hour. If any of the items is defective, they reject the null hypothesis.
   (a) What is the test statistic?
   (b) What is the acceptance region?
   (c) What is the rejection region?
   (d) Is the test a one-tailed test or a two-tailed test?

8.3 "The statistician should set the probability of Type I error as low as possible."
   (a) Do you agree with this statement? Why or why not?
   (b) In Figure 8.2, is the decision rule in accord with this statement?

8.4 Which of the following are two-tailed tests? Which are one-tailed tests?
   (a) Determine whether $\mu = 3{,}000$, the alternative hypothesis being that $\mu \neq 3{,}000$.
   (b) Determine whether $\mu = 10$, the alternative hypothesis being that $\mu < 10$.
   (c) Determine whether $\mu = 0$, the alternative hypothesis being that $\mu > 0$.

[5] For a classic discussion of these and related matters, see R. A. Fisher, "Statistical Inference," reprinted in E. Mansfield, *Statistics for Business and Economics: Readings and Cases* (New York: Norton, 1980). As pointed out in footnote 4, the null hypothesis may sometimes be defined as a population parameter's being at most or at least some value. More is said about this in footnote 9.

8.4 One-Sample Test of a Mean: Large Samples

Having described some of the basic concepts in the statistical theory of hypothesis testing, we are ready now for a detailed consideration of the most important statistical tests. In this and the following sections, we are concerned with the case where data are available concerning a single sample. This section covers the test of a mean; the following section discusses the test of a proportion. In both sections,[6] we assume that the sample is large ($n > 30$).

SETTING UP THE TEST

Although there is no cut-and-dried procedure to be followed, certain steps are essential in testing any hypothesis. The first step is to formulate the null hypothesis (the hypothesis that is being tested) and the alternative hypothesis. Exactly what parameter is relevant for the problem at hand, and what value of this parameter constitutes the null hypothesis? Is the alternative hypothesis a *one-sided alternative,* which means that the decision maker cares about departures from the null hypothesis in one direction? Or is the alternative hypothesis a *two-sided alternative,* which means that the decision maker wants to detect departures from the null hypothesis in both directions?[7]

To be specific, suppose that we take the case of the school system that receives the shipment of light bulbs. In this case, as we know from the previous section, the null hypothesis is that the mean length of life of the bulbs in the shipment is 2,000 hours; and the alternative hypothesis is one-sided (namely, that the mean length of life is less than 2,000 hours). Suppose that the school system knows from past experience that the standard deviation of the length of life of light bulbs in a shipment of this sort is 200 hours, and that it tests a random sample of 100 bulbs. What sort of test procedure or decision rule should the school system use to determine whether, once it obtains the results of the sample, it should accept the null hypothesis (and accept the shipment) or reject the null hypothesis (and reject the shipment)? Clearly, the school system should be more inclined to reject the null hypothesis if the sample mean is relatively low than if it is relatively high, but how low must the sample mean be to justify rejecting the null hypothesis?

To answer this question, the decision maker or statistician must first recognize that any such test procedure or decision rule can make either a Type I error or a Type II error. In this case, a Type I error will

[6] In the next section, it is sometimes assumed that n is considerably in excess of 30. See footnote 14.

[7] Of course, one-sided alternative hypotheses lead to one-tailed tests, and two-sided alternative hypotheses lead to two-tailed tests.

Alternative courses of action	State of nature*	
	$\mu = 2,000$	$\mu < 2,000$
Accept shipment	Correct decision	Type II error
Reject shipment	Type I error	Correct decision

* If $\mu > 2,000$, the correct decision is to accept the shipment. See footnotes 3, 4, and 9.

Table 8.2
Possible Outcomes of the Decision Problem concerning the Shipment of Light Bulbs

arise if the school system rejects a shipment in which the mean length of life of the bulbs is 2,000 hours; a Type II error will arise if the school system accepts a shipment where the mean length of life of the bulbs is less than 2,000 hours. Table 8.2 shows the possibilities in this case, much as Table 8.1 did in the case of the engineers. As stressed in an earlier section, any test procedure or decision rule is characterized by a certain probability of a Type I error and a certain probability of a Type II error, denoted as follows:

> The **probability of a Type I error** is designated as α (alpha), and the **probability of a Type II error** is designated as β (beta).

α **and** β

Whether a particular test procedure or decision rule is appropriate is dependent on whether α and β are set at the right levels from the standpoint of the decision maker. In this case, whether a particular test procedure or decision rule is appropriate depends on whether α and β are set at levels that are satisfactory to the school system receiving the shipment of light bulbs.

The value assigned to α is called the **significance level** of the test. Typically, the significance level is set at .05 or .01, which means that the probability of a Type I error is .05 or .01. The decision maker can set the significance level at any amount that he or she chooses. However, in setting this level it generally is wise to keep in mind the relative costs of committing various types of errors. If a Type I error is very costly, a very low level for α should be chosen; on the other hand, if a Type I error is not very costly, a higher value for α is acceptable. At the same time it should be recognized that for a fixed sample size, the probability of a Type I error cannot be reduced without increasing the probability of a Type II error. (For example, if the school system establishes a decision rule whereby the shipment is rejected if the sample mean is less than 1,600 hours, this rule will have a lower value of α than if it rejects only those where the sample mean is less than 1,800 hours; but the value of β will be higher for the former rule.) Thus, in choosing α, the decision maker should keep β in mind as well, picking a value of α such that α and β bear a sensible relation to the relative costs of Type

I and Type II errors. (In a subsequent part of this section, we shall indicate how the value of β can be calculated.)[8]

DECISION RULES

Once the value of α has been specified, it is relatively simple to determine the decision rule that should be applied to test whether a population mean equals a specified value. For example, in the case of the school system that is testing whether the mean length of life of the light bulbs is 2,000 hours, we know that if the null hypothesis is true, the sampling distribution of the sample mean is as shown in Figure 8.3. Since the sample size is large, the central limit theorem assures us that the sample mean will be approximately normally distributed. And since $\sigma = 200$ and $n = 100$, $\sigma/\sqrt{n} = 20$. Thus, if the null hypothesis is true (that is, if the population mean is 2,000) the sample mean must be (approximately) normally distributed, with a mean of 2,000 and a standard deviation of 20, as shown in Figure 8.3.

Since (for reasons given in the previous section) this is a one-tailed test, the rejection region is entirely in the lower tail of the sampling distribution in Figure 8.3. In other words, our test consists of determining whether the value of the sample mean is below a certain number (labeled c in Figure 8.3). If the sample mean is below this number, the null hypothesis is rejected; if it is not below this number, the null hypothesis is accepted. The value of c that should be used depends on the choice of α, because α equals the probability that the sample mean's value will fall below c when the null hypothesis is true. In other words, the value of α equals the area under the sampling distribution to the left of c in Figure 8.3. Thus, since this sampling distribution is (approximately) normal with a mean of 2,000 and a standard deviation of 20,

$$c = 2,000 - z_\alpha \cdot 20,$$

where z_α is the value of the standard normal variable that is exceeded with a probability of α. Why is this equation correct? Because the area under the standard normal curve to the left of $-z_\alpha$ equals α. Thus, to find c, we must find the value of $\overline{X}$ corresponding to a Z value of $-z_\alpha$. What is the value of $\overline{X}$ corresponding to a Z value of $-z_\alpha$? Based on equation (5.2), the answer is $\mu - z_\alpha\sigma_{\overline{x}}$, or $2,000 - z_\alpha \cdot 20$, which is what is shown in the equation above.

To illustrate the use of this equation, suppose that the school system testing the light bulbs decides to set α at .05. In other words, it

---

[8] For some discussion of the historical origins of the .05 significance level, which is often adopted, see M. Cowles and C. Davis, "On the Origins of the .05 Level of Statistical Significance," *American Psychologist*, May 1982.

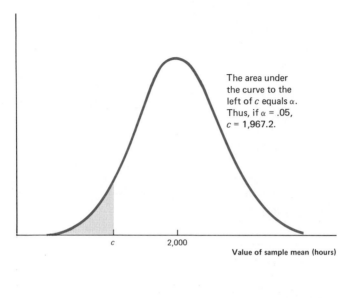

The area under
the curve to the
left of c equals α.
Thus, if $\alpha = .05$,
$c = 1,967.2$.

Figure 8.3
Sampling Distribution
of Mean Length of
Life of Bulbs in Sample
(if Null Hypothesis Is
True)

wants the probability of a Type I error to equal .05. In this case, since $z_{.05} = 1.64$, the null hypothesis should be rejected if the value of the sample mean is less than 1,967.2 hours (because $c = 2,000 - 1.64(20)$, or 1,967.2). If the sample mean is not less than 1,967.2 hours, the null hypothesis should be accepted. Or suppose that the school system decides to set α at .01, which means that it wants the probability of a Type I error to equal .01. In this case, since $z_{.01} = 2.33$, the null hypothesis should be rejected if the value of the sample mean is less than 1,953.4 hours (because $c = 2,000 - 2.33(20)$, or 1,953.4). If the sample mean is not less than 1,953.4 hours, the null hypothesis should be accepted.

In general, the decision rule for testing the null hypothesis that a population mean equals a certain amount μ_0 against the alternative hypothesis that it is less than μ_0 is as follows.

> ***Decision Rule when Alternative Hypothesis is $\mu < \mu_0$:*** *Accept the null hypothesis if $\bar{x} \geq \mu_0 - z_\alpha \sigma / \sqrt{n}$; reject the null hypothesis if $\bar{x} < \mu_0 - z_\alpha \sigma / \sqrt{n}$.*

Test that $\mu = \mu_o$, the Alternative Hypothesis Being that $\mu < \mu_0$.

This is no more than a restatement of what we have already said, since $c = \mu_0 - z_\alpha \sigma / \sqrt{n}$, according to our previous results.[9]

---

[9] This decision rule often is stated in terms of $z = \sqrt{n}(\bar{x} - \mu_0)/\sigma$, rather than $\bar{x}$. In these terms, the rule is: Accept the null hypothesis if $z \geq -z_\alpha$; reject the null hypothesis if $z < -z_\alpha$.

In addition, we should present the decision rule for testing the null hypothesis that a population mean equals a certain amount μ_0 against the alternative hypothesis that it is *more than* μ_0.

Test that $\mu = \mu_0$, the Alternative Hypothesis Being that $\mu > \mu_0$

Decision Rule when Alternative Hypothesis is $\mu > \mu_0$: *Accept the null hypothesis if $\bar{x} \leq \mu_0 + z_\alpha \sigma / \sqrt{n}$; reject the null hypothesis if $\bar{x} > \mu_0 + z_\alpha \sigma / \sqrt{n}$.*

The reasoning leading up to this decision rule is, of course, precisely the same as that leading up to the decision rule in the case where the alternative hypothesis is $\mu < \mu_0$. The only difference is that the rejection region is in the upper tail, not the lower tail, of the sampling distribution of the sample mean.[10]

If the test is two-tailed—that is, if the alternative hypothesis is that the population mean *differs from* μ_0—the decision rule is as follows.

Test that $\mu = \mu_0$, the Alternative Hypothesis Being that $\mu \neq \mu_0$

Decision Rule when Alternative Hypothesis is $\mu \neq \mu_0$: *Accept the null hypothesis if $\mu_0 - z_{\alpha/2} \sigma / \sqrt{n} \leq \bar{x} \leq \mu_0 + z_{\alpha/2} \sigma / \sqrt{n}$; reject the null hypothesis if $\bar{x} < \mu_0 - z_{\alpha/2} \sigma / \sqrt{n}$ or if $\bar{x} > \mu_0 + z_{\alpha/2} \sigma / \sqrt{n}$.*

Since the probability of a Type I error must equal α, the probability that the sample mean will fall in the rejection region in *each* tail of its sampling distribution, given that the null hypothesis is true, is set equal to $\alpha/2$.

Finally, note that there is a close connection between the test procedure described in the previous paragraph and the confidence interval described in Chapter 7. *If the null hypothesis is that the population mean equals μ_0, this hypothesis will be rejected if and only if a confidence interval (with confidence coefficient equal to $1 - \alpha$) for μ does not include μ_0.* For example, if one computes a 95 percent confidence interval for the population mean, and it turns out to be 15 to 20, then it follows that a two-tailed test (with $\alpha = .05$) will reject the null hypothesis that the population mean is any value below 15 or above 20. This intimate connection between hypothesis testing and confidence intervals is present for other tests too.[11]

In footnotes 3 and 4 we noted that the null hypothesis can be defined as $\mu \geq 2,000$ rather than $\mu = 2,000$. If this definition is used, α is the *maximum* probability of a Type I error. That is, if the population mean exceeds 2,000 hours, the probability of a Type I error will be less than α.

[10] As pointed out in footnote 9, decision rules of this sort can be expressed in a variety of equivalent ways. For example, when stated in terms of $z = \sqrt{n}(\bar{x} - \mu_0)/\sigma$, this rule is: Accept the null hypothesis if $z \leq z_\alpha$; reject the null hypothesis if $z > z_\alpha$. Of course, the other decision rules presented below can also be stated in a variety of different ways.

[11] For example, see Exercise 8.36.

OPERATING-CHARACTERISTIC CURVE

Up to now the decision rules presented in this section have been constructed so that the probability of a Type I error—that is, α—is fixed at a predetermined level. But how large is β, the probability of a Type II error? For example, if the school system receiving the shipment of light bulbs sets α equal to .05, how large is the probability that the school system will accept the shipment, given that the mean length of life of the bulbs is less than 2,000 hours? Clearly, the answer depends upon the extent to which the mean length of life of the bulbs falls below 2,000 hours. To see this, recall from the previous part of this section that if $\alpha = .05$ the test procedure is the following: Reject the null hypothesis (and thus the shipment) if $\bar{x}$ is less than 1,967.2 hours; otherwise, accept the null hypothesis. Thus, the probability that the school system will accept the shipment, given that the mean length of life of the light bulbs is less than 2,000 hours, equals the probability that the sample mean will be 1,967.2 hours or more under these circumstances. It is obvious that this probability will decrease as the population mean falls increasingly farther below 2,000 hours.

Figure 8.4 shows the value of β at each value of the population mean. In accord with our argument in the previous paragraph, the value of β decreases as the population mean falls farther and farther below 2,000 hours. To show how the value of β is calculated, suppose that the mean length of life of the bulbs in the shipment is 1,960 hours. The probability that the shipment will be accepted equals the probability that the sample mean will be 1,967.2 hours or more. (See panel B of Figure 8.5.) Since $\bar{X} \geqslant 1,967.2$ if and only if $(\bar{X} - 1,960) \div 20 \geqslant (1,967.2 - 1,960) \div 20$, the probability that $\bar{X} \geqslant 1,967.2$ equals the probability that $(\bar{X} - 1,960) \div 20 \geqslant 0.36$ (because $(1,967.2 - 1,960) \div$

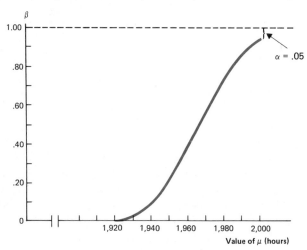

Figure 8.4
Operating-Characteristic Curve of Test that Mean Length of Life of Bulbs Equals 2,000 Hours

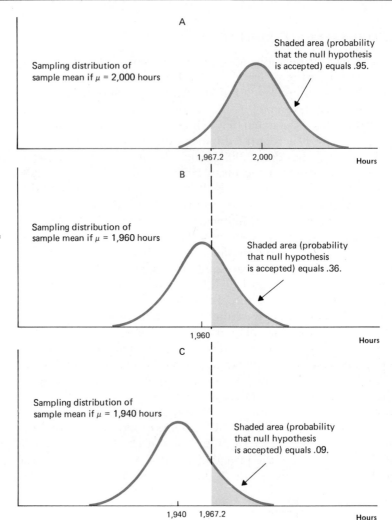

Figure 8.5
Sampling Distribution
of Sample Mean, if $\mu =$
2,000, 1,960, and
1,940 Hours

20 = 0.36). Since $\overline{X}$ is distributed approximately normally with a mean of 1,960 hours and a standard deviation of 20 hours, $(X - 1,960) \div 20$ has the standard normal distribution. Thus, the probability that $(\overline{X} - 1,960) \div 20 \geq 0.36$ equals the area under the standard normal curve to the right of .36, which equals .36. (See Appendix Table 2.)

What if the population mean is 1,940 hours, not 1,960 hours? Under these conditions, what is β? The probability that the shipment will be accepted equals the probability that the sample mean will be 1,967.2 hours or more. (See panel c of Figure 8.5.) Since $\overline{X} \geq 1,967.2$ if and only if $(\overline{X} - 1,940) \div 20 \geq (1,967.2 - 1,940) \div 20$, the probability that $\overline{X} \geq 1,967.2$ equals the probability that $(\overline{X} - 1,940) \div 20 \geq 1.36$ (because $(1,967.2 - 1,940) \div 20 = 1.36$). Since $\overline{X}$ is distributed approximately normally with a mean of 1,940 hours and a standard deviation of 20 hours, $(\overline{X} - 1{,}940) \div 20$ has the standard normal distribution. Thus, the probability that $(\overline{X} - 1{,}940) \div 20 \geqslant 1.36$ equals the area under the standard normal curve to the right of 1.36, which equals .09. (See Appendix Table 2.)

Using the methods described in the previous two paragraphs, we can calculate the value of β at a large number of values of the population mean. (For example, panel A of Figure 8.5 corresponds to the case where the population mean is 2,000 hours.) If we plot each value of β against the corresponding value of the population mean, we get the curve in Figure 8.4, which is called the operating-characteristic curve of this test. The definition of an operating-characteristic curve (or *OC curve*) is as follows.

> OPERATING-CHARACTERISTIC CURVE: *A test's **operating-characteristic curve** shows the value of β (that is, the probability of a Type II error) if the population parameter assumes various values other than that specified by the null hypothesis.*

Sometimes statisticians use the *power curve* rather than the operating-characteristic curve. The power curve shows the value of $(1 - \beta)$—whereas the operating-characteristic curve shows the value of β—at each value of the parameter under test. Clearly, one can easily deduce the power curve from the operating-characteristic curve, and vice versa, since they provide the same information.[12]

A test's operating-characteristic curve can be used to determine the value of α as well as the value of β. At the point where the parameter equals the value specified by the null hypothesis, the OC curve shows the probability that the null hypothesis will be accepted. Thus 1 minus this probability must equal the probability that the null hypothesis will be rejected, given that it is true (which is α, the probability of a Type I error). Consequently, at the point where the parameter equals the value specified by the null hypothesis, the distance between the OC curve and 1 equals α. For example, Figure 8.4 shows that at the point where the mean length of life of the bulbs is 2,000 hours, the distance between the OC curve and 1 equals .05, which we know equals α in this case.[13]

Not all tests have operating-characteristic curves shaped like that in Figure 8.4. If a test is designed to detect positive rather than nega-

---

[12] To derive the power curve of the test in Figure 8.4, we would plot $(1 - \beta)$ on the vertical axis, rather than β.

[13] Recall that we pointed out in an earlier part of this section that if the sample size is fixed, a reduction in the probability of a Type I error can be achieved only at the expense of an increase in the probability of a Type II error. The reason for this can be explained more fully now. Basically, the reason is that, if α is reduced, the OC curve must be shifted upwards, which means an increase in β.

tive departures of the population mean from the value specified by the null hypothesis, its operating-characteristic curve will be shaped like that in panel A of Figure 8.6. If a test is designed to detect departures of the population mean from the value specified by the null hypothesis in either direction (both positive and negative) the operating-characteristic curve will be shaped like that in panel B of Figure 8.6.

Once a test's OC curve is available, one is in a much better position to determine whether the test is appropriate. For example, if the

Figure 8.6
Shape of Operating-Characteristic Curve if Alternative Hypothesis Is $\mu > \mu_0$ or $\mu \neq \mu_0$

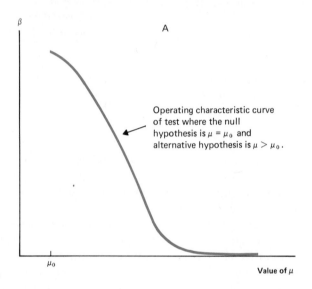

A

Operating characteristic curve of test where the null hypothesis is $\mu = \mu_0$ and alternative hypothesis is $\mu > \mu_0$.

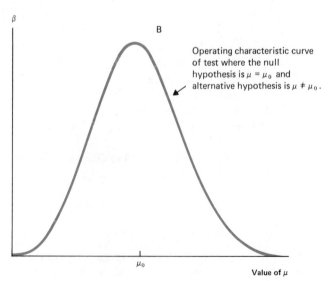

B

Operating characteristic curve of test where the null hypothesis is $\mu = \mu_0$ and alternative hypothesis is $\mu \neq \mu_0$.

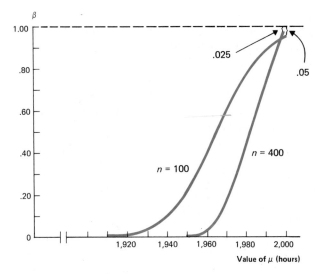

Figure 8.7
Operating-Characteristic Curves of Two Tests

officials of the school system receiving the light bulbs are informed of the OC curve for this test, they may feel that it results in too high a probability of both Type I error and Type II error. Figure 8.7 shows the OC curve for another test, one whereby the school system would choose a sample of 400 bulbs and reject the shipment if the sample mean were less than 1,980.4 hours. As you can see, this test results in a much smaller probability of Type I error (.025 rather than .05) and of Type II error (for practically all values of the population mean) than the earlier test. Why does this test have smaller probabilities of both types of error? Because it is based on a much larger sample. Whether or not the reduction in risk is worth the extra cost of inspecting the larger sample must be determined by the school system's officials. After weighing the relevant costs, the school system can specify the sort of OC curve it wants. Then, by a proper choice of sample size and significance level a test can be chosen that is best suited to its requirements.

EXTENSIONS OF THE TEST PROCEDURES

At this point we must indicate how the test procedures described in this section can be modified to suit circumstances somewhat different from those specified above. In particular, we will describe how a test of this sort can be constructed when the standard deviation is unknown and when the population is not large relative to the sample.

WHERE THE POPULATION STANDARD DEVIATION IS UNKNOWN. If σ is unknown and if the sample size is large ($n > 30$), the sample standard deviation s can be substituted for the population standard deviation σ in the decision rules given above. For example, if the alternative hypothesis is $\mu < \mu_0$, the decision rule says: accept the null hypothesis if $\bar{x} \geq \mu_0 - z_\alpha s/\sqrt{n}$; and reject the null hypothesis if $\bar{x} < \mu_0 - z_\alpha s/\sqrt{n}$. In Section 8.8 we shall indicate how a test can be constructed if σ is unknown and the sample size is small.

WHERE THE POPULATION IS SMALL. If the population is less than 20 times the sample (and if sampling is without replacement), the decision rules given above must be modified. In particular, $\sigma/\sqrt{n}$ (or $s/\sqrt{n}$ if σ is unknown) must be multiplied by the finite population correction factor $\sqrt{(N-n)/(N-1)}$. For example, if the alternative hypothesis is $\mu < \mu_0$ (and σ is known), the decision rule says: accept the null hypothesis if $\bar{x} \geq \mu_0 - z_\alpha(\sigma/\sqrt{n})\sqrt{(N-n)/(N-1)}$; and reject the null hypothesis if $\bar{x} < \mu_0 - z_\alpha(\sigma/\sqrt{n})\sqrt{(N-n)/(N-1)}$.

INTERPRETATION OF TEST RESULTS

When we test the hypothesis that the mean of a particular population equals a specified value μ_0, we base our test on the value of the sample mean. Because of sampling variation, we cannot expect the mean of the sample to be exactly equal to the population mean. Thus, if the sample mean is different from μ_0 we cannot conclude that the population mean does not equal μ_0. What matters is the probability that the size of the difference between the sample mean and μ_0 could have arisen by chance. If this probability is so great that the null hypothesis is not rejected, we say that the difference between the sample mean and μ_0 (the value of the population mean specified by the null hypothesis) is *not statistically significant.* That is, this difference could be attributable to chance. On the other hand, if this probability is so low that the null hypothesis is rejected this difference between the sample mean and μ_0 is termed *statistically significant.*

Statistical Significance versus Practical Significance

The concept of *statistical* significance should not be confused with *practical* significance. Even if there is a statistically significant difference between the sample mean and the postulated value of the population mean, this difference may not really matter. For example, a sociologist wants to determine whether the mean income of families in a certain town is $24,000. Based on a very large sample, he finds that the sample mean is $24,152, which differs significantly (in a statistical sense) from the postulated $24,000. Whether this difference is of any practical significance depends on whether he really cares about a difference of $152. For many purposes, such a difference may not be important. For instance, he may only want to determine whether the mean falls between $23,500 and $24,500, or whether it falls outside this range. If so, the observed difference of $152 between the sample mean and the postulated value of the population mean is of no practical importance even though it is statistically significant.

THE CASE OF QUICKNESS AND DEXTERITY

To measure the effects of a particular training program, a psychologist obtains data concerning the performance of two groups of workers. One group has undergone the training program; the other has not. The psychologist obtains data on about 50 aspects of the workers' performance. With regard to two aspects, he finds that there is a statistically significant difference (at the 0.05 probability level) between the mean performance of the two groups. The results for these two aspects of performance—quickness and dexterity—are as follows:

	Mean quickness	*Mean dexterity*
Workers with training	11.03	16.97
Workers without training	10.98	17.02

With regard to the other aspects of performance, there is not a statistically significant difference between the two groups. The psychologist issues a report saying that this training program has a positive effect on workers' quickness, but a negative effect on their dexterity. Since these differences are statistically significant, he says that they cannot be due to chance, and that they are important. When his report is published, it is criticized severely by his colleagues, who imply (in a polite way) that he is a statistical ignoramus. Where did he go wrong?

SOLUTION

If 50 tests are carried out, and if there is a 5 percent chance that each of them will reject the null hypothesis (that the training has no effect) when it is true, one would expect that about 50 times 0.05, or 2.5 tests will turn out this way even if the training really has no effect. Thus, when one looks at the battery of tests as a whole, the probability of getting two "statistically significant" differences by chance is much larger than if only these two tests had been carried out. After all, if one carries out enough tests, eventually one or two "statistically significant" results will arise by chance. Also, the differences between the means in the table above are less than 1/2 of 1 percent. Even if such differences could not be attributed to chance, they are likely to be of little practical importance.

It is also worth noting that the fact that the null hypothesis is rejected by any of the decision rules given above does not *prove* that the null hypothesis is false. Moreover, the fact that the null hypothesis is accepted by any one of these decision rules does not *prove* that the null hypothesis is true. These test procedures are designed to detect cases where it is unlikely that the sample evidence could have occurred if the null hypothesis was true: In such cases the null hypothesis is rejected. But as long as there is *some chance* that the sample evidence could have occurred if the null hypothesis was true, there is no way to *guarantee* that the results of the test will be absolutely correct. As stressed throughout this section, statistical test procedures are generally designed so that there is a nonzero probability of a Type I or a Type II error. To eliminate such errors entirely would generally be impossible or foolishly expensive.

EXERCISES

8.5 A physician suspects that her patients are being cheated by a local pharmacist. When filling a particular prescription, the pharmacist should give each patient 12 capsules in a small bottle. The physician picks a random sample of 60 bottles, and finds that the mean number in a bottle equals 11.95, the standard deviation being 0.09. She sets α equal to 0.05.
  (a) If the physician is only interested in detecting whether the true mean is under 12, what is the relevant decision rule?
  (b) If the true mean equals 11.9, and if the decision rule in (a) is applied, what is the value of β?
  (c) What should the physician's decision be?

8.6 A firm produces metal wheels. The mean diameter of the wheels should be 4 inches. Because of chance variation and other factors, the diameters of the wheels vary, the standard deviation being 0.05 inches. To test whether the mean is really 4 inches, the firm selects a random sample of 50 wheels and finds that the sample mean diameter equals 3.97 inches. (The sample is less than 1 percent of the population.)
  (a) If the firm is interested in detecting whether the true mean is above or below 4 inches, and if α is set equal to 0.01, what is the relevant decision rule?
  (b) If the true mean equals 3.99 inches, and if the decision rule in (a) is applied, what is the value of β?
  (c) What should the firm's decision be?

8.7 A class contains 60 students. The teacher wants to test whether the mean IQ in the class equals 120. He chooses a random sample of 36 students (without replacement), and finds that the mean and standard deviation of the IQs in the sample are 122.8 and 10.9, respectively.
  (a) If the teacher is only interested in detecting whether the true mean exceeds 120, and if α is set at 0.02, what is the relevant decision rule?

(b) If the true mean equals 121, and if the decision rule in (a) is applied, what is the value of β?

(c) What should the teacher's decision be?

8.8 A statistician is testing whether $\mu = 100$ on the basis of a random sample of 49 observations from a very large population. She knows that $\sigma = 5$, and the test is two-tailed.

(a) If she uses the test procedure described in the text, and if α is set equal to 0.05, what is the probability of a Type II error if $\mu = 96, 98, 100, 102,$ and 104?

(b) Draw five points on the operating-characteristic curve of the test.

(c) Draw five points on the power curve of the test.

8.9 A sociologist at a large state university is testing whether the mean age of employees at the university equals 50 years on the basis of a random sample of 64 employees. He knows that the standard deviation of the ages of the employees at the university equals 10 years. The alternative hypothesis is that $\mu < 50$ years.

(a) If he uses the test procedure described in the text, and if α is set at 0.01, what is the probability of a Type II error if $\mu = 47, 48, 49,$ and 50 years?

(b) Draw four points on the operating-characteristic curve of the test.

(c) Draw four points on the power curve of the test.

8.5 One-Sample Test of a Proportion: Large Samples

Many statistical investigations are aimed at testing whether a population proportion equals a specified value. For example, in acceptance sampling the object is to test whether the proportion of defective items in a shipment equals a specified value. (Recall our discussion of acceptance sampling in Chapter 4.) In this section we will describe the statistical test procedures that can be used in such investigations.

SETTING UP THE TEST

As emphasized in the previous section, the first step in any test procedure (whether of a mean, a proportion, or any other parameter) is to formulate the hypothesis that is being tested. This is a crucial aspect of the work, since, if the null hypothesis and alternative hypothesis are formulated incorrectly, the test procedure will be designed to answer the wrong question. Suppose that a United States senator receives a report from a Washington lobbyist, stating that 70 percent of American voters favor increased nonmilitary aid to Central America as a means of reducing international tensions there. The senator is skeptical of this report since she believes that the percentage favoring such aid is lower than this figure. The senator's assistants draw a simple random sample of 400 voters and ask these voters' preferences in this regard. What is the null hypothesis in this case? It is that the proportion Π of all voters favoring increased aid equals .70. What is the alternative hypothesis in this case? It is that Π is less than .70 since the senator is interested only in determining whether Π is below .70. It makes no difference to the senator whether Π equals .70 or is greater than .70.

Having formulated the null and alternative hypotheses, the next step is to determine the proper values of α (the probability of a Type I error) and of β (the probability of a Type II error). As stressed in the previous section, these probabilities should be chosen on the basis of the relative costs of a Type I and a Type II error. In the present case, a Type I error occurs if the senator rejects the hypothesis that $\Pi = .70$ when in fact this hypothesis is true. A Type II error occurs if the senator accepts the hypothesis that $\Pi = .70$ when in fact Π is less than .70. If a Type I error is much more costly than a Type II error, the value of α should be low relative to the value of β. If a Type II error is much more costly than a Type I error, the value of β should be low relative to the value of α. For each possible value of α one can calculate the test's operating-characteristic curve which shows β. (The method for calculating the operating-characteristic curve is demonstrated later in this section.) The value of α should be chosen so that the OC curve is in line with the relative costs in the case at hand. For the sake of concreteness, suppose that in this case α is set equal to .05.

DECISION RULES

Once the value of α has been specified, we can determine the decision rule that should be applied to test whether a population proportion equals a specified value Π_0. As one might guess, the test statistic is the sample proportion p. If the null hypothesis is true (that is, if $\Pi = \Pi_0$), the sample proportion is approximately normally distributed with a mean of Π_0 and a standard deviation of $\sqrt{\Pi_0(1 - \Pi_0)/n}$, if the sample size n is sufficiently large. (Recall Chapter 7.) Thus, if the null hypothesis is true, the sampling distribution of the sample proportion is as shown in Figure 8.8. In the case of the senator, the rejection region is entirely in the lower tail of this sampling distribution. (It is a one-tailed test.) Thus, our test consists of determining whether the value of the sample proportion is below a certain number, labeled b in Figure 8.8. If the value of the sample proportion is below this number the null hypothesis is rejected; if it is not below this number the null hypothesis is accepted. The value of b depends on the choice of α, because α equals the probability that the sample proportion's value will fall below b

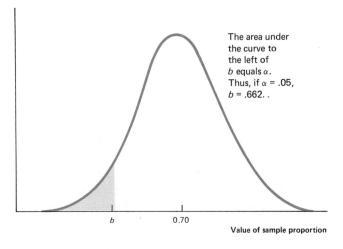

The area under the curve to the left of b equals α. Thus, if $\alpha = .05$, $b = .662$. .

Figure 8.8
Sampling Distribution of Proportion of Voters in Sample Favoring Increased Non-military Aid to Central America, if Null Hypothesis is True

Rejection region   Acceptance region

when the null hypothesis is true. In other words, as shown in Figure 8.8, α equals the value of the area under the sampling distribution to the left of b. Thus,

$$b = \Pi_0 - z_\alpha \sqrt{\frac{\Pi_0(1 - \Pi_0)}{n}},$$

where z_α is the value of the standard normal variable that is exceeded with a probability of α.

In the case of the senator, the null hypothesis is that $\Pi = .70$, which means that $\Pi_0 = .70$. Since $n = 400$, it follows that if the null hypothesis is true, the sample proportion will be approximately normally distributed with a mean of 0.70 and a standard deviation of $\sqrt{(.70)(.30)/400}$ or .023. Given that α is set equal to .05, what must be the value of b? Applying the formula in the previous paragraph, b must equal $.70 - 1.64(.023)$, or .662, since $z_{.05} = 1.64$. Thus, the decision rule that the senator should use is the following: Reject the null hypothesis if the sample proportion is less than .662; accept the null hypothesis if the sample proportion is greater than or equal to .662.

In general, when n is sufficiently large,[14] the decision rule for

---

[14] In Chapters 5 and 7, we discussed how large the sample size must be for the normal approximation to be useful.

testing the null hypothesis that a population proportion equals a certain amount Π_0 against the alternative hypothesis that the population is less than Π_0 is as follows.

Test that $\Pi = \Pi_0$, the Alternative Hypothesis Being that $\Pi < \Pi_0$

Decision Rule when Alternative Hypothesis is $\Pi < \Pi_0$: *Accept the null hypothesis if* $p \geq \Pi_0 - z_\alpha \sqrt{\Pi_0(1 - \Pi_0)/n}$; *reject the null hypothesis if* $p < \Pi_0 - z_\alpha \sqrt{\Pi_0(1 - \Pi_0)/n}$.

This restates what we have already said, since $b = \Pi_0 - z_\alpha\sqrt{\Pi_0(1 - \Pi_0)/n}$.

The decision rule for testing the null hypothesis that a population proportion equals a certain amount Π_0 against the alternative hypothesis that it is *greater* than Π_0 is as follows.

Test that $\Pi = \Pi_0$, the Alternative Hypothesis Being that $\Pi > \Pi_0$

Decision Rule when Alternative Hypothesis is $\Pi > \Pi_0$: *Accept the null hypothesis if* $p \leq \Pi_0 + z_\alpha \sqrt{\Pi_0(1 - \Pi_0)/n}$; *reject the null hypothesis if* $p > \Pi_0 + z_\alpha \sqrt{\Pi_0(1 - \Pi_0)/n}$.

The reasoning behind this decision rule is the same as that for the decision rule in the previous paragraph. The only difference is that the rejection region is in the upper tail, not the lower tail, of the sampling distribution of the sample proportion.

Finally, if the test is two-tailed—that is, if the alternative hypothesis is that the population proportion *differs* from P_0—the decision rule is the following:

Test that $\Pi = \Pi_0$, the Alternative Hypothesis Being that $\Pi \neq \Pi_0$

Decision Rule when Alternative Hypothesis is $\Pi \neq \Pi_0$: *Accept the null hypothesis if* $\Pi_0 - z_{\alpha/2} \sqrt{\Pi_0(1 - \Pi_0)/n} \leq p \leq \Pi_0 + z_{\alpha/2} \sqrt{\Pi_0(1 - \Pi_0)/n}$; *reject the null hypothesis if* $p < \Pi_0 - z_{\alpha/2} \sqrt{\Pi_0(1 - \Pi_0)/n}$ *or if* $p > \Pi_0 + z_{\alpha/2} \sqrt{\Pi_0(1 - \Pi_0)/n}$.

Since the probability of a Type I error must equal α, the probability that the sample proportion will fall in the rejection region in *each* tail of its sampling distribution, given that the null hypothesis is true, is set equal to $\alpha/2$.[15]

OPERATING-CHARACTERISTIC CURVE

The decision rules presented in the previous part of this section are constructed so that the probability of a Type I error equals α. For example, in the case of the senator, α was set equal to .05. But how large is β, the probability of a Type II error? The answer is given by the

---

[15] Of course, if the population is small relative to the sample, $\sqrt{\Pi_0(1 - \Pi_0)/n}$ should be multiplied by $\sqrt{(N - n)/(N - 1)}$ in each of these decision rules.

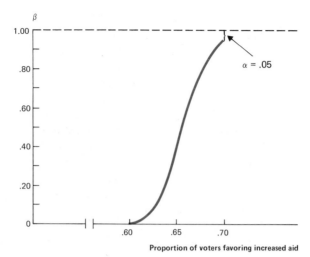

Figure 8.9
Operating-Characteristic Curve of Test that Proportion of Voters Favoring Non-military Aid to Central America Is 0.70

test's operating-characteristic curve. Figure 8.9 shows the operating-characteristic curve of the senator's test. This curve indicates that the probability that the senator will accept the hypothesis that $\Pi = .70$ when in fact $\Pi = 2/3$ is .58, and the probability that the senator will accept the hypothesis that $\Pi = .70$ when in fact $\Pi = .60$ is less than .01. As we pointed out at the beginning of this section, the decision maker should set the value of α (and n) so that the operating-characteristic curve of the test is suitable for the problem at hand.

To illustrate how the OC curve in Figure 8.9 was calculated, consider the point on this curve corresponding to $\Pi = .64$. To determine the value on the OC curve that corresponds to $\Pi = .64$, recall that the null hypothesis ($\Pi = .70$) will be accepted by the senator if the sample proportion p is .662 or greater. Thus, the probability that the senator will accept the null hypothesis equals the probability that the sample proportion will be .662 or greater. If $\Pi = .64$, the sample proportion will be approximately normally distributed with mean equal to .64 and standard deviation equal to $\sqrt{(.64)(.36)/400}$, or .024. Consequently, the probability that the sample proportion will be .662 or more is equal to the probability that a normal variable with a mean of .64 and a standard deviation of .024 will be .662 or more. Applying the techniques we learned in Chapter 5, this probability is .18.[16] Thus, .18 is the value of the OC curve when $\Pi = .64$.

The following example further illustrates the application of these test procedures.

---

[16] The probability that the value of a normal variable with a mean of .64 and a standard deviation of .024 will be at least .662 equals the probability that the value of the standard normal variable will be at least $(.662 - .64) \div .024$, or .92. Using Appendix Table 2, we find that this probability is .18.

EXAMPLE 8.1 In recent years, there has been considerable controversy among criminologists concerning whether or not people in lower social classes are more likely than others to commit crimes. According to John Braithwaite of the Australian Institute of Criminology,[17] 44 out of a sample of 53 studies indicated that this was the case for officially recorded juvenile crime. Assuming that this was a random sample of all such studies, test the hypothesis that half of all such studies have indicated that this was the case. (Assume that $\alpha = .05$ and that you are interested in detecting whether the proportion exceeds or is less than one-half.)

SOLUTION: Let Π be the proportion of all studies that indicate that people in lower social classes are more likely than others to commit crimes. Then H_0: $\Pi = .50$ and H_1: $\Pi \neq .50$. (That is, the null hypothesis is that $\Pi = .50$ and the alternative hypothesis is that $\Pi \neq .50$.) Since $z_{.025} = 1.96$,

$$z_{\alpha/2}\sqrt{\Pi_0(1 - \Pi_0)/n} = 1.96\sqrt{\frac{(.50)(.50)}{53}} = .135$$

$$\Pi_0 + z_{\alpha/2}\sqrt{\Pi_0(1 - \Pi_0)/n} = .50 + .135 = .635$$

$$\Pi_0 - z_{\alpha/2}\sqrt{\Pi_0(1 - \Pi_0)/n} = .50 - .135 = .365.$$

Thus, the null hypothesis should be rejected if the sample proportion is greater than .635 or less than .365. Since the sample proportion is $44/53 = .83$, the null hypothesis should be rejected.[18]

[17] J. Braithwaite, "The Myth of Social Class and Criminality Reconsidered," *American Sociological Review*, February 1981.

[18] Of course, this is only an example illustrating the use of statistical techniques, not a real analysis of this complex issue. The conclusion in the text is not presented as anything more than the outcome of a simple statistical illustration, useful for pedagogical purposes. For discussion of this issue, see Braithwaite's article and the references cited there.

EXERCISES

8.10 Suppose that Massachusetts General Hospital is testing the hypothesis that the proportion of its patients without medical insurance equals 10 percent.
(a) In this case, what is H_0?
(b) Under what circumstances will the hospital incur a Type I error?
(c) Under what circumstances will the hospital incur a Type II error?
(d) In this case, what considerations will determine the proper value of α and β?

8.11 The mayor of Atlantis makes a speech in which he hints that too many errors are being made by the city clerks in their review of property tax forms submitted by the city residents. One of the city clerks, in response to this speech, suggests that a random sample of 20 of the forms handled by each clerk should be inspected each day and that the clerk should be reprimanded if any form is handled incorrectly.

(a) What is the decision rule if this suggestion is accepted?

(b) What is the value of α if this suggestion is accepted (and if the hypothesis being tested is that 1 percent of the forms are handled incorrectly)?

(c) If this suggestion is adopted, what is the value of β if 5 percent of the forms are handled incorrectly?

8.12 The president of the Crooked Arrow National Bank wants to test whether 60 percent of the bank's loans are made to persons who reside in the city where the bank is located. The bank's statistician chooses a random sample of 200 of the people to whom the bank has made loans and finds that 52 percent reside in this city.

(a) If a 5 percent significance level is used, should this hypothesis be accepted or rejected? (Use a two-tailed test).

(b) If a 1 percent significance level is used, should this hypothesis be accepted or rejected? (Use a two-tailed test.)

8.13 A survey is to be carried out to determine whether the proportion of high school students in Chester, Pa., who pass a particular achievement test is different from the past. In previous years, 40 percent passed this test. It is desired to set a 0.01 probability of concluding that a change has occurred when in fact this proportion has not changed. The survey consists of a simple random sample of 120 high school students (out of the 2,500 high school students in Chester).

(a) Set up the decision rule for this test.

(b) What is the probability of concluding that no change has occurred, if the true proportion who pass the test is now 42 percent?

(c) The survey shows that 43 percent of the sample now pass the test. What should one conclude from this result?

8.14 William Malone is responsible for maintaining the quality of parts produced by the factory for which he works. Since inspection destroys the part, he must employ sampling techniques. If the proportion of parts produced that are defective exceeds 8 percent, it is important that he be aware of it. He draws a random sample of 100 parts (from 20,000 parts produced) and finds that 16 are defective.

(a) Is this persuasive evidence that the defective rate exceeds 8 percent? Why, or why not?

(b) If he uses the test described in the text, what is the probability of accepting the hypothesis that the defective rate is 8 percent when in fact it is 12 percent?

8.15 The Internal Revenue Service wants to determine whether the percentage of personal income tax returns filed on time by taxpayers is less than last year's percentage, 81 percent. It selects a random sample of 400 of

this year's tax returns and finds that 317 of them were filed on time.
(a) Can the Internal Revenue Service be reasonably sure that the percentage filed on time has declined?
(b) If the significance level (that is, α) is set at 0.05, is the apparent decline statistically significant?
(c) If the test described in the text is used, what is the probability of accepting the hypothesis of no decline in the percentage, if in fact it now equals 79 percent?

8.16 George Moriarty (the professor's grandson) is testing whether $\Pi = 0.65$ on the basis of a random sample of 100 observations from a very large population. The alternative hypothesis is that $\Pi < 0.65$.
(a) If he uses the test procedure described in the text, and if α is set at 0.025, what is the probability of a Type II error if $\Pi = 0.60, 0.62$, and 0.65?
(b) Draw three points on the operating-characteristic curve of the test.
(c) Draw three points on the power curve of the test.

8.17 A public-interest law firm picks a random sample of 100 hi-fi repair stores and asks them to repair a hi-fi set. In 34 of the cases the customer is undercharged.
(a) Use these results to test (at the 10 percent significance level) the hypothesis that in one-half of all such cases customers are undercharged. (Use a two-tailed test.)
(b) Use a one-tailed test, where the alternative hypothesis is that $\Pi < .50$.

8.18 In October 1981, banks began to offer the All Savers Certificate, a new type of one-year certificate with interest that was tax-free (up to $1,000). A Virginia bank believes that about 70 percent of the people who bought these certificates obtained the money by withdrawing it from their savings accounts. It selects a random sample of 150 people who bought such certificates (which is a very small proportion of all such people).
(a) The bank says that it will reject this belief if more than 114 or less than 96 of the people in the sample obtained the money in this way. What value of α is the bank establishing?
(b) It turns out that 110 of the people in the sample obtained the money in this way. If the bank had set α at 0.05, would the difference between the sample proportion and 70 percent be statistically significant?

8.6 Two-Sample Test of Means: Large Samples

In previous sections, we have described how a statistical procedure can be formulated to test whether a population mean or a population proportion equals a specified value. We turn now to a case where a random sample is drawn from each of two populations. In this section we describe how one can test the hypothesis that the means of two populations are equal. In the next section, we will describe how one can test the hypothesis that two population proportions are equal. In both sections,[19] we assume that the sample size in each population is large (that is, $n > 30$).

SETTING UP THE TEST

Suppose that a market research firm wants to determine whether the mean rating consumers give to their favorite beer differs from that which is given to brand X, when both beers are unidentified. Specifically, the firm picks a random sample of n_1 people, gives each person his or her favorite beer in an unmarked can, and asks the individual to rate it on a scale from 1 (very poor) to 90 (perfect). The firm then picks another random sample of n_2 people, gives each person brand X (also in an unmarked can) and asks the person to rate it on the same scale. The market research firm would like to test whether the mean rating of the favorite beer differs from the mean rating of brand X.

In contrast to the previous two sections, we are concerned here with two populations, not one. The first population is the population of ratings that beer drinkers will give to their favorite beer if it is served to them in an unmarked can. The mean of this population is μ_1 and the standard deviation is σ_1. The second population is the population of ratings that beer drinkers will give to brand X when it is given to them in an unmarked can. The mean of this population is μ_2 and the standard deviation is σ_2. The market research firm draws a random sample of n_1 ratings from the first population. In other words, n_1 individuals are asked to rate their favorite beer in an unmarked can. Also, the market research firm draws a random sample of n_2 ratings from the second population. This means that n_2 people are asked to rate brand X beer in an unmarked can. There is no relation between the individuals chosen in the two samples, which are completely independent. How can the market research firm use these two samples to test whether μ_1 equals μ_2?

As we have stressed before, the first step in constructing a test is to formulate the null hypothesis and the alternative hypothesis. The null hypothesis here is that $\mu_1 = \mu_2$—or that $\mu_1 - \mu_2 = 0$. The alternative hypothesis is that $\mu_1 - \mu_2 \neq 0$, since the firm is interested in rejecting the null hypothesis both when $\mu_1 - \mu_2 < 0$ and when $\mu_1 - \mu_2 > 0$. In other words, the firm wants to reject the null hypothesis both when brand X's mean rating is higher than the mean rating of the favorite beer and when the reverse is true. Stated differently, what the firm wants is a two-tailed test.

[19] In the next section, n may have to be substantially greater than 30 for the normal approximation to apply. As pointed out in Chapters 5 and 7, this depends on the value of Π. In the rest of this chapter, we assume that the population is large relative to the sample (or that sampling is with replacement).

The next step is to specify the value of α, the significance level of the test. As we have emphasized in previous sections, the value of α should be determined on the basis of the relative costs of both Type I and Type II errors. An effort should be made to strike a proper balance between α and β, even if (as is sometimes the case) one has only a vague idea of the relative costs. One important point is that α should always be specified *before the data are examined*. (This is true for any type of test, not just those described in this section.) Otherwise, it would be possible for the investigator to choose a significance level small enough so that the null hypothesis is accepted (if the investigator wants to accept it) or a significance level large enough so that the null hypothesis is rejected (if the investigator wants to reject it). Such a procedure would be a mockery of correct statistical practice. For the sake of concreteness, suppose that α in this case is set equal to .01.

DECISION RULES

Once the value of α has been specified, we can determine the decision rule that should be applied to test whether the difference between the two population means (that is, $\mu_1 - \mu_2$) is zero. If both n_1 and n_2 are large, it can be shown that the sampling distribution of the difference between the sample means (that is, $\overline{X}_1 - \overline{X}_2$) is approximately normal, with a mean of $(\mu_1 - \mu_2)$ and a standard deviation of $\sqrt{\sigma_1^2/n_1 + \sigma_2^2/n_2}$. Thus, if the null hypothesis is true (which means that $\mu_1 - \mu_2 = 0$), the difference between the sample means is approximately normally distributed, with a mean of zero and a standard deviation of $\sqrt{\sigma_1^2/n_1 + \sigma_2^2/n_2}$. Consequently, if the market research firm knows that σ_1 (the standard deviation of the ratings of the favorite beer) is 12 and σ_2 (the standard deviation of the ratings of brand X) is 10, and if $n_1 = 100$ and $n_2 = 200$, the difference between the sample means is approximately normally distributed, with a mean of zero and a standard deviation of $\sqrt{12^2/100 + 10^2/200}$ or 1.39. In other words, if the null hypothesis is true, the sampling distribution of $(\overline{X}_1 - \overline{X}_2)$ is as shown in Figure 8.10.

Since the market research firm wants to reject the null hypothesis either when $\mu_1 - \mu_2 < 0$ or when $\mu_1 - \mu_2 > 0$, it must set up a rejection region under each tail of the sampling distribution of $(\overline{X}_1 - \overline{X}_2)$, as shown in Figure 8.10. The rejection region in *each* tail must be constructed so that the probability is $\alpha/2$ that the difference between the sample means will fall in this region if the null hypothesis is true. (Since there are two tails, the total probability of a Type I error will be α.) If the rejection region in the lower tail is

$$\overline{x}_1 - \overline{x}_2 < -z_{\alpha/2} \sqrt{\frac{\sigma_1^2}{n_1} + \frac{\sigma_2^2}{n_2}} = -2.58(1.39) = -3.59,$$

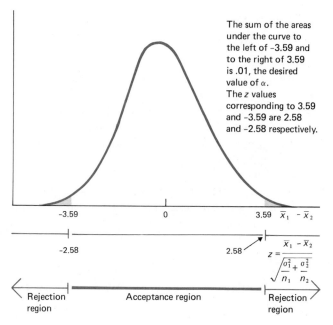

The sum of the areas under the curve to the left of –3.59 and to the right of 3.59 is .01, the desired value of α. The z values corresponding to 3.59 and –3.59 are 2.58 and –2.58 respectively.

Figure 8.10
Sampling Distribution of Difference between Sample Means, if Null Hypothesis Is True

and if the rejection region in the upper tail is

$$\bar{x}_1 - \bar{x} > z_{\alpha/2} \sqrt{\frac{\sigma_1^2}{n_1} + \frac{\sigma_2^2}{n_2}} = 2.58(1.39) = 3.59,$$

this condition is met. That is, the probability is .005 that $(\bar{X}_1 - \bar{X}_2)$ will be less than -3.59 if the two population means are equal, and the probability is .005 that $(\bar{X}_1 - \bar{X}_2)$ will be greater than 3.59 if the two population means are equal. Thus, the market research firm should use the following decision rule: Accept the null hypothesis if $-3.59 \leqslant (\bar{x}_1 - \bar{x}_2) \leqslant 3.59$; reject the null hypothesis if $(\bar{x}_1 - \bar{x}_2) < -3.59$ or if $(\bar{x}_1 - \bar{x}_2) > 3.59$. This decision rule has the desired probability of a Type I error, namely .01.

Like any random variable, the difference between the sample means can be expressed in standard units (that is, in z values). Since its mean is zero (if the null hypothesis is true) and its standard deviation is $\sqrt{\sigma_1^2/n_1 + \sigma_2^2/n_2}$, its observed value in standard units is

$$z = (\bar{x}_1 - \bar{x}_2) \div \sqrt{\frac{\sigma_1^2}{n_1} + \frac{\sigma_2^2}{n_2}}.$$

Stated in terms of z, the rejection region in the lower tail is

$$z < -z_{\alpha/2} = -2.58$$

and the rejection region in the upper tail is

$$z > z_{\alpha/2} = 2.58.$$

(Why? Because if $(\bar{x}_1 - \bar{x}_2) < -3.59$, it follows that $z < -3.59 \div 1.39$, or -2.58. Similarly, if $(\bar{x}_1 - \bar{x}_2) > 3.59$, it follows that $z > 3.59 \div 1.39$, or 2.58. This is because z equals $(\bar{x}_1 - \bar{x}_2) \div 1.39$, since $\sqrt{\sigma_1^2/n_1 + \sigma_2^2/n_2} = 1.39$.) Thus, the market research firm should accept the null hypothesis if $-2.58 \le z \le 2.58$, and reject it otherwise.

In general, the decision rules for testing the null hypothesis that $\mu_1 - \mu_2 = 0$ are as follows.

Test that $\mu_1 - \mu_2 = 0$

Decision Rules:[20] *When the alternative hypothesis is $\mu_1 - \mu_2 \neq 0$, reject the null hypothesis if z exceeds $z_{\alpha/2}$ or is less than $-z_{\alpha/2}$. When the alternative hypothesis is $\mu_1 - \mu_2 < 0$, reject the null hypothesis if z < $-z_\alpha$. When the alternative hypothesis is $\mu_1 - \mu_2 > 0$, reject the null hypothesis if z > z_α.*

The following example is provided to further illustrate the application of these tests.

[20] If the population standard deviations are unknown, the sample standard deviations can be substituted in these decision rules if both samples are large.

EXAMPLE 8.2 A university, which administers two hospitals, wants to determine whether the mean length of stay of patients in the one hospital equals the mean length of stay of patients in the other hospital. If the mean length of stay at either hospital is larger than at the other hospital, the hypothesis that the mean lengths of stay are equal at the two hospitals will have to be rejected. At both hospitals the standard deviation of the length of stay of patients is 5 days. A random sample of 100 patients is chosen from each hospital, and the sample mean of their lengths of stay is found to be 7.3 days at the one hospital and 8.2 days at the other. If the significance level is .05, should the university reject the hypothesis that the means are equal?

SOLUTION: Let μ_1 be the mean length of stay of patients at the one hospital and μ_2 be the mean length of stay of patients at the other hospital. The null hypothesis is that $\mu_1 - \mu_2 = 0$, and the alternative hypothesis is that $\mu_1 - \mu_2 \neq 0$. According to the decision rule given above, the null hypothesis should be rejected if $z > 1.96$ or if $z < -1.96$, since $z_{.025} = 1.96$. Because $z = (7.3 - 8.2) \div \sqrt{5^2/100 + 5^2/100} = -0.9 \div .7071 = -1.27$, the university should not reject the null hypothesis that the mean lengths of stay are equal.

8.7 Two-Sample Test of Proportions: Large Samples

In this section we will describe a statistical procedure to test the hypothesis that two population proportions are equal. This test procedure is based on two independent samples, one from each population. It is assumed that each of these samples is large.

SETTING UP THE TEST

To illustrate this test, we take up an actual study published in the *American Sociological Review* of trends in men's and women's sex-role attitudes. The authors, Andrew Cherlin and Pamela Walters of Johns Hopkins University, investigated a variety of topics, including changes in men's attitudes toward a woman being president of the United States.[21] They obtained data from a sample of 638 men in 1972 and a sample of 559 men in 1978.[22] For each sample, they determined the proportion of men who said that they would vote for a woman for president if she were qualified and nominated by their party. Using these data, how can we test whether the proportion of all American men who said that they would vote for a woman for president was greater in 1978 than in 1972?

As we have stressed, the first step in setting up a test is to formulate the null hypothesis and the alternative hypothesis. Clearly, the null hypothesis is that Π_1, the proportion of men who said they would vote for a woman for president in 1978, equals Π_2, the proportion of men who said they would do so in 1972. Let's assume that the alternative hypothesis is that $\Pi_1 > \Pi_2$. In other words, let's assume that we want to detect an increase over time in this proportion, but that we do not care whether (if no increase has occurred) the proportion decreased or remained constant.

The next step, as we know, is to set a significance level α. The considerations influencing this choice are no different here than in earlier sections. Suppose that the decision is made to set α equal to .05.

DECISION RULES

Given that the value of α has been chosen, we can determine the decision rule that should be applied to determine whether the differ-

---

[21] A. Cherlin and P. B. Walters, "Trends in United States Men's and Women's Sex-Role Attitudes: 1972 to 1978," *American Sociological Review,* August 1981.

[22] A different sample was selected each year. The data used here pertain to whites. Other simplifications are made, but they are of no importance for present purposes.

Figure 8.11
Sampling Distribution
of Difference between
Sample Proportions, if
Null Hypothesis Is
True

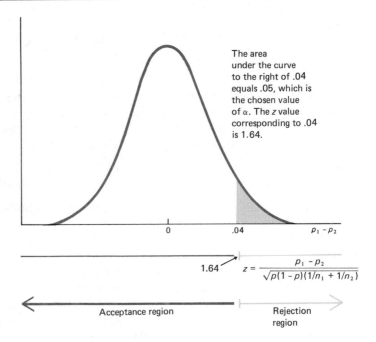

The area
under the curve
to the right of .04
equals .05, which is
the chosen value
of α. The z value
corresponding to .04
is 1.64.

ence between the two population proportions (that is, $\Pi_1 - \Pi_2$) is zero.
If the null hypothesis is true and if both n_1 and n_2 are large, the difference between the sample proportions ($p_1 - p_2$) is approximately normally distributed with a mean of zero and a standard deviation of
$\sqrt{\Pi(1 - \Pi)(1/n_1 + 1/n_2)}$, where Π is the value of the population proportion in both populations. (In other words, if the null hypothesis is
true, the sampling distribution of the difference between the sample
proportions is as shown in Figure 8.11.)

If the null hypothesis is true, a good estimate of Π is

$$p = \frac{n_1 p_1 + n_2 p_2}{n_1 + n_2}$$

Clearly, p is the weighted mean of the sample proportions in the two
samples, each being weighted by the sample size. If n_1 and n_2 are large,
we can substitute the above estimate for Π in the expression in
the previous paragraph. Thus, if the null hypothesis is true, the difference between the sample proportions is approximately normally distributed, with a mean of zero and a standard deviation of
$\sqrt{p(1 - p)(1/n_1 + 1/n_2)}$.

Since we want to reject the null hypothesis only if $\Pi_1 > \Pi_2$, the
rejection region must be set up under the upper tail of the sampling

distribution of ($p_1 - p_2$), as shown in Figure 8.11. This rejection region must be constructed so that the probability is α that the difference between the sample proportions will fall in this region, if the null hypothesis is true. If this rejection region is

$$p_1 - p_2 > z_\alpha \sqrt{p(1-p)\left(\frac{1}{n_1} + \frac{1}{n_2}\right)},$$

this condition is met. Thus, since $\alpha = .05$, we should use the following decision rule: Accept the null hypothesis if $p_1 - p_2 \leq 1.64 \sqrt{p(1-p)(1/n_1 + 1/n_2)}$; but if $p_1 - p_2 > 1.64 \sqrt{p(1-p)(1/n_1 + 1/n_2)}$, reject the null hypothesis.

Like any random variable, the difference between the sample proportions can be expressed in standard units. Since its mean is zero (if the null hypothesis is true) and its standard deviation is $\sqrt{p(1-p)(1/n_1 + 1/n_2)}$, its observed value in standard units is

$$z = (p_1 - p_2) \div \sqrt{p(1-p)(1/n_1 + 1/n_2)}.$$

Stated in terms of z, we should reject the null hypothesis if $z > 1.64$, and accept it otherwise. (Why? Because the z value exceeds 1.64 if $p_1 - p_2 > 1.64 \sqrt{p(1-p)(1/n_1 + 1/n_2)}$ which, according to the previous paragraph, is the condition for rejecting the null hypothesis.)

In fact, Cherlin and Walters found that 73 percent of the men in 1972 and 83 percent of the men in 1978 said that they would vote for a woman for president. Thus,

$$p = \frac{559(.83) + 638(.73)}{559 + 638} = \frac{930}{1197} = .78,$$

and

$$z = (.83 - .73) \div \sqrt{.78(.22)\left(\frac{1}{559} + \frac{1}{638}\right)}$$

$$= .10 \div .024 = 4.17.$$

Consequently, we should reject the null hypothesis (that $\Pi_1 = \Pi_2$) because $z > 1.64$. Using this sort of test, Cherlin and Walters concluded that there was an increase between 1972 and 1978 in the proportion of men who said that they would vote for a woman for president.

GETTING DOWN
TO CASES

Who Is Prone to Alcoholism?

Alcoholism is a serious problem in American society. Many psychologists (and others in the health professions) have spent considerable time trying to determine the sorts of people who are prone toward alcohol abuse.[23] For example, one study compared men who had several relatives dependent on alcohol with men who had no relatives dependent on alcohol. The proportion of men in each group who were dependent on alcohol was as follows:

Number of relatives dependent on alcohol	Number of men in sample	Proportion of men dependent on alcohol
None	178	10/178
Several	71	34/71

(a) If each sample is randomly chosen, can we use these data to test whether the incidence of alcohol dependence is higher among men with several alcohol-dependent relatives than among men with no such relatives?

(b) What is the null hypothesis?

(c) What is the alternative hypothesis?

(d) If the significance level is set at .05, does this evidence indicate that there is a real difference in this regard between these two groups of men?

(e) Does this prove that alcoholism is hereditary? Why or why not?

(f) Does this prove that alcoholism is due to cultural factors? Why or why not?

[23] G. Vaillant and E. Milofsky, "The Etiology of Alcoholism," *American Psychologist*, May 1982.

In general, when n_1 and n_2 are sufficiently large, the decision rules for testing the null hypothesis that $\Pi_1 - \Pi_2 = 0$ are as follows.

Test that
$\Pi_1 - \Pi_2 = 0$

Decision Rules: *When the alternative hypothesis is $\Pi_1 - \Pi_2 \neq 0$, reject the null hypothesis if z exceeds $z_{\alpha/2}$ or is less than $-z_{\alpha/2}$. When the alternative hypothesis is $\Pi_1 - \Pi_2 < 0$, reject the null hypothesis if $z < -z_\alpha$. When the alternative hypothesis is $\Pi_1 - \Pi_2 > 0$, reject the null hypothesis if $z > z_\alpha$.*

A further illustration of the application of these tests follows.

EXAMPLE 8.3 A polling organization is interested in determining whether candidate Jones will run better in urban or in rural areas. A random sample is drawn of 400 urban voters and 400 rural voters, and it is determined that 55 percent of the urban voters and 49 percent of the rural voters prefer candidate Jones over the others. Should the polling organization conclude that the popularity of this candidate differs between urban and rural areas? (Assume that $\alpha = .05$.)

SOLUTION: Let Π_1 be the proportion of urban voters who prefer candidate Jones, and let Π_2 be the proportion of rural voters who prefer Jones. Since the polling organization wants to reject the null hypothesis (that $\Pi_1 = \Pi_2$) when either $\Pi_1 > \Pi_2$ or $\Pi_2 > \Pi_1$, the correct decision rule is to reject the null hypothesis if z is greater than 1.96 or less than -1.96 (because $z_{\alpha/2} = 1.96$). Since $n_1 = n_2 = 400$,

$$p = \frac{(400)(.55) + (400)(.49)}{800} = .52,$$

and

$$z = (.55 - .49) \div \sqrt{.52(.48)(.005)}$$

$$= .06 \div .035 = 1.71.$$

Thus, the polling organization should not reject the null hypothesis (that $\Pi_1 = \Pi_2$), since z is not greater than 1.96 or less than -1.96.

EXERCISES

8.19 (a) A university psychologist decides to test whether the mean score on a particular aptitude test is different among students majoring in mathematics than among those majoring in history. She knows that the standard deviation of the scores among the mathematics students is 10 points and that the same is true for the history students. A random sample is taken of 100 mathemetics students and of 100 history students. The mean score of the mathematics students is 80 and the mean score of the history students is 78.

    (i) What is the null hypothesis? What is the alternative hypothesis?

    (ii) What is the appropriate decision rule?

    (iii) If $\alpha = .10$, should the psychologist accept or reject the null hypothesis?

(iv) If $\alpha = .05$, should she accept or reject the null hypothesis?

(v) If $\alpha = .01$, should she accept or reject the null hypothesis?

(b) Suppose that the psychologist wants to reject the hypothesis that the mean scores of mathematics and history students are the same only if the mean mathematics score exceeds the mean history score.

(i) What is the null hypothesis? What is the alternative hypothesis?

(ii) What is the appropriate decision rule?

(iii) If $\alpha = .10$, should the psychologist accept or reject the null hypothesis?

(iv) If $\alpha = .05$, should she accept or reject the null hypothesis?

(v) If $\alpha = .01$, should she accept or reject the null hypothesis?

(c) Suppose that the psychologist wants to reject the hypothesis that the mean scores of mathematics and history students are the same only if the mean history score exceeds the mean mathematics score.

(i) What is the null hypothesis? What is the alternative hypothesis?

(ii) What is the appropriate decision rule?

(iii) If $\alpha = .10$, should the null hypothesis be accepted or rejected?

(iv) If $\alpha = .05$, should the null hypothesis be accepted or rejected?

(v) If $\alpha = .01$, should the null hypothesis be accepted or rejected?

8.20 (a) A government official wants to determine whether the proportion of tool and die firms now using numerically controlled machine tools is different in Canada than in the United States. The official draws a random sample of 81 tool and die firms in Canada and 100 tool and die firms in the United States, and finds that 20 of the Canadian firms and 30 of the American firms have introduced numerically controlled machine tools.

(i) What is the null hypothesis? What is the alternative hypothesis?

(ii) What is the appropriate decision rule?

(iii) If $\alpha = .10$, should the null hypothesis be accepted or rejected?

(iv) If $\alpha = .05$, should the null hypothesis be accepted or rejected?

(v) If $\alpha = .01$, should the null hypothesis be accepted or rejected?

(b) Suppose that the official wants to reject the hypothesis that the proportion of tool and die firms now using numerically controlled machine tools is the same in Canada as in the United States only if the proportion is higher in Canada.

(i) What is the null hypothesis? What is the alternative hypothesis?

(ii) What is the appropriate decision rule?

(iii) If $\alpha = .10$, should the null hypothesis be accepted or rejected?

(iv) If $\alpha = .05$, should the null hypothesis be accepted or rejected?

(v) If $\alpha = .01$, should the null hypothesis be accepted or rejected?

(c) Suppose that the official wants to reject the hypothesis that the proportion of tool and die firms now using numerically controlled machine tools is the same in Canada as in the United States only if the proportion of users is higher in the United States.

(i) What is the null hypothesis? What is the alternative hypothesis?

(ii) What is the appropriate decision rule?

(iii) If $\alpha = .10$, should the null hypothesis be accepted or rejected?

     (iv) If $\alpha = .05$, should the null hypothesis be accepted or rejected?

     (v) If $\alpha = .01$, should the null hypothesis be accepted or rejected?

8.21 It is claimed that men are better than women at a certain clerical task. To see whether this is the case, a psychologist chooses a random sample of 100 men and a random sample of 100 women. Each person is given this task to do, and his or her performance is graded from zero to 100. For the men, the sample mean is 60.8 and the sample standard deviation is 9.9. For the women, the sample mean is 58.4 and the sample standard deviation is 8.7.

    (a) If the significance level is set at 0.05, does the evidence indicate that men are better than women at this task?

    (b) Does this test depend on the assumption that the standard deviation of the grades is the same among men as among women?

8.22 A chemical firm has two very large research laboratories, one in the United States and one in Europe. It wants to determine whether the chemists at its U.S. laboratory tend to be older or younger than those at its European laboratory. It selects a random sample of 80 chemists from each laboratory, the results being as follows:

	Mean age (years)	Standard deviation (years)
United States	50.6	8.7
Europe	47.3	7.6

Each laboratory hires 1700 chemists.

    (a) If the significance level is set at 0.01, does the evidence indicate that the mean age is the same at the two laboratories?

    (b) If the significance level is set at 0.05 instead, what does the evidence indicate?

8.23 A medical school is evaluating a new type of diaper. It wants to determine whether the proportion of mothers preferring this new product to existing products is different in the east than in the west. Based on a random sample of 100 eastern mothers and a random sample of 100 western mothers, the results are as follows:

	East	West
Number preferring new product	71	56
Number not preferring new product	29	44
Total	100	100

    (a) If the significance level is set at 0.02, does the evidence indicate that the proportion of mothers preferring the new product is the same in the east as in the west?

    (b) If the significance level is set at 0.05 instead, what does the evidence indicate?

8.8 One-Sample Test of a Mean: Small Samples

In previous sections of this chapter we have assumed that the sample was large. We now turn to the case where the sample is small ($n \leq 30$). In this section, we indicate how one can test the hypothesis that the population mean equals a specified value. In the next section, we will indicate how one can test the hypothesis that two population means are equal. In both sections we assume that the populations are normal or approximately normal. In neither section do we assume that any population standard deviation is known.

SETTING UP THE TEST

Suppose that the beer manufacturer (discussed briefly in Section 8.3) wants to test whether the mean rating for a new brand of its beer equals 60 (which is known to be the population mean rating for its old brand). If the mean rating for the new brand is higher than 60, the beer manufacturer will substitute the new brand for the old. If the mean rating for the new brand is less than 60, the implications are the same as if it equals 60. That is, in either of these two latter cases the new brand will be dropped. Suppose that the beer manufacturer samples 16 beer drinkers and obtains ratings on a scale of 1 to 90 for the new brand of beer, the results being those in Table 8.3. Should the manufacturer accept or reject the hypothesis that the mean rating for the new brand is 60?

As usual, we begin by formulating the null hypothesis and the alternative hypothesis. In this case, the null hypothesis is that the population mean rating for the new brand μ equals 60. The alternative hypothesis is that this mean exceeds 60. (The beer manufacturer will not reject the null hypothesis if $\mu < 60$, since if $\mu < 60$ the implications are the same as if μ equals 60.) To sum up, H_0 (the null hypothesis) is that $\mu = 60$, and H_1 (the alternative hypothesis) is that $\mu > 60$. We also need to know the significance level: Assume that the beer manufacturer wants α to equal .05.

Table 8.3
Ratings of a New
Brand of Beer by 16
Randomly Selected
Consumers

Person	Rating	Person	Rating	Person	Rating
1	46	7	71	12	54
2	64	8	77	13	64
3	62	9	69	14	63
4	58	10	67	15	57
5	54	11	59	16	68
6	65				

DECISION RULES

In contrast to our earlier discussion (in Section 8.4) of tests that μ equals a specified value, we do not know the population standard deviation, and the sample size is small. The decision rules in Section 8.4 are therefore not appropriate here. Nonetheless, if the population is normal we can obtain the decision rule we want by using the t distribution. As pointed out in Chapter 7, $(\overline{X} - \mu) \div s/\sqrt{n}$ has the t distribution (with $n-1$ degrees of freedom) if the population is normal. (Of course, s is the sample standard deviation, and n is the sample size.) Thus, if $\mu = \mu_0$, the value specified by the null hypothesis, $(\overline{X} - \mu_0) \div s/\sqrt{n}$ has the t distribution. To test the null hypothesis, we compute

$$t = (\overline{x} - \mu_0) \div s/\sqrt{n},$$

which is used as follows.

Decision Rules: *When the alternative hypothesis is $\mu \neq \mu_0$, reject the null hypothesis if* t *exceeds* $t_{\alpha/2}$ *or is less than* $-t_{\alpha/2}$. *When the alternative hypothesis is $\mu < \mu_0$, reject the null hypothesis if* t $< -t_\alpha$. *When the alternative hypothesis is $\mu > \mu_0$, reject the null hypothesis if* t $> t_\alpha$.

Test that $\mu = \mu_0$, where n ≤ 30

Since the alternative hypothesis in the beer manufacturer's case is that $\mu > 60$, the third decision rule is the applicable one. Using the data in Table 8.3, we find that the sample standard deviation s is 7.60. Thus, the decision rule is: Reject the null hypothesis if t is greater than 1.753, since, as shown in Appendix Table 6, $t_{.05} = 1.753$. (Note that since $n = 16$, there are $16 - 1$, or 15, degrees of freedom.) Because $\overline{x} = 62.38$, $t = (62.38 - 60) \div 7.6/\sqrt{16}$, or 1.25. Since t is not greater than 1.753, there is no reason to reject the null hypothesis that the mean rating of the new brand of beer is 60.

The example below provides a further application of these tests.

<u>EXAMPLE 8.4</u> An ambulance service says that it takes, on the average, 8.0 minutes to get to its destination under emergency conditions. To check this claim, a hospital timed the service on 9 emergency calls (chosen at random), the results (in minutes) being as follows:

7.7	8.0	7.4
7.8	8.4	7.2
7.9	8.2	7.6

The hospital wants to reject the hypothesis that the mean equals 8.0 minutes if the mean is greater than 8.0 minutes, but not if it is less than 8.0 minutes. The time taken by the ambulance service is believed to be normally distributed. Should the hospital accept the service's claim? (Let the significance level equal .05.)

SOLUTION: The null hypothesis is that the mean time interval μ equals 8.0 minutes, and the alternative hypothesis is that $\mu >$ 8.0 minutes. The sample standard deviation s is .38 minutes. The null hypothesis should be rejected if $t > 1.86$, since Appendix Table 6 shows that $t_{.05}$ equals 1.86 when there are 8 degrees of freedom. Because $\mu_0 = 8.0$ and $\bar{x} = 7.8$,

$$t = (7.8 - 8.0) \div .38/\sqrt{9}$$

$$= -1.57.$$

Since the observed value of t is not more than 1.86, the null hypothesis should not be rejected. That is, the hospital should not reject the ambulance service's claim.[24]

[24] For further discussion of the use of statistics in health administration, see R. Broyles and C. Lay, *Statistics in Health Administration.*

WHAT COMPUTERS CAN DO

HYPOTHESIS TESTING AND p VALUES

Computer packages like Minitab can be used to test hypotheses. Suppose, for example, that one wants to test the hypothesis that a population mean equals 19. The sample is 12, 18, 15, 16, 20, 21, and the population standard deviation is unknown. Since these numbers are already in column C2 of the worksheet that Minitab maintains in the computer (recall page 274), all that one has to do is to type the following command.

```
– TTEST OF POPULATION MU=19, DATA IN C2
```

After this command is typed, the computer automatically prints out the following:

```
C2     N = 6     MEAN =     17.000     ST.DEV. =     3.35

TEST OF MU =     19.0000 VS. MU N.E.     19.0000
T = −1.464
THE TEST IS SIGNIFICANT AT 0.2031
CANNOT REJECT AT ALPHA = 0.05
```

The first line contains the number of observations and the sample mean and standard deviation. The second line says that this is a test that $\mu = 19$, where the alternative hypothesis is that μ is not equal to 19. The third line provides the value of t. The fourth line says that the null hypothesis (that $\mu = 19$) should be rejected only if α is 0.2031 or more. The last line says that, if $\alpha = 0.05$, the null hypothesis should not be rejected.

The number shown on the fourth line of the printout is often called the *p value* of the test. *The p value is the lowest significance level at which the test would result in the rejection of the null hypothesis.* Thus, in this case the observed value of t (that is, -1.464) is such that the null hypothesis should be rejected if α is 0.2031 or more. Thus, the p value is 0.2031. The advantage of reporting the p value is that the reader of the printout can use whatever significance level he or she wants. If the computer reports only that the null hypothesis was rejected at a particular significance level, the reader cannot tell whether the same conclusion would result if some other significance level (which the reader believes to be more appropriate) were used instead. For this reason, statisticians often report p values in research reports.

p Values

8.9 Two-Sample Test of Means: Small Samples

We will now describe how one can test the hypothesis that the means of two populations are equal, if the samples are small and the population standard deviations are unknown. (The tests assume that the population standard deviations in the two populations are equal.)

SETTING UP THE TEST

Suppose that the market research firm in Section 8.6 wants to test whether the mean rating given by consumers to their favorite beer differs from that given to brand Y. The firm samples nine people and asks them to rate a beer (in reality their favorite brand) in an unmarked can. Suppose that the ratings are as shown in Table 8.4. Also, the firm samples another nine people and asks them to rate a beer (in reality brand Y) in an unmarked can. These ratings are also shown in Table 8.4. What should the firm conclude from these results?

Again, the first step is to specify the null and alternative hypotheses. The null hypothesis is that μ_1, the population mean of the ratings for the favorite beer, equals μ_2, the population mean of the ratings for brand Y. The alternative hypothesis is $\mu_1 - \mu_2 \neq 0$. The next step is to set the significance level, α. Suppose that the firm sets α at .05.

Table 8.4
Results of Market-Research Firm's Study

Person	Rating of favorite brand in unmarked can	Person	Rating of brand Y in unmarked can
1	58	10	62
2	60	11	60
3	62	12	60
4	60	13	62
5	60	14	63
6	62	15	59
7	58	16	58
8	59	17	64
9	61	18	61

$$\bar{x}_1 = 60$$

$$s_1 = \sqrt{\frac{18}{8}} = 1.5$$

$$\bar{x}_2 = 61$$

$$s_2 = \sqrt{\frac{30}{8}} = 1.94$$

DECISION RULES

In contrast to our earlier discussion in Section 8.6 of tests of whether two population means are equal, we do not know the standard deviation of each population, and the sample size is small. Nonetheless, if both populations are normal, and if their standard deviations are equal, we can obtain the decision rules we need. First, an estimate of the variance in each population is

$$s^2 = \frac{(n_1 - 1)s_1^2 + (n_2 - 1)s_2^2}{n_1 + n_2 - 2}.$$

If the null hypothesis is true, it can be shown that $(\bar{X}_1 - \bar{X}_2) \div \sqrt{s^2(1/n_1 + 1/n_2)}$ has the t distribution with $n_1 + n_2 - 2$ degrees of freedom. Thus, to test the null hypothesis, we compute

$$t = (\bar{x}_1 - \bar{x}_2) \div \sqrt{s^2(1/n_1 + 1/n_2)},$$

which is used as follows.

Decision Rules: *When the alternative hypothesis is* $\mu_1 - \mu_2 \neq 0$, *reject the null hypothesis if* t *exceeds* $t_{\alpha/2}$ *or is less than* $-t_{\alpha/2}$. *When the alternative hypothesis is* $\mu_1 - \mu_2 < 0$, *reject the null hypothesis if* t $< -t_\alpha$. *When the alternative hypothesis is* $\mu_1 - \mu_2 > 0$, *reject the null hypothesis if* t $> t_\alpha$.

Test that $\mu_1 - \mu_2 = 0$, *where* $n \leqslant 30$

Because the alternative hypothesis is $\mu_1 - \mu_2 \neq 0$, the first decision rule is the appropriate one for the market research firm. Using the data in Table 8.4, we must calculate s^2, which is

$$\frac{(n_1 - 1)s_1^2 + (n_2 - 1)s_2^2}{n_1 + n_2 - 2} = \frac{8(1.5)^2 + 8(1.94)^2}{16} = 3.$$

Thus,

$$t = (60 - 61) \div \sqrt{3(1/9 + 1/9)}$$

$$= -1 \div .82 = -1.22.$$

Since $t_{.025} = 2.12$ when there are 16 degrees of freedom (see Appendix Table 6), the observed value of t does not exceed $t_{.025}$ or fall below $-t_{.025}$. Consequently, the null hypothesis should not be rejected.

The above illustration is based on an actual study carried out by the director of research of Carling Brewing Company.[25] The actual study was more complicated and involved larger samples than those considered here, but the principles are essentially the same. Also, it is interesting to note that the results were much the same as the hypothetical results analyzed here. In particular, the actual study concluded that "participants, in general, did not appear to be able to discern the taste differences among the various beer brands. . . ."[26]

The example below again illustrates the application of these tests.

[25] R. Allison and K. Uhl, "Influence of Beer Brand Identification on Taste Perception," *Journal of Marketing Research,* August 1964, pp. 36–39.
[26] Ibid., p. 39.

EXAMPLE 8.5 The Department of Defense tests two types of aircraft, type A and type B, to determine whether their mean speed under a particular set of conditions is the same. In four tests of type A aircraft the mean speed under the prescribed conditions was 590 miles per hour, and the sample standard deviation was 100 miles per hour. In four tests of type B aircraft the mean speed under these conditions was 750 miles per hour, and the standard deviation was 80 miles per hour. The Defense Department wants to detect differences in mean speed, regardless of which type of aircraft may be faster. There is good reason to believe that the speeds of each type of aircraft are normally distributed with the same standard deviation. If $\alpha = .05$, should the Department of Defense accept or reject the hypothesis that the mean speeds are equal?

SOLUTION: Let μ_1 be the population mean speed of type A aircraft, and let μ_2 be the population mean speed of type B aircraft. Since the alternative hypothesis is $\mu_1 - \mu_2 \neq 0$, the first decision rule given above is the appropriate one. Since,

$$s^2 = \frac{(n_1 - 1)s_1^2 + (n_2 - 1)s_2^2}{n_1 + n_2 - 2} = \frac{3(100^2) + 3(80^2)}{6} = 8{,}200,$$

it follows that

$$t = (\bar{x}_1 - \bar{x}_2) \div \sqrt{s^2\left(\frac{1}{n_1} + \frac{1}{n_2}\right)} = (590 - 750) \div \sqrt{8200\left(\frac{1}{4} + \frac{1}{4}\right)}$$

$$= -160 \div 64.03 = -2.50.$$

As shown in Appendix Table 6, $t_{.025} = 2.447$ (because there are $4 + 4 - 2$, or 6, degrees of freedom). Thus, the Department of Defense should reject the null hypothesis if the observed value of t exceeds 2.447 or is less than -2.447. Hence, since $t = -2.50$, it follows that the null hypothesis (that $\mu_1 = \mu_2$) should be rejected.

EXERCISES

8.24 The Internal Revenue Service wants to determine the mean amount of income that waiters fail to report on their income tax. A random sample of 12 waiters is selected, and very detailed and accurate records are kept of their actual incomes by IRS agents. In this sample, the mean amount of unreported income is $4,896, and the standard deviation is $1,592. The Service would like to test whether the true mean is $6,000. If the significance level is set at 0.05, and a two-tailed test is used, are these data consistent with the Service's hypothesis?

8.25 The California Bar Association, in an investigation of the legal costs of a particular kind of antitrust suit, selects 16 such antitrust suits at random. The mean legal cost is $2.3 million, and the standard deviation is $1.4 million. Based on a one-tailed test with α equal to 0.05, are these data consistent with the statement that the mean legal cost of such antitrust cases is $2 million? (The alternative hypothesis is that it exceeds $2 million.)

8.26 If the sample size in the previous exercise had been 160 rather than 16, would you have reached a different conclusion? (Assume that the total number of such antitrust suits is 4,000.)

8.27 A Cincinnati physician wants to determine whether a weight-reducing drug has a different effect on adults over 40 years old than on adults that are no more than 40 years old. Twelve people over 40 are given the drug; their mean weight loss is 8.9 pounds, and the standard deviation is 4.1 pounds. Twelve people no more than 40 are given the drug; their mean weight loss is 11.3 pounds, and the standard deviation is 3.8 pounds.

(a) Based on a two-tailed test with α equal to 0.05, are these data consistent with the hypothesis that the average effect of this drug is the same in both age groups?

(b) Would your answer to (a) change if the significance level were 0.01 rather than 0.05?

(c) What assumptions are you making in (a) and (b)?

8.28 The Bona Fide Washing Machine Company chooses a random sample of 25 motors received from supplier I. It determines the length of life of each of the motors, the results (in thousands of hours) being as follows:

4.1	4.6	4.6	4.6	5.1
4.3	4.7	4.6	4.8	4.8
4.5	4.2	5.0	4.4	4.7
4.7	4.1	3.8	4.2	4.6
3.9	4.0	4.4	4.0	4.5

Suppose that the firm wants to test the hypothesis that the mean length of life of motors received from supplier I equals 4,900 hours. If the Bona Fide Washing Machine Company does not know the standard deviation of the length of life of motors received from supplier I, should it accept this hypothesis (if it sets α equal to .05)? (Use a one-tailed test, the alternative hypothesis being that the mean is less than 4,900 hours.)

8.29 A New England college mails out a great many letters and bulletins to alumni and prospective students. Because of the importance of its mailing costs, the college continually scrutinizes the average weight of the letters it mails. A simple random sample of 12 letters mailed by the college shows a mean weight of 2.7 ounces and a standard deviation of 1.1 ounces.

(a) Using a two-tailed test with α equal to 0.02, are these data consistent with the college's belief that the mean weight of all letters it mails is 2.0 ounces?

(b) Suppose that the sample size had been 120 rather than 12. Would you have reached the same conclusion as in (a)?

The Chapter in a Nutshell

1. The basic hypothesis that is being tested for possible rejection is called the *null hypothesis, H_0*. In testing any null hypothesis, there is an *alternative hypothesis, H_1*, that the decision maker accepts if the null hypothesis is rejected. A Type I error occurs when the null hypothesis

is rejected when it is true. A Type II error occurs when the null hypothesis is accepted when it is not true. A test procedure specifies a *decision rule* indicating the conditions under which the null hypothesis should be accepted or rejected. This decision rule is based on the value of a test statistic, which is a statistic computed from a sample. All possible values of the test statistic are divided into two mutually exclusive regions: the *acceptance region* and the *rejection region*. If the test statistic's value falls within the acceptance region, the null hypothesis is accepted; if it falls within the rejection region, the null hypothesis is rejected.

2. The probability of a Type I error is designated as α, and the probability of a Type II error is designated as β. Whether or not a particular test procedure is appropriate depends on whether α and β are set at the right levels from the point of view of the decision maker. The value of α that is chosen is called the *significance level* of the test. A test's *operating-characteristic curve* shows the value of β if the population parameter assumes various values other than that specified by the null hypothesis. The shape of a test's operating-characteristic curve is influenced by the sample size and by the chosen level of α. With increases in sample size, it is possible to reduce the probability of both types of error (that is, both α and β).

3. If the alternative hypothesis is that the value of the parameter differs in either direction from the value specified by the null hypothesis, the test is a *two-tailed test*. If the alternative hypothesis is that the value of the parameter differs in only one direction from the value specified by the null hypothesis, the test is a *one-tailed test*.

4. For a two-tailed test of the null hypothesis that a population mean equals a specified value μ_0, the decision rule is the following: Accept the null hypothesis if $\mu_0 - z_{\alpha/2}\sigma/\sqrt{n} \leqslant \bar{x} \leqslant \mu_0 + z_{\alpha/2}\sigma/\sqrt{n}$; reject the null hypothesis if $\bar{x} < \mu_0 - z_{\alpha/2}\sigma/\sqrt{n}$ or if $\bar{x} > \mu_0 + z_{\alpha/2}\sigma/\sqrt{n}$. If the population standard deviation σ is unknown, it can be replaced with s, the sample standard deviation. This decision rule applies to large samples (that is, when $n > 30$). For small samples, the t distribution (with $n - 1$ degrees of freedom) must be used in place of the standard normal distribution. For small samples the decision rule is: Accept the null hypothesis if $-t_{\alpha/2} \leqslant t \leqslant t_{\alpha/2}$; reject it otherwise. For purposes of this test, $t = (\bar{x} - \mu_0) \div s/\sqrt{n}$. This small-sample test assumes that the population is approximately normal.

5. For a two-tailed test of the null hypothesis that a population proportion equals a specified value Π_0, the decision rule is: Accept the null hypothesis if $\Pi_0 - z_{\alpha/2}\sqrt{\Pi_0(1 - \Pi_0)/n} \leqslant p \leqslant \Pi_0 + z_{\alpha/2}\sqrt{\Pi_0(1 - \Pi_0)/n}$; reject the null hypothesis if $p < \Pi_0 - z_{\alpha/2}\sqrt{\Pi_0(1 - \Pi_0)/n}$ or if $p > \Pi_0 + z_{\alpha/2}\sqrt{\Pi_0(1 - \Pi_0)/n}$. For a two-tailed test of the null hypothesis that two population proportions are equal, the decision rule is: Accept the null hypothesis if $-z_{\alpha/2} \leqslant z \leqslant z_{\alpha/2}$; re-

ject it otherwise. For purposes of this test, $z = (p_1 - p_2) \div \sqrt{p(1-p)(1/n_1 + 1/n_2)}$. Both tests assume that the samples are large.

6. For a two-tailed test of the null hypothesis that the mean of one population μ_1 equals the mean of another population μ_2, the decision rule is: Accept the null hypothesis if $-z_{\alpha/2} \leqslant z \leqslant z_{\alpha/2}$; reject it otherwise. For purposes of this test, $z = (\bar{x}_1 - \bar{x}_2) \div \sqrt{\sigma_1^2/n_1 + \sigma_2^2/n_2}$. This rule assumes that the samples are large and that the standard deviation of each population is known. If neither of these assumptions is true, the decision rule is: Accept the null hypothesis if $-t_{\alpha/2} \leqslant t \leqslant t_{\alpha/2}$; reject it otherwise. For purposes of this test, $t = (\bar{x}_1 - \bar{x}_2) \div \sqrt{s^2(1/n_1 + 1/n_2)}$, where

$$ s^2 = \frac{(n_1 - 1)s_1^2 + (n_2 - 1)s_2^2}{n_1 + n_2 - 2}. $$

The latter test assumes that both populations are normal and that their standard deviations are equal. (The number of degrees of freedom of the t distribution is $n_1 + n_2 - 2$.)

Chapter Review Exercises

8.30 (a) John Jerome finds that the difference between the mean IQ of a sample of students at the local high school and 100 is not statistically significant. Explain what this means.

(b) Is the finding in (a) independent of the level at which α is set?

(c) John Jerome also finds that the difference between the mean IQ of the sample of students and 140 is statistically significant. Explain what this means.

(d) Is the finding in (c) independent of the level at which α is set?

8.31 The Environmental Protection Agency (EPA) tests the hypothesis that the mean number of miles per gallon achieved by a particular make of car is 28 against the alternative hypothesis that it is not 28. ($\alpha = 0.05$.)

(a) If the standard deviation of the number of miles per gallon achieved by this make of car is 6, and if EPA decides to base its test on a random sample of 100 cars of this make, provide a suitable test procedure.

(b) Suppose that the mean number of miles per gallon for the sample of 100 cars is 26.2. On the basis of this result, should EPA reject the hypothesis that the population mean is 28? Why, or why not?

(c) Suppose that EPA is interested in rejecting the null hypothesis only if the mean number of miles per gallon achieved by this make of car is less than 28. Provide the EPA with a suitable test procedure for this set of circumstances.

(d) If the EPA is interested in rejecting the null hypothesis only if the mean number of miles per gallon achieved by its cars is less than 28, should it reject the null hypothesis [based on the sample in part (b)]? Why, or why not?

8.32 To test whether its police officers are performing adequately, the city of Atlantis takes a random sample of 10 of the traffic tickets made out by each police officer each month. If 1 or more of a certain police officer's traffic tickets contain an error, the police officer is reprimanded.

(a) What is the decision rule in this case?

(b) If the hypothesis being tested is that 1 percent of the traffic tickets contain errors, what is α in this case? (*Hint:* The normal distribution should *not* be used here.)

(c) What is the probability of a Type II error if in fact 2 percent of the traffic tickets contain errors?

(d) What is the probability of a Type II error if 5 percent of the traffic tickets contain errors?

(e) Draw the operating-characteristic curve of this test.

8.33 An Atlantic City gambling casino wants to make sure that the probability of winning on a certain slot machine is 0.4 (no more, no less). The machine is operated 200 times, and the operator wins 89 times.

(a) If α is set at 0.01, is the difference between the sample proportion and 0.4 statistically significant?

(b) If the test described in the text is used, what is the probability of accepting the hypothesis that $\Pi = 0.4$ when in reality $\Pi = 0.38$, 0.40, and 0.42?

(c) Draw three points on the operating-characteristic curve of the test.

8.34 In a poll of television viewers in Los Angeles, 130 out of 200 Democrats disliked a particular news program, while 158 out of 300 Republicans disliked it. Is there a real difference of opinion along party lines on this matter?

8.35 In an agricultural experiment to determine the effects of a particular insecticide, a field was planted with corn. Half of the plants were sprayed with the insecticide, and half were not sprayed with it. Several weeks later, a random sample of 200 sprayed plants was selected, and a random sample of 200 unsprayed plants was selected. The number of healthy plants in each sample was as follows:

	Sprayed	*Unsprayed*
Healthy	121	109
Not healthy	79	91
Total	200	200

(a) If the significance level is set at 0.05, does the evidence indicate that a higher proportion of sprayed than of unsprayed plants was healthy?

(b) If the significance level is set at 0.02 instead, what does the evidence indicate?

8.36 Recall from Exercise 7.34 that the Mercer Company has no idea of the standard deviation of the life of the motors it receives from a particular

supplier, and that it chose a random sample of 9 motors received from this supplier and determined the length of life of each. The results (in thousands of hours) were as follows:

4.3	4.6	3.8
4.2	4.3	3.9
4.1	3.9	4.0

(a) Use these results to test the hypothesis that the mean life of the motors received from this supplier equals 4,200 hours (against the alternative that it does not equal 4,200 hours) at the 5 percent level of significance. Do the same at the 10 percent level of significance. (Use two-tailed tests.)

(b) Compare your results in (a) with your results in Exercise 7.34. Can you see how the results you obtained in Exercise 7.34 could have been used to provide the answer to (a)?

8.37 A North Carolina sociologist wants to test whether the mean income of black workers in a particular community equals $15,000, the alternative hypothesis being that the mean income does not equal $15,000. She takes a random sample of 36 black workers in this community and finds that their incomes in 1985 (in dollars) were as follows:

20,100	8,200	13,200	5,200	11,100	51,100
19,400	8,900	14,300	10,100	10,800	9,600
10,100	10,300	15,800	12,300	9,100	10,900
23,000	26,000	16,100	14,000	7,200	12,000
24,200	11,400	17,200	15,100	4,300	13,200
25,100	12,900	18,900	16,000	38,000	15,100

(a) If the sociologist sets a 5 percent significance level, should she accept or reject this hypothesis? (Use a two-tailed test.)

(b) If she sets a 1 percent significance level, should she accept or reject this hypothesis? (Use a two-tailed test.)

8.38 The sociologist in the previous exercise believes that the mean income of white workers in this community equals $20,000. She wants to test this hypothesis against the alternative that this mean is less than $20,000. A random sample of 9 white workers in this community is chosen, and their incomes (in dollars) turn out to be

24,000	13,400	18,400
22,900	13,800	8,200
11,100	9,300	14,600

(a) If the sociologist sets a 5 percent significance level, should she accept or reject this hypothesis?

(b) If she sets a 1 percent significance level, should she accept or reject this hypothesis?

8.39 The sociologist in the previous two exercises is asked to test whether the mean income of black workers is equal to the mean income of white workers in this community.

(a) If the significance level is set at 5 percent, should she accept or reject this hypothesis, based on the data in Exercises 8.37 and 8.38? (Use a two-tailed test.)

(b) If the significance level is set at 1 percent, should she accept or reject this hypothesis, based on the data in Exercises 8.37 and 8.38? (Use a two-tailed test.)

*8.40 As pointed out in Exercise 7.43, a random sample of ten women who had legal abortions in 1986 indicated that their ages (in years at the time of the abortion) were 21, 18, 28, 16, 22, 26, 31, 21, 19, 30. Minitab is used to test the hypothesis that the mean age of women having legal abortions in 1986 was 20 years. Interpret the computer printout, which is as follows:

```
TTEST OF POPULATION MU=20,DATA IN C2

C2  N = 10  MEAN = 23.200  ST.DEV. = 5.22

TEST OF MU = 20.0000 VS. MU N.E. 20.0000
T = 1.937
THE TEST IS SIGNIFICANT AT 0.0847
CANNOT REJECT AT ALPHA = 0.05
```

* Exercise requiring some familiarity with Minitab.

PART IV

Chi-Square Tests, Nonparametric Techniques, and the Analysis of Variance

OBJECTIVES

Chi-square (denoted by χ^2) tests are very frequently used in a variety of fields; so are nonparametric techniques. The overall purpose of this chapter is to indicate how χ^2 tests and nonparametric techniques can be applied. Among the specific objectives are:

1 To show how χ^2 tests are used to test whether a number of proportions are equal, and to analyze contingency tables.

2 To describe how χ^2 tests indicate whether an observed frequency distribution corresponds to a theoretical distribution.

3 To show how the sign test, the Mann-Whitney test, and the runs test (all nonparametric techniques) are carried out.

* Some instructors may want to skip the latter part of this chapter, which deals with nonparametric techniques. Subsequent chapters have been written so that this may be done without loss of continuity.

Chi-Square Tests and Nonparametric Techniques*

9.1 Introduction

The procedures described in the previous chapter are only a small (albeit very important) sample of the many statistical tests that are available for coping with a wide variety of problems. In this chapter we will present some additional tests which have a wide application. These tests are based on the principles described in Chapter 8, but they deal with different hypotheses and assumptions than those encountered there.

The chi-square (denoted by χ^2) tests taken up in this chapter are based on a very important sampling distribution that we have not encountered before—the χ^2 distribution. (Note that the "chi" in "chi-square" is pronounced like "ki" in "kite.") Besides describing these tests, we will discuss the χ^2 distribution itself. Then we will take up nonparametric techniques, which make fewer assumptions about the nature of the population being sampled than the tests described in the previous chapter. Like the χ^2 tests, nonparametric techniques are very widely used in practical work.

9.2 Target Practice and Pearson's Test Statistic

To begin our discussion of χ^2 tests, let's consider the following sporting situation. Suppose that a marksman fires 200 shots at a target, and that

359

each shot can result in three outcomes: (1) a bull's-eye; (2) a hit on the target, but not a bull's-eye; and (3) a miss of the target. If the probability that a shot results in each of these outcomes is Π_1, Π_2, and Π_3, respectively, and if the outcomes of successive shots are statistically independent, what is the expected number of bull's-eyes? Clearly, that answer is $200\Pi_1$. What is the expected number of hits (but not bull's-eyes)? Clearly, $200\Pi_2$. What is the expected number of misses? Clearly, $200\Pi_3$.

Given the outcomes of the 200 shots, we want to test the null hypothesis that Π_1, Π_2, and Π_3 equal specified values (denoted by Π_1^0, Π_2^0, and Π_3^0). To test this hypothesis, it seems reasonable to compare the actual number of shots having each outcome with the number that would be expected, given that these specified values of Π_1, Π_2, and Π_3 are true. For example, if Π_1^0 (the hypothesized value of Π_1) is .2, we would expect about .2(200), or 40, shots to result in bull's-eyes. If the actual number of bull's-eyes is only 14, there is a seemingly large difference (14–40) between the actual and expected number of bull's-eyes. To tell whether we should view this difference as evidence that the hypothesis that $\Pi_1 = .2$ is untrue, we must know the probability of a difference of this size, given that $\Pi_1 = .2$.

In 1900 Karl Pearson, an English statistician who was one of the fathers of modern statistics, proposed that in a situation of this sort the following procedure be used to test the null hypothesis that Π_1, Π_2, and Π_3 equal the specified values Π_1^0, Π_2^0, and Π_3^0. First, we determine the expected number of shots that would be bull's-eyes if $\Pi_1 = \Pi_1^0$. This expected number (which equals $200\Pi_1^0$) is denoted by e_1. Second, we determine the expected number of shots that would be hits (but not bull's-eyes) if $\Pi_2 = \Pi_2^0$. This expected number (which equals $200\Pi_2^0$) is denoted by e_2. Third, we determine the expected number of shots that would be misses if $\Pi_3 = \Pi_3^0$. This expected number (which equals $200\Pi_3^0$) is denoted by e_3. Then we calculate the following test statistic:

The Chi-Square Test Statistic

$$\sum_{i=1}^{3} \frac{(f_i - e_i)^2}{e_i},$$
(9.1)

where f_1 is the actual number of shots that are bull's-eyes, f_2 is the actual number that are hits (but not bull's-eyes), and f_3 is the actual number of misses. Pearson showed that *if the null hypothesis is true (that is, if $\Pi_1 = \Pi_1^0$, $\Pi_2 = \Pi_2^0$ and $\Pi_3 = \Pi_3^0$), and if the sample size is large enough so that the smallest value of e_i is at least 5, this test statistic will have a sampling distribution which can be approximated adequately by the χ^2 distribution,* a very important probability distribution to which we now turn.

9.3 The Chi-Square Distribution

The χ^2 distribution, which is the probability distribution of a χ^2 random variable, is defined as follows.

> χ^2 DISTRIBUTION: *The χ^2* **distribution,** *with v (nu) degrees of freedom, is the probability distribution of the sum of squares of v independent standard normal variables.*

To clarify this definition, let's begin by taking a single standard normal variable. (Recall from Chapter 5 that the standard normal variable is $(X - \mu)/\sigma$, where X is a normal variable with mean equal to μ and standard deviation equal to σ.) Instead of considering the probability distribution of its value, let's consider the probability distribution of the *square* of its value. If one knows that this variable has a standard normal distribution, it should be possible to figure out the distribution of the square of its value. This distribution, derived many years ago, is a χ^2 distribution with 1 degree of freedom. Next, suppose that we consider two independent standard normal variables. Let's square the value of each and add the squares together. The distribution of this sum is a χ^2 distribution with 2 degrees of freedom. Finally, suppose that we consider four independent standard normal variables. Let's square the value of each and add the squares together. The distribution of this sum, shown in panel A of Figure 9.1, is a χ^2 distribution with 4 degrees of freedom.

Like the t distribution, the χ^2 distribution is a family of distributions, each of which is characterized by a certain number of *degrees of freedom.*[1] The number of degrees of freedom is the number of squares of standard normal variables that are summed up. Thus, in Figure 9.1, panel A shows a χ^2 distribution with 4 degrees of freedom; panel B shows a χ^2 distribution with 10 degrees of freedom; and panel C shows a χ^2 distribution with 20 degrees of freedom.

As is evident from Figure 9.1, the χ^2 distribution generally is skewed to the right. Since the χ^2 random variable is a sum of squares, the probability that it is negative is zero. If there are v degrees of freedom, it can be shown that

$$E(\chi^2) = v, \qquad (9.2)$$
$$\sigma(\chi^2) = \sqrt{2v}. \qquad (9.3)$$

Mean and Standard Deviation of a χ^2 Random Variable

[1] There is a close relationship between the t and χ^2 distributions. It can be shown that t equals the ratio of a standard normal variable to the square root of a χ^2 variable divided by its number of degrees of freedom. For further explanation one should consult a more advanced statistics text.

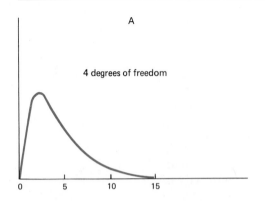

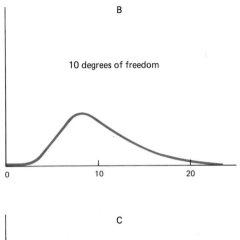

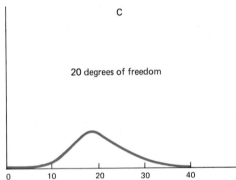

Figure 9.1
χ^2 Probability Density
Functions with 4, 10,
and 20 Degrees of
Freedom

In other words, the mean of a χ^2 random variable equals its number of degrees of freedom, and its standard deviation equals the square root of twice its number of degrees of freedom.

In applying the test procedures described in the following sections, it is essential that we be able to find the value of χ^2 that is exceeded with a probability of .05, .10, or some other such amount. That is, we need to be able to calculate χ^2_α, which is the value of χ^2 that is exceeded with a probability of α. Appendix Table 8 shows the value of χ^2_α for various numbers of degrees of freedom and for various values of α. Each row in this table corresponds to a certain number of degrees of freedom. For example, the tenth row shows that, if there are 10 degrees of freedom, the probability is .10 that the value of χ^2 will exceed 15.987, that the probability is .05 that it will exceed 18.307, and that the probability is .01 that it will exceed 23.2093. As the number of degrees of freedom becomes very large, the χ^2 distribution can be approximated by the normal distribution.

9.4 Test of Differences among Proportions

One very important application of the χ^2 distribution is to problems where the decision maker wants to determine whether a number of proportions are equal. In the previous chapter, we showed how one can test whether two population proportions are equal. Now we show how one can test whether any number of population proportions are equal. To illustrate the sort of situation to which this test is applicable, we will describe an actual case study involving the relationship between parents' culture and dependence on alcohol. According to Harvard's George Vaillant and Eva Milofsky, cultural factors play an important role in influencing whether or not a person becomes an alcoholic. In their view, it is important to "contrast cultures that forbid drinking in children but condone drunkenness in adults with those cultures that teach children how to drink responsibly but that forbid adult drunkenness."[2]

Specifically, they point out that the non-Moslem Mediterranean countries have no sanctions against children's drinking (in moderation),but have strong sanctions against adult drunkenness. On the other hand, the Irish culture forbids drinking in adolescence,but gives explicit or tacit approval to drunkenness in adult males. The North American culture is regarded as being rather similar in this respect to the Irish, whereas the Northern European culture is assigned an intermediate position between the Mediterraneans and the Irish. To see whether culture is related in this way to alcohol dependence, data were obtained concerning the proportion of men whose parents were raised in three cultural groups (Irish, North American and North European, and Mediterranean) that are alcohol dependent. Table 9.1 shows the results for their sample of 398 men. One question that Vaillant and Milofsky wanted to answer was: Is the proportion of men who are alcohol dependent constant, regardless of the cultural group in which their parents were raised, or does it vary?

In a problem of this sort, *the null hypothesis is that the population proportions are all equal.* Thus, in this case study, the null hypothesis is that $\Pi_1 = \Pi_2 = \Pi_3$, and *the alternative hypothesis is that these proportions are not all equal.* (Of course, Π_1 is the population proportion that is alcohol dependent among the Irish, Π_2 is the population proportion that

---

[2] G. Vaillant and E. Milofsky, "The Etiology of Alcoholism: A Prospective Viewpoint," *American Psychologist*, May 1982, p. 499. Of course, many factors other than these cultural considerations influence the incidence of alcoholism. The purpose of this brief discussion is to illustrate the use of the χ^2 test, not to provide a full treatment of this complex topic. The hazards of generalizing about cultural, national, and racial groups are well known, and need not be repeated here.

Table 9.1
Number of Alcohol
Dependent Men,
Three Cultural Groups

	Irish	*Cultural group in which man's parents were raised:* *North American or* *North European*	*Mediterranean*
Number alcohol dependent	21	44	5
Number not alcohol dependent	54	149	125
Sample size	75	193	130

Source: Vaillant and Milofsky, "Etiology of Alcoholism," p. 499.

is alcohol dependent among the North Americans and North Europeans, and Π_3 is the population proportion that is alcohol dependent among the Mediterraneans.) If the null hypothesis is true, the population proportion that is alcohol dependent can be estimated by pooling the data for all cultures. Clearly, the common proportion that is alcohol dependent equals

$$\sum_{i=1}^{3} x_i \bigg/ \sum_{i=1}^{3} n_i,$$

where x_i is the number of alcohol dependent men in the ith cultural group, and n_i is the size of the total sample in the ith cultural group. Thus, the expected proportion that is alcohol dependent in all cultural groups combined is

$$\frac{21 + 44 + 5}{75 + 193 + 130} = .176$$

To test whether the null hypothesis is correct, we calculate the expected number of alcoholics and the expected number of nonalcoholics in each cultural group. Since there are 75 Irish in the sample, we would expect that $75(.176) = 13.2$ of them would be alcoholics, and that $75(.824) = 61.8$ would not be alcoholics. Since there are 130 Mediterraneans in the sample, we would expect that $130(.176) = 22.9$ of them would be alcoholics, and that $130(.824) = 107.1$ would not be alcoholics. To test the null hypothesis, we compare these theoretical, or expected, frequencies with the actual ones. Clearly, the greater the difference between the theoretical frequencies and the actual ones, the less likely it is that the null hypothesis is true.

Specifically, the test procedure is as follows. Having calculated each expected frequency (shown in Table 9.2), we must compute the following

Cultural group	Number alcohol dependent			Number not alcohol dependent		
	Actual (f)	Expected (e)	$\frac{(f-e)^2}{e}$	Actual (f)	Expected (e)	$\frac{(f-e)^2}{e}$
Irish	21	13.2	4.61	54	61.8	0.98
North American or North European	44	34.0	2.94	149	159.0	0.63
Mediterranean	5	22.9	13.99	125	107.1	2.99

$$\sum \frac{(f-e)^2}{e} = 26.14$$

Table 9.2
Expected Frequencies, and Calculation of
$$\sum \frac{(f-e)^2}{e}$$

test statistic,

$$\sum \frac{(f-e)^2}{e}, \qquad\qquad (9.4)$$

where f is the actual frequency, e is the corresponding expected frequency, and the summation is over all items in Table 9.2.[3] If the null hypothesis is true, this test statistic has a sampling distribution that can be approximated by the χ^2 distribution with degrees of freedom equal to $(r - 1)$, where r is the number of population proportions that are being compared.

Why is $(r - 1)$ the appropriate number of degrees of freedom? The general rule is that *the appropriate number of degrees of freedom equals the number of comparisons between actual and expected frequencies, less the number of independent linear restrictions placed upon the frequencies.* Because the number of such restrictions varies from one case to another in this chapter, the number of degrees of freedom will be given by different formulas from one case to another. Here the appropriate formula is $(r - 1)$, since the sum of the frequencies in each cultural group is given and the total number of alcoholics in the sample is also given. Thus, there are $(r + 1)$ restrictions on the frequencies.[4] Subtracting $(r + 1)$ from $2r$ (the number of comparisons between actual and expected frequencies), we get $(r - 1)$, the number of degrees of freedom.

As we just noted, if the null hypothesis is true, the distribution of $\Sigma(f - e)^2/e$ can be approximated by the χ^2 distribution with $(r - 1)$ de-

---

[3] Note that this test statistic is the same as that in expression (9.1).

[4] What are the $(r + 1)$ restrictions? For each of the r populations (in this case, 3 cultural groups), the number of alcoholics plus the number of non-alcoholics must equal the sample size. Besides these r restrictions, the total number of alcoholics is taken as given. (That is, $\sum_{i=1}^{3} x_i = 70$.) Thus, there are $(r + 1)$ restrictions in all.

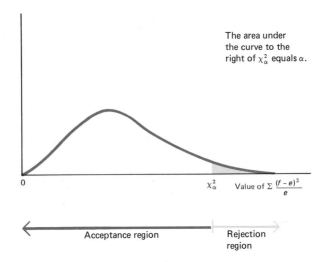

Figure 9.2
Sampling Distribution of $\sum \frac{(f-e)^2}{e}$, if Null Hypothesis is True

The area under the curve to the right of χ^2_α equals α.

χ^2_α Value of $\sum \frac{(f-e)^2}{e}$

Acceptance region

Rejection region

grees of freedom, which is shown in Figure 9.2. The rejection region is under the upper tail of this distribution, since large values of this test statistic indicate large discrepancies between the actual and expected frequencies. In general, to test whether *r* population proportions are equal, the appropriate decision rule is as follows.

Test that $\Pi_1 = \Pi_2 = \ldots = \Pi_r$

Decision Rule: *Reject the null hypothesis that* $\Pi_1 = \Pi_2 = \cdots = \Pi_r$ *if* $\Sigma(f-e)^2/e > \chi^2_\alpha$ *where* α *is the desired significance level of the test (and the number of degrees of freedom is* r − 1*). Accept the null hypothesis if* $\Sigma(f-e)^2/e \leq \chi^2_\alpha$.

Table 9.2 shows that the value of $\Sigma(f-e)^2/e$ is approximately 26.14. Appendix Table 8 shows that $\chi^2_{.05} = 5.99147$ if the number of degrees of freedom equals 2. (Why are there 2 degrees of freedom? Because the number of degrees of freedom equals the number of population proportions to be compared—3 in this case—minus 1.) Since 26.14 exceeds 5.99147, it is clear that the null hypothesis should be rejected, if the significance level is set equal to .05. Even if the decision maker sets the significance level at .01, the null hypothesis should be rejected since $\chi^2_{.01} = 9.21034$ if the number of degrees of freedom equals 2. Thus, based on these data, it appears that the proportion of men who are alcohol dependent varies from one cultural group to another, in accord with the views of Vaillant and Milofsky.

The test of differences among proportions that we have just described is commonly employed. The following is another illustration of how it is used.

EXAMPLE 9.1 Psychologists have noted that mothers who have access to their babies immediately after birth prefer to hold the baby on their left side, whereas mothers who do not have early contact with their babies prefer to hold them on their right side. It is speculated that the heartbeat of the mother is comforting to the newborn infant, and that the sound of the heartbeat is more pronounced if the child is held on the mother's left side. According to one study, the data for a sample of 200 mothers were as follows:[5]

Mother's preference	Immediately after birth	No early contact
Left side	75	40
Right side	25	60
Total	100	100

Do these data indicate a difference in preference between mothers who have immediate access to their babies and those that have no early contact with their infants? (Assume that the significance level equals .05.)

SOLUTION: In all, 115 of the 200 mothers showed a preference for the left side. Thus, the common or overall proportion is 115/200, or .575. If there is no difference between the preferences of the two groups of mothers, the expected numbers would be as follows:

Mother's preference	Immediately after birth	No early contact
Left side	57.5	57.5
Right side	42.5	42.5
Total	100	100

Thus,

$$\sum \frac{(f-e)^2}{e} = \frac{(75-57.5)^2}{57.5} + \frac{(40-57.5)^2}{57.5} + \frac{(25-42.5)^2}{42.5} + \frac{(60-42.5)^2}{42.5}$$

$$= 25.064.$$

[5] L. Salk, "The Role of the Heartbeat in the Relations between Mother and Infant," *Scientific American*, May 1973.

Since there are two populations being compared, the number of degrees of freedom is $2 - 1$, or 1. According to Appendix Table 8, $\chi^2_{.05} = 3.84146$ when there is 1 degree of freedom. Thus, since the observed value of $\Sigma(f - e)^2/e$ exceeds 3.84146, the null hypothesis (that the proportion of mothers preferring the left side is the same in both groups) should be rejected. In other words, there does seem to be a difference in preference between mothers in the two groups.[6]

[6] As pointed out in note 9, when there is only one degree of freedom, a slightly different formula should be used. But in this case the results are almost precisely the same as those shown above. (Prove this is true as an exercise.)

9.5 Changes in Sex-Role Attitudes in the United States: A Case Study

Because χ^2 tests of this sort are so important, it is worthwhile to present another actual case illustrating their use. In 1981, Andrew Cherlin and Pamela B. Walters published a study of trends in sex-role attitudes in the United States.[7] In particular, based on data from random samples of men in 1972, 1975, and 1978, they found that the proportion who approved of a married woman earning money (if she has a husband capable of supporting her) increased from 0.62 in 1972 to 0.70 in 1975 to 0.73 in 1978. Table 9.3 shows the number of men in the sample who approved and disapproved in each year. The proportion of men in the three samples combined (for 1972, 1975, and 1978) who approved equals $(410 + 412 + 409) \div (410 + 252 + 412 + 176 + 409 + 151) = 0.68$.

To test whether the observed differences in these proportions are due to chance, we can calculate the number of men in the sample who would be expected to approve each year, based on the proportion of the total sample (in 1972, 1975, and 1978 combined) who approved. Since the previous paragraph showed that this proportion is 0.68, the expected number in 1972 is .68 (662) or 450; the expected number in 1975 is .68 (588) or 400; and the expected number in 1978 is .68 (560) or 381. As indicated in Table 9.3, the value of $\Sigma(f - e)^2/e$ is 18.68.

[7] A. Cherlin and P. B. Walters, "Trends in United States Men's and Women's Sex-Role Attitudes: 1972 to 1978," *American Sociological Review,* August 1981. To simplify the analysis, we assume that all men in the sample either approved or disapproved; this simplification allows us to use their published data. For present purposes (which are solely illustrative), this simplification is of no importance. The data used here pertain to whites; for blacks, the results seem to be different.

	Number Approving			Number Disapproving			Table 9.3
Year	Actual (f)	Expected (e)	$\frac{(f-e)^2}{e}$	Actual (f)	Expected (e)	$\frac{(f-e)^2}{e}$	
1972	410	450	3.56	252	212	7.55	
1975	412	400	0.36	176	188	0.77	
1978	409	381	2.06	151	179	4.38	

$$\sum \frac{(f-e)^2}{e} = 18.68$$

Table 9.3
χ^2 Test of Changes Over Time in the Proportion of Men Approving of a Married Woman Earning Money if She Has a Husband Capable of Supporting Her, 1972–78.

Since we are comparing three proportions, the number of degrees of freedom is $(3 - 1)$, or 2. Appendix Table 8 shows that $\chi^2_{.01} = 9.21034$ if the number of degrees of freedom equals 2. Thus, if we set α equal to .01, we should reject the null hypothesis that the proportion who approve has not changed over time. In other words, we should reject the hypothesis that the observed differences among the proportions in 1972, 1975, and 1978 are due to chance.

EXERCISES

9.1 Find the value of $\chi^2_{.05}$ when there are (a) 5 degrees of freedom; (b) 10 degrees of freedom; (c) 20 degrees of freedom.

9.2 Find the value of $\chi^2_{.01}$ when there are (a) 8 degrees of freedom; (b) 14 degrees of freedom; (c) 26 degrees of freedom.

9.3 If the coefficient of variation of a χ^2 random variable equals 1, how many degrees of freedom does it have?

9.4 If $\chi^2_{.01} = 29.1413$, what is the number of degrees of freedom? If $\chi^2_{.95} = 10.8508$, what is the number of degrees of freedom?

9.5 We add up the squared values of 15 independent standard normal variables. (a) What sort of probability distribution does this sum have? (b) What is the expected value of this sum? (c) What is the variance of this sum?

9.6 A large city hospital selects a random sample of 100 of its patients each day. Each patient is asked whether he or she has any complaints about the food. The number of patients having such complaints is as follows for a period of 10 consecutive days: 9, 10, 8, 6, 12, 15, 12, 9, 8, 6. Test whether the proportion of patients that have complaints about the food is the same each day. (Let $\alpha = .05$.)

9.7 Four large state universities each select a random sample of 200 applications for admission in 1986. The number of these applications containing falsified information is 6, 8, 9, and 12. Test whether the proportion of applications containing falsified information is the same at each university. (Let $\alpha = .05$.)

9.8 The Alpha Corporation, a bicycle maker, obtains tires from two suppliers, firm A and firm B. During 1985, it received the following number of acceptable and defective tires from each supplier:

Supplier	Number of acceptable tires	Number of defective tires	Total number of tires received
A	940	89	1,029
B	780	32	812

Is the probability that the Alpha Corporation would receive a defective tire the same, regardless of the supplier? Test this hypothesis, based on a significance level of .05.

9.9 A psychologist wants to determine whether the proportion of children who watch television more than 5 hours a day is the same in four quite different communities. He chooses a random sample of 80 children in each community and finds out whether each child watches television more than 5 hours a day, the results being as follows:

Number of hours per day watching television	Community A	B	C	D
	(Number of children)			
More than 5	9	15	20	21
No more than 5	71	65	60	59
Total	80	80	80	80

(a) Can the observed differences among communities be attributed to chance variation? (Set $\alpha = .05$.)

(b) If α is set at .01, can the observed differences be attributed to chance variation?

9.6 Contingency Tables

Decision makers are often concerned with problems involving contingency tables. A *contingency table* indicates whether two characteristics or variables are dependent on one another. In other words, a contingency table contains two variables of classification, and the point is to determine whether these two variables are related. For example, Table 9.4 shows a simple contingency table where the vertical columns represent Republicans and Democrats and the horizontal rows show the number of persons earning more or no more than $30,000 per year. Based on a random sample of 300 individuals in a particular Pennsylvania town, the results are as shown in Table 9.4. This is a 2 × 2 contingency table since there are two rows and two columns. *A table of this sort can be used to test whether a person's income is independent of his or her political affiliation.*

The null hypothesis here is that the probability that a Republican will have an income above $30,000 is the same as the probability that a

Income	Republicans	Democrats	Total
More than $30,000	30(20)	30(40)	60
No more than $30,000	70(80)	170(160)	240
Total	100	200	300

Table 9.4
A 2 × 2 Contingency Table: Income and Political Affiliation in a Pennsylvania Town

Democrat will have an income above $30,000. That is, the null hypothesis is that income is *independent* of political party. To test this hypothesis, we compute the expected frequencies in Table 9.4, assuming that the hypothesis is true. Since the probability of a person making over $30,000 is 60 ÷ 300 or .20 in the sample as a whole, we would expect that 20 percent of the Republicans—that is, .20(100), or 20 Republicans—would earn over $30,000 per year, and that 20 percent of the Democrats—that is, .20(200), or 40 Democrats—would earn over $30,000 per year, if the null hypothesis is true. By the same token, we would expect that 80 percent of the Republicans—that is, .80(100), or 80 Republicans—would make no more than $30,000 per year, and that 80 percent of the Democrats—that is, .80(200), or 160 Democrats—would make no more than $30,000 per year, if the null hypothesis is true.

Given these expected frequencies, which are shown in parentheses in Table 9.4, we test the null hypothesis by computing $\Sigma(f-e)^2/e$. In other words $\Sigma(f-e)^2/e$ is the test statistic. If α is the desired significance level, the decision rule is as follows.

> ***Decision Rule:*** *Reject the null hypothesis (of independence) if* $\Sigma(f-e)^2/e > \chi_\alpha^2$, *where there are* $(r-1)(c-1)$ *degrees of freedom (r being the number of rows and c being the number of columns). Accept the null hypothesis if* $\Sigma(f-e)^2/e \leq \chi_\alpha^2$.

Test of Indepencence in a Contingency Table

Why is the appropriate number of degrees of freedom equal to $(r-1)(c-1)$? As noted in the previous section, the appropriate number of degrees of freedom equals the number of comparisons between actual and expected frequencies less the number of restrictions placed upon these frequencies. Since the number of entries in a contingency table equals cr, there are cr actual frequencies to be compared with the corresponding expected frequencies. Because the sum of the frequencies in each row and each column is a given quantity, there are $r + c - 1$ such restrictions.[8] Thus, the appropriate number of degrees of freedom is $cr - (r + c - 1)$, or $(r-1)(c-1)$.

---

[8] There is a restriction for each row and column, since the sum of the frequencies in each row or column is given. However, one of these restrictions must hold if all the others are met, so the number of independent restrictions is $r + c - 1$.

In this case the value of the test statistic is

$$\Sigma \frac{(f-e)^2}{e} = \frac{(30-20)^2}{20} + \frac{(30-40)^2}{40} + \frac{(70-80)^2}{80} + \frac{(170-160)^2}{160}$$

$$= \frac{100}{20} + \frac{100}{40} + \frac{100}{80} + \frac{100}{160} = 9.375.$$

Suppose that the significance level is set at .05. If so, the null hypothesis should be rejected, since $\Sigma(f-e)^2/e$ exceeds $\chi^2_{.05}$, which is 3.84. (Note that there is only 1 degree of freedom here, since $(r-1) \cdot (c-1) = 1$ in a 2×2 contingency table.)[9] In other words based on this evidence, the probability that a person's income is above \$30,000 does not seem to be independent of his or her political affiliation.

9.7 Tests of Goodness of Fit

Still another important application of the χ^2 distribution is where the decision maker wants to determine whether an observed frequency distribution conforms to a theoretical distribution. For example, suppose that a physician believes that the probability is .50 that a person exposed to a particular disease will contract the illness. In each of 160 hospitals, there are four persons who have been exposed to this disease. The physician obtains data from each hospital concerning the number (zero, 1, 2, 3, or 4) who contracted the illness. The results are shown in the last column of Table 9.5.

Let's assume that the physician, after collecting these data, wants to determine whether the binomial distribution is an accurate repre-

---

[9] When there is only 1 degree of freedom, one should use a slightly different formula:

$$\Sigma \frac{(|f-e|-1/2)^2}{e}.$$

In this case, the result would be

$$\frac{(9.5)^2}{20} + \frac{(9.5)^2}{40} + \frac{(9.5)^2}{80} + \frac{(9.5)^2}{160} = 90.25\left(\frac{3}{32}\right) = 8.46.$$

Thus, the results are essentially the same as in the text. Note that this so-called *continuity correction factor* also applies to the χ^2 tests in Sections 9.4 and 9.7, as well as to contingency tables, when there is only 1 degree of freedom.

Number of persons contracting the disease	Theoretical probability	Theoretical frequency	Actual frequency
0	1/16	10	12
1	1/4	40	35
2	3/8	60	60
3	1/4	40	45
4	1/16	10	8

Table 9.5
Goodness-of-Fit Test

sentation of the actual distribution. The theoretical frequency distribution is provided in the third column of Table 9.5. To obtain this
distribution, we use the binomial distribution to calculate the theoretical probabilities that zero, 1, 2, 3, or 4 persons will contract the illness.
These probabilities are shown in the second column of Table 9.5. Then
we multiply each of these numbers by 160. Of course, the actual and
theoretical distributions do not coincide exactly, but this does not
prove that the theoretical distribution is inappropriate, since some discrepancy between the two distributions would be expected due to
chance. The question facing the physician is: Are the discrepancies between the actual and theoretical distributions so large that they cannot
reasonably be attributed to chance?

To answer this kind of question, statisticians use a procedure
quite similar to those described in the previous two sections. *The null
hypothesis is that the actual distribution can in fact be represented by the
theoretical distribution, and that the discrepancies between them are due to
chance.* To test this hypothesis, we calculate $\Sigma(f - e)^2/e$, where f is the
observed frequency in a particular class interval of the frequency distribution and e is the theoretical, or expected, frequency in the same
class interval of the frequency distribution. If the null hypothesis is
true, it can be shown that $\Sigma(f - e)^2/e$ has approximately a χ^2 distribution, the number of degrees of freedom being 1 less than the number
of values of $(f - e)^2/e$ that are summed up. (If some of the parameters
of the theoretical distribution are estimated from the sample, the
number of degrees of freedom is less than this amount by the number
of parameters that are estimated.[10])

Once this test has been calculated, the decision rule is as follows.

---

[10] In this case there are two parameters (n and Π) of the theoretical probability
distribution, and their values (4 and 0.5) are given. However, often we want to test
whether the data conform to a binomial distribution without specifying the value of Π,
since we have no a priori information concerning its value. More will be said about this
below.

Test that
Discrepancies
between Actual and
Theoretical
Distributions Are
Due to Chance

Decision Rule: *Reject the null hypothesis that the discrepancy between the actual and theoretical frequency distributions is due to chance if* $\Sigma(f - e)^2/e > \chi_\alpha^2$ *where the number of degrees of freedom equals the number of class intervals of the frequency distribution minus* $(h + 1)$, *where* h *is the number of parameters estimated from the sample. Accept the null hypothesis if* $\Sigma(f - e)^2/e \leqslant \chi_\alpha^2$.

Given the actual and theoretical frequency distributions in Table 9.5, the value of the test statistic is

$$\Sigma \frac{(f-e)^2}{e} = \frac{(12-10)^2}{10} + \frac{(35-40)^2}{40} + \frac{(60-60)^2}{60}$$

$$+ \frac{(45-40)^2}{40} + \frac{(8-10)^2}{10}$$

$$= 0.4 + 0.625 + 0 + 0.625 + 0.4 = 2.05$$

Since no parameter of the theoretical distribution is estimated from the data, the number of degrees of freedom is 1 less than the number of values of $(f-e)^2/e$ that are summed up; that is, $5 - 1 = 4$. Thus, if the .05 significance level is chosen, the null hypothesis should be rejected if $\Sigma(f-e)^2/e$ exceeds 9.488, since this is the value of $\chi_{.05}^2$ when there are 4 degrees of freedom. (See Appendix Table 8.) Since $\Sigma(f-e)^2/e$ does not exceed 9.488, it follows that there is no reason to reject the null hypothesis. In other words, the probability is greater than .05 that the observed discrepancies between the actual distribution and the binomial distribution could be due to chance.

In carrying out goodness-of-fit tests, statisticians sometimes force the mean of the theoretical distribution to equal the mean of the actual distribution. Frequently, the mean of the theoretical distribution is not stipulated on a priori grounds, and therefore this seems the sensible thing to do. For example, if the physician wanted to test whether the actual distribution in Table 9.5 was binomial, but was not willing to specify that the mean of the binomial distribution was 2.0, then the mean of the theoretical distribution would be set equal to the mean of the actual distribution. Since this would enable us to specify the entire theoretical distribution, we could calculate $\Sigma(f-e)^2/e$ and see whether it exceeds $\chi_{.05}^2$. But it is important to note that χ^2 would have one less degree of freedom, because we estimated an additional parameter of the theoretical distribution (namely, the mean) from the actual distribution.[11]

[11] In testing whether a frequency distribution conforms to a normal population, statisticians frequently force the population mean to equal the sample mean and the population standard deviation to equal the sample standard deviation. Under these circum-

Finally, in carrying out all such goodness-of-fit tests, one should define the class intervals of the frequency distribution so that the *theoretical, or expected, frequency in each and every class interval is at least 5.* The reason is that the χ^2 distribution is not a good approximation of the distribution of $\bar{\Sigma}(f-e)^2/e$ when the null hypothesis is correct, if the theoretical frequencies in particular class intervals are very small. To make sure that this rule of thumb is met, all that one has to do is to combine adjacent class intervals when any of them has a theoretical frequency of less than 5. (Example 9.2 illustrates this procedure.)

This rule of thumb should also be observed in the other applications of the χ^2 distribution described in the preceding sections. In the case of tests of differences among proportions, each expected or theoretical frequency must equal 5 or more; in contingency tables, the theoretical frequency in each cell must equal 5 or more. Recall that, when we discussed Karl Pearson's findings in Section 9.2, we pointed out that his results assumed that the smallest expected or theoretical frequency (that is, the smallest value of e_i) is at least 5. This rule of thumb is important and should be observed.

Goodness-of-fit tests are of considerable importance in a wide variety of fields. The following example illustrates how they can be used by government agencies.

EXAMPLE 9.2 Let's return to the replacement of parts on Polaris submarines, an actual case study presented in the appendix to Chapter 5. Suppose that the Department of Defense believes that the probability distribution of the number of submarine parts of a certain type that will fail during a mission is as follows:[12]

Number of failures per mission	Theoretical probability
0	.368
1	.368
2	.184
3	.061
4 or more	.019

Data for 500 missions indicate that the number of these missions in which each number of failures occurred was as follows:

---

stances, the number of degrees of freedom is the number of class intervals in the frequency distribution minus 3 since two parameters are estimated.

[12] Readers who have covered the appendix to Chapter 5 should note that this probability distribution is a Poisson distribution in which the mean number of failures per mission is 1.0.

Number of failures per mission	Number of missions
0	190
1	180
2	90
3	30
4 or more	10

Test the hypothesis that the discrepancies between the actual and theoretical frequency distributions are due to chance. (Let the significance level equal .05.)

SOLUTION: The theoretical frequency distribution can be obtained by multiplying the theoretical probabilities by 500, the results being as follows:

Number of failures per mission	Expected number of missions
0	.368(500) = 184.0
1	.368(500) = 184.0
2	.184(500) = 92.0
3	.061(500) = 30.5
4 or more	.019(500) = 9.5

Note that we have lumped together all numbers of failures exceeding 3 in one class interval to hold the theoretical frequency in the class interval to a level of five or more.

The value of the test statistic is

$$\sum \frac{(f-e)^2}{e} = \frac{(190-184)^2}{184} + \frac{(180-184)^2}{184} + \frac{(90-92)^2}{92} +$$

$$\frac{(30-30.5)^2}{30.5} + \frac{(10-9.5)^2}{9.5}$$

$$= .196 + .087 + .043 + .008 + .026 = .360.$$

Since no parameters of the theoretical frequency distribution are estimated from the sample, the number of degrees of freedom is $5 - 1$, or 4. Appendix Table 8 shows that if there are 4 degrees of freedom, $\chi^2_{.05} = 9.488$. Since the observed value of $\Sigma(f-e)^2/e$ is less than 9.488, the null hypothesis should not be rejected. The probability is much higher than .05 that the observed value of the test statistic could have arisen, given that the theoretical distribution was valid. In other words, the Defense Department should not reject the hypothesis that the data conform to the theoretical distribution.

TESTING FOR NORMALITY AT THE AMERICAN STOVE COMPANY

A number of years ago, the American Stove Company collected data concerning the heights of a metal piece that the firm manufactured. The firm measured the heights of 145 of these metal pieces, and the results are shown in the following frequency distribution:[13]

Height of metal piece (inches)	Number of metal pieces
Less than .8215	9
.8215 and under .8245	5
.8245 and under .8275	14
.8275 and under .8305	21
.8305 and under .8335	55
.8335 and under .8365	23
.8365 and under .8395	7
.8395 and under .8425	6
.8425 and over	5
Total	145

For many purposes it is important to test whether a certain random variable is normally distributed. For example, as we saw in Chapter 8, some statistical tests assume normality. Suppose that the American Stove Company had given you the task of using the above data to test whether the heights of the metal pieces are normally distributed.

(a) Express the upper limit of each class interval in the frequency distribution as a deviation from the sample mean (.8314 inches), and divide this deviation by the sample standard deviation (.0059 inches).

(b) Use the results in (a)—which correspond to points on the standard normal distribution—to determine the theoretical number of metal pieces that should fall into each class interval if the heights are normally distributed.

(c) Calculate the difference between the actual and expected frequency in each class, square this difference, divide it by the theoretical frequency, and sum the results for all classes.

(d) Use the χ^2 distribution to test the hypothesis that the heights are normally distributed. (Let $\alpha = .05$.)

(e) Explain your choice of the number of degrees of freedom.

(f) Write a three-sentence memorandum summarizing your results.

[13] See A. J. Duncan, *Quality Control and Industrial Statistics.*

9.8 Tests and Confidence Intervals Concerning the Variance

Before turning to nonparametric techniques, one final application of the χ^2 distribution should be discussed. If the population is normal, the χ^2 distribution can be used to test hypotheses concerning the variance or to construct confidence intervals for the variance. For example, a psychologist is interested in the variability in the time it takes a person to react to a red traffic light. Specifically, the psychologist wants to test the hypothesis that the variance, measured in (seconds)^2, of the reaction times equals .0001. A random sample of 11 people is chosen, and it is found that the sample variance s^2 equals .0003. If the significance level is set at .05, should the psychologist reject the hypothesis that the population variance equals .0001?

The first step is to formulate the null hypothesis and the alternative hypothesis. In this case, the null hypothesis is that the population variance equals .0001, and the alternative hypothesis is that the population variance is either less than or greater than .0001. If the population is normal, and if its variance equals σ_0^2, it can be shown that $(n - 1)s^2 \div \sigma_0^2$ has the χ^2 distribution with $(n - 1)$ degrees of freedom, where n is the sample size. Thus, to test the null hypothesis that the variance equals σ_0^2, we compute

$$\frac{(n - 1)s^2}{\sigma_0^2},$$

which is used as follows.

Test that $\sigma^2 = \sigma_0^2$

Decision Rule: *When the alternative hypothesis is $\sigma^2 \neq \sigma_0^2$, reject the null hypothesis if $(n - 1)s^2/\sigma_0^2$ exceeds $\chi_{\alpha/2}^2$ or is less than $\chi_{1-\alpha/2}^2$. When the alternative hypothesis is $\sigma^2 > \sigma_0^2$, reject the null hypothesis if $(n - 1)s^2/\sigma_0^2$ exceeds χ_α^2. When the alternative hypothesis is $\sigma^2 < \sigma_0^2$, reject the null hypothesis if $(n - 1)s^2/\sigma_0^2$ is less than $\chi_{1-\alpha}^2$.*

The alternative hypothesis in this case is that $\sigma^2 \neq \sigma_0^2$. Appendix Table 8 shows that $\chi_{.025}^2 = 20.48$ and $\chi_{.975}^2 = 3.25$ because the number of degrees of freedom equals $(n - 1) = 11 - 1 = 10$. Thus, the decision rule is: Reject the null hypothesis if $(n - 1)s^2/\sigma_0^2$ is greater than 20.48 or less than 3.25. In fact,

$$\frac{(n - 1)s^2}{\sigma_0^2} = \frac{10(.0003)}{.0001} = 30,$$

The Case of the Canine Life Preservers

According to some observers, the presence of a dog may help to extend the life of coronary patients and others.[14] A medical researcher wants to test whether this is true. He studies the condition of 300 coronary patients six months after their release from the hospital. These patients are divided into three groups: (1) owners of dogs, (2) nonowners of dogs who live in a family where there is a dog, (3) nonowners of dogs who live without a dog. In each group, the number of people who are alive or dead at the end of this 6-month period is as follows:

	Owner of dog	Nonowner of dog	
		In family with dog	Without dog
Alive	99	97	94
Dead	1	3	6

The researcher carries out a χ^2 test to determine whether the three groups differ with regard to the proportion of patients who survive. The results are:

$$\sum \frac{(f-e)^2}{e} = \frac{1}{96.67}\{(99-96.67)^2 + (97-96.67)^2 + (94-96.67)^2\}$$

$$+ \frac{1}{3.33}\{(1-3.33)^2 + (3-3.33)^2 + (6-3.33)^2\}$$

Based on 1 degree of freedom, $\chi^2_{.05} = 3.84146$. Thus, the researcher rejects the hypothesis that the proportion who survive is the same in each group. Do you agree with this statistical procedure? If not, what mistakes were made?

SOLUTION: At least two important mistakes were made. First, as stressed in Section 9.7, the theoretical, or expected, frequency should always be at least 5. This is not the case here. In three cases, the expected frequency is only 3.33. Second, the wrong number of degrees of freedom is used. Since this is a contingency table with 3 columns and 2 rows, the correct number of degrees of freedom is $(3-1)(2-1) = 2$. For these reasons, the analysis is incorrect.

[14] R. Runyon and A. Haber, *Fundamentals of Behavioral Statistics,* 4th edition (Reading, Mass.: Addison-Wesley, 1980), p. 316.

because $n = 11$, $s^2 = .0003$, and $\sigma_0^2 = .0001$. Thus, the psychologist should reject the null hypothesis that the variance of the reaction times equals .0001.

In many circumstances, statisticians must estimate the variance of a normal population based on the results of a random sample of n observations. *If the confidence coefficient is $(1 - \alpha)$, the confidence interval for the population variance is*

Confidence Interval for σ^2

$$\frac{(n-1)s^2}{\chi_{\alpha/2}^2} < \sigma^2 < \frac{(n-1)s^2}{\chi_{1-\alpha/2}^2},$$

(9.5)

where $\chi_{\alpha/2}^2$ is the value of a χ^2 variable (with n $- 1$ *degrees of freedom) that is exceeded with a probability of $\alpha/2$, and $\chi_{1-\alpha/2}^2$ is the value that is exceeded with a probability of $1 - \alpha/2$.* Thus, if a sample of 20 observations is chosen, and the sample variance equals 5, the 95 percent confidence interval for the population variance is

$$\frac{19(5)}{32.85} < \sigma^2 < \frac{19(5)}{8.91},$$

since Appendix Table 8 shows that for 19 degrees of freedom, $\chi_{.025}^2 = 32.85$ and $\chi_{.975}^2 = 8.91$. Simplifying terms, it follows that the confidence interval in this case is 2.89 to 10.66.

EXERCISES

9.10 Aversion conditioning is a way of treating alcoholism.[15] A hospital uses aversion conditioning on a random sample of 100 patients from each of four occupational groups: (1) employed, (2) self-employed, (3) housewife, and (4) unemployed. The results are as follows:

Occupational status	Unsuccessful	Partially successful	Successful	Total
Employed	30	10	60	100
Self-employed	41	21	38	100
Housewife	28	7	65	100
Unemployed	32	14	54	100
Total	131	52	217	400

Test whether there are differences among occupational groups in the probability that this treatment is (1) unsuccessful, (2) partially successful, (3) successful. (Let $\alpha = .01$.)

[15] A. Wiens and C. Menustik, "Treatment Outcome and Patient Characteristics in an Aversion Therapy Program for Alcoholism," *American Psychologist,* October 1983.

9.11 A publisher is interested in determining which of three book covers is most attractive. It interviews 400 people in each of three states (California, Illinois, and New York), and asks each person which of the covers he or she prefers. The number preferring each cover is as follows:

	California	*Illinois*	*New York*
First cover	81	60	182
Second cover	78	93	95
Third cover	241	247	123
Total	400	400	400

Do these data indicate that there are regional differences in people's preferences concerning these covers?
(a) Use the .05 level of significance.
(b) Use the .01 level of significance.

9.12 A soap manufacturer is trying to determine whether or not to market a new type of soap. It chooses a random sample of 300 people in the United States, and samples of equivalent size in England and France. It asks each of the people in each sample to try the new soap, and see whether he or she likes it better than other soaps. The results are as follows:

	United States	*England*	*France*
Prefer new soap	81	43	26
Do not prefer it	219	257	274
Total	300	300	300

Does it appear that there are international differences in the proportion of people who prefer the new soap?
(a) Use the .01 level of significance.
(b) Use the .05 level of significance.

9.13 A gambler wants to determine whether a die is true. He throws the die 100 times, with the following results:

Number on face of die	*Number of outcomes*
1	15
2	18
3	20
4	17
5	13
6	17
Total	100

Is there any evidence that the die is not true? (Let $\alpha = .05$.)

9.14 A detective wants to determine whether a pair of dice is "loaded." He rolls the dice 200 times, with the following results:

Number on dice	Number of outcomes
2	0
3	2
4	2
5	16
6	30
7	100
8	30
9	14
10	2
11	4
12	0
Total	200

Is there any evidence that these dice are not true? (Let $\alpha = .01$.)

9.15 A teacher claims that over the long run the grades that she gives are normally distributed and that she gives 8 percent As, 25 percent Bs, 34 percent Cs, 25 percent Ds, and 8 percent Fs. Last year, she gave the 110 students in her class the following grades:

Grade	A	B	C	D	F
Number of students	5	22	50	27	6

Are these data consistent with her claim?
(a) Use the .05 significance level.
(b) Use the .01 significance level.

9.16 The following table shows the number of daughters in 50 families, each of which contains five children.

Number of daughters	Number of families
0	7
1	12
2	13
3	8
4	6
5	4
Total	50

(a) Use the binomial distribuion with $n = 5$ and $\Pi = 0.5$ to compute the expected number of families with each number of daughters.

(b) Based on a χ^2 test, does it appear that these data conform to the binomial distribution? (Let $\alpha = .05$.)

9.17 A random sample of 30 observations is taken from a normal population. The sample variance equals 10.
   (a) Test the hypothesis that the population variance equals 20, the alternative hypothesis being that $\sigma^2 \neq 20$. Use the .05 significance level.
   (b) Calculate a 98 percent confidence interval for the population variance.

9.18 If a population is normal, and if the sample standard deviation (based on a sample of 18 observations) equals 3.2, test the hypothesis that $\sigma = 4$.
   (a) Use a two-tailed test with $\alpha = .05$.
   (b) Use a two-tailed test with $\alpha = .02$.

9.19 An engineer draws a random sample of 18 ball bearings from a large shipment. The sample standard deviation of their diameters is 0.003 inches.
   (a) Construct a 95 percent confidence interval for the population standard deviation.
   (b) Construct a 90 percent confidence interval for the population standard deviation.
   (c) What assumptions underlie your answers to (a) and (b)?

9.20 Find a 95 percent confidence interval for the variance if $s = 8$ and if
   (a) $n = 10$.
   (b) $n = 20$.
   (c) $n = 41$.

9.9 Nonparametric Techniques

Many of the tests discussed here and in the previous chapter are based on the assumption that the population or populations under consideration are normal. Although this assumption frequently is close enough to the truth so that these tests are good approximations, cases often arise where statisticians prefer instead to use various kinds of nonparametric, or distribution-free, tests. The hallmark of these tests is that they avoid the assumption of normality. For practically every test discussed in this and the previous chapter, there is a nonparametric analogue. Besides having the advantage that normality is not assumed, nonparametric tests often are easier to carry out because the computations are simpler. In addition, some experiments yield responses that can only be ranked, not measured along a cardinal scale. Nonparametric techniques are designed to analyze data of this sort, which is another important reason for their use.

In the following sections we shall describe three of the most commonly used nonparametric techniques: the sign test, the Mann-Whitney test, and the runs test. Each of these has important applications in a wide variety of fields.

9.10 The Sign Test

The sign test is used to test the null hypothesis that the population median equals a certain amount. For example, suppose that the Federal Trade Commission has 36 tires from a manufacturer tested to determine their length of life. The purpose is to find out whether the manufacturer's advertising claim that the tires average 25,000 miles of use is correct. If the median length of life is 25,000 miles, the probability that a tire chosen at random will last for less than 25,000 miles is 1/2, and the probability that it will last for more than 25,000 miles is also 1/2. If the results for the 36 tires are as shown in Table 9.6,

Table 9.6
Length of Life of
Sample of 36 Tires

27,100(+)	24,500(−)	24,800(−)	25,100(+)
24,200(−)	25,600(+)	24,600(−)	24,800(−)
23,300(−)	24,600(−)	25,500(+)	24,700(−)
23,800(−)	24,800(−)	26,100(+)	25,900(+)
22,600(−)	24,100(−)	23,900(−)	23,900(−)
24,800(−)	23,900(−)	24,100(−)	24,800(−)
24,100(−)	24,900(−)	24,300(−)	24,700(−)
24,800(−)	24,600(−)	24,700(−)	24,900(−)
25,500(+)	26,100(+)	24,900(−)	25,100(+)

we can place a minus sign to the right of the figure for each tire that lasts less than 25,000 miles. By the same token, we place a plus sign to the right of the figure for each tire that lasts more than 25,000 miles. If the null hypothesis holds (that is, if the median life is 25,000 miles), the number of pluses will have a binomial distribution:

$$P(x) = \frac{36!}{x!(36-x)!}\left(\frac{1}{2}\right)^x\left(\frac{1}{2}\right)^{36-x},$$

where x is the number of pluses. (Why? Because there is a probability of 1/2 that each tire in the sample will last more than 25,000 miles and thus will have a plus beside it.) Since the sample size is relatively large ($n = 36$), we know from Chapter 5 that this binomial distribution can be approximated by a normal distribution with a mean of 18 and a stan-

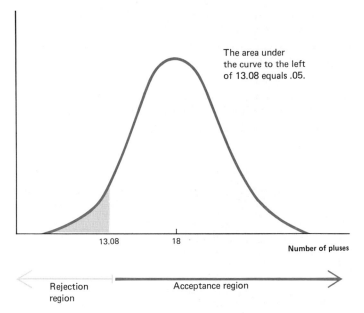

The area under
the curve to the left
of 13.08 equals .05.

13.08   18

Number of pluses

Rejection
region

Acceptance region

Figure 9.3
Sampling Distribution
of Number of Pluses, if
Null Hypothesis Is
True

dard deviation of 3. (Because $n = 36$ and $\Pi = 1/2$, $n\Pi = 36(1/2)$ or 18, and $\sqrt{n\Pi(1 - \Pi)} = \sqrt{36(1/2)(1/2)}$, or 3.) Thus, the sampling distribution of the number of pluses is as shown in Figure 9.3.

The null hypothesis, as noted above, is that the median length of life is 25,000 miles. The alternative hypothesis is that the median length of life is less than 25,000 miles. (Why is the alternative hypothesis one-sided? Because the FTC is interested in taking action against the advertising claims only if the median is less than 25,000 miles.) Thus, the FTC's statisticians will reject the null hypothesis only if the number of pluses—that is, x—is so small that it is unlikely that the sampling distribution in Figure 9.3 is valid. If the significance level is set at .05, it is clear that the null hypothesis should be rejected if the number of pluses is less than

$$18 - 1.64(3) = 13.08,$$

since there is a .05 probability that the number of pluses will be less than 13.08, given that the null hypothesis is true.[16] (See Figure 9.3.) Based on the data in Table 9.6, it follows that the FTC should reject the hypothesis that the median length of life is 25,000 miles, because the number of pluses is 9, which is considerably less than 13.08.

---

[16] We have ignored the continuity correction discussed in Chapter 5 because it has no material effect on our results.

The sign test is based on the fact that if the population median (*M*) is equal to the value specified by the null hypothesis (say, M_0), *x*, the number of observations in the sample above M_0, has a binomial distribution with Π equal to $1/2$ and *n* equal to the sample size. Thus, if the sample size is large, the following decision rules can be used.

Test that $M = M_0$

Decision Rule if the Alternative Hypothesis is that M < M₀: *Accept the null hypothesis if* $x \geq n/2 - z_\alpha \sqrt{n/4}$; *reject the null hypothesis if* $x < n/2 - z_\alpha \sqrt{n/4}$.

Decision Rule if the Alternative Hypothesis is that M > M₀: *Accept the null hypothesis if* $x \leq n/2 + z_\alpha \sqrt{n/4}$; *reject the null hypothesis if* $x > n/2 + z_\alpha \sqrt{n/4}$.

Decision Rule if the Alternative Hypothesis is that M ≠ M₀: *Accept the null hypothesis if* $n/2 - z_{\alpha/2}\sqrt{n/4} \leq x \leq n/2 + z_{\alpha/2}\sqrt{n/4}$; *reject the null hypothesis if* $x < n/2 - z_{\alpha/2}\sqrt{n/4}$ *or if* $x > n/2 + z_{\alpha/2}\sqrt{n/4}$.

If the sample size is small, the binomial distribution, not the normal distribution, must be used; thus, these decision rules must be altered.

Note that the sign test does not depend on any assumptions about the nature or shape of the population. It rests only on the fact that the probability that a randomly chosen observation falls above the median is $1/2$; and this is true for any population, normal or otherwise. (Basically, this is why the sign test is nonparametric.) Also, note that the computations involved in this test are comparatively simple. As pointed out previously, nonparametric tests are often used for this reason.

The sign test is useful in a wide variety of circumstances. The following example illustrates how it is employed in a sample consisting of *matched pairs.* (Recall our discussion of matched pairs in Chapter 7.)

EXAMPLE 9.3 An experiment is conducted in which a subject takes a test when he or she is 18 years old and the same test two years later. The sample consists of 100 college students. The investigator wants to test the null hypothesis that the median difference in test score equals zero. For 62 students, the test score is higher when they take the test the second time; for 38 students the reverse is true. Use the sign test to test the null hypothesis. (Let the significance level equal .05.)

SOLUTION: If the null hypothesis is true (that is, if the median difference is zero), there is a 50–50 probability that a student will obtain a higher score the second time than the first time. Thus if we put a plus next to each student where the second test score exceeds the first, the probability distribution of the number of pluses should be binomial with $n = 100$ and $\pi = 1/2$, if the null hypothesis is true. Since n is large, the binomial distribution can be approximated by the normal distribution, and the null hypothesis should be rejected if the number of pluses x is less than $n/2 - z_{.025} \sqrt{n/4}$ or greater than $n/2 + z_{.025} \sqrt{n/4}$. Since $n = 100$, the null hypothesis should be rejected if $x < 50 - 1.96(5)$, or 40.2, or if $x > 50 + 1.96(5)$, or 59.8. Since $x = 62$, the null hypothesis should be rejected. In other words, the evidence seems to indicate that students obtain higher test scores the second time than the first time.

Note that any student who obtained the same score both times should be omitted from the analysis.

9.11 The Mann-Whitney Test

Another important nonparametric technique is the Mann-Whitney test, also known as the Wilcoxon test or the U test. This technique is used to find out whether two populations are identical. For example, the Federal Trade Commission might need to determine whether there is a difference between the population of lifetimes of firm C's tires and the population of lifetimes of firm D's tires. A random sample of 14 of firm C's tires and a random sample of 14 of firm D's tires are taken, the results being shown in Table 9.7. In this case, *the null hypothesis is that there is no difference between the two populations. The alternative hypothesis is that they are different.* Note that in contrast to Example 9.3, the two samples are entirely independent; *they are not matched pairs.*

The first step in applying the Mann-Whitney test is to rank all observations in the two combined samples. Table 9.7 shows the 28 observations ranked from lowest to highest, with the letter under each observation (a C or a D) indicating the firm from which it came. The members of firm C's sample are ranked 1, 2, 3, 4, 5, 7, $\cdots$, and the members of firm D's sample are ranked 6, 8, 9, 11, 14, 16, 17 $\cdots$.[17] If the null hypothesis is true, we would expect that the average rank for the items from one sample would be approximately equal to the average rank for the items from the other sample. On the other hand, if the alternative hypothesis is true, we would expect a marked difference between the average ranks in the two samples.

[17] If there are ties, we must assign the tied items the mean of the ranks they have in common. For example, if two items are tied for 3d and 4th place, each would receive the rank of 3.5.

Table 9.7
Lengths of Life of
Sample of Tires from
Two Firms

Firm C		Firm D	
19,800	19,700	19,850	20,850
19,900	19,600	31,250	20,650
20,100	19,500	20,380	21,050
20,200	20,400	19,950	19,820
31,300	20,300	20,260	20,350
18,700	21,000	19,750	31,150
18,900	21,100	20,450	20,750

Observations Ranked from Lowest to Highest

Observation	18,700	18,900	19,500	19,600	19,700	19,750	19,800	19,820,
Firm	C	C	C	C	C	D	C	D
Rank	1	2	3	4	5	6	7	8

Observation	19,850	19,900	19,950	20,100	20,200	20,260	20,300	20,350,
Firm	D	C	D	C	C	D	C	D
Rank	9	10	11	12	13	14	15	16

Observation	20,380	20,400	20,450	20,650	20,750	20,850	21,000	21,050,
Firm	D	C	D	D	D	D	C	D
Rank	17	18	19	20	21	22	23	24

Observation	21,100	31,150	31,250	31,300
Firm	C	D	D	C
Rank	25	26	27	28

To carry out the Mann-Whitney test, it is more convenient to use the *rank-sum* instead of the average rank in a sample. Let R_1 be the sum of the ranks for one of the samples. (It doesn't matter which sample one chooses.) Then we compute

$$U = n_1 n_2 + \frac{n_1(n_1 + 1)}{2} - R_1, \qquad (9.6)$$

where n_1 is the number of observations in the sample on which R_1 is based and n_2 is the number of observations in the other sample. If the null hypothesis is true, the sampling distribution of U has a mean equal to

$$E_u = \frac{n_1 n_2}{2}, \qquad (9.7)$$

and a standard deviation equal to

$$\sigma_U = \sqrt{\frac{n_1 n_2 (n_1 + n_2 + 1)}{12}}. \tag{9.8}$$

And if n_1 and n_2 are at least 10, the sampling distribution of U can be approximated quite well by the normal distribution.

Based on these facts concerning the sampling distribution of U (if the null hypothesis is true), it is easy to formulate an appropriate decision rule.

Decision Rule: *Accept the null hypothesis that the two populations are*   *Mann-Whitney Test* *identical if* $-z_{\alpha/2} \leq (U - E_u) \div \sigma_u \leq z_{\alpha/2}$; *reject the null hypothesis if* $(U - E_u) \div \sigma_u < -z_{\alpha/2}$ *or if* $(U - E_u) \div \sigma_u > z_{\alpha/2}$.

In applying this decision rule, note that

$$\frac{U - E_u}{\sigma_u} = \frac{n_1 n_2 + \dfrac{n_1(n_1 + 1)}{2} - R_1 - \dfrac{n_1 n_2}{2}}{\sqrt{n_1 n_2 (n_1 + n_2 + 1)/12}}. \tag{9.9}$$

Also, it should be recognized that this decision rule applies only to a two-tailed test. (The extension to a one-tailed test is straightforward.) Further, this decision rule assumes that both n_1 and n_2 are at least 10; if not, the test must be based on special tables, and the procedure described here is not appropriate.

To illustrate the use of the Mann-Whitney test, let's return to the data in Table 9.7. To compute R_1, let's use the sample of firm C's tires. Since the observations in this sample have ranks of 1, 2, 3, 4, 5, 7, 10, 12, 13, 15, 18, 23, 25, and 28, it is clear that $R_1 = 166$. Thus, since $n_1 = n_2 = 14$,

$$U = 14(14) + \frac{14(15)}{2} - 166 = 135,$$

and

$$E_u = \frac{14(14)}{2} = 98$$

$$\sigma_u = \sqrt{\frac{14(14)(29)}{12}} = 21.8.$$

Consequently,

$$\frac{U - E_u}{\sigma_u} = \frac{135 - 98}{21.8} = 1.70.$$

If the .05 significance level is chosen, it appears that the null hypothesis should not be rejected, since $z_{.025} = 1.96$, which is more than $(U - E_u) \div \sigma_u$. Thus, the Federal Trade Commission should conclude that the available evidence does not indicate that there is a difference between the population of lifetimes of firm C's tires and the population of lifetimes of firm D's tires.

Finally, it is worth noting that this test does not assume that the populations in question are normal. That is, there is no need to assume that the population of lifetimes of firm C's tires is normal, or that the population of lifetimes of firm D's tires is normal. Thus, this test avoids the assumption of normality underlying the t test, which was used in the previous chapter to test a similar kind of hypothesis. Moreover, this test requires far less computation than the t test, an important advantage under some circumstances.

The Mann-Whitney test is useful as a possible substitute for the t test. The following example further illustrates how it is used.

EXAMPLE 9.4 A psychologist gives an aptitude test to 12 men and 12 women, the scores being as follows:

Men   80, 79, 92, 65, 83, 84, 95, 78, 81, 85, 73, 52
Women 82, 87, 89, 91, 93, 76, 74, 70, 88, 99, 61, 94

Use these data to test the null hypothesis that the distribution of scores on this test is the same for men as for women. (Let the significance level equal .05.)

SOLUTION: Rank the scores from lowest to highest, and put an M under each score if it is a man's and a W under each score if it is a woman's:

52 61 65 70 73 74 76 78 79 80 81 82 83 84 85 87 88
M W M W M W W M M M M W M M M W W

89 91 92 93 94 95 99
W W M W W M W

The sum of the ranks for men is $1 + 3 + 5 + 8 + 9 + 10 + 11 + 13 + 14 + 15 + 20 + 23 = 132$. Thus $U = 12(12) + [12(13)/2] - 132 = 90$. Also, $E_u = 12(12)/2 = 72$, and $\sigma_u = \sqrt{12(12)(25)/12} = \sqrt{300} = 17.3$. Thus, since $z_{\alpha/2} = 1.96$, the null hypothesis should be rejected if $(U - E_u) \div \sigma_u$ is less than -1.96 or greater than 1.96. In fact, $(U - E_u) \div \sigma_u = (90 - 72) \div 17.3 = 1.0$. Thus, the null hypothesis should not be rejected. The probability is greater than .05 that these results could have occurred, given that the distribution of scores on this test was the same for men as for women.

9.12 The Runs Test

Another important nonparametric technique is the runs test, which is designed to test the hypothesis that a sequence of numbers, symbols, or objects is in random order. A tennis fan interested in the outcomes of the matches between John McEnroe and Ivan Lendl finds that these outcomes are as follows (where M stands for a McEnroe victory and L stands for a Lendl victory):

L M L M L M L M L M L M L M.

Does this sequence appear to be in a random order? No, since the winner seems to alternate back and forth. (If this were the case, one might suspect that tennis players, like some phoney wrestlers, decide in advance who will win and alternate the winner.) On the other hand, suppose that the sequence is

M M M M M M M L L L L L L L.

Does this sequence appear to be in a random order? No, since the winner of all the early matches is McEnroe, and the winner of all the later matches is Lendl. (If this were the case, we might suspect that Lendl had greatly improved—or that McEnroe's play had deteriorated—at about the middle of the sequence of matches.)

To understand the runs tests, you must know the statistician's definition of a run, which is the following:

> **A run** *is a sequence of identical numbers, symbols, objects, or events preceded and followed by different numbers, symbols, objects, or events (or by nothing at all).*

In the first version of the McEnroe-Lendl matches given above, there are 14 runs; in the second version there are only 2 runs. Suppose that a third version is as follows:

M L M M M M L L L M L M L L.

How many runs are there? The answer is 8. (The lines under the letters designate each run.)

The runs test is based on the idea that if the probability of the occurrence of one number, symbol, object, or event is constant throughout the sequence, then there should be neither a very large number of runs (as in our first "history" of the McEnroe-Lendl matches) nor a very small number of runs (as in our second such "history"). If there are two possible outcomes (such as L and M), and if the probability of each outcome remains constant throughout the sequence (and if the successive outcomes are independent), it can be shown that the expected number of runs is:

$$E_r = \frac{2n_1n_2}{n_1 + n_2} + 1,$$ (9.10)

and the standard deviation of the number of runs is

$$\sigma_r = \sqrt{\frac{2n_1n_2(2n_1n_2 - n_1 - n_2)}{(n_1 + n_2)^2(n_1 + n_2 - 1)}},$$ (9.11)

where n_1 is the number of outcomes of one type (such as L) and n_2 is the number of outcomes of the other type (such as M). If either n_1 or n_2 is larger than 20, it can be shown that the number of runs r is approximately normally distributed. When both n_1 and n_2 exceed 10, this approximation is good.

The runs test is designed to test the null hypothesis that a given sequence of outcomes is in random order—that is, the probability of each outcome remains constant throughout the sequence and that successive outcomes are independent. If the significance level is set equal to α, it follows from the previous paragraph that the following decision rule can be applied.

Runs Test   **Decision Rule:** *Accept the null hypothesis (of randomness) if* $E_r - z_{\alpha/2}\sigma_r \leq r \leq E_r + z_{\alpha/2}\sigma_r$; *reject the null hypothesis if* $r < E_r - z_{\alpha/2}\sigma_r$ *or if* $r > E_r + z_{\alpha/2}\sigma_r$.

This assumes that n_1 and n_2 meet the requirements in the previous paragraph.

To illustrate the application of this test, suppose that McEnroe and Lendl play 37 times, with the following results:

MLMMLMMLMMLLLMLLMMMLMLMMLMMLLMMMLLMML.

If n_1 is the number of McEnroe victories and n_2 is the number of Lendl victories, $n_1 = 21$ and $n_2 = 16$. Thus,

$$E_r = \frac{2(21)(16)}{37} + 1 = 19.16$$

and

$$\sigma_r = \sqrt{\frac{2(21)(16)[2(21)(16) - 21 - 16]}{37^2(36)}} = \sqrt{\frac{672(635)}{49{,}284}} = \sqrt{8.7} = 2.94.$$

Thus, if α is set equal to .05,

$$E_r - z_{\alpha/2}\sigma_r = 19.16 - 1.96(2.94) = 13.4$$

$$E_r + z_{\alpha/2}\sigma_r = 19.16 + 1.96(2.94) = 24.9$$

Since the actual number of runs r equals 22, the null hypothesis should not be rejected (r being neither less than 13.4 nor more than 24.9).

The runs test has many applications. The example below shows how it can be used in the field of environmental science.

EXAMPLE 9.5 An environmental scientist measures daily the air quality in Galveston, Texas. Each day during a 32-day period is classified A (above-average air quality) or B (below-average air quality),the results being as follows:

B A A A A B B A B B A A A A B B B B A B A A A A A A A A B A A A A.

Use the runs test to determine whether there is a departure from randomness in the sequence. (Let the significance level equal .05.)

SOLUTION: Since $n_1 = 21$ and $n_2 = 11$,

$$E_r = \frac{2(21)(11)}{32} + 1, \text{ or } 15.4,$$

and

$$\sigma_r = \sqrt{\frac{2(21)(11)[2(21)(11) - 21 - 11]}{32^2(31)}}$$

$$= \sqrt{\frac{198{,}660}{31{,}744}} = \sqrt{6.26}, \text{ or } 2.50.$$

Thus, the null hypothesis should be rejected if the number of runs is less than

$$E_r - z_{\alpha/2}\sigma_r = 15.4 - 1.96(2.50) = 10.5,$$

or greater than

$$E_r + z_{\alpha/2}\sigma_r = 15.4 + 1.96(2.50) = 20.3.$$

Since the number of runs equals 12, the environmental scientist should not reject the null hypothesis. In other words, he should not reject the hypothesis that the sequence is random.

9.13 Nonparametric Techniques: Pros and Cons

Based on the discussion in the previous sections, nonparametric techniques are clearly a very important statistical tool. Frequently, as we have seen, the statistician has the choice of using a nonparametric technique or a parametric procedure of the sort discussed in the previous chapter. For example, the sign test is to some extent a substitute for the test (in the previous chapter) of a specified value of the mean, and the Mann-Whitney test is to some extent a substitute for the parametric test (also in the previous chapter) of equality between two population means. The following factors should guide the choice between nonparametric and parametric techniques in situations where both can be used.

SIMPLICITY OF CALCULATIONS. Under some circumstances, it is very important that the computations in a statistical test be simple and rapid. For example, it may be necessary for a person with a relatively limited mathematical background to carry out the computations; or there may be a premium on speed. In such cases, nonparametric techniques like the sign test may be preferred over parametric tests.

REALISM OF ASSUMPTIONS. Many parametric tests assume that the relevant populations are normal. Two such tests are the small-sample test that a population mean equals a specified value and the small-sample test for equality between two population means. Although these procedures are dependable in the face of moderate departures from normality, they can be quite misleading in the face of gross departures of this sort. Thus, if the populations are very far from normality, statisticians often prefer nonparametric techniques like the sign test and Mann-Whitney test.

POWER OF THE TEST. If the assumptions underlying the parametric tests in the previous chapter are reasonably close to reality, these tests generally are more powerful than nonparametric techniques. In other words, the parametric techniques are less likely to produce a Type II error, given that the probability of a Type I error is the same for both tests. This is an important advantage of parametric tests, as long as their assumptions are reasonably valid. Nonparametric techniques, as we have seen, often use rankings or orderings as distinct from the actual values of the observations, thus losing a certain amount of the information in the sample.

EXERCISES

9.21 A North Carolina sociologist chooses a random sample of 36 black workers in a particular community and finds that the 1985 income of each is as follows:

Dollars

20,100	8,200	13,200	5,200	11,100	51,100
19,400	8,900	14,300	10,100	10,800	9,600
10,100	10,300	15,800	12,300	9,100	10,900
23,000	26,000	16,100	14,000	7,200	12,000
24,200	11,400	17,200	15,100	4,300	13,200
25,100	12,900	18,900	16,000	38,000	15,100

Use the sign test to test the hypothesis that the median income of the black workers in this community equals $15,000, the alternative hypothesis being that it does not equal $15,000.

(a) Carry out this test at the 5 percent significance level.

(b) Carry out this test at the 1 percent significance level.

(c) Compare your results with those obtained in Exercise 8.37. Are the results consistent? Are the two tests aimed at testing the same hypothesis? Are the two tests based on the same assumptions? If not, what are the differences in the assumptions?

9.22 According to a famous criminologist, the murder rate in a particular type of city averages 3.5 per 100,000 people. To test this statement, a statistician selects a random sample of 100 such cities, and finds that 58 percent of them have murder rates exceeding 3.5 per 100,000 people.

(a) Test whether the median murder rate in such cities is 3.5 per 100,000 people. Let $\alpha = .05$.

(b) Do your results change if α is set equal to .01 rather than .05?

9.23 A psychologist claims that the length of time a person takes to do a particular job is not influenced by whether the person is male or female. To test this claim a statistician asks 11 men and 11 women to do this job, and measures each one's time. The 22 times are ranked (from lowest to highest), with the following results:

Men: 1, 2, 4, 6, 8, 9, 10, 11, 14, 15, 16
Women: 3, 5, 7, 12, 13, 17, 18, 19, 20, 21, 22

(a) Are the data consistent with the psychologist's claim, if $\alpha = .05$?

(b) What are the results if $\alpha = .01$ rather than .05?

9.24 Having installed a new procedure to reduce clerical errors, the Social Security Administration wants to determine whether this new procedure has resulted in a difference in the distribution of the number of clerical errors per day in one of its departments. The Social Security Administration chooses a random sample of 11 days prior to the installation of the new procedure and finds that the number of errors per day was as follows: 7, 8, 10, 6, 5, 8, 9, 11, 10, 7, 9. It also chooses a random sample of 11 days following the installation of the new procedure and finds that the number of errors per day was as follows: 6, 5, 8, 7, 9, 6, 5, 4, 9, 7, 6.

(a) Use the Mann-Whitney test to determine whether the population distribution of the number of errors per day was different after vs. before the installation of the new procedure. (Let $\alpha = .05$.)

(b) Use a t test to test this hypothesis, based on the data given above.

(c) Are the results consistent?

(d) Are the two tests aimed at testing the same hypothesis?

(e) Are the two tests based on the same assumptions? If not, what are the differences in the assumptions?

9.25 The Social Security Administration decides that a better way to test error reduction under its new procedure is to choose 10 employees randomly and determine how many errors each made in a week before the introduction of the new procedure and how many each made in a week after its introduction. The results are as follows:

Type of procedure	*Person*									
	1	*2*	*3*	*4*	*5*	*6*	*7*	*8*	*9*	*10*
Old	40	38	37	34	39	37	42	33	39	40
New	37	37	32	40	37	35	39	32	37	38

Use a nonparametric technique to test the hypothesis that there is no difference between the old and new procedure in the incidence of errors committed. (Let $\alpha = .05$.)

9.26 There is a sequence containing 3 Ss and 17 Ts. (a) What is the minimum number of runs that can occur in this sequence? (b) What is the maximum number?

9.27 A table of one-digit random numbers begins as follows:

```
0 1 3 4 6 8 9 7 6 5 4 1 0 0 9 8 7 6 5 4 3 2 1
1 4 3 8 7 6 6 7 5 4 5 0 2 5 8 7 6 9 4 3 2 1 0
9 7 6 5 4 3 2 2 1 1 4 5 6 7 7 8 9 2 4 4 5
```

(a) If an even number (or zero) is denoted by E and an odd number is denoted by O, how many runs of Es or Os are there in this sequence?

(b) Based on the number of runs of even (including zero) and odd numbers in the sequence, is there any evidence that the sequence of numbers given there is not random? (Let $\alpha = .05$.)

(c) Let any number from zero to 4 be denoted by *A*, and any number from 5 to 9 be denoted by *B*. How many runs of *A*s or *B*s are there in the sequence?

(d) Based on the number of runs of *A*s and *B*s in the sequence, is there any evidence that the sequence of numbers given there is not random? (Let $\alpha = .01$.)

9.28 The price of a share of U.S. Steel common stock behaves in the following way over a 40-day period. (A "+" indicates that the price rose; a "−" indicates that it did not rise.)

$$+ + + − − + − + + + − − + − + − + − + + + − − + − + −$$
$$+ − + − − + + − − + − + −$$

(a) If $\alpha = .05$, is there evidence of a departure from randomness in this sequence?

(b) Does your answer to (a) change if α equals .01 rather than .05?

9.29 The following sequence indicates whether or not a robbery occurred in a particular residential neighborhood in a 24-day period. (A "1" means that a robbery occurred on a particular day; a "0" means that no robbery occurred on this day.)

$$1\ 0\ 1\ 1\ 0\ 1\ 1\ 0\ 1\ 1\ 1\ 0\ 0\ 0\ 0\ 1\ 0\ 1\ 0\ 1\ 0\ 0\ 1\ 1$$

(a) If $\alpha = .02$, is there evidence of a departure from randomness in this sequence?

(b) Does your answer to (a) change if α equals .01 rather than .02?

The Chapter in a Nutshell

1. The χ^2 *distribution* (with v degrees of freedom) is the probability distribution of the sum of squares of v independent standard normal variables. The χ^2 distribution is a continuous probability distribution and is generally skewed to the right; but as the number of degrees of freedom increases, it approaches normality. There is a zero probability that a χ^2 random variable will be negative.

2. To test whether a number of population proportions are equal (against the alternative hypothesis that they are unequal), the sample proportion (in the samples from all populations combined) should be computed. This proportion should then be multiplied by the sample size from each population to obtain the expected frequency of "successes" in this sample. The test statistic is $\Sigma(f-e)^2/e$, where f is the actual frequency of "successes" or "failures" in each sample and e is the corresponding expected frequency. If the test statistic exceeds χ^2_α (where the number of degrees of freedom is one less than the number of populations being compared), the null hypothesis should be rejected.

3. A *contingency table* contains a certain number of rows and columns, and the decision maker wants to know whether the probability distribution in one column differs from that in another column. In other words, is the variable represented by the rows independent of that represented by the columns? Assuming such independence, we can compute the expected frequencies in the table. The test statistic is $\Sigma(f-e)^2/e$, where f is an actual frequency and e is the corresponding expected frequency. If the test statistic exceeds χ_α^2 (where the number of degrees of freedom equals $[r-1][c-1]$, r being the number of rows and c being the number of columns), the null hypothesis (of independence) should be rejected.

4. To test whether an *observed frequency distribution* conforms to a *theoretical frequency distribution*, we calculate the test statistic $\Sigma(f-e)^2/e$, where f is the observed frequency in a particular class interval and e is the theoretical, or expected, frequency in the same class interval. If the test statistic exceeds χ_α^2—where the number of degrees of freedom equals the number of class intervals minus $(1+h)$, and h is the number of parameters estimated from the sample—the null hypothesis (that the observed frequency distribution conforms to the theoretical one) should be rejected.

5. To test whether the variance of a normal population equals σ_0^2, we compute the test statistic: $(n-1)s^2 \div \sigma_0^2$, where n is the sample size and s^2 is the sample variance. If the test statistic exceeds $\chi_{\alpha/2}^2$ or is less than $\chi_{1-\alpha/2}^2$ (where $n-1$ is the number of degrees of freedom), the null hypothesis should be rejected if the alternative hypothesis is two-sided.

6. The *sign test* is used to test the null hypothesis that the population median equals a certain amount. A plus sign is placed next to a sample observation exceeding this amount; a minus sign is placed next to one that falls below it. The number of plus signs x has a binomial distribution (where $\Pi = 1/2$) if the null hypothesis is true. For large samples, the null hypothesis should be rejected if $x < n/2 - z_{\alpha/2}\sqrt{n/4}$ or if $x > n/2 + z_{\alpha/2}\sqrt{n/4}$, assuming that a two-tailed test is appropriate.

7. The *Mann-Whitney test* is used to test the null hypothesis that two samples come from the same population. The first step is to rank all observations in the two samples combined and to compute R_1, the sum of the ranks for the first sample. One then computes the test statistic

$$\frac{n_1 n_2 + \dfrac{n_1(n_1+1)}{2} - R_1 - \dfrac{n_1 n_2}{2}}{\sqrt{n_1 n_2(n_1+n_2+1)/12}},$$

where n_1 is the number of observations in the first sample and n_2 is the number of observations in the second sample. If this test statistic is greater than $z_{\alpha/2}$ or less than $-z_{\alpha/2}$ the null hypothesis should be rejected. (This assumes that n_1 and n_2 are at least 10.)

8. The *runs test* is designed to test the null hypothesis that a sequence of numbers, symbols, or objects is in random order. A *run* is a sequence of *identical* numbers, symbols, objects, or events preceded and followed by *different* numbers, symbols, objects, or events, or by nothing at all. If a sequence is in random order, the expected number of runs E_r equals $[(2n_1n_2)/(n_1 + n_2)] + 1$, and the standard deviation of the number of runs σ_r, equals

$$\sqrt{\frac{2n_1n_2(2n_1n_2 - n_1 - n_2)}{(n_1 + n_2)^2(n_1 + n_2 - 1)}},$$

where n_1 is the number of observations in the sequence of one type and n_2 is the number of the other type. If both n_1 and n_2 exceed 10, the null hypothesis should be rejected if the number of runs r is less than $E_r - z_{\alpha/2}\sigma_r$ or if it exceeds $E_r + z_{\alpha/2}\sigma_r$.

Chapter Review Exercises

9.30 A clinic wants to evaluate three methods to help people quit smoking. A random sample of 100 persons who want to quit smoking is subjected to the first method; another such random sample is subjected to the second method; and still another such random sample is subjected to the third method. The results are as follows:

	Method I	*Method II*	*Method III*
Did not quit smoking	81	72	60
Successfully quit smoking	19	28	40
Total	100	100	100

(a) Can the observed differences among methods be attributed to chance variation? (Set $\alpha = .01$.)

(b) If α is set equal to .05, can the observed differences be attributed to chance variation?

9.31 The Department of Transportation wants to determine whether six types of cars differ with respect to the probability of stalling when the temperature is minus 40 degrees. The Department chooses a random sample of 100 cars of each type, and operates them under this set of circumstances, the results being:

	Type 1	*Type 2*	*Type 3*	*Type 4*	*Type 5*	*Type 6*
Stall	31	22	29	33	23	24
No stall	69	78	71	67	77	76
Total	100	100	100	100	100	100

Test whether these types of cars differ significantly with respect to the probability of stalling under these circumstances. (Set $\alpha = 0.10$.)

9.32 According to a theory put forth by plant geneticists, the number of peas of each of four types should be as shown in the table below. The actual numbers are also given.

Type of pea	Expected	Actual
Smooth green	321	303
Wrinkled green	105	109
Smooth yellow	105	98
Wrinkled yellow	36	57
Total	567	567

Are the data consistent with the theory?
(a) Use the .01 level of significance.
(b) Use the .05 level of significance.

9.33 The mayor of a large city wants to determine the proportion of voters that support an increase in schoolteachers' salaries. He asks that a random sample of 100 voters be chosen from each of the city's four districts, and that each person in these samples be asked whether he or she supports such an increase. The results are as follows:

	District I	District II	District III	District IV
Supports an increase	38	43	78	53
Does not support an increase	62	57	22	47
Total	100	100	100	100

Can the differences among districts in the proportion supporting such an increase be attributed to chance?
(a) Use the .10 level of significance.
(b) Use the .05 level of significance.
(c) Use the .01 level of significance.

9.34 A researcher claims that the median increase in SAT scores due to a new coaching technique is 20 points. Another researcher tries the new technique on 100 randomly chosen students and finds that in 41 of these cases there is an increase of more than 20 points. In 59 of these cases there is an increase of less than 20 points.
(a) Use the sign test to indicate whether the researcher's claim is correct, against a two-sided alternative. (Let $\alpha = .05$.)
(b) Use the sign test to indicate whether the researcher's claim is correct, against the alternative hypothesis that the median increase is less than 20 points. (Let $\alpha = .05$.)

9.35 Test the hypothesis that $\sigma = 10$, given that $s = 8$ for a sample of 15.
(a) The alternative hypothesis is $\sigma > 10$, and $\alpha = .05$.
(b) The alternative hypothesis is $\sigma < 10$, and $\alpha = .01$.

9.36 The following sequence indicates whether the murder rate in the city of Atlantis is above average or below average on a particular day. (An *A* means that it is above average; a *B* means that it is below average.) The data pertain to 25 consecutive days.

   A B B A A B B B A B A B A A A B B B A A B A A B A B A

   (a) If $\alpha = .01$, is there any evidence of a departure from randomness in this sequence?
   (b) Why should the law enforcement officers in the city of Atlantis care whether a sequence of this sort is random?

9.37 The Department of Defense wants to determine whether alloy *A* withstands heat better or worse than alloy *B*. This information is important in deciding which alloy should be used in a new weapons system. Thirty pieces of alloy *A* are matched against thirty pieces of alloy *B*. In each of these 30 cases, one piece of each type of alloy is subjected to the same intense heat, and engineers determine which type of alloy fares better. In 19 of the 30 cases, alloy *B* fares better than alloy *A*.
   (a) Use these results to test whether there is any difference between alloy *A* and alloy *B* in how well they withstand heat. Let $\alpha = .05$.
   (b) Do your results change if α is set equal to .01 rather than .05?

9.38 A scientist is interested in comparing the accuracy of two methods of measuring weight. An object is weighed 10 times based on the first method and 10 times based on the second method. The size of the error is recorded in each of the 20 cases. When these 20 errors are ranked (from smallest to largest), the results are as follows:

   Method 1: 2, 3, 5, 6, 9, 11, 13, 14, 15, 16
   Method 2: 1, 4, 7, 8, 10, 12, 17, 18, 19, 20

   (a) Test whether the accuracy of the two methods is the same. (Let $\alpha = .05$.)
   (b) Does the assumption of normality of the errors underlie your results in (a)?

9.39 Students at an eastern university are asked whether they favor increased military aid to El Salvador. The sample is divided into fraternity members, sorority members, and unaffiliated students. The results are as follows:

	Fraternity	Sorority	Unaffiliated
Oppose Strongly	20	12	19
Oppose Somewhat	24	28	21
Favor Strongly	16	20	20
Favor Somewhat	16	20	20
Total	76	80	80

Use Minitab (or some other computer package) to test whether student opinion on this issue is independent of whether a student belongs to a fraternity or sorority or is unaffiliated. (Let $\alpha = .05$.) If Minitab (or some other computer package) is not available, do the computations by hand.

*9.40 In a high school in suburban Los Angeles, 200 students were chosen at random. The sample was divided at random into three groups, one of which spent one hour per week with a tutor, one of which spent 1/2 hour per week, and one of which spent no time with the tutor. The number of students in each group whose grades improved, deteriorated, or remained unchanged was as follows:

Change in grades	Hours per week with tutor		
	1	1/2	0
Improved	32	16	10
Deteriorated	11	20	31
Unchanged	27	34	19
Total	70	70	60

Minitab was used to carry out a χ^2 test to see whether the change in grades was independent of the amount of time spent with the tutor. Interpret the results, given below.

```
READ THE FOLLOWING DATA INTO COLUMNS C4,C5,C6

32 16 10
11 20 31
27 34 19
CHISQUARE ANALYSIS OF THE TABLE IN COLUMNS C4,C5,C6
EXPECTED FREQUENCIES ARE PRINTED BELOW OBSERVED
FREQUENCIES
           I    C4   I    C5   I    C6   I  TOTALS
     ------I--------I--------I--------I------
       1   I   32   I   16   I   10   I    58
           I  20.3  I  20.3  I  17.4  I
     ------I--------I--------I--------I------
       2   I   11   I   20   I   31   I    62
           I  21.7  I  21.7  I  18.6  I
     ------I--------I--------I--------I------
       3   I   27   I   34   I   19   I    80
           I  28.0  I  28.0  I  24.0  I
     ------I--------I--------I--------I------
    TOTALS I   70   I   70   I   60   I   200

TOTAL CHI SQUARE =

         6.74 + 0.91 + 3.15 +

         5.28 + 0.13 + 8.27 +

         0.04 + 1.29 + 1.04 +

         = 26.84

DEGREES OF FREEDOM = ( 3-1) X ( 3-1) = 4
```

* Exercise requiring some familiarity with Minitab.

*9.41 An ambulance service has two ambulances. It wants to determine whether one makes the trip from the center of town to the local hospital quicker than the other. Over a one-week period, the time (in minutes) that it takes each ambulance to make this trip is shown below:

Ambulance A   6.3, 7.2, 5.2, 3.5, 4.8, 5.7, 5.2, 6.4, 5.9, 6.1, 6.3, 6.8
Ambulance B   7.4, 6.5, 7.4, 8.2, 5.6, 5.4, 6.3, 7.4, 6.2, 7.3, 4.6, 5.9

Minitab is used to carry out a Mann-Whitney test to determine whether the median time for the two ambulances is the same. (Let $\alpha = .05$.) Interpret the results shown below.

```
READ AMBULANCE A INTO C1, AMBULANCE B INTO C2
6.3 7.4
7.2 6.5
5.2 7.4
3.5 8.2
4.8 5.6
5.7 5.4
5.2 6.3
6.4 7.4
5.9 6.2
6.1 7.3
6.3 4.6
6.8 5.9
MANN-WHITNEY ON DATA IN C1,C2

C1  N = 12  MEDIAN = 6.0000
C2  N = 12  MEDIAN = 6.4000

TEST OF ETA1 = ETA2 VS. ETA1 N.E. ETA2
THE TEST IS SIGNIFICANT AT 0.1190

CANNOT REJECT AT ALPHA = 0.05
```

* Exercise requiring some familiarity with Minitab.

10

OBJECTIVES

One of the most important functions of statistics is to help promote the proper design of experiments (both laboratory experiments and a variety of other kinds). The overall purpose of this chapter is to describe some experimental designs that are commonly used, and to discuss the analysis of variance, the statistical procedure used to analyze the results of such experiments. Among the specific objectives are:

1   To describe some major principles of experimental design.

2   To describe the F distribution, and the conditions under which it arises.

3   To show how the one-way analysis of variance can be used to test whether the means of a number of populations are equal.

4   To indicate how the two-way analysis of variance is carried out.

* Some instructors may want to skip the last sections of this chapter, which deal with the two-way analysis of variance. Others may want to take up only the first three sections, or skip this chapter entirely. Any of these options is feasible, since an understanding of this chapter is not required in subsequent chapters.

Experimental Design and the Analysis of Variance*

10.1 Design of Experiments

Although the neophyte commonly believes that it is a simple matter to design an experiment for testing a certain hypothesis, often a considerable amount of ingenuity is required to achieve a design that really tests the hypothesis one wants to test—and does so at something near minimum cost. As we know from Chapter 1, one of the commonest pitfalls is an experimental design in which the effect of the variable one wants to estimate is inextricably entangled with the effect of some other factor. (In other words, more than one factor may be responsible for a particular observed experimental result.) When this occurs, the effect of the variable one wants to estimate is said to be *confounded* with the effect of another factor or factors.

To illustrate this, recall our discussion in Chapter 1 of the results of a nationwide test of the effectiveness of the Salk antipolio vaccine. These data (reproduced in Table 10.1) seem perfectly adequate for measuring the effects of the vaccine on the incidence of polio in children, but in fact they contain a major bias: Only those second-graders who received their parent's permission received the vaccine, whereas *all* first- and third-graders were taken into account in estimating what the incidence of polio would be without the vaccine. The problem was that the incidence of polio was substantially lower among nonvaccinated children who did not receive permission than among those who did. (Recall from Chapter 1 that children of lower-income parents are less likely to receive permission— and because they grow up in less hy-

Table 10.1
Incidence of Polio in
Two Groups of
Children

School grade	Treatment	Number of children	Number afflicted with polio	Number afflicted per 100,000
Second grade	Salk vaccine	222,000	38	17
First and third grades	No vaccine	725,000	330	46

Table 10.2
Randomized
Experiment with Salk
Vaccine

Permission given	Treatment	Number of children	Number afflicted with polio	Number afflicted per 100,000
Yes	Salk vaccine	201,000	33	16
Yes	Placebo	201,000	115	57
No	None	339,000	121	36

gienic conditions, they are less likely to get polio.) Table 10.2 shows the results of a much more adequate experimental design in which children who received permission are differentiated from those who did not. Those who received permission were assigned at random to either the group receiving the vaccine or the group receiving the placebo (similar in appearance to the vaccine, but of no medical significance). As you can see, the results indicate that the reduction in the incidence of polio due to the vaccine is much larger than is shown in Table 10.1[1]

In general, *an experiment dictates that one group of people (or machines, materials, or other experimental units) be treated differently from another group; and the effect of this difference in treatment is estimated by comparing certain measurable characteristics of the two groups.* However, unless the people (or machines, materials, or other experimental units)

Randomization   are assigned to one group or the other *at random,* all sorts of confounding can occur. For example, in the case of the Salk vaccine, if doctors had decided which children were to receive the vaccine and which were not, there might have been a tendency to give the vaccine to children where the consent of the parents was easy to obtain, and such children might differ from the others in the likelihood that they would contract polio. The importance of randomization cannot be overemphasized. The experimental units or subjects should be assigned at random to the groups receiving different treatments.

[1] This example, including the data in Tables 10.1 and 10.2, is from W. J. Youden, "Chance, Uncertainty, and Truth in Science," *Journal of Quality Technology,* 1972.

10.2 Should the Effects of Several Factors Be Studied One at a Time?

In many experiments, the statistician or decision maker is interested in estimating the effects of more than one factor on a certain characteristic of the the relevant experimental unit. For example, psychologists have carried out considerable research to estimate the effects of differences in method of treatment and differences in type of patient on the effectiveness of anti-smoking programs.[2] The traditional way of estimating these effects is the one-at-a-time method. That is, the psychologist would hold constant the type of patient and observe the effects of differences in the method of treatment. For example, if there are three types of patients (adult men, adult women, and teenagers) and if there are three methods of reducing smoking (I, II, and III), the psychologist might use each method on one type of patient (say, adult men) to observe how the effectiveness of the antismoking program varies with the method used. Similarly, to estimate the effect of the type of patient on the effectiveness of the program, the psychologist might use only one method (say, method I), and observe how the effectiveness of the program varies among the types of patients.

An important disadvantage of the one-at-a-time approach is that the results may be too narrowly focused. For example, the experiments described in the previous paragraph will provide information concerning the effects of the method of reducing smoking on the effectiveness of the program *if the patients are adult men.* However, the differences among methods in this regard may *not* be independent of the type of patient. Thus, although method I may be more effective than the others for adult men patients, it may be less effective than the others for teenager patients. Similarly, the above experiments will provide information concerning the effects of the type of patient on the effectiveness of the program when method I is used, but the differences among types of patient in this regard may not be independent of the method used. Thus, although adult men may show a greater reduction in smoking than adult women or teenagers when method I is used, they may show a smaller reduction than adult women or teenagers when method III is used.

In general, modern statisticians tend to emphasize the advantages of not controlling an experiment too closely. Even if the factors are independent, it frequently is less expensive to conduct an experiment where the factors are allowed to vary, rather than hold one or more

[2] I. Janis, "The Role of Social Support in Adherence to Stressful Decisions," *American Psychologist,* February 1983.

constant, because fewer observations often are required to obtain the same precision. For example, in the above case, a better experimental design may be to obtain data concerning the reduction in smoking when each combination of type of patient and method is used. Thus, the psychologist might use method I with adult men, adult women, and teenagers. Similarly, each of the other methods (II and III) might be used with adult men, adult women, and teenagers. If the effects of the method are independent of the type of patient, and if the effects of the type of patient are independent of the method, the resulting nine observations could be used to estimate the effect of each method and each type of patient. (If they are not independent, more observations are needed, as we shall see below.) More will be said about this in subsequent sections of this chapter.

10.3 Randomized Blocks: An Example of an Experimental Design

*Treatment Effects
and Block Effects*

There are many kinds of experimental designs commonly used by statisticians. The type recommended in the preceding paragraph is known as a randomized block design. In a *randomized block design* there are two kinds of effects, *treatment effects* and *block effects.* These terms are derived from agricultural research, where a field may be split into several blocks, and various treatments (such as fertilizers, pesticides, or some other factor whose effects the researcher is interested in) may be randomly assigned to plots in each block. Each block is constructed so that it contains relatively homogeneous experimental conditions. In an agricultural experiment each block may be a piece of land which has relatively homogeneous soil, sunlight, rainfall, and so forth. In the case of the psychologist, either the types of patients or the methods of reducing smoking can be regarded as treatments, and the other factor can be considered blocks. If each of the methods of reducing smoking is regarded as a treatment, each type of patient can then be regarded as a block because the use of this (and only this) type of patient results in relatively homogeneous experimental conditions for comparing the effects of the treatments (the methods of reducing smoking).

In a randomized block design, the statistician obtains data concerning the effect of each treatment in each block. Thus, in this case data are obtained concerning the reduction in smoking resulting from the use of each method with each type of patient. The results might be as shown in Table 10.3. Note that only one observation is obtained concerning the effect of each treatment in each block. That is, no attempt is made to obtain more than one measurement of the reduction

THE CASE OF FLUORESCEIN

MISTAKES TO AVOID

A number of years ago, the claim was made by some agricultural scientists that plants grow better when watered with a dilute solution of fluorescein. To test this hypothesis, the agricultural scientists watered one group of plants with water and an adjacent group with fluorescein. The spatial distribution of plants was as shown below:

The results indicated a large difference in growth between the two groups. Unfortunately, however, these results were not supported by subsequent research because the experimental design was faulty. What fundamental mistake was made? How could the design have been improved?

SOLUTION

The biggest mistake was to put the plants watered with fluorescein on a different plot of land than the plants watered with water alone. The effects of fluorescein are confounded with whatever difference exists between the plots in fertility. In fact, the soil to which the fluorescein was applied contained more nutrients than the soil to which plain water was applied. This, rather than fluorescein, was responsible for the observed difference in plants' growth rates. A more effective test of the effect of fluorescein might have been accomplished by the spatial distribution of plants shown below:

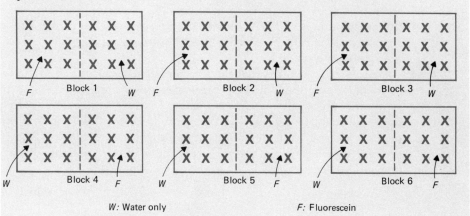

According to this design, the entire field is divided into 6 relatively homogeneous blocks, each of which contains 18 plants. Each block is then divided in half, and a coin is flipped to determine which half (that is, which nine plants) will be watered with fluorescein and which half will receive plain water. This is a randomized block with nine observations concerning the effect of each treatment in each block. The difference between the average rate of growth of the plants receiving flourescein and the average rate of growth of those receiving plain water is a measure of the net effect of the treatment. (The net block effects—that is, the differences in rate of growth among various blocks—are of subsidiary importance in this case.)[3]

[3] W. J. Youden, "Chance, Uncertainty, and Truth in Science."

Table 10.3
Results of Experiment To Measure Effectiveness of Antismoking Programs*

| Type of patient | Method of Reducing Smoking | | |
| | I | II | III |
		(reduction in cigarettes smoked daily)	
Adult Men	50	40	45
Adult Women	48	39	45
Teenagers	52	44	48

* Data show reduction in number of cigarettes smoked daily. Before treatment, each of the subjects smoked 60 cigarettes per day.

in smoking resulting from the use of a particular method with a particular type of patient. As shown in Table 10.3, data are obtained concerning only 9 patients. In cases of this sort, the statistician must assume that the treatment effects and block effects are independent; if this assumption is violated, the *experimental errors*—the errors due to chance variation in the effect of each treatment in each block—will tend to be overstated.

To avoid assuming that the treatment and block effects are independent, and to obtain more precise estimates of these effects, statisticians often specify that *replications* occur. In other words, they ask that the experiment be repeated so that more than one observation is obtained concerning the effect of each treatment in each block. Thus, in the case above, each method might be applied to two patients of each type, and the results might be as shown in Table 10.4.

Type of patient	Method of Reducing Smoking		
	I	II (reduction in cigarettes smoked daily)	III
Adult Men	50,52	40,38	45,46
Adult Women	48,47	39,39	45,44
Teenagers	52,53	44,45	48,47

Table 10.4
Results of Experiment
To Measure
Effectiveness of
Antismoking Programs,
Where Replication
Occurs*

* See note, Table 10.3.

EXERCISES

10.1 Many studies have been carried out to measure the effect of a particular surgical technique. Of the studies that showed marked enthusiasm for the technique, none had randomized controls. Of those that showed no enthusiasm for it, half had randomized controls. (By randomized controls, we mean that patients were assigned at random to this treatment rather than to another treatment.) How can you explain this?

10.2 A placebo is "any inactive substance or procedure used with a patient under the guise of an effective treatment." Placebos often make people feel that they have recovered.[4] Given that this is the case, suppose that a new headache remedy is administered to 1,000 patients, and 700 say that it reduces the severity of their headaches. Is this proof of the effectiveness of this remedy? Why, or why not?

10.3 In the early 1960s, "gastric freezing" was used to treat duodenal ulcers. This treatment meant lowering a small balloon into the stomach and filling it with very cold alcohol to temporarily freeze the ulcer. Ulcer patients seemed to experience dramatic improvements after receiving this treatment.
(a) Is this proof of the effectiveness of this treatment? Why, or why not?
(b) To see whether "gastric freezing" really worked, researchers divided all patients entering a certain medical center into two groups: those who would undergo "gastric freezing," and those who would receive a placebo. (The latter patients were given the balloon treatment, but with alcohol that was not frigid enough to freeze the ulcer.) The latter group showed as much improvement as the former group. How do you account for this?

10.4 An industrial psychologist wants to test whether method A results in higher output per worker in his plant than method B. To find out, he produces 10 percent of the plant's output, using method A. Since method

[4] See J. Critelli and K. Neumann, "The Placebo: Conceptual Analysis of a Construct in Transition," *American Psychologist,* January 1984.

A is still experimental, he is forced to use it only on the night shift, when supervision is less strict. Comment on this experimental design.

10.5 In the 1950s, the United States Army carried out tests to determine whether large amounts of vitamins C and B complex would increase the physical performance of soldiers involved in a high-activity program in a cold environment.

(a) One way that this study could have been carried out is by picking a sample of 100 soldiers at random and sending them to a cold place to engage in calisthenics. All 100 would receive large amounts of the vitamins mentioned. Then their average calisthenics performance might be compared with the average performance of all U.S. Army recruits. What problems can you detect in this design?

(b) Another way in which the experiment might be designed is as follows: Half of the 100 soldiers in the study might be given large amounts of vitamins C and B complex while the other half might receive a placebo. Then the average performance of one group might be compared with that of the other. In determining which group a soldier would join, the medical record of each individual would be inspected to see whether he had taken any vitamin supplements before. Only those who had not done so would be assigned to the group receiving the placebo. What problems can you detect in this design?

(c) As still another way of designing the experiment, half of the 100 soldiers in the study would be given large amounts of the vitamins, while the other half would receive nothing. Soldiers would be allocated at random to the two groups. Those receiving the vitamins would be told what they were being given, and those receiving nothing would be told that they were receiving nothing. Then the average performance of one group might be compared with that of the other. What problems can you detect in this design?

(d) The army actually carried out this experiment in the following way: A random sample of soldiers was drawn from four platoons. The soldiers chosen from each platoon were randomly divided into two equal groups, one of which received large amounts of the vitamins, the other of which received a placebo. However, neither group knew that the latter was a placebo. In each platoon the average performance of one group was compared with that of the other.

  (i) Is this a randomized block? Why, or why not?
  (ii) If so, what are the blocks, and what are the treatments?
  (iii) Does this design contain replications?

10.6 The following table shows the results of 9 different measurements of the viscosity of silicone gum rubber.[5] As can be seen, 3 of the measurements pertain to the first batch of this material, 3 pertain to the second batch, and 3 to the third batch. Also, 3 of the measurements were made by William Jones, 3 were made by John Beam, and 3 were made by Joan Read.

---

[5] These data are from A. Duncan, *Quality Control and Industrial Statistics*, p. 747. They are coded; that is, a constant amount is subtracted from each measurement to make the data easier to work with.

Three of the measurements were made with type A viscosity measuring jars, 3 were made with type B measuring jars, and 3 were made with type C measuring jars.

Batch	Jones	Measurer Beam	Read
I	9 (A)	8 (B)	−3 (C)
II	17 (B)	−2 (C)	7 (A)
III	−2 (C)	41 (A)	2 (B)

(a) Is this a randomized block? Why, or why not?
(b) If so, what are the blocks, and what are the treatments?
(c) Does the design contain replications?
(d) Which of the batches has the highest average rating?
(e) Which type of measuring jar seems to result in the highest average rating?
(f) Without carrying out the appropriate tests of significance, can we be sure that these differences among batches and among types of measuring jars are not due to chance?

10.4 The *F* Distribution

Earlier in this chapter we described particular experimental designs, but we have not yet indicated how one can test whether the observed differences among treatment means (that is, differences among the means of the observations resulting from various treatments) are statistically significant. For example, the results in Table 10.3 seem to indicate that the mean reduction in number of cigarettes smoked daily is greater for patients treated by method I than for those treated by methods II or III. But is this observed difference statistically significant? In other words, is it quite unlikely that it could have arisen due to chance? To answer this question, we must take up the *F* distribution, another of the most important probability distributions in statistics.

The *F* distribution is a continuous probability distribution. It was named for R. A. Fisher, the great British statistician who developed it in the early 1920s. Its definition is as follows.

> F DISTRIBUTION: *If a random variable Y_1 has a χ^2 distribution with v_1 degrees of freedom, and if a random variable Y_2 has a χ^2 distribution with v_2 degrees of freedom, then $Y_1/v_1 \div Y_2/v_2$ has an* **F distribution** *with v_1 and v_2 degrees of freedom (if Y_1 and Y_2 are independent).*

Figure 10.1
The *F* Probability
Density Function, with
2 and 9 Degrees of
Freedom

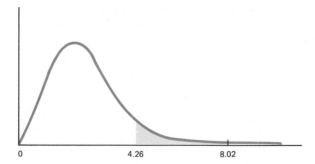

In other words, the ratio of two independent χ^2 random variables, each divided by its number of degrees of freedom, is an *F* random variable.

Like the *t* and χ^2 distributions, the *F* distribution is in reality a family of probability distributions, each corresponding to certain numbers of degrees of freedom. But unlike the *t* and χ^2 distributions, the *F* distribution has two numbers of degrees of freedom, not one. Figure 10.1 shows the *F* distribution with 2 and 9 degrees of freedom. As you can see, the *F* distribution is skewed to the right. However, as both numbers of degrees of freedom become very large, the *F* distribution tends toward normality. As in the case of the χ^2 distribution, the probability that an *F* random variable is negative is zero. This must be true since an *F* random variable is a ratio of two nonnegative numbers. (Y_1/v_1 and Y_2/v_2 are both nonnegative.) Once again, it should be emphasized that any *F* random variable has *two* numbers of degrees of freedom. Be careful to keep these numbers of degrees of freedom in *the correct order,* because an *F* distribution with v_1 and v_2 degrees of freedom is *not* the same as an *F* distribution with v_2 and v_1 degrees of freedom.

Tables are available which show the values of *F* that are exceeded with certain probabilities, such as .05 and .01. Appendix Table 9 shows, for various numbers of degrees of freedom, the value of *F* that is exceeded with probability equal to .05. For example, if the numbers of degrees of freedom are 2 and 9, the value of *F* that is exceeded with probability equal to .05 is 4.26. Similarly, Appendix Table 10 shows, for various numbers of degrees of freedom, the value of *F* that is exceeded with probability equal to .01. For example, if the numbers of degrees of freedom are 2 and 9, the value of *F* exceeded with probability equal to .01 is 8.02.

In the following sections of this chapter, we shall show how the *F* distribution can be used to test whether the differences among various treatment means are statistically significant. For now, it is important to become familiar with Appendix Tables 9 and 10. The following examples show how these tables are used.

Smoking and Lung Cancer

In the past thirty years, a great change has occurred in the attitudes of the public and of the government toward smoking. To a considerable extent, this change was due to a succession of statistical studies linking the use of tobacco to cancer, heart attacks, and other diseases.

(a) In 1939, studies were conducted in which cases of lung cancer were identified, and some other persons comparable to the lung-cancer patients in sex, age, and other characteristics were selected. It was found that the persons with lung cancer smoked much more heavily than the other persons with similar characteristics. Was this a randomized experiment? What are the disadvantages of this experimental procedure?

(b) Two British statisticians, Richard Doll and A. Bradford Hill, sent questionnaires to about 60,000 men and women to obtain data concerning their smoking habits, age, and sex. Then they determined how many of these people died in each subsequent year. Their findings indicated that heavy smokers had a lung-cancer death rate 24 times higher than nonsmokers. Was this a randomized experiment? What are the disadvantages of this experimental procedure?

(c) R. A. Fisher argued that if people with a hereditary predilection for smoking also have a hereditary tendency to contract cancer, the results would be like those obtained in the above studies. How can one determine whether smoking is associated with lung cancer when heredity is held constant?[6]

[6] For further discussion, see B. Brown, "Statistics, Scientific Method, and Smoking," in J. Tanur (ed.), *Statistics: A Guide to the Unknown* (San Francisco: Holden-Day, 1972).

EXAMPLE 10.1 A random variable has the *F* distribution with 10 and 12 degrees of freedom. What is the value of this random variable that is exceeded with a probability of .05? With a probability of .01?

SOLUTION: Appendix Table 9 shows that the answer to the first question is 2.75; Appendix Table 10 shows that the answer to the second question is 4.30.

EXAMPLE 10.2 A random variable has the F distribution with 6 and 18 degrees of freedom. What is the value of this random variable that is exceeded with a probability of .05? With a probability of .01?

SOLUTION: Appendix Table 9 shows that the answer to the first question is 2.66; Appendix Table 10 shows that the answer to the second question is 4.01.

10.5 Analysis of a Completely Randomized Design

The procedure of one of the simplest experimental designs is to divide a set of people or other experimental units into groups, and subject each group to a different treatment. (The people or other experimental units are allocated to the groups at random.) Then the differences among the mean responses of the various groups are used to measure the net effects of the various treatments. This is called a *completely randomized design*. For example, suppose that a market researcher picks a random sample of 20 beer drinkers whose favorite brand of beer is known to be brand W. These beer drinkers are then divided randomly into four groups of five individuals each. The first group is asked to rate a beer in an unmarked can which is in reality their favorite (brand W). The second group is asked to rate a beer in an unmarked can which is in reality brand X. The third group is asked to rate a beer in an unmarked can which is in reality brand Y. The fourth group is asked to rate a beer in an unmarked can which is in reality brand Z. If the results are as shown in Table 10.5, the mean ratings provided by the four groups are 60, 61, 58, and 61, respectively. The market researcher would like to test the hypothesis that these differences among the four means are due to chance.

In this case, the null hypothesis is that $\mu_1 = \mu_2 = \mu_3 = \mu_4$, where μ_1 is the population mean rating for the favorite beer, μ_2 is the population mean rating for brand X, μ_3 is the population mean rating for brand Y, and μ_4 is the population mean rating for brand Z. The alternative hypothesis is that μ_1, μ_2, μ_3, and μ_4 are not all equal. Clearly, the null hypothesis tends to be supported if the four sample means are close together, whereas the alternative hypthesis tends to be supported if they are far apart. A reasonable measure of how close together or far apart the four sample means are is their variance—that is, the square of their standard deviation. Specifically, their variance equals

	Respondent's favorite beer	*Ratings*		
		Brand X	*Brand Y*	*Brand Z*
	60	61	58	61
	59	60	54	57
	61	66	58	61
	55	62	58	61
	65	56	62	65
$\bar{x}_j$	60	61	58	61
$\sum_{i=1}^{5} (x_{ij} - \bar{x}_j)^2$	52	52	32	32
s_j^2	13	13	8	8

Table 10.5
Results of Market Researcher's Survey

$$s_{\bar{x}}^2 = \sum_{j=1}^{4} \frac{(\bar{x}_j - \bar{\bar{x}})^2}{3},$$

where $\bar{x}_j$ is the mean of the jth sample and $\bar{\bar{x}}$ is the mean of the four sample means.[7] If the numerical values in Table 10.5 are substituted in this equation, we have

$$s_{\bar{x}}^2 = \frac{(60 - 60)^2 + (61 - 60)^2 + (58 - 60)^2 + (61 - 60)^2}{3}$$

$$= 2.$$

Suppose that each of the four populations of ratings can be approximated by a normal distribution, and that the standard deviation of each population σ is the same. Then if the null hypothesis is true (that is, if $\mu_1 = \mu_2 = \mu_3 = \mu_4$), our four samples in Table 10.5 are four samples from the same population. Consequently, since the variance of the sample means drawn from the same population equals σ^2/n (as we know from Chapter 6), it follows that $s_{\bar{x}}^2$ is an estimate of σ^2/n where n is the size of each sample. (In this case, $n = 5$.) Thus, $ns_{\bar{x}}^2$ is an estimate of σ^2, the common variance of these four populations, if the null hypothesis and our other assumptions are true. In particular, in the case

[7] The denominator is 3 because this is a sample variance, where the denominator is the sample size minus 1. There are four sample means, so the denominator is $(4 - 1) = 3$.

in Table 10.5, 5(2), or 10, is an estimate of σ^2 if the null hypothesis is true (since $n = 5$ and $s_{\bar{x}}^2 = 2$).

To test whether the null hypothesis is true, we compare this estimate of σ^2 with another estimate of σ^2 that is valid regardless of whether or not the null hypothesis is true. This latter estimate of σ^2 is the mean of the four variances within the samples, s_1^2, s_2^2, s_3^2, and s_4^2. As we know from Chapter 7, each of these sample variances is an unbiased estimate of the population variance (assumed to be the same in the four populations). Thus, the mean of these variances is also an unbiased estimate of the population variance; and this is true regardless of whether or not μ_1, μ_2, μ_3, and μ_4 are equal. In the case shown in Table 10.5, the mean of the sample variances equals

$$\frac{s_1^2 + s_2^2 + s_3^2 + s_4^2}{4} = \frac{13 + 13 + 8 + 8}{4} = \frac{42}{4} = 10.5.$$

The derivation of s_1^2, s_2^2, s_3^2, and s_4^2 is shown in Table 10.5.

At this point, we have two estimates of σ^2, the first being $ns_{\bar{x}}^2$ (in this case, 10), the second being $(s_1^2 + s_2^2 + s_3^2 + s_4^2)/4$ (in this case, 10.5). To test the null hypothesis, we form the ratio of these two estimates:

The Variance Ratio

$$F = \frac{ns_{\bar{x}}^2}{(s_1^2 + s_2^2 + s_3^2 + s_4^2)/4} \qquad (10.1)$$

This ratio is called a *variance ratio*. If the null hypothesis is true, this ratio should be fairly close to 1, since both the numerator and the denominator should be approximately equal to σ^2. If the null hypothesis is not true, $ns_{\bar{x}}^2$ should be much greater than σ^2, since it will also reflect the variation among μ_1, μ_2, μ_3, and μ_4. Consequently, this variance ratio, F, is used as a test statistic. High values of F are evidence that the null hypothesis should be rejected.

But how large can F be expected to be by chance if the null hypothesis is true? The answer is given by the F distribution, since the variance ratio has an F distribution with $(k - 1)$ and $k(n - 1)$ degrees of freedom (where k is the number of population means being compared and n is the sample size in each population), if the null hypothesis is true. Thus, in the case of Table 10.5, the variance ratio has an F distribution with 3 and 16 degrees of freedom, since $k = 4$ and $n = 5$. As shown in Appendix Table 9, there is a .05 chance that an F variable with 3 and 16 degrees of freedom will exceed 3.24. Thus, if the .05 significance level is appropriate, the null hypothesis should be rejected if the variance ratio exceeds 3.24. In fact, since $ns_{\bar{x}}^2 = 10$ and $(s_1^2 + s_2^2 + s_3^2 + s_4^2)/4 = 10.5$, the variance ratio in Table 10.5 equals 0.95, and there is no reason to reject the null hypothesis.

In general, the decision rule for this test procedure is as follows.

Decision Rule: *Reject the null hypothesis that the population means are all equal if*

$$\frac{ns_{\bar{x}}^2}{\sum_{j=1}^{k} s_j^2/k} > F_\alpha,$$

Test that $\mu_1 = \mu_2 = \ldots = \mu_k$

where F_α *is the value of an F random variable—with degrees of freedom equal to* (k − 1) *and* k(n − 1)—*that is exceeded with probability equal to* α *(the desired level of significance), where* k *is the number of population means being compared, and* n *is the sample size in each population. Otherwise, accept the null hypothesis.*

In the next section, we shall show how this test can be presented in another format which is used in the design and analysis of experiments.

10.6 One-Way Analysis of Variance

The analysis of variance is a technique designed to divide the total variation in a set of data into its component parts, each of which can be ascribed to a particular source. For example, in the problem presented in the previous section, the total variation might be split into two parts, one representing the differences among beer brands in their average ratings, the other representing the variation among consumers in their rating of a particular brand. In this section, we will show how an analysis of variance can be carried out for this type of problem. Although this is merely another way of presenting the test described in the previous section, it is useful to know this format since it is often employed and since it can be adapted for more complex problems, as we shall see in Section 10.8.

NOTATION

In any one-way analysis of variance, the purpose is to test whether the means of k populaions are equal. Since n observations are chosen from each population, the data can be arrayed as shown in Table 10.6. Each (vertical) column contains observations from the same population. Thus, x_{11} is the first observation from the first population, x_{21} is the second observation from the first population, x_{12} is the first observation from the second population, and so on. In general, x_{ij} is the *i*th observation from the *j*th population. We can apply this notation to

Table 10.6
Array of Data for
Analysis of Variance

x_{11}	x_{12}	$\cdot$	x_{1j}	$\cdot$	x_{1k}		
x_{21}	x_{22}	$\cdot$	x_{2j}	$\cdot$	x_{2k}		
x_{31}	x_{32}	$\cdot$	x_{3j}	$\cdot$	x_{3k}		
$\cdot$	$\cdot$						
x_{i1}	x_{i2}	$\cdot$	x_{ij}	$\cdot$	x_{ik}		
$\cdot$	$\cdot$	$\cdot$		$\cdot$			
x_{n1}	x_{n2}	$\cdot$	x_{nj}	$\cdot$	x_{nk}		

Means $\bar{x}_{.1}$ $\bar{x}_{.2}$ $\cdot$ $\bar{x}_{.j}$ $\cdot$ $\bar{x}_{.k}$

Table 10.5, where $k = 4$ and $n = 5$. Clearly, in Table 10.5, $x_{11} = 60$, $x_{21} = 59$, $x_{12} = 61$, and so on.

The mean of the observations in the jth column is denoted by $\bar{x}_{.j}$. Thus, in Table 10.5, $\bar{x}_{.1} = 60$, $\bar{x}_{.2} = 61$, $\bar{x}_{.3} = 58$, and $\bar{x}_{.4} = 61$. On the other hand, the mean of the observations in the ith (horizontal) row is denoted by $\bar{x}_{i.}$. The dot indicates the subscript over which the averaging takes place. Since the first subscript designates the row of the observation, $\bar{x}_{.j}$ indicates that the averaging is over the observations in all rows in the jth column. Similarly, since the second subscript designates the column of the observation, $\bar{x}_{i.}$ indicates that the averaging is over the observations in all columns in the ith row. Thus, in Table 10.5, $\bar{x}_{1.} = (60 + 61 + 58 + 61)/4$, or 60.

BASIC IDENTITY

Using this notation, we can define the total variation in a set of data as follows:

Total Sum of Squares

$$TSS = \sum_{i=1}^{n} \sum_{j=1}^{k} (x_{ij} - \bar{x}_{..})^2, \tag{10.2}$$

where $\bar{x}_{..}$ is the overall mean of all $n \times k$ observations. The total variation is generally referred to as the **total sum of squares,** which accounts for the TSS on the left-hand side of equation (10.2). With a bit of algebraic manipulation, it can be shown that the total sum of squares is identically equal to the sum of two terms:

Total Sum of Squares
= Between-Group
Sum of Squares +
Within-Group Sum
of Squares

$$\sum_{i=1}^{n} \sum_{j=1}^{k} (x_{ij} - \bar{x}_{..})^2 = n \sum_{j=1}^{k} (\bar{x}_{.j} - \bar{x}_{..})^2 + \sum_{i=1}^{n} \sum_{j=1}^{k} (x_{ij} - \bar{x}_{.j})^2. \tag{10.3}$$

This is the basic identity underlying the one-way analysis of variance.

To understand the one-way analysis of variance, it is essential to interpret each of the terms in equation (10.3). The total sum of squares, which is on the left-hand side of this equation, is a measure of how much variation exists among all of the $n \times k$ observations in the sample. According to equation (10.3), this total variation can be split into two parts, one reflecting differences among the means of the observations taken from different populations, the other reflecting differences among the observations taken from the same population. Holding n and k constant, the larger the first part (relative to the second part), the less likely it would seem that the population means are all equal.

The first term on the right-hand side of equation (10.3) is often called the **between-group sum of squares** since it reflects differences between the populations in the sample means. To satisfy yourself that it does reflect such differences, note that it equals $(k - 1)$ times $ns_{\bar{x}}^2$. Another frequently used name for this term is the **treatment sum of squares** since the differences among the population means are often called the net effects of different treatments. *Treatment* is a completely general term used to characterize each of the populations being compared. For example, each brand of beer can be called a different treatment in Table 10.5.

The second term on the right-hand side of equation (10.3) is often referred to as the **within-group sum of squares,** since it measures the variation within the populations. To see that it does measure such variation, note that it is equal to $k(n - 1)$ times the denominator of the variance ratio in equation (10.1). (In equation (10.1), $k = 4$ and $n = 5$.) It is also frequently called the **error sum of squares,** since the within-group variation is often interpreted as being due to experimental error.

To illustrate that the total sum of squares is identically equal to the between-group sum of squares plus the within-group sum of squares, let's return to the data in Table 10.5. Clearly, the total sum of squares equals

$$
\begin{aligned}
TSS = {} & (60 - 60)^2 + (59 - 60)^2 + (61 - 60)^2 + (55 - 60)^2 \\
& + (65 - 60)^2 + (61 - 60)^2 + (60 - 60)^2 + (66 - 60)^2 \\
& + (62 - 60)^2 + (56 - 60)^2 + (58 - 60)^2 + (54 - 60)^2 \\
& + (58 - 60)^2 + (58 - 60)^2 + (62 - 60)^2 + (61 - 60)^2 \\
& + (57 - 60)^2 + (61 - 60)^2 + (61 - 60)^2 + (65 - 60)^2 \\
= {} & 198.
\end{aligned}
$$

The between-group sum of squares equals

$$BSS = 5[(60 - 60)^2 + (61 - 60)^2 + (58 - 60)^2 + (61 - 60)^2]$$

$$= 30.$$

And the within-group sum of squares equals

$$WSS = (60 - 60)^2 + (59 - 60)^2 + (61 - 60)^2 + (55 - 60)^2$$
$$+ (65 - 60)^2 + (61 - 61)^2 + (60 - 61)^2 + (66 - 61)^2$$
$$+ (62 - 61)^2 + (56 - 61)^2 + (58 - 58)^2 + (54 - 58)^2$$
$$+ (58 - 58)^2 + (58 - 58)^2 + (62 - 58)^2 + (61 - 61)^2$$
$$+ (57 - 61)^2 + (61 - 61)^2 + (61 - 61)^2 + (65 - 61)^2$$
$$= 168.$$

Since $198 = 30 + 168$, it is obvious that the total sum of squares does equal the sum of the between-group sum of squares and the within-group sum of squares.

ANALYSIS-OF-VARIANCE TABLE

The test procedure used in the one-way analysis of variance is precisely the same as that described in Section 10.5. What is different is that the computations are presented in the format of an analysis of variance table. The general form of a one-way analysis of variance table is shown in Table 10.7. The first column shows the source or type of variation, and the second column shows the corresponding sum of squares. The third column shows the number of degrees of freedom corresponding to each sum of squares, these numbers being the figures that were used in equation (10.1) to divide each sum of squares to ob-

Table 10.7
General One-Way
Analysis-of-Variance
Table

Source of variation	Sum of squares	Degrees of freedom	Mean square	F
Between groups	BSS	$k - 1$	$\dfrac{BSS}{k - 1}$	$\dfrac{BSS}{k - 1} \div \dfrac{WSS}{k(n - 1)}$
Within groups	WSS	$k(n - 1)$	$\dfrac{WSS}{k(n - 1)}$	
Total	TSS	$nk - 1$		

tain an estimate of σ^2, the variance of each population.[8] The fourth column shows each mean square, which is the sum of squares divided by the number of degrees of freedom. The last column shows the ratio of the two mean squares. If the null hypothesis is true, this ratio—which is precisely the same as the variance ratio in equation (10.1)—has an F distribution with $(k - 1)$ and $k(n - 1)$ degrees of freedom. (Note that these are the numbers of degrees of freedom in the first and second rows of the table.) Thus, in a one-way analysis of variance table, the decision rule is as follows.

> **Decision Rule:** *Reject the null hypothesis that the population means are all equal if the ratio of the between-group mean square to the within-group mean square exceeds* F_α, *where* α *is the desired significance level. Otherwise, accept the null hypothesis.*

Test that $\mu_1 = \mu_2 = \ldots = \mu_k$

Of course, this decision rule is essentially the same as the decision rule in Section 10.5.

To illustrate the application of a one-way analysis of variance, let's go back to the experiment carried out by the market researcher. Based on the data in Table 10.5, we know from previous paragraphs that the between-group sum of squares equals 30 and that the within-group sum of squares equals 168. These figures constitute the second column of Table 10.7. And since $k = 4$ and $n = 5$, it is evident that the numbers of degrees of freedom are 3 and 16. Thus, dividing 30 by 3, we get the between-group mean square, which is 10; and dividing 168 by 16, we get the within-group mean square, which is 10.5. The ratio of the mean squares is $10 \div 10.5 = 0.95$. Since $F_{.05} = 3.24$ when there are 3 and 16 degrees of freedom, the ratio of the mean squares is not greater than $F_{.05}$, and there is no reason (at the .05 significance level) to reject the null hypothesis that the mean ratings for the four brands of beer are equal. The analysis-of-variance table is given in Table 10.8.

Source of variation	Sum of squares	Degrees of freedom	Mean square	F
Between groups	30	3	10	0.95
Within groups	168	16	10.5	
Total	198	19		

Table 10.8
One-Way Analysis-of-Variance Table, Beer Example

[8] In the numerator of the right-hand side of equation (10.1), the between-group sum of squares is divided by 3. (Note that $ns_{\bar{x}}^2 = BSS \div 3$.) In the denominator of the right-hand side of equation (10.1), the within-group sum of squares is divided by $4 \times (n - 1)$. (Note that $(s_1^2 + s_2^2 + s_3^2 + s_4^2) \div 4 = WSS \div [4 \times (n - 1)]$.) Since $k = 4$ in equation (10.1), the numbers of degrees of freedom are indeed the numbers used in equation (10.1) to divide the sums of squares to get estimates of σ^2.

The following example illustrates further how the one-way analysis of variance is used.

EXAMPLE 10.3 A psychologist (encountered earlier in this chapter) is studying the effectiveness of three methods of reducing smoking. He wants to determine whether the mean reduction in the number of cigarettes smoked daily differs from one method to another among men patients. Sixteen men are included in the experiment. Each smoked 60 cigarettes per day before treatment. Four randomly chosen members of the group pursue method I; four pursue method II; and so on. The results are as follows:

	Method	
I	*II*	*III*
50	41	49
51	40	47
51	39	45
52	40	47

Use a one-way analysis of variance to test whether the mean reduction in the number of cigarettes smoked daily is equal for the three methods. (Let the significance level equal .05.)

SOLUTION: The mean reduction for the first method is 51; for the second method it is 40; and for the third method it is 47. The mean for all methods combined is 46. Thus, the between-group sum of squares equals $4 \times [(51-46)^2 + (40-46)^2 + (47-46)^2]$, or 248. The within-group sum of squares equals $(50-51)^2 + (51-51)^2 + (51-51)^2 + (52-51)^2 + (41-40)^2 + (40-40)^2 + (39-40)^2 + (40-40)^2 + (49-47)^2 + (47-47)^2 + (45-47)^2 + (47-47)^2 = 12$. Thus, the analysis-of-variance table is

Source of variation	Sum of squares	Degrees of freedom	Mean square	F
Between groups	248	2	124	93
Within groups	12	9	1.33	
Total	260	11		

Since there are 2 and 9 degrees of freedom, $F_{.05} = 4.26$. Since the observed value of F far exceeds this amount, the psychologist should reject the null hypothesis that the mean reduction in the number of cigarettes smoked daily is the same for the three methods.[9]

---

[9] See Appendix 10.2 for some formulas that are useful in calculating the required sums of squares in more complicated cases.

10.7 Confidence Intervals for Differences among Means

In the previous section, we saw how the one-way analysis of variance can be used to test whether several population means are all equal. However, in most statistical investigations the purpose is to find out the *extent* to which these means differ, not just *whether* they differ. For example, in the case of the experiment carried out by the psychologist in Example 10.3, it is important to estimate the differences between the mean reductions in the number of cigarettes smoked daily resulting from various methods of reducing smoking. The psychologist wants to answer questions like: What is the difference between the mean reduction resulting from method I and method II? In this section we will show how confidence intervals can be constructed for *all* differences among the population means. The confidence coefficient attached to these intervals is the probability that *all* these intervals will include the respective differences among the population means *simultaneously*.

In the case of the psychologist, there are three differences between the population means, since there are three population means. If μ_1 is the mean reduction resulting from method I, μ_2 is the mean reduction resulting from method II, and μ_3 is the mean reduction resulting from method III, the probability is $(1 - \alpha)$ that *all* the following statements hold true *simultaneously:*

$$\bar{x}_{.1} - \bar{x}_{.2} - \sqrt{F_\alpha}s_w\sqrt{\frac{2(k-1)}{n}} < \mu_1 - \mu_2 < \bar{x}_{.1} - \bar{x}_{.2} + \sqrt{F_\alpha}s_w\sqrt{\frac{2(k-1)}{n}}$$

$$\bar{x}_{.1} - \bar{x}_{.3} - \sqrt{F_\alpha}s_w\sqrt{\frac{2(k-1)}{n}} < \mu_1 - \mu_3 < \bar{x}_{.1} - \bar{x}_{.3} + \sqrt{F_\alpha}s_w\sqrt{\frac{2(k-1)}{n}}$$

$$\bar{x}_{.2} - \bar{x}_{.3} - \sqrt{F_\alpha}s_w\sqrt{\frac{2(k-1)}{n}} < \mu_2 - \mu_3 < \bar{x}_{.2} - \bar{x}_{.3} + \sqrt{F_\alpha}s_w\sqrt{\frac{2(k-1)}{n}}$$

Confidence Intervals for Differences among Means

$$(10.4)$$

where s_w is the square root of the within-group mean square defined as $WSS/k(n-1)$, F_α is the value of an F random variable [with $(k-1)$ and $k(n-1)$ degrees of freedom] exceeded with a probability of α, k is the number of means being compared, and n is the size of the sample taken from each population.[10]

---

[10] This result and that in Section 10.9 are from H. Scheffe, *The Analysis of Variance* (New York: Wiley, 1959).

Since $\bar{x}_{.1} = 51, \bar{x}_{.2} = 40, \bar{x}_{.3} = 47, F_{.05} = 4.26, s_w = 1.15, k = 3$, and $n = 4$, the 95 percent confidence interval for all of the differences between the population means is as follows:

$$6.63 < \mu_1 - \mu_2 < 11.37$$

$$1.63 < \mu_1 - \mu_3 < 6.37$$

$$-9.37 < \mu_2 - \mu_3 < -4.63.$$

Clearly, method I seems to result in a larger reduction in smoking than methods II or III, and method III seems to result in a larger reduction than method II.

Expression (10.4) can be used to construct confidence intervals for the differences between the population means, no matter what the values of k and n may be. The following is an example of how these intervals are constructed.

EXAMPLE 10.4 Using the data in Table 10.5, construct a 95 percent confidence interval for the six differences between the mean ratings of the four beers (the favorite, brand X, brand Y, and brand Z).

SOLUTION: In this case, $\bar{x}_{.1} = 60, \bar{x}_{.2} = 61, \bar{x}_{.3} = 58, \bar{x}_{.4} = 61$, $F_{.05} = 3.24$ (since there are 3 and 16 degrees of freedom), $s_w = \sqrt{10.5}, k = 4$, and $n = 5$. Thus, $\sqrt{F_\alpha}\, s_w\sqrt{2(k-1)/n} = \sqrt{3.24} \times \sqrt{10.5} \times \sqrt{(3/5)} \times 2 = 6.39$, and the confidence intervals are as follows:

$$-7.39 < \mu_1 - \mu_2 < 5.39$$

$$-4.39 < \mu_1 - \mu_3 < 8.39$$

$$-7.39 < \mu_1 - \mu_4 < 5.39$$

$$-3.39 < \mu_2 - \mu_3 < 9.39$$

$$-6.39 < \mu_2 - \mu_4 < 6.39$$

$$-9.39 < \mu_3 - \mu_4 < 3.39$$

As you can see, all these confidence intervals include zero, which is what would be expected, given that the analysis of variance in the previous section concluded that none of these differences is statistically significant.

EXERCISES

10.7 If an *F* random variable has 40 and 30 degrees of freedom, what is the probability that it will exceed (a) 1.79; (b) 2.30?

10.8 Suppose that *X* has a χ^2 distribution with 4 degrees of freedom and *Y* has a χ^2 distribution with 7 degrees of freedom. (a) What is the distribution of $X/4 \div Y/7$? (b) What is the distribution of $Y/7 \div X/4$?

10.9 Suppose that a random variable has an *F* distribution with 10 and 12 degrees of freedom. (a) What is the value of this variable that is exceeded with a probability of .05? (b) What is the value of this variable that is exceeded with a probability of .01?

10.10 Fill in the blanks in the following analysis of variance table:

Source of variation	Sum of squares	Degrees of freedom	Mean square	F
Between groups	_____	2	_____	_____
Within groups	_____	11	14	
Total	200	_____		

10.11 The Department of Transportation runs an experiment in which four of one U.S. auto firm's cars are chosen at random, four of another U.S. firm's cars are chosen at random, four of a German firm's cars are chosen at random, and four of a Japanese firm's cars are chosen at random. Each of the 16 cars is operated under identical conditions for a month, and the mileage per gallon of gasoline is determined. The results are as follows:

Miles per gallon			
U.S. firm I	U.S. firm II	German firm	Japanese firm
18	22	25	29
20	21	27	28
19	24	26	24
17	20	28	25

(a) Test the hypothesis that the mean number of miles per gallon is the same for all four firms' cars (using $\alpha = .05$).
(b) Construct an analysis-of-variance table summarizing the results.
(c) Construct 95 percent confidence intervals for the differences among the four types of cars with respect to the mean number of miles per gallon.

10.12 An agricultural experiment station treats three plots of ground with fertilizer I, another three plots of ground with fertilizer II, and still another three plots of ground with fertilizer III. Then wheat is grown on the nine plots, the yield on each plot being as follows:

Fertilizer I	Fertilizer II	Fertilizer III
9.1	8.2	9.3
8.3	5.9	8.8
6.4	7.0	7.9

(a) Which fertilizer seems to result in the highest average yield? The lowest average yield?

(b) Construct the relevant analysis of variance table.

(c) Can the observed differences among fertilizers in the mean yield be attributed to chance? (Let $\alpha = .05$.)

(d) What assumptions underlie your answer to (c)?

(e) Construct 95 percent confidence intervals for the differences among the means.

(f) Criticize the design of this experiment.

(g) If the station determines that the mean yield with one of these fertilizers is higher than with the others, does this mean that farmers should replace the other fertilizers with this one?

10.13 An industrial engineer identifies four ways that a certain job can be done. To determine how long it takes workers to do the job when each of these methods is used, he asks four workers to do the job using method A, another four workers to do this job using method B, and so on. Each worker's time (in seconds) is shown below:

Method A	Method B	Method C	Method D
19	18	21	22
17	16	20	23
22	15	19	21
20	14	19	20

(a) Which method seems fastest? Slowest?

(b) Construct the relevant analysis of variance table.

(c) Can the differences among methods in average time be attributed to chance? (Let $\alpha = 0.01$.)

(d) Construct 99 percent confidence intervals for the differences among the means.

(e) Criticize the design of this experiment.

(f) If the engineer finds that one of these methods is faster than the others, does this mean that only this method should be used?

10.8 Two-Way Analysis of Variance

The one-way analysis of variance is the simplest type; there are a variety of more complicated types which are taken up in more specialized textbooks. For present purposes, it is sufficient to discuss a single extension, the two-way analysis of variance. This technique varies from the one-way analysis because two, not one, sources of variation (other than the error sum of squares) are singled out for attention. In particular, two-way analysis is the technique used to test whether the differences among treatment means in a randomized block without replications are statistically significant.

To illustrate two-way analysis of variance, let's return to the market researcher discussed previously. Suppose that, when the study described in Section 10.5 is presented to the market researcher's clients, they suggest to the market researcher that a somewhat different experimental design be used. In particular, it is suggested that the same five people be asked to rate all four of the brands of beer, rather than employing a different sample for each brand. In this way, the differences among brands in sample means will not be clouded by differences in the composition of the sample. The result is a randomized block design in which each person constitutes a block.

Person	Favorite brand	Rating Brand X	Rating Brand Y	Rating Brand Z	Block mean
Jones	63	59	62	61	61.25
Smith	61	62	57	63	60.75
Klein	61	64	60	58	60.75
Carlucci	62	62	60	62	61.50
Weill	58	63	61	61	60.75
Treatment mean	61	62	60	61	

Table 10.9
Results of Survey
Regarding Beer
Preferences

Suppose that the market researcher carries out the experiment suggested and the results are as shown in Table 10.9. In this new experimental design, the market researcher is interested in the differences among brands of beer and the differences among people. The differences among brands of beer in their mean ratings are called *differences among treatment means,* as in the previous section. The differences among persons in their mean ratings are called *differences among block means,* and they reflect the fact that some individuals may tend to rate all these brands of beer more highly than would other individuals. Our primary interest here is in whether or not the observed differences among treatment means are due to chance. The differences among the block means are of only secondary interest in this case; but since they contribute to the total sum of squares, they must be included in the analysis.

As in the case of the one-way analysis of variance, the total sum of squares equals

Total Sum of Squares

$$TSS = \sum_{i=1}^{n} \sum_{j=1}^{k} (x_{ij} - \bar{x}..)^2. \qquad (10.5)$$

In this case, however, the total sum of squares can be split into three parts in the following way:

$$\sum_{i=1}^{n} \sum_{j=1}^{k} (x_{ij} - \bar{x}..)^2 = n \sum_{j=1}^{k} (\bar{x}_{.j} - \bar{x}..)^2 + k \sum_{i=1}^{n} (\bar{x}_{i.} - \bar{x}..)^2 \qquad (10.6)$$

+ error sum of squares,

where the error sum of squares (denoted by *ESS*) can be obtained by subtraction. That is,

Error Sum of Squares

$$ESS = \sum_{i=1}^{n} \sum_{j=1}^{k} (x_{ij} - \bar{x}..)^2 - n \sum_{j=1}^{k} (\bar{x}_{.j} - \bar{x}..)^2 \qquad (10.7)$$

$$- k \sum_{i=1}^{n} (\bar{x}_{i.} - \bar{x}..)^2.$$

The identity in equation (10.6) is the basis for the two-way analysis of variance. The first term on the right-hand side of equation (10.6) is called the **treatment sum of squares,** as in previous sections. Clearly, it reflects differences in the treatment means (for example, differences in the average rating of various brands of beer in Table 10.9). We denote this term by *BSS*. The second term on the right-hand side of equation (10.6) is called the **block sum of squares,** since it reflects differences in the block means (for example, differences among the persons in Table 10.9 in their average ratings of all beers). We denote the block sum of squares by *RSS*. The identity in equation (10.6) says that

The Basic Identity

total sum of squares = treatment sum of squares + block

sum of squares + error sum of squares

In other words, $TSS = BSS + RSS + ESS$.

This identity is used in the two-way analysis-of-variance table shown in Table 10.10. The first column of the table shows the source of variation (treatment, block, or error), and the second column shows

Table 10.10
General Two-Way
Analysis-of-Variance
Table

Source of variation	Sum of squares	Degrees of freedom	Mean square	F
Treatments	BSS	$k-1$	$\dfrac{BSS}{k-1}$	$\dfrac{BSS}{k-1} \div \dfrac{ESS}{(k-1)(n-1)}$
Blocks	RSS	$n-1$	$\dfrac{RSS}{n-1}$	$\dfrac{RSS}{n-1} \div \dfrac{ESS}{(k-1)(n-1)}$
Error	ESS	$(k-1)(n-1)$	$\dfrac{ESS}{(k-1)(n-1)}$	
Total	TSS	$nk-1$		

the corresponding sum of squares. The third column shows the number of degrees of freedom for each sum of squares—$(k-1)$ for the treatment sum of squares, $(n-1)$ for the block sum of squares, and $(k-1)(n-1)$ for the error sum of squares. The fourth column shows the mean square for treatments, blocks, and errors, each being the relevant sum of squares divided by the relevant degrees of freedom (that is, the second column divided by the third column). The last column shows (1) the treatment mean square divided by the error mean square and (2) the block mean square divided by the error mean square. As explained in the next paragraph, these two ratios are the pay dirt of the entire analysis.

In a two-way analysis of variance, we can test two different null hypotheses, not just one. The first null hypothesis is that *the treatment means are all equal*. In terms of our example, this hypothesis says that the mean rating for each brand of beer is the same. To test this hypothesis, we use the ratio of the treatment mean square to the error mean square. If α is the significance level, the decision rule is the following.

> ***Decision Rule:*** *Reject the above null hypothesis if the ratio of the treatment mean square to the error mean square exceeds* F_α *where there are* (k − 1) *and* (k − 1)(n − 1) *degrees of freedom. Accept the above null hypothesis if this ratio does not exceed* F_α.

Test that the Treatment Means Are All Equal

The second null hypothesis is that *the block means are all equal*. In terms of our example, this hypothesis says that the mean rating is the same for all people. To test this second hypothesis, we use the ratio of the block mean square to the error mean square. If α is the significance level, the decision rule is the following.

Decision Rule: *Reject the above null hypothesis if the ratio of the block mean square to the error mean square exceeds* F_α *where there are* $(n-1)$ *and* $(k-1)(n-1)$ *degrees of freedom. Accept the above null hypothesis if this ratio does not exceed* F_α.

To illustrate the application of two-way analysis of variance, consider the data concerning the beers in Table 10.9. As shown in Table 10.11, the total sum of squares equals 66, the treatment sum of squares equals 10, the block sum of squares equals 2, and the error sum of squares equals 54. (Each of these numbers is derived under the table.)

Table 10.11
Analysis of Results of
Beer Survey

Source of variation	Sum of squares	Degrees of freedom	Mean square	F
Treatments (beers)	10	3	3.33	$\dfrac{3.33}{4.50} = 0.74$
Blocks (people)	2	4	0.50	$\dfrac{0.50}{4.50} = 0.11$
Error	54	12	4.50	
Total	66	19		

As shown in Table 10.9, the treatment means are 61, 62, 60, and 61. Thus, since $\bar{x}.. = 61$,

$$\text{Treatment sum of squares} = 5[(61-61)^2 + (62-61)^2 + (60-61)^2 + (61-61)^2]$$
$$= 5(2) = 10$$

As shown in Table 10.9, the block means are 61.25, 60.75, 60.75, 61.50, and 60.75. Thus,

$$\text{Block sum of squares} = 4[(61.25-61)^2 + (60.75-61)^2 + (60.75-61)^2$$
$$+ (61.50-61)^2 + (60.75-61)^2] = 4(.5) = 2.$$

Based on the data in Table 10.9,

$$\text{Total sum of squares} = (63-61)^2 + (59-61)^2 + (62-61)^2 + (61-61)^2$$
$$+ (61-61)^2 + (62-61)^2 + (57-61)^2 + (63-61)^2$$
$$+ (61-61)^2 + (64-61)^2 + (60-61)^2 + (58-61)^2$$
$$+ (62-61)^2 + (62-61)^2 + (60-61)^2 + (62-61)^2$$
$$+ (58-61)^2 + (63-61)^2 + (61-61)^2 + (61-61)^2$$
$$= 66.$$

Using the previous results,

$$\text{Error sum of squares} = 66 - 10 - 2 = 54.$$

Since $k = 4$ and $n = 5$, the degrees of freedom for each sum of squares is as shown in the third column of the table; and dividing each sum of squares by its number of degrees of freedom, we get the mean squares in the fourth column. Finally, dividing the treatment mean square by the error mean square, we get 0.74. This is less than 3.49, which is $F_{.05}$ when there are 3 and 12 degrees of freedom. Thus, there is no reason to reject the hypothesis that the treatment means are equal. Dividing the block mean square by the error mean square, we get 0.11. This is less than 3.26, which is $F_{.05}$ when there are 4 and 12 degrees of freedom. Thus, we should not reject the hypothesis that the block means are equal. Overall, the results of this analysis indicate that the individuals in the sample do not exhibit different average ratings for all included beer brands, and there is no evidence of differences in the average ratings of the four brands of beer.

The two-way analysis of variance is of great practical importance because, as we know from earlier sections, randomized blocks are a commonly used experimental design. The following is a further example of how two-way analysis of variance is carried out.

EXAMPLE 10.5 Use the data in Table 10.3 to construct a two-way analysis of variance to test whether the mean reduction in the number of cigarettes smoked daily differs from one method to another. Then test whether the mean reduction differs from one type of patient to another. Set the significance level of each test equal to .05.

SOLUTION: Let x_{ij} be the reduction in the number of cigarettes smoked daily by the ith type of patient pursuing the jth method of reducing smoking, where adult men are the first type of patient, adult women are the second type, and so on. Based on Table 10.3, it is clear that

$$\bar{x}_{.1} = 50 \qquad \bar{x}_{1.} = 45$$

$$\bar{x}_{.2} = 41 \qquad \bar{x}_{2.} = 44$$

$$\bar{x}_{.3} = 46 \qquad \bar{x}_{3.} = 48.$$

If the various methods are regarded as treatments, and if the types of patients are regarded as blocks, the treatment sum of squares is

$$3 \times \left[\left(50 - 45\frac{2}{3} \right)^2 + \left(41 - 45\frac{2}{3} \right)^2 + \left(46 - 45\frac{2}{3} \right)^2 \right] = 122,$$

and the block of squares is

$$3 \times \left[\left(45 - 45\frac{2}{3} \right)^2 + \left(44 - 45\frac{2}{3} \right)^2 + \left(48 - 45\frac{2}{3} \right)^2 \right] = 26,$$

since $\bar{x}.. = 45\ 2/3$. Thus, the error sum of squares is

$$\left(50 - 45\frac{2}{3} \right)^2 + \left(48 - 45\frac{2}{3} \right)^2 + \left(52 - 45\frac{2}{3} \right)^2$$

$$+ \left(40 - 45\frac{2}{3} \right)^2 + \left(39 - 45\frac{2}{3} \right)^2 + \left(44 - 45\frac{2}{3} \right)^2$$

$$+ \left(45 - 45\frac{2}{3} \right)^2 + \left(45 - 45\frac{2}{3} \right)^2 + \left(48 - 45\frac{2}{3} \right)^2$$

$$- 122 - 26 = 150 - 122 - 26 = 2.$$

The analysis-of-variance table is as follows:

Source of variation	Sum of squares	Degrees of freedom	Mean square	F
Methods (treatments)	122	2	61	$61/0.5 = 122$
Types of patients (blocks)	26	2	13	$13/0.5 = 26$
Error	2	4	0.5	
Total	150	8		

Since $F_{.05} = 6.94$ when there are 2 and 4 degrees of freedom, both values of F in the table exceed $F_{.05}$. Thus, the psychologist should reject both the null hypothesis that the mean reduction in smoking is the same for all methods and the null hypothesis that the mean reduction in smoking is the same for all types of patients.[11]

[11] See Appendix 10.2 for some formulas that are useful in calculating the required sums of squares in more complicated cases.

10.9 Confidence Intervals for Differences among Means

In the previous section, we saw how the two-way analysis of variance can be used to test whether the treatment means are all equal. In most experiments, however, the purpose is to find out the *extent* to which the treatment means differ, not just *whether* they differ. In this section, we will show how confidence intervals can be constructed for all differences between treatment means. The confidence coefficient attached to these intervals is the probability that *all* these intervals will include the true differences among treatment means *simultaneously*.

In the case of the psychologist, there are three differences among the treatment (that is, method) means. Assuming that the type of patient is held constant, let μ_1 once again be the mean reduction in number of cigarettes smoked daily resulting from method I, μ_2 be the mean reduction in number of cigarettes smoked daily resulting from method II, and so on. The probability is $(1 - \alpha)$ that *all* the following statements hold true *simultaneously*:

$$\bar{x}_{.1} - \bar{x}_{.2} - \sqrt{F_\alpha} s_E \sqrt{\frac{2(k-1)}{n}} < \mu_1 - \mu_2 < \bar{x}_{.1} - \bar{x}_{.2} + \sqrt{F_\alpha} s_E \sqrt{\frac{2(k-1)}{n}}$$

$$\bar{x}_{.1} - \bar{x}_{.3} - \sqrt{F_\alpha} s_E \sqrt{\frac{2(k-1)}{n}} < \mu_1 - \mu_3 < \bar{x}_{.1} - \bar{x}_{.3} + \sqrt{F_\alpha} s_E \sqrt{\frac{2(k-1)}{n}}$$

$$\bar{x}_{.2} - \bar{x}_{.3} - \sqrt{F_\alpha} s_E \sqrt{\frac{2(k-1)}{n}} < \mu_2 - \mu_3 < \bar{x}_{.2} - \bar{x}_{.3} + \sqrt{F_\alpha} s_E \sqrt{\frac{2(k-1)}{n}}$$

Confidence Intervals for Differences Among Treatment Means

$$(10.8)$$

where k is the number of treatments, n the number of blocks, F_α is the value of F with $(k-1)$ and $(k-1)(n-1)$ degrees of freedom exceeded with a probability of α, and s_E is the square root of the error mean square. Since $\bar{x}_{.1} = 50$, $\bar{x}_{.2} = 41$, $\bar{x}_{.3} = 46$, $F_{.05} = 6.94$, $k = 3$, $n = 3$, and $s_E = .71$, the 95 percent confidence interval for all the differences between treatment (method) means when the block (type of patient) is held constant, is as follows:

$$6.84 < \mu_1 - \mu_2 < 11.16$$

$$1.84 < \mu_1 - \mu_3 < 6.16$$

$$-7.16 < \mu_2 - \mu_3 < -2.84.$$

It is also possible to construct confidence intervals of this sort for the differences among the block means when the treatment is held constant. Thus, in this case, when the method is held constant, one can construct confidence intervals for the differences among the means corresponding to various types of patients. (Since this entails only a slight modification of the procedures covered in the previous paragraph, we describe this procedure in a footnote.)[12]

The following is a further illustration of the construction of confidence intervals of this sort.

EXAMPLE 10.6 Use the data in Table 10.9 to construct a 95 percent confidence interval for the six differences between the mean ratings of the four beers, when the person making the ratings is held constant.

SOLUTION: In this case, $\bar{x}_{.1} = 61, \bar{x}_{.2} = 62, \bar{x}_{.3} = 60, \bar{x}_{.4} = 61$, $F_{.05} = 3.49$, $s_E = \sqrt{4.50}$, $k = 4$,   and $n = 5$.   Thus, $\sqrt{F_\alpha}\, s_E \sqrt{2 \times [(k-1)/n]} = \sqrt{3.49} \times \sqrt{4.50} \times \sqrt{6/5} = 4.34$, and the confidence intervals are as follows:

$$-5.34 < \mu_1 - \mu_2 < 3.34$$

$$-3.34 < \mu_1 - \mu_3 < 5.34$$

$$-4.34 < \mu_1 - \mu_4 < 4.34$$

$$-2.34 < \mu_2 - \mu_3 < 6.34$$

$$-3.34 < \mu_2 - \mu_4 < 5.34$$

$$-5.34 < \mu_3 - \mu_4 < 3.34.$$

All the above confidence intervals include zero, which would be expected, since the analysis of variance in the previous section concluded that none of the differences among the treatment means is statistically significant.

10.10 A Final Caution

Before concluding this chapter, it is important to note the assumptions underlying the analysis of variance. If these assumptions are not met,

---

[12] To obtain confidence intervals for the block means, all that one has to do is substitute k for n and vice versa in expression (10.8). Of course, F_α must also differ from that in expression (10.8) because there are $(n-1)$ and $(k-1)(n-1)$ degrees of freedom in this case.

the use of the analysis of variance may be misleading. One assumption is that the populations being compared are normally distributed. Fortunately, this assumption is not as stringent as it sounds, since studies have shown that the validity of the analysis of variance is not significantly affected by moderate departures from normality. In the jargon of the statistician, the analysis of variance is a "robust test" in this regard. Another assumption underlying the analysis of variance is that the variances of the populations are equal. (Appendix 10.1 shows how the *F* distribution can be used to test the hypothesis that the variances of two populations are equal.) If this assumption is not met, trouble can result. Statisticians have devised techniques for handling this problem in some cases; these techniques are more properly discussed in more advanced texts. Still another important assumption is that the observations are statistically independent. To repeat, the analysis of variance should not be used unless the relevant assumptions are at least approximately fulfilled.

EXERCISES

10.14 A psychologist wants to determine which of three teaching methods results in the highest scores on a particular test. She picks 3 girls and 3 boys at random. One of each sex is taught by each method, after which all are given the test. Their grades are as follows:

	Method A	*Method B*	*Method C*
Boys	80	78	64
Girls	91	85	70

(a) Which method seems to result in the highest scores? In the lowest scores?

(b) Do boys seem to do better on this test than girls?

(c) Construct the relevant analysis of variance table.

(d) Can the differences in average score among methods be attributed to chance? (Let $\alpha = 0.05$.)

(e) Can the difference in average score between boys and girls be attributed to chance? (Let $\alpha = 0.05$.)

(f) Criticize the design of the experiment.

(g) If one method can be shown to result in a higher average score than the other methods, should this method replace the others?

10.15 A chemist wants to determine how four catalysts differ in yield. The chemist runs the experiment in eight manufacturing plants. In each plant, the yield is measured with each catalyst. The yields are as follows:

Plant	Catalyst 1	2	3	4
A	35	40	41	40
B	36	37	39	41
C	38	41	42	40
D	38	40	43	39
E	39	41	40	39
F	37	36	39	38
G	35	37	38	37
H	36	37	36	36

(a) Which catalyst seems to have the highest yield? The lowest yield?

(b) Which plant seems to have the highest yield?

(c) Construct the relevant analysis of variance table.

(d) Can the differences among catalysts be attributed to chance? (Let $\alpha = 0.01$.)

(e) Can the difference in average yield among the plants be attributed to chance? (Let $\alpha = 0.01$.)

(f) Calculate 99 percent confidence intervals for the differences among catalysts in mean yield.

10.16 Suppose that the Department of Transportation designs the following experiment: It picks at random four of one U.S. auto firm's cars, four of its U.S. competitor's cars, four of a German firm's cars, and four of a Japanese firm's cars. Then it has one of each firm's cars drive in city traffic; one of each firm's cars drive under suburban conditions; one of each firm's cars drive under mountainous conditions; and one of each firm's cars drive in flat, open country. The results are as follows:

Manufacturer	City	Suburbs	Miles per gallon Mountains	Flat country
U.S. firm I	14	20	21	24
U.S. firm II	15	21	25	26
German firm	18	24	26	28
Japanese firm	19	25	25	27

(a) Construct an analysis-of-variance table summarizing the results.

(b) What are the treatment effects here? What are the block effects?

(c) Does it appear that each make of car gets the same number of miles per gallon? If not, which seems to get the most, and which seems to get the least?

(d) Does it appear that the number of miles per gallon is the same for all types of driving conditions? If not, which type of driving condition seems to result in the largest number of miles per gallon, and which type seems to result in the smallest number of miles per gallon?

(e) Construct 95 percent confidence intervals for the differences among the four types of cars with respect to the mean number of miles per gallon when the type of driving condition is held constant.

The Chapter in a Nutshell

1. An *experiment* dictates that one group of people, machines, or other experimental units be treated differently from another group; and the effect of this difference in treatment is estimated by comparing certain measurable characteristics of the two groups. In an attempt to prevent confounding, statisticians recommend that the people, machines, or other experimental units be assigned to one group or the other *at random*.

2. A frequently used experimental design is the *randomized block design*, in which the experimental units are classified into blocks, and the statistician obtains data concerning the effect of each *treatment* in each *block*. In this design it is possible to estimate both the differences among the treatment means (holding the block constant) and the differences among the blocks means (holding the treatment constant).

3. If Y_1 has a χ^2 distribution with v_1 degrees of freedom, and if Y_2 has a χ^2 distribution with v_2 degrees of freedom, then $Y_1/v_1 \div Y_2/v_2$ has an F distribution with v_1 and v_2 of freedom (if Y_1 and Y_2 are independent). The F distribution is generally skewed to the right, but as both degrees of freedom become very large, the F distribution tends toward normality.

4. In a *completely randomized design*, experimental units are classified at random into groups, and each group is subjected to a different treatment. Then the sample mean in each group is used as an estimate of the population mean, and a test is made of the null hypothesis that the population means are all equal. The null hypothesis should be rejected if $ns_{\bar{x}}^2 \div \Sigma s_j^2/k > F_\alpha$, where k is the number of population means being compared, n is the sample size in each population, $s_{\bar{x}}^2$ is the variance of the k sample means, s_j^2 is the variance of the sample from the jth population, and F_α is the value of an F random variable with $(k-1)$ and $k(n-1)$ degrees of freedom that is exceeded with a probability of α.

5. An *analysis of variance* is a technique designed to analyze the total variation in a set of data, the object being to split up this total variation into component parts, each of which can be ascribed to a particular source. The *one-way analysis of variance* splits the total sum of squares into two parts, the *between-group* (or *treatment*) *sum of squares* and the *within-group* (or *error*) *sum of squares*. Using this breakdown of the total sum of squares, the one-way analysis of variance tests the null hypothesis that the population means in a completely randomized design are all equal. The test procedure is precisely the same as that given in the previous paragraph. What is different is that the computations are presented in the format of an analysis-of-variance table.

6. In most experiments, the purpose is to find out the *extent to which* the population means (corresponding to the various treatments)

differ, not just *whether* they differ. Confidence intervals can be constructed for all differences among the population means. The *confidence coefficient* attached to these intervals is the probability that all these intervals will include the respective differences among the population means simultaneously.

7. The *two-way analysis of variance* recognizes not one, but two sources of variation other than the error sum of squares. The two-way technique is used to test whether the differences among treatment means in a randomized block (without replications) are statistically significant. The two-way analysis of variance splits the total sum of squares into three parts, the *treatment sum of squares,* the *block sum of squares,* and the *error sum of squares.* The null hypothesis that the treatment means are equal should be rejected if the ratio of the treatment mean square to the error mean square exceeds F_α, where there are $(k - 1)$ and $(k - 1)(n - 1)$ degrees of freedom. The null hypothesis that the block means are equal should be rejected if the ratio of the block mean square to the error mean square exceeds F_α, where there are $(n - 1)$ and $(k - 1)(n - 1)$ degrees of freedom.

8. Confidence intervals can be constructed for all differences among the treatment means (holding the block constant). The confidence coefficient attached to these intervals is the probability that all these intervals will include the respective differences among the treatment means simultaneously. Confidence intervals of this sort can also be constructed for the differences among the block means (holding the treatment constant).

Chapter Review Exercises

10.17 The physical fitness department at a Midwestern college wants to determine how long it takes male students under 150 pounds, 150 to 190 pounds, and over 190 pounds to perform a particular physical fitness task. Fifteen students are chosen at random from each of these three weight classes, and the time it takes each student to perform the task is measured. The mean and standard deviation of the students' times (in minutes) are shown below:

	Under 150 lbs.	*150–190 lbs.*	*Over 190 lbs.*
Mean	15.1	16.2	13.4
Standard deviation	3.2	4.0	3.1

(a) In which weight class does the time seem highest? In which one does it seem lowest?

(b) Construct the relevant analysis-of-variance table.

(c) Can the observed differences among the weight classes in the mean time be attributed to chance? (Let $\alpha = 0.05$.)

(d) What assumptions underlie your answer to (c)?

(e) Construct 95 percent confidence intervals for the differences among the means.

10.18 A medical researcher wants to compare the effects of four drugs. He picks a sample of 48 people, and gives drug *A* to 12 of them, drug *B* to the next 12 of them, and so on. The drugs are meant to induce weight loss. The following table shows the mean weight loss (in pounds) and the sample standard deviation of the weight loss for those given each drug:

	Drug			
	A	*B*	*C*	*D*
Mean	10.1	11.4	9.3	8.8
Standard deviation	3.6	3.2	2.9	2.8

(a) Which drug seems most effective? Which drug seems least effective?

(b) Construct the appropriate analysis of variance table.

(c) Can the observed differences in the mean effects of these drugs be attributed to chance? (Let $\alpha = 0.05$.)

(d) What assumptions underlie your answer to (c)?

10.19 Professor Malone assigns her students a new statistics textbook. She finds at the end of the year that 70 percent of the class do well on the final examination, and none fails it. Does this prove that the new book is effective?

10.20 When studying the effects of a new drug, it is advisable to give the drug to some patients (the treatment group) but not others (the control group). Moreover, the experiment should be run *double-blind;* that is, neither the doctors who measure the results nor the subjects should know who is in which group. Why are these practices desirable?

10.21 A consumer research organization tests three brands of tires to see how many miles they can be driven before they should be replaced. One tire of each brand is tested in each of five types of cars. The results (in thousands of miles) are as follows:

Type of car	Brand A	Brand B	Brand C
I	26	29	24
II	30	32	27
III	27	30	26
IV	28	28	25
V	29	31	28

(a) Which brand seems to wear best? Which seems to wear most poorly?

(b) Do tires seem to wear better in one type of car than in another type?

(c) Construct the relevant analysis of variance table.

(d) Can the differences among brands be attributed to chance? (Let $\alpha = 0.05$.)

(e) Can the differences among types of cars be attributed to chance? (Let $\alpha = 0.05$.)

(f) Construct 95 percent confidence intervals for the differences among the brand means.

10.22 Fill in the blanks in the following analysis of variance table:

Source of variation	Sum of squares	Degrees of freedom	Mean square	F
Treatments	———	2	19	———
Blocks	122	———	———	———
Error	———	———	40	
Total	400	11		

*10.23 The following data show the number of words per minute which a secretary types (on a number of occasions) on five different typewriters.

Typewriter A: 67, 72, 64, 70, 67
Typewriter B: 68, 73, 70, 73, 69
Typewriter C: 71, 63, 72, 75, 72
Typewriter D: 70, 68, 74, 69, 74
Typewriter E: 64, 59, 68, 66, 75

Minitab is used to carry out an analysis of variance to determine whether the differences among the typewriters in this regard can be attributed to chance. Interpret the results, shown below.

```
READ THE FOLLOWING DATA INTO COLUMNS C1,C2,C3,C4,C5
67 68 71 70 64
72 73 63 68 59
64 70 72 74 68
70 73 75 69 66
67 69 72 74 75
AOVONEWAY ON DATA IN C1,C2,C3,C4,C5

ANALYSIS OF VARIANCE

DUE TO      DF        SS       MS=SS/DF     F-RATIO
FACTOR       4       81.8        20.5         1.32
ERROR       20      309.6        15.5
TOTAL       24      391.4
```

* Exercise requiring some familiarity with Minitab.

LEVEL	N	MEAN	ST. DEV.
C1	5	68.00	3.08
C2	5	70.60	2.30
C3	5	70.60	4.51
C4	5	71.00	2.83
C5	5	66.40	5.86

10.24 Four students are given three different forms of a standardized mathematics achievement test. Although they are different, the test developer claims that (on the average) they produce equivalent results. The scores for the four students are as follows:

	Student 1	Student 2	Student 3	Student 4
Form A	69	88	63	74
Form B	73	85	60	71
Form C	75	86	59	69

Use Minitab (or some other computer package) to determine whether the test developer's claim is correct. (Let $\alpha = .05$.) If Minitab (or some other computer package) is not available, do the calculations by hand.

Appendix 10.1

COMPARING TWO POPULATION VARIANCES

The F distribution can be used to test the hypothesis that the variance of one normal population equals the variance of another normal population. This test is often useful because a decision maker wants to determine whether one population is more variable than another. For example, a production manager may want to determine whether the variability of the errors made by one measuring instrument is less than the variability of those made by another measuring instrument. In addition, this test is often used to determine whether the assumptions underlying other statistical tests are valid. For example, in carrying out the t test to determine whether the means of two populations are the same, we assume that the variances of the two populations are equal. (Recall Chapter 8.) This assumption can be checked by carrying out the test described below.

The null hypothesis is that the variance of one normal population σ_1^2 equals the variance of the other normal population σ_2^2. To test this hypothesis, a sample of n_1 observations is taken from the first population and a sample of n_2 observations is taken from the second population. The test statistic is $s_1^2 \div s_2^2$, where s_1^2 is the sample variance of the observations taken from the first population and s_2^2 is the sample variance of the observations taken from the second population. If the null hypothesis (that $\sigma_1^2 = \sigma_2^2$) is true, this test statistic has the F distribution with $(n_1 - 1)$ and $(n_2 - 1)$ degrees of freedom.

Decision Rule: *When the alternative hypothesis[13] is $\sigma_1^2 > \sigma_2^2$, reject the null hypothesis if the test statistic exceeds F_α. When the alternative hypothesis is $\sigma_1^2 \neq \sigma_2^2$, let the population with the larger sample variance be the first population (that is, the one whose sample variance is in the numerator of the test statistic); and reject the null hypothesis if the test statistic exceeds $F_{\alpha/2}$.*

To illustrate the use of this test, suppose that a school psychologist wants to test whether the variance of scores on a particular test is the same among eighth graders as among seventh graders. (The significance level is set at .02.) A random sample of 25 eighth graders is selected, as is a random sample of 25 seventh graders, and it is found that the sample variance is 50 for eighth graders and 80 for seventh graders. Since the alternative hypothesis is that the two variances are unequal (regardless of which is bigger), we let the seventh graders constitute the first population (since its sample variance is larger than the sample variance of the eighth graders). Thus, $s_1^2 \div s_2^2 = 80 \div 50$, or 1.6. The null hypothesis should be rejected if this test statistic exceeds $F_{.01}$ (the number of degrees of freedom being 24 and 24). According to Appendix Table 10, $F_{.01} = 2.66$. Since the test statistic does not exceed $F_{.01}$, the null hypothesis should not be rejected.

[13] When the alternative hypothesis is one-sided (that is, when the variance of one population is larger than that of the other population, according to the alternative hypothesis), the population with the larger variance according to the alternative hypothesis should be designated as the first population.

Appendix 10.2

FORMULAS FOR COMPUTATIONS IN THE ANALYSIS OF VARIANCE

In calculating the sums of squares that are required by the analysis of variance, it is usually best to use formulas that require finding only sums and sums of squares of the observations. Such formulas are given below for the sums of squares in Table 10.7:

$$BSS = \frac{1}{n} \sum_{j=1}^{k} T_j^2 - \frac{1}{nk} T^2$$

$$TSS = \sum_{i=1}^{n} \sum_{j=1}^{k} x_{ij}^2 - \frac{1}{kn} T^2,$$

where T_j is the total of the observations from the jth population (or the jth treatment), and T is the total of all observations. Once these two sums of squares are calculated, we can obtain WSS by subtraction. That is,

$$WSS = TSS - BSS.$$

An additional formula that applies to Table 10.10 is

$$RSS = \frac{1}{k} \sum_{i=1}^{n} T_i^2 - \frac{1}{kn} T^2,$$

where T_i is the total of the observations in the ith block.

To illustrate the use of these formulas, consider once more the data in Example 10.3. Clearly, $T = 552$, $T_1 = 204$, $T_2 = 160$, and $T_3 = 188$. Thus,

$$BSS = \frac{1}{4}(204^2 + 160^2 + 188^2) - \frac{1}{4(3)} 552^2 = 25{,}640 - 25{,}392 = 248$$

$$TSS = 25{,}652 - \frac{1}{4(3)} 552^2 = 25{,}652 - 25{,}392 = 260$$

$$WSS = 260 - 248 = 12.$$

Comparing these results with those in the solution to Example 10.3, we find that they are identical. The advantage of the formulas given here is that they are easier and more efficient to calculate. Although this is not obvious in this case (since the numbers were intentionally chosen so that the calculations would be simple), this is generally true.

PART V

Regression and Correlation

11

OBJECTIVES

Regression and correlation techniques are concerned with how one variable is related to another variable. The overall purpose of this chapter is to describe the nature of regression and correlation techniques and how they are applied. Among the specific objectives are:

1 To show how one calculates a sample regression line.

2 To indicate how the sample regression line can be used to predict the value of the dependent variable on the basis of knowledge of the independent variable.

3 To calculate and interpret the sample correlation coefficient, and to test whether the population correlation coefficient equals zero.

Regression and Correlation Techniques

11.1 Introduction

Statisticians frequently must estimate how one variable is related to, or affected by, another variable. A political scientist may need to determine how a city's voting turnout is related to its voting registration, or a psychologist may need to know how a child's score on one form of IQ test is related to his or her score on another form of such a test. To estimate such relationships, statisticians use regression techniques; and to determine how strong such relationships are, they use correlation techniques. Regression and correlation are among the most important and most frequently used methods of statistics. In this chapter, we begin the study of regression and correlation. The final chapter will provide further information concerning these topics.

11.2 Relationship among Variables, and the Scatter Diagram

DETERMINISTIC AND STATISTICAL RELATIONSHIPS

The statistical techniques presented in previous chapters were concerned with a single variable, X. For example, in Chapter 7 we described how to estimate the mean of X; and in Chapter 8 we described how to test whether this mean equals a specified value. In many important practical situations, statisticians must be concerned with more than a single variable; in particular, they must be concerned with the relationships among variables. For example, they may want to determine whether changes in one variable, X, tend to be associated with changes in another variable, Y. (Does Y tend to increase when X increases?) The techniques presented in previous chapters are powerless to handle such a problem.

Deterministic Relationship

When we say that statisticians are interested in the relationships among variables, it is important to note at the outset that these relationships seldom are deterministic. To see what we mean by a *deterministic relationship,* suppose that Y is the variable we want to estimate and X is the variable whose value will be used to make this estimate. If the relationship between Y and X is *exact*, we say that it is a *deterministic relationship.* For example, if Y is the perimeter of a square and X is the length of a side of the square, it is evident that $Y = 4X$. This is a deterministic relationship since, once we are given X, we can predict Y *exactly.* (For example, in this case if X equals 2, we know that Y must equal 8—no more, or less.)

Statistical Relationship

Statisticians are generally interested in statistical, not deterministic, relationships. If a *statistical relationship* exists between Y and X, the average value of Y tends to be related to the value of X, but it is impossible to predict with certainty the value of Y on the basis of the value of X. For example, suppose that X is a family's annual income and Y is the amount the family saves per year. On the average, the amount saved by a family tends to increase as its income increases; and this relationship can be used to predict how much a family will save, if we know the amount of its income. However, this relationship is far from exact. Since families with the same income do not all save the same amount, it is impossible to predict with certainty the amount a family will save on the basis of its income alone.

REGRESSION ANALYSIS

Regression analysis describes the way in which one variable is related to another. (As we shall see in Chapter 12, regression and correlation techniques can handle more than two variables, but only two are considered in this chapter.) Regression analysis derives an equation which can be used to estimate the unknown value of one variable on the basis of the known value of the other variable. Suppose that all secretaries at a large private university are given performance ratings (from 0 to 10) by the university administration, and that each secretary takes an aptitude test (scored from 0 to 10) before being hired. A university administrator, John Kellogg, must decide whether or not to hire a particular job applicant, Jane Cosgrove, who scores 4 on this test.

To help make this decision, he would like to estimate what Ms. Cosgrove's performance rating will be, if she is hired. Although her test score is known, her performance rating cannot be known in advance. Regression analysis can be used to estimate her performance rating on the basis of her known test score.

The term *regression analysis* comes from studies carried out by the English statistician Francis Galton about 80 years ago. Galton compared the heights of parents with the heights of their offspring and found that very tall parents tended to have offspring who were shorter than their parents, while very short parents tended to have offspring who were taller than their parents. In other words, the heights of the offspring of unusually tall or unusually short parents tended to "regress" toward the mean height of the population. Because Galton used the height of the parent to predict the height of the offspring, this type of analysis came to be called regression analysis, even though subsequent applications actually had very little to do with Galton's "regression" of heights toward the mean.

SCATTER DIAGRAM

Since regression analysis is concerned with how one variable is related to another, an analysis of this sort generally begins with data concerning the two variables in question. Take the case of the university administrator who wants to estimate Ms. Cosgrove's performance rating on the basis of her test score. Clearly, a sensible first step for the administrator is to obtain data concerning the performance ratings and test scores of a sample of secretaries. Suppose that he collects such data for a sample of nine secretaries, the results being shown in Table 11.1. It is convenient to plot data of this sort in a so-called **scatter diagram.**

Test score*	Performance rating*
1	2
2	3
4	4
8	7
6	6
5	5
8	8
9	8
7	6

Table 11.1
Test Score and Performance Rating, Sample of Nine Secretaries

* Both test scores and performance ratings range from 0 to 10, where 0 is lowest and 10 is highest.

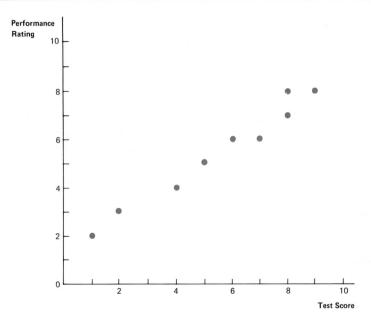

Figure 11.1
Scatter Diagram of
Data on Performance
Ratings and Test
Scores

In this diagram, the known variable—in this case the test score—is plotted along the horizontal axis and is called the *independent variable*. The unknown variable—the performance rating, in this case—is plotted along the vertical axis and is called the *dependent variable*. Of course, during the period to which the data pertain, *both* variables are known, but when the regression analysis is used to estimate what a job applicant's performance rating will be, only the test score will be known.

Figure 11.1 shows the scatter diagram based on the data in Table 11.1. Clearly, this diagram provides a useful visual portrait of the relationship between the dependent and independent variables. Based on such a diagram, one can form an initial impression concerning the following three important questions.

IS THE RELATIONSHIP DIRECT OR INVERSE? The relationship between X and Y is *direct* if increases in X tend to be associated with increases in Y, and decreases in X tend to be associated with decreases in Y. On the other hand, the relationship between X and Y is *inverse* if increases in X tend to be associated with decreases in Y, and decreases in X tend to be associated with increases in Y. Figure 11.1 indicates that the relationship between a secretary's test score and performance rating is direct, as would be expected. Panel A of Figure 11.2 shows a case where the relationship between X and Y is inverse. Not all scatter diagrams indicate either a direct or inverse relationship. Some, like panel C, indicate no correlation at all between X and Y. That is, changes in X do not seem to have any effect on the value of Y.

IS THE RELATIONSHIP LINEAR OR NONLINEAR? A relationship between X and Y is *linear* if a straight line provides an adequate representation of the average relationship between the two variables. On the other hand, if the points in the scatter diagram fall along a curved line or depart in some other way from a linear relationship, the relationship between X and Y is *nonlinear*. Figure 11.1 suggests that the relationship between test score and performance rating is linear (at least in this range). Panel D of Figure 11.2 shows a case where the relationship between X and Y is nonlinear.

HOW STRONG IS THE RELATIONSHIP? The relationship between X and Y is relatively strong if the points in the scatter diagram lie close to the line of average relationship. For example, panel E of Figure 11.2 shows a case where the relationship is strong enough so that one can

Figure 11.2
Scatter Diagrams of Various Types of Relationships between X and Y

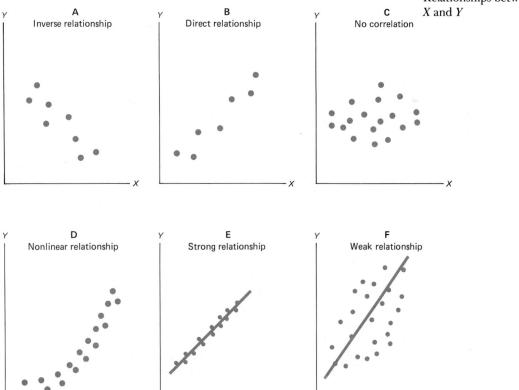

predict the value of *Y* quite accurately on the basis of the value of *X*. This is evidenced by the fact that all the points lie very close to the line. On the other hand, panel F of Figure 11.2 shows a case where the relationship is so weak that one cannot predict the value of *Y* at all well on the basis of the value of *X*. This follows from the fact that the points are scattered widely around the line.

CORRELATION ANALYSIS

Correlation analysis is concerned with the strength of the relationship between two variables. As we have seen, some relationships among variables are much stronger than others. For example, the size of a person's left foot is ordinarily very strongly related to the size of his or her right foot. On the other hand, there is some relationship between a firm's size and how rapidly it adopts new techniques, but this relationship may be rather weak. Correlation analysis is an important and useful complement to regression analysis. Whereas regression analysis describes the *type* of relationship between the two variables, correlation analysis describes the *strength* of this relationship.

11.3 Aims of Regression and Correlation Analysis

Basically, there are four principal goals of regression and correlation analysis. First, *regression analysis provides estimates of the dependent variable for given values of the independent variable.* If the university administrator discussed above wants to estimate the performance rating of the job applicant with a test score of 4, he can use regression analysis to provide such an estimate. His estimate will be based on the *regression line,* which is discussed in detail in subsequent sections. This line estimates the mean value of *Y* for each value of *X*. Thus, in this case the regression line would estimate the mean value of the performance rating for each value of the test score.

Second, *regression analysis provides measures of the errors that are likely to be involved in using the regression line to estimate the dependent variable.* For example, in the case of the university administrator, it clearly would be useful to know how much faith one can put in the estimated performance rating based on the regression line. To answer such questions, statisticians construct confidence intervals, which are described in Section 11.9.

Third, *regression analysis provides an estimate of the effect on the mean value of Y of a one-unit change in X.* For example, the university administrator might well be interested in how much of an increase in performance rating can be expected if the test score increases by 1 point.

In other words, how sensitive does a secretary's performance rating seem to be to differences in the test score? Regression analysis enables us to answer this question, as shown in Section 11.13.

Fourth, *correlation analysis provides estimates of how strong the relationship is between the two variables.* The *coefficient of correlation* and the *coefficient of determination* are two measures generally used for this purpose. These will be discussed in detail in Sections 11.10 and 11.11.

11.4 Linear Regression Model

A *model* is a simplified or idealized representation of the real world. All scientific inquiry is based to some extent on the use of models. In this section, we describe the model—that is, the set of simplifying assumptions—on which regression analysis is based. To begin with, the statistician visualizes a population of all relevant pairs of observations of the independent and dependent variables. For example, in the case of the university administrator, the statistician would visualize a population of pairs of observations concerning test score and performance rating. This population would include all the performance ratings corresponding to all the test scores of all the university's secretaries.

Holding constant the value of X (the independent variable), the statistician assumes that each corresponding value of Y (the dependent variable) is drawn at random from the population. For example, the second pair of observations in Table 11.1 is a case where the test score equals 2 and the performance rating equals 3. The statistician views this pair of observations as arising in the following way. The value of the test score (the independent variable) is fixed at 2. The value of the performance rating (the dependent variable) is the result of a random choice of all levels of the performance rating corresponding to a test score of 2. Thus, the value of the dependent variable is a random variable, which happens in this case to equal 3.

What determines the shape of the probability distribution of the dependent variable Y when the value of the independent variable is fixed at its specified value? For example, suppose that the probability distribution of the performance rating, given that the test score equals 2, is as shown in Figure 11.3. According to this figure, this probability distribution is bell-shaped with a mean of 2.95. Why does the probability distribution have this shape? Because in the population as a whole the values of the performance rating (the dependent variable), when the test score is fixed at 2, have a bell-shaped distribution with a mean of 2.95.

The probability distribution of Y, given a specified value of X, is called the ***conditional probability distribution of Y.*** Thus, Figure 11.3

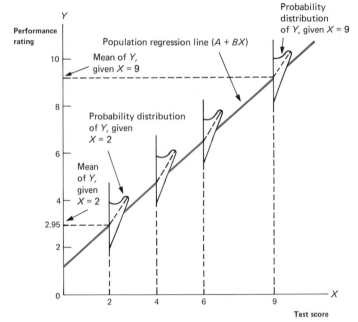

Figure 11.3
Conditional Probability
Distribution of
Performance Rating,
Given that Test Score
Equals 2, 4, 6, or 9

shows the conditional probability distribution of the performance rating, given that the test score equals 2 (in other words, under the *condition* that the test score equals 2).[1] Figure 11.3 also shows the conditional probability distribution of the performance rating, given that the test score equals 4, 6, and 9. The conditional probability distribution of Y, given the specified value of X, is denoted by

$$P(Y|X),$$

where Y is the value of the dependent variable and X is the specified value of the independent variable. The mean of this conditional probability distribution is denoted by $\mu_{Y \cdot X}$, and the standard deviation of this probability distribution is denoted by $\sigma_{Y \cdot X}$. For example, in Figure 11.3, $\mu_{Y \cdot X}$ equals 2.95 if the test score equals 2.

Regression analysis makes the following assumptions about the conditional probability distribution of Y. First, it assumes that *the mean value of* Y, *given the value of* X, *is a linear function of* X. In other words, the mean value of the dependent variable is assumed to be a linear function of the independent variable. Put still differently, the means of

---

[1] The concept of a conditional probability distribution has already been touched on in Appendix 4.3. The present discussion is self-contained and does not presume any knowledge of the material in Appendix 4.3.

the conditional probability distributions are assumed to lie on a straight line, the equation of this line being

$$\mu_{Y \cdot X} = A + BX.$$

Figure 11.3 shows a case of this sort, as evidenced by the fact that the means lie on a straight line. This straight line is called the ***population regression line*** or the ***true regression line.***

Second, regression analysis assumes that *the standard deviation of the conditional probability distribution is the same, regardless of the specified value of the independent variable.* Thus, in Figure 11.3, the spread of each of the conditional probability distributions is the same. For example, the standard deviation of the probability distribution of the performance rating, given that the test score is 2, is the same as the standard deviation of the performance rating, given that the test score is 9. This characteristic (of equal standard deviations) is called *homoscedasticity.*

Third, regression analysis assumes that *the values of* Y *are independent of one another.* For example, if one observation lies below the mean of its conditional probability distribution, it is assumed that this will not affect the chance that some other observation in the sample will lie below the mean of its conditional probability distribution. For example, if one secretary's performance rating is below average, it is assumed that this will not affect the chance that some other secretary's performance rating is below average.

Fourth, regression analysis assumes that the *conditional probability distribution of* Y *is normal.* Actually, as pointed out below, not all aspects of regression analysis require this assumption, but some do. It is also worth noting that in regression analysis only *Y* is regarded as a random variable. The values of *X* are assumed to be fixed. Thus, when regression analysis is used to estimate *Y* on the basis of *X,* the true value of *Y* is subject to error, but the value of *X* is known. For example, if regression analysis is used to estimate Ms. Cosgrove's performance rating when her test score is 4, her actual performance rating can be predicted only subject to error; but her test score (4) is known precisely.

The four assumptions underlying regression analysis can be stated somewhat differently. Together they imply that

$$Y_i = A + BX_i + e_i, \tag{11.1}$$

where Y_i is the ith observed value of the dependent variable, X_i is the ith observed value of the independent variable, and e_i is a normally distributed random variable with a mean of zero and a standard deviation equal to σ_e. Essentially, e_i is an *error term,* that is, a random amount that

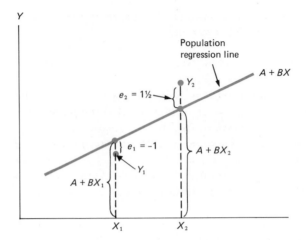

Figure 11.4
The Regression Model

is added to $A + BX_i$ (or subtracted from it if e_i is negative). Because of the presence of this error term, the observed values of Y_i fall around the population regression line, not on it. Thus, as shown in Figure 11.4, if e_1 (the value of the error term for the first observation) is -1, Y_1 will lie 1 below the population regression line. And if e_2 (the value of the error term for the second observation) is $+1.50$, Y_2 will lie 1.50 above the population regression line. Regression analysis assumes that the values of e_i are independent.[2]

Although the assumptions underlying regression analysis are unlikely to be met completely, they are close enough to the truth in a sufficiently large number of cases so that regression analysis is a powerful technique. Nonetheless, it is important to recognize at the start that if these assumptions are not at least approximately valid, the results of a regression analysis can be misleading. In the next chapter we shall indicate how tests can be carried out to check these assumptions, and we will describe how violations of these assumptions may affect the results.

11.5 Sample Regression Line

To carry out a regression analysis, we must obtain the mathematical equation for a line that describes the average relationship between the dependent and independent variable. This line is calculated from the

[2] In this chapter (and in Chapter 12), we use capital letters X and Y to denote both random variables and realized values. Which is referred to should be obvious from the context. (Practically all texts use the same symbols for both at this point. There seems to be general agreement that there is little danger of confusion.)

sample observations and is called the *sample* or *estimated regression line*. It should not be confused with the *population regression line* discussed in the previous section. Whereas the population regression line is based on the entire population, the sample regression line is based only on the sample.

The general expression for the sample regression line is

$$\hat{Y} = a + bX,$$

where $\hat{Y}$ is the value of the dependent variable predicted by the regression line, and a and b are estimators of A and B, respectively. Since this equation implies that $\hat{Y} = a$ when $X = 0$, it follows that a is the value of Y at which the line intersects the Y axis. Thus, a is often called the Y *intercept* of the regression line. And b, which clearly is the *slope* of the line, measures the change in the predicted value of Y associated with a one-unit increase in X.

Figure 11.5 shows the estimated regression line for the data concerning performance ratings and test scores of secretaries. The equation for this regression line is

$$\hat{Y} = 1.266 + 0.752X,$$

where $\hat{Y}$ is the performance rating and X is the test score. What is 1.266? It is the value of a, the estimator of A. What is 0.752? It is the value of b, the estimator of B. We are not interested here in how this equation was determined. (The methods used and their rationale are described in detail in the next section.) What we do want to consider is how this equation should be interpreted.

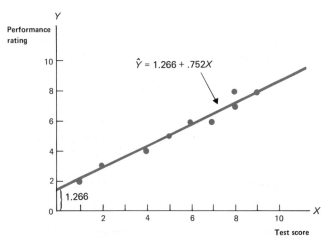

Figure 11.5
Sample Regression Line

To begin with, note the difference between Y and $\hat{Y}$. Whereas Y denotes an *observed* value of the performance rating, $\hat{Y}$ denotes the *computed* or *estimated* value of the performance rating, based on the regression line. For example, the first row of Table 11.1 shows that for the first secretary in the sample, the actual performance rating was 2 and the test score was 1. Thus, $Y = 2$ when $X = 1$. In contrast, the regression line indicates that $\hat{Y} = 1.266 + 0.752(1)$, or 2.018 when $X = 1$. In other words, while the regression line predicts that the performance rating will equal 2.018 when the test score equals 1, the actual performance rating under these circumstances (for the first secretary) was 2.

It is important to be able to identify and interpret the Y intercept and slope of a regression line. What is the Y intercept of the regression line in the case of the secretaries? It is 1.266. This means that if the test score is zero, the estimated performance rating would be 1.266. (As shown in Figure 11.5, 1.266 is the value of the dependent variable at which the regression line intersects the vertical axis.) What is the slope of the regression line in this case? It is 0.752. This means that the estimated performance rating increases by 0.752 when the test score increases by 1.

11.6 Method of Least Squares

In this section, we describe how a sample regression line is calculated. To illustrate how this is done, suppose that we want to estimate a regression line to represent the relationship beween performance rating and test score in Fig. 11.1. Since the equation for this line is

$$\hat{Y} = a + bX,$$

the estimation of a regression line really amounts to the choice of numerical values of a and b. There are an infinite number of possible values of a and b we could choose. One possibility, resulting in the line shown in panel A of Figure 11.6, is that $a = 1/2$ and $b = 1$. Another possibility, resulting in the line in panel B of Figure 11.6, is that $a = 0$ and $b = 1$. Still another possibility, resulting in the line in panel C of Figure 11.6, is that $a = -1$ and $b = 1$. How can we figure out which is best?

The *method of least squares* answers this question in the following way: We should take each possible line (that is, each possible value of a and b) and measure the deviation of each point in the sample from this line. Thus, in panel A of Figure 11.6 the deviation of each point from the line is measured by the broken vertical line from the point to the line. Then, according to the method of least squares, we square each of

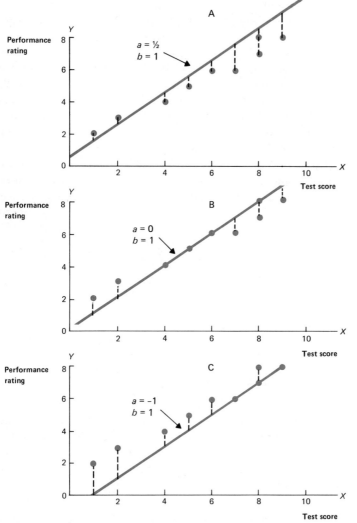

Figure 11.6
Alternative Values of *a*
and *b*

these deviations and add them up. Thus, in panel A of Figure 11.6 the
sum of the squared deviations of the points from the line shown there is
8.25. In panel B the sum of squared deviations is 5, and in panel C the
sum of squared deviations is 12.[3] Then *the method of least squares dictates that we choose the line where the sum of the squared deviations of the
points from the line is a minimum.*

[3] In panel A the deviations of the points from the line are $1/2, 1/2, -1/2, -1/2,$
$-1/2, -3/2, -3/2, -1/2, -3/2$. Thus, the sum of squared deviations is 8.25. In panel
B the deviations are $1, 1, 0, 0, 0, -1, -1, 0, -1$. Thus, the sum of squared deviations is 5.
In panel C the deviations are $2, 2, 1, 1, 1, 0, 1, 0, 0$. Thus, the sum of squared deviations is
12.

Certainly, on an intuitive level, it makes sense to choose the line (that is, the values of *a* and *b*) that minimizes the sum of squared deviations of the data in the sample from the line. Why? Because the bigger the sum of squared deviations of the data from the line, the poorer the line fits the data. Thus, one should minimize the sum of squared deviations if one wants to obtain a line that fits the data as well as possible. This is illustrated in Figure 11.6. Clearly, the line in panel c does not fit the data as well as the line in panel A, which in turn does not fit the data as well as the line in panel B. This fact is reflected in the differences in the sum of the squared deviations. As would be expected, the sum of squared deviations is higher for the line in panel c than for the line in panel A, and higher for the line in panel A than for the line in panel B.

Mathematically, it can be shown that if Y_i and X_i are the *i*th pair of observations concerning the dependent and independent variables, the values of *a* and *b* that result in the minimization of the sum of squared deviations from the regression line satisfy the following equations:

$$\sum_{i=1}^{n} Y_i = na + b \sum_{i=1}^{n} X_i$$

$$\sum_{i=1}^{n} X_i Y_i = a \sum_{i=1}^{n} X_i + b \sum_{i=1}^{n} X_i^2,$$

where *n* is the number of values of X_i (and Y_i) on which the calculation of the sample regression line is based. Solving for *a* and *b*, which are generally referred to as the *least-squares estimators of* A *and* B, we obtain the following important formulas:

Least-Squares Estimators of A and B

$$b = \frac{\sum_{i=1}^{n} (X_i - \overline{X})(Y_i - \overline{Y})}{\sum_{i=1}^{n} (X_i - \overline{X})^2} \qquad (11.2a)$$

$$a = \overline{Y} - b\overline{X}. \qquad (11.2b)$$

The value of *b* in equation (11.2a) is often called the **estimated regression coefficient**.

From the standpoint of computational ease, it frequently is preferable to use a somewhat different formula for *b* than the one given in

	X_i	Y_i	X_i^2	Y_i^2	X_iY_i
	1	2	1	4	2
	2	3	4	9	6
	4	4	16	16	16
	8	7	64	49	56
	6	6	36	36	36
	5	5	25	25	25
	8	8	64	64	64
	9	8	81	64	72
	7	6	49	36	42
Total	50	49	340	303	319

Table 11.2
Computation of ΣX_i, ΣY_i, ΣX_i^2, ΣY_i^2, and ΣX_iY_i

$$\overline{X} = \frac{50}{9} = 5.556$$

$$\overline{Y} = \frac{49}{9} = 5.444$$

equation (11.2a). This alternate formula, which yields the same answer as equation (11.2a), is

$$b = \frac{n \sum_{i=1}^{n} X_iY_i - \left(\sum_{i=1}^{n} X_i\right)\left(\sum_{i=1}^{n} Y_i\right)}{n \sum_{i=1}^{n} X_i^2 - \left(\sum_{i=1}^{n} X_i\right)^2}.$$

In the case of the secretaries, Table 11.2 shows the calculation of ΣX_iY_i, ΣX_i^2, ΣX_i, and ΣY_i. Based on these calculations,

$$b = \frac{9(319) - (50)(49)}{9(340) - 50^2} = \frac{2871 - 2450}{3060 - 2500}$$

$$= \frac{421}{560} = .752.$$

Thus, the value of b, the least-squares estimator of B, is .752, which is the result given in the previous section. In other words, an increase in test score of 1 point is associated with an increase in estimated performance rating of 0.752 points.

Having calculated b, we can readily determine the value of a, the least-squares estimator of A. According to equation (11.2b),

$$a = \bar{Y} - b\bar{X},$$

where $\bar{Y}$ is the mean of the values of Y, and $\bar{X}$ is the mean of the values of X. Since, as shown in Table 11.2, $\bar{Y} = 5.444$ and $\bar{X} = 5.556$, it follows that

$$a = 5.444 - .752(5.556)$$

$$= 1.266.$$

Thus, the least-squares estimate of A is 1.266. Recall that this is the result given in the previous section.

Given a and b, it is a simple matter to specify the average relationship in the sample between performance rating and test score for the secretaries. This relationship is

$$\hat{Y} = 1.266 + 0.752X. \tag{11.3}$$

As we know, this line is often called the *sample regression line* or the *regression of Y on X*. It is the line that we presented in the previous section and that we plotted in Figure 11.5. Now we have shown how this line is derived.

A regression line of this sort can be of great practical importance. For example, let's return to the case of the university administrator who wants to predict the performance rating of a job applicant, Jane Cosgrove, whose test score is 4. Using equation (11.3), the administrator would predict that her performance rating would be

$$1.266 + 0.752(4) = 4.274. \tag{11.3a}$$

Equipped with this prediction, the administrator is better able to judge whether Ms. Cosgrove will perform adequately, and thus whether she should be hired.

The following example illustrates further how one calculates a least-squares regression line.

EXAMPLE 11.1 A sociologist wants to estimate the relationship in an Appalachian community between a family's annual income and the amount that the family saves. The following data from nine families are obtained:

Annual income (thousands of dollars)	Annual savings (thousands of dollars)
12	0.0
13	0.1
14	0.2
15	0.2
16	0.5
17	0.5
18	0.6
19	0.7
20	0.8

Calculate the least-squares regression line, where annual savings is the dependent variable and annual income is the independent variable.

SOLUTION: Letting X_i be the income (in thousands of dollars) of the ith family, and Y_i be the saving (in thousands of dollars) of the ith family, we find that

$$\sum_{i=1}^{9} X_i Y_i = 63.7 \quad \sum_{i=1}^{9} Y_i = 3.6 \quad \bar{Y} = 0.4,$$

$$\sum_{i=1}^{9} X_i^2 = 2364 \quad \sum_{i=1}^{9} X_i = 144 \quad \bar{X} = 16.$$

Thus, substituting these values in the alternate formula for b, we obtain

$$b = \frac{9(63.7) - (144)(3.6)}{9(2364) - 144^2} = \frac{573.3 - 518.4}{21,276 - 20,736} = .1017.$$

Consequently,

$$a = \bar{Y} - b\bar{X} = 0.4 - .1017(16) = -1.2272.$$

Thus, the regression line is

$$\hat{Y} = -1.2272 + .1017X,$$

where both X and Y are measured in thousands of dollars.

EXERCISES

11.1 In aircraft manufacture, rivets are used to join parts. The following table, taken from a study published in *Industrial Quality Control,* shows the number of oversize rivet holes and the number of minor repairs on ten sections of an airplane.

Oversize rivet holes	Minor repairs
45	22
52	26
49	21
60	28
67	33
61	32
70	33
54	25
52	34
67	35

(a) Construct a scatter diagram of these data.

(b) Calculate the sample regression line, where the number of minor repairs is the dependent variable and the number of oversize rivet holes is the independent variable.

(c) Use this regression line to predict the average number of minor repairs if the number of oversize rivet holes is 50.

(d) Predict the average number of minor repairs if the number of oversize rivet holes is 70.

11.2 An environmental scientist samples the water in 15 lakes in the United States. For each lake, he estimates the number of factories per mile of lake front (X) and the percent of impurities in the water (Y). The results are shown below.

X	Y
5.00	0.00
8.90	0.43
7.15	0.00
6.10	0.24
7.70	0.00
8.20	0.91
1.01	0.00
0.61	0.03
0.72	0.01
0.68	0.00
1.14	0.00
1.18	0.01
0.73	0.00
0.87	0.20
1.47	0.00

(a) Construct the scatter diagram of these data.
(b) Calculate the sample regression line, where Y is the dependent variable and X is the independent variable.
(c) Predict the percent of impurities in the water if there are 6 factories per mile of lake front.
(d) Is the relationship between X and Y a causal relationship? Why, or why not?

11.3 A psychologist obtains the IQ of 14 individuals, each of which is asked to take a particular test. The results are as follows:

IQ	Test score
99	54
110	70
141	100
108	61
123	83
129	86
132	98
88	51
101	64
105	68
97	52
96	55
118	71
113	76

(a) Construct a scatter diagram of these data.
(b) Calculate the sample regression line, where the test score is the dependent variable and the IQ is the independent variable.
(c) Predict the average test score if a person's IQ is 130.
(d) If the maximum test score is 100, do you think that the relationship between IQ and test score is linear regardless of the level of IQ? Explain.

11.7 Characteristics of Least-Squares Estimates

Optimally, in their calculations the decision maker and the statistician would like to know the population regression line. For example, the university administrator would like to know the population regression line relating a secretary's performance rating to his or her test score—the regression based on *all* the possible observations of performance ratings and test scores. However, this regression line cannot be calculated because the administrator has only the sample of observations to work with. Therefore, the best that the administrator can do is to calculate the sample regression line and to use it as an estimate of the population regression line. The statistics—*a* and *b*—defined in the previous section are estimators of *A* and *B*, the constants in the population regression. *Whether or not the conditional probability distribution of the dependent variable is normal,* these estimators have the following desirable properties (if the other assumptions in Section 11.4 are met):

1. *Unbiasedness.* It can be shown that *a* is an unbiased estimator of *A*, and that *b* is an unbiased estimator of *B*. In other words, if we were to draw one sample after another, and calculate the least-squares estimators *a* and *b* from each sample, the mean value of *a* would equal *A*, and the mean value of *b* would equal *B*, if we were to draw a very large number of samples. (Recall from Chapter 7 that unbiasedness is one of the criteria statisticians use for choosing among estimators.)

2. *Efficiency.* It can also be shown that of all estimators which are unbiased (and which are linear functions of the dependent variables), *a* and *b* have the smallest standard deviation. In other words, they are the most efficient estimators of this type. This very important result was proved by the so-called *Gauss-Markov theorem.* (Recall from Chapter 7 that efficiency is one of the criteria used by statisticians to choose among estimators.)

3. *Consistency.* We also find that *a* is a consistent estimator of *A*, and that *b* is a consistent estimator of *B*. In other words, as the sample size becomes larger and larger, the value of *a* homes in on *A*, and the value of *b* homes in on *B*. (Recall from Chapter 7 that consistency is one of the criteria used by statisticians for choosing among estimators.)

Because of these very desirable properties, the least-squares estimators a *and* b *are the standard estimators used by statisticians to estimate the constants in the population regression line,* A *and* B. The fact that *a* and *b* have these desirable properties provides fundamental and strong support for the intuitive judgment in the previous section that the method of least squares is a good way to fit a sample regression line.

11.8 Standard Error of Estimate

In previous sections we have shown how regression analysis provides estimates of the dependent variable for given values of the independent variable (the first goal cited in Section 11.3). We will now describe how regression analysis provides measures of the errors that are likely to be involved in this estimation procedure (the second goal cited in Section 11.3). First, it is essential to recall that the standard deviation of the conditional probability distribution of the dependent variable is assumed to be the same, regardless of the value of the independent variable. *This standard deviation, which we denote by* σ_e, *is a measure of the amount of scatter about the regression line in the population.* If σ_e is large, there is much scatter; if σ_e is small, there is little scatter.

The sample statistic used to estimate σ_e is the standard error of estimate. It is defined as

$$s_e = \sqrt{\frac{\sum_{i=1}^{n}(Y_i - \hat{Y}_i)^2}{n-2}},$$ (11.4)

Formula for s_e

where Y_i is the ith value of the dependent variable, n is the sample size, and $\hat{Y}_i$ is the estimate of Y_i from the regression line (that is, $\hat{Y}_i = a + bX_i$). Clearly, the value of s_e rises with increases in the amount of scatter about the regression line in the sample. If there is no scatter at all (which means that all the points are on the regression line), s_e equals zero, since Y_i and $\hat{Y}_i$ are always the same. But if there is much scatter, Y_i often differs greatly from $\hat{Y}_i'$, with the result that s_e will be large.

Another formula used frequently for the standard error of estimate is

$$s_e = \sqrt{\frac{\sum_{i=1}^{n}Y_i^2 - a\sum_{i=1}^{n}Y_i - b\sum_{i=1}^{n}X_iY_i}{n-2}}.$$ (11.5)

This expression is often easier to calculate than the one given in equation (11.4). Of course, equations (11.4) and (11.5) will always give the same result. In each of these equations $(n-2)$ is the denominator because this results in s_e^2 being an unbiased estimate of σ_e^2.

To illustrate the use of equation (11.5), let's return to the case of the university secretaries. Since Table 11.2 shows that $\Sigma Y_i^2 = 303$, $\Sigma Y_i = 49$, and $\Sigma X_iY_i = 319$, it follows that

$$s_e = \sqrt{\frac{303 - (1.266)(49) - (0.752)(319)}{7}} = \sqrt{.154} = 0.392.$$

Thus, if we knew the true values of A and B, the standard deviation of the errors in prediction based on this true regression line would be about .392. In other words, the situation would be as shown in Figure 11.7.

The standard error of estimate will be used frequently in subsequent sections. The example below further illustrates how it is calculated.

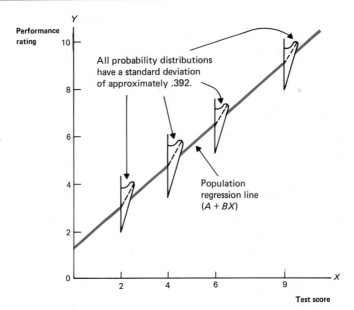

Figure 11.7
Conditional Probability
Distribution of
Performance Rating,
Given that Test Score
Equals 2, 4, 6, or 9

All probability distributions have a standard deviation of approximately .392.

Population regression line ($A + BX$)

Test score

EXAMPLE 11.2 Based on the data in Example 11.1, the sociologist wants to calculate the standard error of estimate. What is its value?

SOLUTION: From equation (11.5), it follows that

$$s_e = \sqrt{\frac{\sum_{i=1}^{9} Y_i^2 - a \sum_{i=1}^{9} Y_i - b \sum_{i=1}^{9} X_i Y_i}{7}}$$

$$= \sqrt{\frac{2.08 + (1.2272)(3.6) - (.1017)(63.7)}{7}}$$

$$= \sqrt{\frac{2.08 + 4.4179 - 6.4783}{7}} = \sqrt{.0028}$$

$$= .053.$$

Thus, the standard error of estimate is .053 thousands of dollars, or $53.

11.9 Estimators of (1) the Conditional Mean and (2) an Individual Value of *Y*

ESTIMATING THE CONDITIONAL MEAN

In this section we show how one can estimate the conditional mean of *Y*. In contrast to the previous section, it is not assumed that we know *A* and *B;* instead, we assume that both the regression line and the estimate are based on least-squares estimates of *A* and *B*. To be specific, let's return again to the example of the university secretaries. Suppose that the university administrator is interested in predicting the *mean* performance rating of *all* secretaries whose test scores are 4. In other words, the administrator is interested in estimating the vertical coordinate of the point on the population regression line corresponding to a test score of 4. In Figure 11.8, this conditional mean is denoted by $\mu_{y.4}$.

Since the population regression line is unknown, the best that the university administrator can do is to substitute the sample regression line. That is, to estimate $A + B(4)$, one uses $a + b(4)$, or 4.274. This, of course, is the point on the sample regression line corresponding to $X = 4$, as shown in Figure 11.8. In general, in order to estimate the conditional mean of *Y* (that is, the vertical coordinate of the point on the true

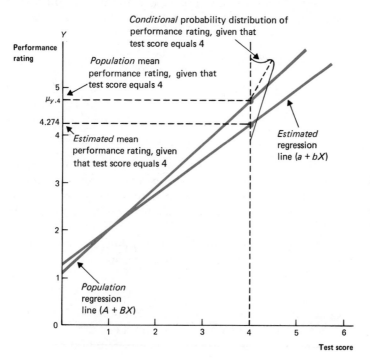

Figure 11.8
Estimated (and Population) Mean Performance Rating and Conditional Probability Distribution of Performance Rating, if Test Score Equals 4

regression line) when the independent variable equals $X*$, one should use

$$a + bX*.$$

A confidence interval for the conditional mean of Y (given that the independent variable equals $X*$) is

Confidence Interval for Conditional Mean

$$(a + bX*) \pm t_{\alpha/2} s_e \sqrt{\frac{1}{n} + \frac{(X* - \overline{X})^2}{\sum_{i=1}^{n} X_i^2 - n\overline{X}^2}} \qquad (11.6)$$

If the conditional probability distribution of the dependent variable is normal, the probability is $(1 - \alpha)$ that this interval will include this conditional mean. Note that this confidence interval becomes wider as the value of $X*$ lies farther and farther from $\overline{X}$. This makes sense because as one moves farther and farther from $\overline{X}$, an error in the estimated slope of the regression line will result in an increasingly larger error in the location of the regression line.

To illustrate the use of the confidence interval in expression (11.6), let's consider the prediction of the mean performance rating for all secretaries whose test score is 4. In this case, the 95 percent confidence interval for this mean rating would be

$$4.274 \pm 2.365(0.392) \sqrt{\frac{1}{9} + \frac{(4 - 5.556)^2}{62.2}}$$

or

$$4.274 \pm .359.$$

(Note that $t_{\alpha/2}$ is based on $(n - 2)$, or 7, degrees of freedom.) Thus, if the university administrator is interested in estimating the point on the population regression line corresponding to a test score of 4, the 95 percent confidence interval for this conditional mean is 3.915 to 4.633.

The following example further illustrates how a confidence interval is computed for a conditional mean.

EXAMPLE 11.3 Based on the data in Example 11.1, calculate a 95 percent confidence interval for the mean amount of family savings among families with an income of $20,000.

SOLUTION: Substituting in equation (11.6), we obtain

$$-1.2272 + .1017(20) \pm 2.365(.053) \sqrt{\frac{1}{9} + \frac{(20-16)^2}{2364 - 9(16)^2}},$$

or

$$.8068 \pm 2.365(.053)(.615),$$

or

$$0.730 \text{ to } 0.884.$$

Thus, the 95 percent confidence interval for the mean amount of savings among families with an income of $20,000 is .730 to .884 thousands of dollars, or $730 to $884.

PREDICTING AN INDIVIDUAL VALUE OF *Y*

Sample regression lines are often used to predict an individual value of *Y*. As we have seen, the university administrator might want to predict a secretary's performance rating if his or her test score equals 4. This kind of prediction can be made with the use of the sample regression line. The vertical coordinate of the point on the sample regression line corresponding to $X = 4$ can be used as a predictor. As shown in Figure 11.8 (and as we already know from Section 11.6), the prediction in this case would be that $Y = 4.274$. In general, to predict an individual value of Y when the independent variable equals X^*, one should use

$$a + bX^*.$$

A confidence interval for the value of *Y* that will occur if the independent variable is set at X^* is

$$(a + bX^*) \pm t_{\alpha/2} s_e \sqrt{\frac{n+1}{n} + \frac{(X^* - \overline{X})^2}{\sum\limits_{i=1}^{n} X_i^2 - n\overline{X}^2}} \qquad (11.7)$$

Confidence Interval for an Individual Value of Y

If the conditional probability distribution of the dependent variable is normal, the probability is $(1 - \alpha)$ that this interval will include the true value of Y. A comparison of this confidence interval with the one shown in expression (11.6) indicates that this one is wider than the other. This, of course, is reasonable because the sampling error in predicting an indi-

vidual value of the dependent variable will be greater than the sampling error in estimating the *conditional mean value* of the dependent variable.

To illustrate the use of this confidence interval, let's return once more to the case of the university administrator who is interested in predicting the performance rating of a job applicant, Jane Cosgrove, whose test score is 4. Based on expression (11.7), the 95 percent confidence interval for this performance rating is

$$4.274 \pm 2.365(0.392) \sqrt{\frac{10}{9} + \frac{(4 - 5.556)^2}{62.2}}$$

or

$$4.274 \pm 0.994.$$

(Again, $t_{\alpha/2}$ is based on $(n - 2)$, or 7, degrees of freedom.) Thus, the 95 percent confidence interval for this rating is 3.280 to 5.268.

It is important to recognize that the estimate of the *conditional mean* of Y covered above is quite different from the prediction of an *individual value* of Y being discussed here. An estimate of the *average* performance rating of *all* secretaries whose test score is 4 is not the same as a prediction of the performance rating of a *single* secretary with this test score. In the latter case, we are trying to predict the value of the random variable Y whose distribution is shown in Figure 11.8; in the former case, we are attempting to estimate the conditional mean of this distribution ($\mu_{Y \cdot 4}$), also shown in Figure 11.8. Put still differently, in the latter case we are attempting to predict an individual value of Y, while in the former we are trying to estimate a point on the population regression line. Although these two things are not the same, we use the same point estimate, $a + b(4)$, for both. However, as we have seen, we use a different confidence interval for each.

Before leaving the topic of the confidence intervals in expressions (11.6) and (11.7), it is important to spell out the underlying assumptions once more. First, *it is assumed that the population regression is linear; that is, that the conditional mean of the dependent variable is a linear function of the independent variable.* Second, *it is assumed that the conditional probability distribution of the dependent variable is normal and that its standard deviation is the same, regardless of the value of the independent variable.* If any one of these assumptions is violated, the results will be in error to at least some degree. It is also worth noting that the confidence interval in expresssion (11.7) relates only to a single prediction. In other words, the confidence interval pertains to the entire process of gathering a sample of size n and making a single prediction. It does not pertain to more than one prediction based on a single sample.

Example 11.4 shows how a confidence interval is computed for a predicted value of the dependent variable.

EXAMPLE 11.4 Based on the data in Example 11.1, predict the savings of a family chosen at random from among families with an income of $20,000, and calculate a 95 percent confidence interval for this prediction.

SOLUTION: The prediction is $-1.2272 + .1017(20)$, or .807 thousands of dollars (that is, $807). Using expression (11.7), the confidence interval is

$$-1.2272 + .1017(20) \pm 2.365(.053) \sqrt{\frac{10}{9} + \frac{(20-16)^2}{2364 - 9(16^2)}},$$

or

$$.8068 \pm 2.365(.053)1.174,$$

or

$$.660 \text{ to } .954.$$

Thus, the 95 percent confidence interval for the savings of a family chosen at random from among families with an income of $20,000 is .660 to .954 thousands of dollars, or $660 to $954.

EXERCISES

11.4 A sample of ten psychology textbooks (all paperbacks) shows the following relationship between their price and length (prices are rounded off):

Price (dollars)	Length (pages)
16	520
20	680
21	740
9	200
10	400
22	800
20	750
15	500
10	300
10	350

(a) Construct a scatter diagram of these data.
(b) Compute the sample regression line, where price is the dependent variable and length is the independent variable.
(c) Calculate the standard error of estimate.
(d) Compute a 95 percent confidence interval for the conditional mean price, if the book's length is 600 pages.
(e) Compute a 95 percent confidence interval for the price of a book, if its length is 600 pages.
(f) Is length causally related to price?

11.5 For eight families, the amount spent annually on food, and their annual incomes, are given below:

Food expenditure	Income
(thousands of dollars)	
4	20
6	40
3	11
5	30
2	9
2	12
3	15
3	21

(a) Compute the sample regression line, where income is the independent variable and amount spent on food is the dependent variable.
(b) Calculate the standard error of estimate.
(c) Compute a 90 percent confidence interval for the conditional mean food expenditure if family income is $20,000.
(d) Compute a 90 percent confidence interval for a family's food expenditure if its income is $20,000.

11.6 (a) In Exercise 11.1, what is the standard error of estimate?
(b) Compute a 99 percent confidence interval for the conditional mean number of minor repairs if the number of oversize rivet holes is 50.

11.7 (a) In Exercise 11.2, what is the standard error of estimate?
(b) Compute a 50 percent confidence interval for the percent of impurities in the water if there are 6 factories per mile of lake front.

11.8 (a) In Exercise 11.3, what is the standard error of estimate?
(b) Compute a 90 percent confidence interval for a person's test score if his IQ is 130.

11.9 The United States Department of Agriculture has published data concerning the strength of cotton yarn and the length of the cotton fibers that make up the yarn.[4] Results for 10 pieces of yarn are as follows:

[4] U.S. Department of Agriculture, *Results of Fiber and Spinning Tests for Some Varieties of Upland Cotton Grown in the U.S.*

Strength of yarn (pounds)	Fiber length (hundredths of an inch)
99	85
93	82
99	75
97	74
90	76
96	74
93	73
130	96
118	93
88	70

(a) Construct a scatter diagram of these data.

(b) Based on this scatter diagram, does the relationship between these two variables seem to be direct or inverse? Is this in accord with common sense? Why, or why not? Does the relationship seem to be linear?

(c) Assume that the conditional mean value of yarn strength is a linear function of fiber length. Calculate the least-squares estimates of the parameters (A and B) of this linear function.

(d) What is the sample regression line for these data? Use this regression line to predict the average strength of yarn made from fibers of length equal to 0.80 inches. Use this regression line to predict the average strength of yarn made from fibers of length equal to 0.90 inches.

(e) Calculate the standard error of estimate.

(f) Compute a 90 percent confidence interval for the conditional mean strength of yarn corresponding to a fiber length of 0.80 inches.

(g) Compute a 90 percent confidence interval for the strength of a piece of yarn, if the fiber length is 0.80 inches. Why is this confidence interval wider than the confidence interval in (f)?

11.10 Coefficient of Determination

In previous sections, we have shown how a regression line can be calculated. Once the regression line has been found, the statistician wants to know how well this line fits the data. There can be vast differences in how well a regression line fits a set of data, as shown in Figure 11.9. Clearly, the regression line in panel B of Figure 11.9 provides a better fit than the regression line in panel A of the same figure. How can we measure how well a regression line fits the data?

As a first step toward answering this question, we must discuss the concept of *variation*, which refers to a sum of squared deviations. The total variation in the dependent variable Y equals

$$\sum_{i=1}^{n} (Y_i - \overline{Y})^2.$$

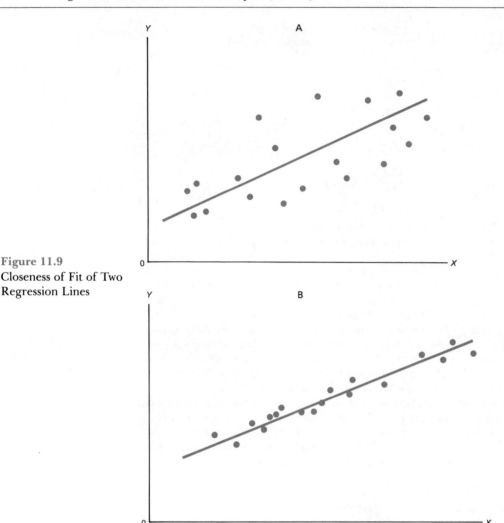

Figure 11.9
Closeness of Fit of Two
Regression Lines

In other words, the total variation equals the sum of the squared deviations of Y from its mean. (In Chapter 10, this was often called the
total sum of squares.)

To measure how well a regression line fits the data, we divide the
total variation in the dependent variable into two parts: (1) the variation that *can* be explained by the regression line; and (2) the variation
that *cannot* be explained by the regression line. To divide the total variation in this way, we must note that for the ith observation,

$$(Y_i - \overline{Y}) = (Y_i - \hat{Y}_i) + (\hat{Y}_i - \overline{Y}),$$ (11.8)

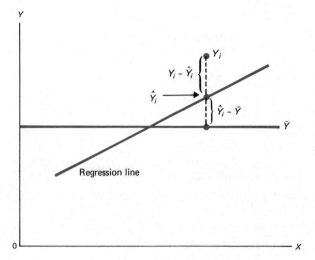

Figure 11.10
Division of $(Y_i - \overline{Y})$ into Two Parts: $(Y_i - \hat{Y}_i)$ and $(\hat{Y}_i - \overline{Y})$

where $\hat{Y}_i$ is the value of Y_i that would be predicted on the basis of the regression line. In other words, as shown in Figure 11.10, the discrepancy between Y_i and the mean value of Y can be split into two parts: the discrepancy between Y_i and the point on the regression line directly below (or above) Y_i and the discrepancy between the point on the regression line directly below (or above) Y_i and $\overline{Y}$.

If we square both sides of equation (11.8) and sum the result over all values of i, we find that[5]

$$\sum_{i=1}^{n} (Y_i - \overline{Y})^2 = \sum_{i=1}^{n} (Y_i - \hat{Y}_i)^2 + \sum_{i=1}^{n} (\hat{Y}_i - \overline{Y})^2. \qquad (11.9)$$

Total Variation = Unexplained Variation + Explained Variation

The term on the left-hand side of this equation shows the *total variation* in the dependent variable. The first term on the right-hand side measures the *variation in the dependent variable that is not explained by the regression.* This is a reasonable interpretation of this term since it is the sum of squared deviations of the actual observations from the regres-

[5] To derive this result, note that

$$\sum_{i=1}^{n} (Y_i - \overline{Y})^2 = \sum_{i=1}^{n} [(Y_i - \hat{Y}_i) + (\hat{Y}_i - \overline{Y})]^2$$

$$= \sum_{i=1}^{n} (Y_i - \hat{Y}_i)^2 + \sum_{i=1}^{n} (\hat{Y}_i - \overline{Y})^2 + 2 \sum_{i=1}^{n} (Y_i - \hat{Y}_i)(\hat{Y}_i - \overline{Y}).$$

The last term on the right-hand side equals zero, so this result follows.

sion line. Clearly, the larger the value of this term, the poorer the regression equation fits the data.

The second term on the right-hand side of the equation measures the *variation in the dependent variable that is explained by the regression.* This is a reasonable interpretation of this term since it shows how much the dependent variable would be expected to vary on the basis of the regression alone. Putting it differently, this second term shows the reduction in unexplained variation due to the use of the regression instead of $\bar{Y}$ as an estimator of Y. Using $\bar{Y}$ as an estimator, the total variation is unexplained, whereas the first term on the right is unexplained when predictions based on the regression are used. Thus, the second term on the right shows the reduction in unexplained variation due to the use of predictions based on the regression instead of $\bar{Y}$.

To measure the closeness of fit of a regression line, statisticians use the **coefficient of determination:**

Formula for r²

$$r^2 = 1 - \frac{\sum\limits_{i=1}^{n} (Y_i - \hat{Y}_i)^2}{\sum\limits_{i=1}^{n} (Y_i - \bar{Y})^2}. \tag{11.10}$$

In other words,

$$r^2 = 1 - \frac{\text{Variation not explained by regression}}{\text{Total variation}}$$
$$= \frac{\text{Variation explained by regression}}{\text{Total variation}} \tag{11.11}$$

Clearly, the coefficient of determination is a reasonable measure of the closeness of fit of the regression line, since it equals *the proportion of the total variation in the dependent variable that is explained by the regression line.*

In practical work, a more convenient formula for the coefficient of determination is

Alternative Formula for r²

$$r^2 = \frac{\left[n \sum\limits_{i=1}^{n} X_i Y_i - \left(\sum\limits_{i=1}^{n} X_i \right) \left(\sum\limits_{i=1}^{n} Y_i \right) \right]^2}{\left[n \sum\limits_{i=1}^{n} X_i^2 - \left(\sum\limits_{i=1}^{n} X_i \right)^2 \right] \left[n \sum\limits_{i=1}^{n} Y_i^2 - \left(\sum\limits_{i=1}^{n} Y_i \right)^2 \right]}. \tag{11.12a}$$

To illustrate the computation of the coefficient of determination, Table 11.2 shows the various quantities needed in equation (11.12a) in the case of the university secretaries. Substituting these quantities into equation (11.12a), we have

$$r^2 = \frac{[9(319) - 50(49)]^2}{[9(340) - 50^2][9(303) - 49^2]} = \frac{421^2}{560(326)}$$

$$= 0.97.$$

Thus, the coefficient of determination between performance rating and test score for the secretaries is 0.97. In other words, the regression line in Figure 11.5 can explain about 97 percent of the variation in performance ratings.

Still another formula that is sometimes even more convenient is

$$r^2 = \frac{a \sum_{i=1}^{n} Y_i + b \sum_{i=1}^{n} X_i Y_i - \frac{1}{n} \left(\sum_{i=1}^{n} Y_i \right)^2}{\sum_{i=1}^{n} Y_i^2 - \frac{1}{n} \left(\sum_{i=1}^{n} Y_i \right)^2}. \tag{11.12b}$$

The advantage of this formula over equation (11.12a) is that if one has already calculated the regression line, the values of a and b are already available. If this formula is used in the case of the university secretaries, the result is

$$r^2 = \frac{1.266(49) + .752(319) - 266.778}{36.222} = \frac{62.034 + 239.888 - 266.778}{36.222}$$

$$= .97,$$

which, of course, is the answer we obtained in the previous paragraph.

11.11 The Correlation Coefficient

As pointed out at the beginning of this chapter, the purpose of correlation analysis is to measure the strength of the relationship between two variables, X and Y. The assumptions (or model) underlying correlation analysis are as follows: First, both X and Y are assumed to be normally distributed random variables. This is different from regression analysis where Y is assumed to be a random variable but X is not.

Second, the standard deviation of the Ys is assumed to be constant for all values of X, and the standard deviation of the Xs is assumed to be constant for all values of Y.

The correlation coefficient is commonly used as a measure of the strength of the relationship between two variables. The **correlation coefficient** r is simply the square root of the coefficient of determination. That is,

$$r = \sqrt{r^2}.$$

The sign of r must equal the sign of the slope of the regression line. Thus, the positive square root of r^2 is taken if $b > 0$, and the negative square root is taken if $b < 0$. For ease of computation, the following formula is often used:[6]

Formula for r

$$r = \frac{n \sum_{i=1}^{n} X_i Y_i - \sum_{i=1}^{n} X_i \sum_{i=1}^{n} Y_i}{\sqrt{n \sum_{i=1}^{n} X_i^2 - \left(\sum_{i=1}^{n} X_i \right)^2} \sqrt{n \sum_{i=1}^{n} Y_i^2 - \left(\sum_{i=1}^{n} Y_i \right)^2}} \tag{11.13}$$

The correlation coefficient cannot be greater than 1 or less than -1. *If* $r = 1$, *there is a perfect linear relationship between the independent and dependent variables, and the relationship is direct.* In other words, the situation is like that shown in panel A of Figure 11.11. On the other hand, *if* $r = -1$, *there is a perfect linear relationship between the independent and dependent variables, and the relationship is inverse.* The situation is like that shown in panel B of Figure 11.11. In each case the relationship is perfect in the sense that the regression explains all the variation in the dependent variable (since all the points fall on the regression line). Why does the correlation coefficient equal either $+1$ or -1 if all the

---

[6] The correlation coefficient in the text is the correlation coefficient *unadjusted* for degrees of freedom, and is a biased estimate of the population correlation coefficient. (It is biased away from zero.) An unbiased estimate is the *adjusted* correlation coefficient, which is

$$r = \sqrt{1 - \frac{\sum_{i=1}^{n} (Y_i - \hat{Y}_i)^2 \div (n - 2)}{\sum_{i=1}^{n} (Y_i - \bar{Y})^2 \div (n - 1)}},$$

where n is the number of observations. (*Adjusted* here means adjusted for degrees of freedom.)

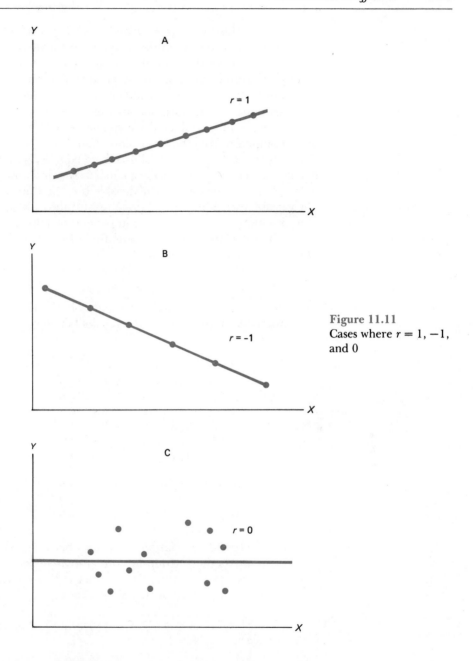

Figure 11.11
Cases where $r = 1, -1$, and 0

variation in Y is explained by the regression line? Because under these circumstances, the actual value of Y must always be equal to the value computed from the regression, which means that $\Sigma(Y_i - \hat{Y}_i)^2 = 0$. In other words, none of the variation in Y is unexplained by the regression. Hence, it follows from equation (11.11) that r^2 must equal 1, which means that r must equal $+1$ or -1.

If r = 0, *there is zero correlation between the independent and dependent variables.* In this case, the least-squares estimate of B will turn out to be zero, indicating that, on the average, changes in the independent variable have no effect on the dependent variable. Under such circumstances, the correlation coefficient is zero because the regression explains none of the variation in the dependent variable. In other words, $\Sigma(\hat{Y}_i - \overline{Y})^2 = 0$ since the mean value of Y is always equal to the value computed from the regression. Also, $\Sigma(Y_i - \hat{Y}_i)^2 = \Sigma(Y_i - \overline{Y})^2$ since $\hat{Y}_i$ always equals $\overline{Y}$. Hence, it follows from equation (11.10) that r^2 must equal 0, which means that r must equal 0. In a situation of this sort, the best estimate of the dependent variable is $\overline{Y}$; the value of the independent variable provides no additional useful information on this score. The situation is like that shown in panel c of Figure 11.11.

The calculation and interpretation of the sample correlation coefficient are illustrated below.

EXAMPLE 11.5 An industrial psychologist obtains the IQ score and productivity of 10 workers, the results being

IQ score	Productivity (output per hour)
110	5.2
120	6.0
130	6.3
126	5.7
122	4.8
121	4.2
103	3.0
98	2.9
80	2.7
97	3.2

Compute the correlation coefficient between IQ score and productivity. Is the relationship direct or inverse?

SOLUTION: Letting the productivity of the *i*th worker equal Y_i and the IQ score of the *i*th worker equal X_i,

$$\sum_{i=1}^{10} X_i Y_i = 5042.6 \quad \sum_{i=1}^{10} X_i = 1107 \quad \sum_{i=1}^{10} Y_i = 44.0$$

$$\sum_{i=1}^{10} Y_i^2 = 210.84 \quad \sum_{i=1}^{10} X_i^2 = 124{,}823 \quad n = 10.$$

Thus, from equation (11.13), it follows that

$$r = \frac{10(5{,}042.6) - (1107)(44)}{\sqrt{10(124{,}823) - (1107)^2}\ \sqrt{10(210.84) - (44)^2}}$$

$$= \frac{50{,}426 - 48{,}708}{\sqrt{1{,}248{,}230 - 1{,}225{,}449} \; \sqrt{2108.4 - 1936}} = \frac{1718}{\sqrt{22{,}781} \; \sqrt{172.4}}$$

$$= \frac{1718}{(150.93)(13.13)} = \frac{1718}{1982} = .867.$$

The value of the correlation coefficient is about .87. Since this value is positive, the relationship seems to be direct.

EXAMPLE 11.6 Walter Mischel of Columbia University has studied the correlation between how long a preschool child is willing to delay gratification and his or her characteristics as an adolescent.[7] For each of 77 four-year-old children, he measured the number of minutes they were willing to wait by themselves for a preferred but delayed gratification. (Specifically, they could have two marshmallows when they waited until the experimenter returned of her own accord or one marshmallow immediately whenever the child decided to end the delay.) When each of the children was about 16 years old, Mischel obtained measurements of how attentive (and able to concentrate) each one was. The correlation coefficient between the number of minutes a person was willing to delay gratification at age 4 and how attentive he or she was at age 16 equaled 0.49. Is the relationship between these two variables direct or inverse? What percent of the variation in attentiveness can be explained by how long a person was willing to delay gratification?

SOLUTION: Since the correlation coefficient is positive, the relation seems to be direct. The proportion of variation in attentiveness that can be explained in this way equals $r^2 = 0.49^2$ or .24. Thus, it appears that 24 percent of this variation can be explained in this way.

[7] W. Mischel, "Convergences and Challenges in the Search for Consistency," *American Psychologist*, April 1984.

11.12 Inference Concerning the Population Correlation Coefficient[8]

The correlation coefficient described in the previous section is a *sample* correlation coefficient, and it varies from one sample to another. In

[8] Some instructors may prefer to take up Chapter 11 before Chapter 8. Chapter 11 has been written so this can be done; but Sections 11.12 and 11.13 should be taken up after Chapter 8 has been covered.

contrast, the ***population correlation coefficient,*** ρ (rho), pertains to the entire population. The definition of ρ is the same as that of r, the only difference being that r is based on the sample data whereas ρ is based on all the data in the population. In most cases, statisticians are much more interested in the correlation coefficient in the population than in the sample, and they use r as an estimate of ρ.

Frequently, statisticians are interested in testing whether the population correlation coefficient is zero. *If one variable is independent of another variable, the population correlation coefficient equals zero.* For example, if the number of days a patient with acute upper respiratory infection stays in the hospital is completely unrelated to his or her age, the correlation coefficient in the population between a patient's length of hospital stay and age will be zero. A hospital administrator may want to determine whether length of hospital stay and age really are uncorrelated for such patients. Having measured the length of hospital stay and age for such patients, the administrator may want to test the null hypothesis that $\rho = 0$.

Test that $\rho = 0$

To carry out such a test, the first step is to specify the null hypothesis and the alternative hypothesis. The null hypothesis is that $\rho = 0$, and the alternative hypothesis is that $\rho \neq 0$. To test the null hypothesis, we compute

$$t = \frac{r}{\sqrt{(1 - r^2)/(n - 2)}}, \qquad (11.14)$$

which has the t distribution with $(n - 2)$ degrees of freedom if the null hypothesis is true. Thus, the decision rule used to test this null hypothesis is: *Reject the null hypothesis that $\rho = 0$ if* t *is greater than* $t_{\alpha/2}$ *or less than* $-t_{\alpha/2}$; *accept the null hypothesis otherwise.*

To illustrate the way in which this test is carried out, suppose that in a sample of 18 patients, the sample correlation coefficient between length of hospital stay and age is 0.42. To test the null hypothesis that the population correlation coefficient equals zero, we compute

$$t = \frac{0.42}{\sqrt{(1 - .42^2)/16}} = \frac{0.42}{\sqrt{.2059}} = 0.93.$$

If the significance level of the test is set at .05, the null hypothesis should be rejected if $t > 2.12$ or if $t < -2.12$, since Appendix Table 6 shows that $r_{.025} = 2.12$ if there are 16 degrees of freedom. Because the observed value of t does not exceed 2.12 or fall below -2.12, the null hypothesis should not be rejected.

EXAMPLE 11.7 Pamela Jackson of Rhode Island College and Leo Carroll of the University of Rhode Island carried out a study of the determinants of municipal police expenditures in 90 nonsouthern cities.[9] Among other things, they found that the coefficient of correlation between a city's 1971 police expenditures (per capita) and its 1968–70 crime rate was 0.54. Test whether the population correlation coefficient equals zero. (Let $\alpha = .05$.)

SOLUTION: Since $r = 0.54$ and $n = 90$,

$$t = \frac{0.54}{\sqrt{(1 - .54^2)/88}} = \frac{0.54}{\sqrt{.00805}} = 6.02.$$

Appendix Table 6 shows that $t_{.025}$ is about 1.99, since there are 88 degrees of freedom. Because the observed value of t exceeds 1.99, the null hypothesis that $\rho = 0$ should be rejected.

---

[9] P. Jackson and L. Carroll, "Race and the War on Crime: The Sociopolitical Determinants of Municipal Police Expenditures in 90 Non-Southern U.S. Cities," *American Sociological Review*, June 1981.

ADMISSIONS TESTING ON TRIAL

GETTING DOWN TO CASES

In recent years, there has been a great deal of controversy over the role of tests like the Law School Admissions Test (LSAT) in universities' admission processes. According to critics like A. Nairn (and Ralph Nader), LSAT scores are of little or no use in predicting the relative performance of students. To test such claims, hundreds of studies have been carried out. Among schools where the standard deviation of the LSAT scores is about 100, the correlation coefficient between a student's LSAT score and his or her first-year grades in law school have been found to be about 0.51. In seven states, Carlson and Werts found that the coefficient of correlation between a student's grade on the essay portions of the bar examination and his or her LSAT score are: .37, .36, .35, .37, .31, .43, .20.[10]

(a) What proportion of the variation among students in first-year law grades can be explained by LSAT scores?

(b) About what proportion of the variation among students in their grades on the essay portions of the bar examination can be explained by LSAT scores?

[10] R. Lynn, "Admission Testing on Trial," *American Psychologist*, March 1982.

(c) According to Carlson and Werts, the coefficient of correlation between a student's cumulative law school grade point average and his or her grade on the essay portions of the bar examination is about 0.48. Is the grade on the essay portions of the bar examination more closely related to a student's cumulative law school grade point average or to a student's LSAT score?

(d) Based on the fact that the correlation coefficient between a student's SAT score and his or her grade rank is .345, Nairn concludes that "for 88% of the applicants (though it is impossible to predict which ones) an SAT score will predict their grade rank no more accurately than a pair of dice."[11] Can you provide a more accurate interpretation?

[11] A. Nairn and Associates, *The Reign of ETS: The Corporation That Makes Up Minds* (Washington, D.C.: Ralph Nader, 1980).

11.13 Inference Concerning the Value of *B*

In regression analysis, the slope of the sample regression line *b* varies from one sample to another. Like any sample statistic, it has a sampling distribution; and an estimate of the standard deviation of this sampling distribution is

$$s_b = s_e \div \sqrt{\sum_{i=1}^{n} X_i^2 - n\overline{X}^2},$$

which is often called the *standard error of* b. There are many occasions when the statistician wants to use the observed value of *b* to calculate a confidence interval for *B*, the slope of the population regression line. Such a confidence interval is

Confidence Interval for **B**

$$b \pm t_{\alpha/2} s_b.$$

(11.15)

If the conditional probability distribution of the dependent variable is normal, the probability is $(1 - \alpha)$ *that these limits will include the true value of* B. To illustrate the use of this formula, we can calculate the 95 percent confidence interval for *B* in the case of the university secretaries. Since $b = .752$, $t_{.025} = 2.365$, $s_e = .392$, and $\Sigma X_i^2 - n\overline{X}^2 = 62.2$, it follows that the confidence interval is

$$.752 \pm 2.365(.392) \div \sqrt{62.2},$$

or

.752 ± .118.

Thus, the 95 percent confidence interval for the increase in perform-ance rating associated with a 1-point increase in test score is .634 to .870.

In addition to estimating the value of *B*, the statistician frequently wants to test the hypothesis that *B* equals zero. If this hypothesis is true, the mean of the dependent variable is the same, regardless of the value of the independent variable. (Specifically, the mean equals *A*, since *BX* = 0.) Thus, a knowledge of the independent variable is of no use in predicting the dependent variable since the conditional probability distribution of the dependent variable is not influenced by the value of the independent variable. In other words, if this hypothesis is true (and if the assumptions given in Section 11.4 hold), there is *no relationship* between the dependent and the independent variable.

If *B* equals zero, it does not follow that *b* must equal zero. On the contrary, it is quite likely that *b* will be nonzero because of random fluc-tuations. The decision rules for testing the null hypothesis that $B = 0$ are given below. These rules assume that the conditional probability distribution of the dependent variable is normal, and which rule is ap-propriate depends on the nature of the alternative hypothesis.

Test that **B = 0**

Alternative Hypothesis: *B > 0. Reject the null hypothesis if* $b \div s_b >$ t_α; *accept the null hypothesis if* $b \div s_b \leq t_\alpha$. *(The number of degrees of freedom is* $n - 2$, *and* α *is the significance level.)*

Alternative Hypothesis: *B < 0. Reject the null hypothesis if* $b \div s_b <$ $-t_\alpha$; *accept the null hypothesis if* $b \div s_b \geq -t_\alpha$.

Alternative Hypothesis: *B ≠ 0. Reject the null hypothesis if* $b \div s_b >$ $t_{\alpha/2}$ *or* $< -t_{\alpha/2}$; *accept the null hypothesis if* $-t_{\alpha/2} \leq b \div s_b \leq t_{\alpha/2}$.

As an illustration of how this test is carried out, suppose that the university administrator wants to test whether the slope of the popula-tion regression line relating a secretary's performance rating to test score equals zero (against the alternative hypothesis that $B \neq 0$.) The significance level is .05. From previous sections, we know that in this case $b = 0.752$, $s_e = 0.392$, $\Sigma X_i^2 - n\overline{X}^2 = 62.2$, and $n = 9$. Thus,

$$b \div s_b = \frac{.752}{.392 \div \sqrt{62.2}} = 15.1.$$

Since $n = 9$, there are 7 degrees of freedom, and $t_{.025} = 2.365$. Since the value of the test statistic (15.1) exceeds 2.365, the null hypothesis should be rejected.

Below is a further illustration of how one can test the hypothesis that $B = 0$.

EXAMPLE 11.8 Based on the data in Example 11.1, test the hypothesis that the slope of the true regression line relating savings to income equals zero. The alternative hypothesis is that this slope exceeds zero. Use the .05 significance level.

SOLUTION: From the results in Examples 11.1 and 11.2, we know that $b = .1017$, $s_e = .053$, $\Sigma X_i^2 - n\overline{X}^2 = 60$, and $n = 9$. Thus,

$$b \div s_b = \frac{.1017}{.053 \div \sqrt{60}} = 15.$$

Since $t_{.05} = 1.895$ when there are 7 degrees of freedom, the null hypothesis should be rejected. That is, the evidence seems to indicate that the slope of the true regression is positive, not zero.

11.14 Hazards and Problems in Regression and Correlation

There are a number of pitfalls in regression and correlation analyses that should be emphasized. First, *it is by no means true that a high coefficient of determination (or correlation coefficient) between two variables means that one variable causes the other variable to vary.* For example, if one regresses the size of a person's left foot on the size of his or her right foot, this regression is bound to fit very well, since the size of a person's left foot is closely correlated with the size of his or her right foot. But this does not mean that the size of a person's right foot *causes* a person's left foot to be as large or small as it is. Two variables can be highly correlated without causation being implied.

Second, *even if an observed correlation is due to a causal relationship, the direction of causation may be the reverse of that implied by the regression.* For example, suppose that we regress a firm's profits on its R and D (research and development) expenditure, the firm's profits being the dependent variable and its R and D expenditures being the independent variable. If the correlation between these two variables turns out to be high, does this imply that high R and D expenditures produce high profits? Obviously not. The line of causation could run the other way: High profits could result in high R and D expenditures.

Thus, in interpreting the results of regression and correlation studies, it is important to ask oneself whether the line of causation assumed in the studies is correct.

Third, *regressions are sometimes used to forecast values of the dependent variable corresponding to values of the independent variable lying beyond the sample range.* For example, in Figure 11.12, the scatter diagram shows that the data for the independent variable range from about 1 to 7. But the regression may be used to forecast the dependent variable when the independent variable assumes a value of 9, which is outside the sample range. This procedure, known as **extrapolation,** is dangerous because the available data provide no evidence that the true regression is linear beyond the range of the sample data. For example, the true regression may be as shown in Figure 11.12, in which case a forecast based on the estimated regression may be very poor.

Hazards of Extrapolation

Fourth, *it is important to recognize that a regression based on past data may not be a good predictor, due to shifts in the regression line.* For example, suppose that the university changes the standards on which its ratings of secretaries' performance are based. If the standards are lowered, the regression line relating performance rating to test score is likely to shift upward and to the left; and predictions based on historical data are likely to underestimate future performance ratings.

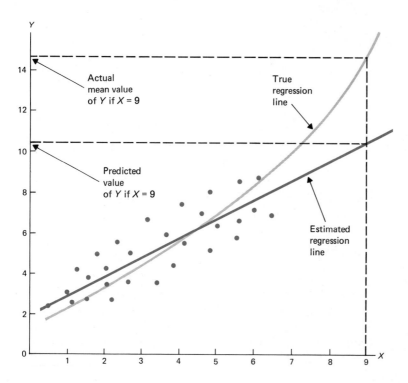

Figure 11.12
Dangers of Extrapolation

NOTE: If the curved line is the true (that is, the population) regression line, and if the estimated regression line is extrapolated to forecast the value of Y when $X = 9$, the result will be very inaccurate. Whereas the forecast is that Y will equal about 10.5, the mean value of Y(if $X = 9$) is really about 14.5

Finally, *when carrying out a regression, it is important to try to make sure that the assumptions in Section 11.4 are met.* Statisticians often plot the regression line on the scatter diagram and look for evidence of departures from the assumptions. For example, the scatter diagram may indicate that the relationship between the variables is curvilinear, not linear. (See panel D of Figure 11.2.) If this is so, multiple regression (described in the next chapter) may be more appropriate than the simple linear regression techniques discussed here.

MISTAKES TO AVOID

THE CASE OF COPPER AND LEAD

An analyst at the Department of State is interested in whether a country's copper output is related to its lead output. He is particularly interested in three countries, *A, B,* and *C.* To normalize for differences among the sizes of the countries, he uses copper output *per capita* as one variable and lead output *per capita* as the other. The scatter diagram, shown below, indicates that there is a strong positive correlation between these variables.

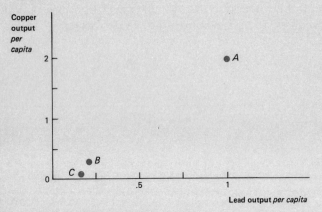

He shows this scatter diagram to his boss, who says that it is extremely misleading. What was his mistake?

SOLUTION

A spurious correlation can be created by dividing both the independent and dependent variables by the same quantity. In this case, suppose that the populations of countries *A, B,* and *C* are 1 million, 10 million, and 20 million, respectively. Then this procedure (of using copper output per capita and lead output per capita) results in a high correlation, as shown above. But in fact there is no relationship whatever between a country's copper output and its lead output, as shown in the graph below. The moral is that one should be very careful to avoid creating a spurious correlation by dividing both the independent and dependent variables by the same quantity.

In addition, one must be very careful about drawing conclusions from such a small sample. Even if the sample correlation coefficient is quite high, it may be due to chance. (For a relevant test procedure, see Section 11.12.)

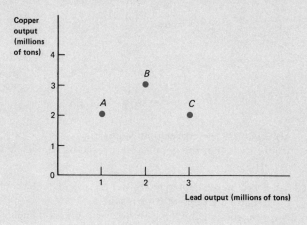

EXERCISES

11.10 Data are collected by the Department of Transportation concerning the relationship between a car's speed and the distance it travels before stopping (after being braked). Would you expect the correlation coefficient to be positive or negative? Is there a causal relationship in this case?

11.11 According to actual published data, the relationship between detergency (Y) and concentration (X) was as follows for a sample of eight detergents:

Y	X
37	10
42	20
46	30
48	40
53	10
62	10
79	30
84	40

(a) Calculate the correlation coefficient.
(b) Test whether the population correlation coefficient is zero. (Let $\alpha =$ .05.)

11.12 For a random sample of ten students, the relationship between their college grade-point average (Y) and their high school grade-point average (X) is as follows:

Y	X
2.5	3.1
2.0	2.7
3.1	3.6
3.4	3.7
3.9	4.0
2.1	3.0
2.8	3.3
2.9	3.5
3.0	3.6
3.8	4.0

(a) Calculate the correlation coefficient.
(b) Test whether the population correlation coefficient is zero. (Let $\alpha =$ .01.)

11.13 (a) Based on the data in Exercise 11.1, calculate the coefficient of determination. Interpret your result.
(b) Test whether the slope of the true regression line in Exercise 11.1 is zero. (Let $\alpha = .05$, and use a two-tailed test.)

11.14 (a) Based on the data in Exercise 11.2, calculate the coefficient of determination. Interpret your result.
(b) Test whether the slope of the true regression line in Exercise 11.2 is zero. (Let $\alpha = .05$, and use a one-tailed test where the alternative hypothesis is that the slope is positive.)

11.15 (a) Based on the data in Exercise 11.4, calculate the standard error of the slope of the sample regression line.
(b) Calculate a 90 percent confidence interval for the slope of the true regression line in Exercise 11.4.

11.16 (a) Based on the data in Exercise 11.5, calculate the standard error of the slope of the sample regression line.
(b) Calculate a 95 percent confidence interval for the slope of the true regression line in Exercise 11.5.

11.17 (a) Using the data in Exercise 11.9, calculate the sample correlation coefficient between fiber length and yarn strength.
(b) What proportion of the variation in yarn strength in the sample can be explained by fiber length?

11.18 Based on the data in Example 11.5, test whether the population correlation coefficient between IQ score and productivity is zero. (Let $\alpha = .05$, and use a two-tailed test.)

11.19 On the basis of the data in Table 11.1 and the regression equation derived therefrom, a university administrator estimates that if the aptitude test score equals 10, the performance rating should equal 8.786. Comment on this estimate.

11.20 If a regression line cannot explain 36 percent of the variation in the dependent variable, what is the correlation coefficient?

11.21 If the standard deviation of the conditional probability distribution of the dependent variable varies with the value of the independent variable, *heteroscedasticity* is said to occur.
(a) Can you think of any reasons why heteroscedasticity might occur in the case of the relationship between savings and income in Example 11.1?
(b) What factors might cause heteroscedasticity in the case of the relationship between a firm's costs and its level of output?

11.22 If two variables X and Y are statistically independent, their correlation coefficient is zero. But is it true that if the population correlation coefficient between X and Y is zero they must be statistically independent? Explain.

The Chapter in a Nutshell

1. *Regression analysis* indicates how one variable is related to another. A first step in describing such a relationship is to plot a scatter diagram. The *regression line* shows the average relationship between the dependent variable and the independent variable. The method of least squares is the standard technique used to fit a regression line to a set of data. If the regression line is $\hat{Y} = a + bX$, and if a and b are calculated by least squares,

$$b = \frac{\sum_{i=1}^{n} (X_i - \overline{X})(Y_i - \overline{Y})}{\sum_{i=1}^{n} (X_i - \overline{X})^2}$$

$$a = \overline{Y} - b\overline{X}.$$

This value of b is often called the *estimated regression coefficient*.

2. The regression line calculated by the method of least squares is generally based on a sample, not on the entire population. Suppose that the means of the conditional probability distributions of the dependent variable fall on a straight line: $\mu_{Y \cdot X} = A + BX$. Also, suppose

that the standard deviation of these conditional probability distributions is the same for all values of X, that the observations in the sample are statistically independent, and that the values of X are known with certainty. Then a is an unbiased and consistent estimator of A, and is the most efficient estimator of A (among those unbiased estimators that are linear functions of the dependent variable). Similarly, b is an unbiased and consistent estimator of B, and is the most efficient estimator of B (among those unbiased estimators that are linear functions of the dependent variable).

3. The *standard deviation of the conditional probability distribution of the dependent variable is* σ_e. It is a measure of the amount of scatter about the regression line in the population. The sample statistic which is used to estimate σ_e is the standard error of estimate, defined as

$$s_e = \sqrt{\frac{\sum\limits_{i=1}^{n} (Y_i - \hat{Y}_i)^2}{n-2}},$$

where $\hat{Y}_i$ is the estimate of Y_i based on the regression line.

4. If we can assume that the conditional probability distribution of the dependent variable is normal, we can calculate a confidence interval for the conditional mean of the dependent variable when the independent variable equals X^*. This confidence interval is

$$(a + bX^*) \pm t_{\alpha/2} s_e \sqrt{\frac{1}{n} + \frac{(X^* - \overline{X})^2}{\sum\limits_{i=1}^{n} X_i^2 - n\overline{X}^2}}$$

if the confidence coefficient is $(1 - \alpha)$. Also, we can calculate a confidence interval for the value of the dependent variable that will occur if the independent variable is set at X^*. This confidence interval is

$$(a + bX^*) \pm t_{\alpha/2} s_e \sqrt{\frac{n+1}{n} + \frac{(X^* - \overline{X})^2}{\sum\limits_{i=1}^{n} X_i^2 - n\overline{X}^2}}$$

if the confidence coefficient is $(1 - \alpha)$. The number of degrees of freedom is $n - 2$.

5. Assuming that the conditional probability distribution of the dependent variable is normal, a confidence interval for the slope B of the population regression line is

$$b \pm t_{\alpha/2} s_b,$$

where

$$s_b = s_e \div \sqrt{\sum_{i=1}^{n} X_i^2 - n\overline{X}^2}$$

and where $(1 - \alpha)$ is the confidence coefficient. To test the null hypothesis that $B = 0$ (against the alternative hypothesis that $B \neq 0$), reject the null hypothesis if $b \div s_b > t_{\alpha/2}$ or $< -t_{\alpha/2}$; accept the null hypothesis if $-t_{\alpha/2} \leq b \div s_b \leq t_{\alpha/2}$. The number of degrees of freedom is $(n - 2)$, and α is the significance level.

6. To measure the closeness of fit of a regression line, statisticians often use the *coefficient of determination,* defined as

$$r^2 = \frac{\left[n \sum_{i=1}^{n} X_i Y_i - \left(\sum_{i=1}^{n} X_i \right)\left(\sum_{i=1}^{n} Y_i \right) \right]^2}{\left[n \sum_{i=1}^{n} X_i^2 - \left(\sum_{i=1}^{n} X_i \right)^2 \right]\left[n \sum_{i=1}^{n} Y_i^2 - \left(\sum_{i=1}^{n} Y_i \right)^2 \right]}.$$

The *coefficient of determination* equals the proportion of the total variation in the dependent variable that is explained by the regression line.

7. *Correlation analysis* is concerned with measuring the strength of the relationship between two variables. The correlation coefficient, which is the square root of the coefficient of determination, is often used for this purpose. If $r = 1$, there is a perfect linear relationship between the two variables, and the relationship is direct. If $r = -1$, there is a perfect linear relationship between the two variables, and the relationship is inverse. If two variables are statistically independent, $r = 0$. To test whether the population correlation coefficient ρ equals zero, one can calculate

$$t = \frac{r}{\sqrt{\dfrac{(1 - r^2)}{(n - 2)}}}$$

The null hypothesis that $\rho = 0$ should be rejected if this test statistic exceeds $t_{\alpha/2}$ or is less than $-t_{\alpha/2}$, where α is the significance level and the number of degrees of freedom equals $n - 2$. The alternative hypothesis is that $\rho \neq 0$.

8. There are many pitfalls in regression and correlation analysis. A high correlation between two variables does not necessarily mean that the variables are causally related. And even if they are causally related, the direction of causation may be different from that presumed in the analysis. It is extremely dangerous to extrapolate a regression line beyond the range of the data. One should be careful to avoid causing spurious correlation by dividing both the independent and the de-

pendent variable by the same quantity. A regression based on past data may not be a good predictor, due to shifts in the regression line. It is important to try to determine whether the assumptions underlying regression analysis are met.

Chapter Review Exercises

11.23 An urban sociologist obtains data for seven communities concerning the percent of people over 25 years old who are current users of marijuana and the percent who are current users of heroin. The results are as follows:

Percent using marijuana	Percent using heroin
5.7	0.27
6.7	0.12
0.2	0.00
0.6	0.04
3.8	0.05
12.5	0.46
0.5	0.00

(a) Construct a scatter diagram of these data.
(b) Compute the sample regression line, where the percent using heroin is the dependent variable and the percent using marijuana is the independent variable.
(c) Predict the percent using heroin in a community where the percent using marijuana is 2 percent.

11.24 *Business Week* has presented the following data concerning the research and development (R and D) expenditures and sales of six telecommunications firms:

Firm	Sales	R and D
	(millions of dollars)	
A T & T	50,790	419
Comsat	300	12
GTE	9,980	162
Rolm	201	13
United	1,904	3
Western Union	794	5

(a) Construct a scatter diagram of these data.
(b) Compute the sample regression line, where R and D expenditure is the dependent variable and sales is the independent variable.

(c) Predict the average R and D expenditure if sales equal $10 billion.

(d) Do you think that this regression can be applied to other industries? Explain.

11.25 For the following 10 countries, the numbers of telephones and television receivers are as follows:

Country	Telephones (millions)	Televisions (millions)
Argentina	2.6	5.1
Belgium	3.6	3.5
Canada	16.5	11.3
France	24.7	19.0
Germany	28.6	20.8
Italy	19.3	22.0
Japan	53.6	63.0
Spain	11.9	9.4
Turkey	1.9	3.3
United Kingdom	26.7	22.6

Source: Statistical Abstract of the U.S., 1982–83.

Use Minitab (or some other computer package) to construct the scatter diagram. If no computer package is available, do it by hand.

11.26 A firm examines a random sample of 10 spot welds of steel. In each case, the shear strength of the weld and the diameter of the weld are determined, the results being as follows:[12]

Shear strength (pounds)	Weld diameter (thousandths of a inch)
680	190
800	200
780	209
885	215
975	215
1,025	215
1,100	230
1,030	250
1,175	265
1,300	250

(a) Construct a scatter diagram of these data.

(b) Based on this scatter diagram, does the relationship between these two variables seem to be direct or inverse? Does this accord with

[12] A. Duncan, *Quality Control and Industrial Statistics*, p. 651.

common sense? Why, or why not? Does the relationship seem to be linear?

(c) Assume that the conditional mean value of a weld's shear strength is a linear function of its diameter. Calculate the least-squares estimates of the parameters (A and B) of this linear function.

(d) Plot the regression line. Use this regression line to predict the average shear strength of a weld $1/5$ inch in diameter. Use the regression line to predict the average shear strength of a weld $1/4$ inch in diameter.

(e) Compute the standard error of estimate. What does this number mean?

(f) Compute the 95 percent confidence interval for the mean shear strength among welds with a diameter of $1/4$ inch.

(g) Compute the 95 percent confidence interval for the shear strength of a weld if its diameter is $1/4$ inch. What assumptions underlie the calculation of this confidence interval?

11.27 (a) Using the data in Exercise 11.26, calculate the sample correlation coefficient between weld strength and weld diameter.

(b) What proportion of the variation in weld strength in the sample can be explained by weld diameter?

11.28 If the sample correlation coefficient is .30 and n equals 20, test whether the population coefficient is zero. (Let $\alpha = .05$, and use a two-tailed test.)

11.29 Using the data in Exercise 11.9, test the hypothesis that the slope of the true regression line relating yarn strength (the dependent variable) and fiber length (the independent variable) equals zero. The alternative hypothesis is that the slope exceeds zero. (Use the .01 significance level.) Can we be sure that variation in fiber length *causes* variation in yarn strength? Why, or why not?

11.30 Using the data in Exercise 11.26, test the hypothesis that the slope of the true regression line relating a weld's strength (the dependent variable) to its diameter (the independent variable) equals zero. The alternative hypothesis is that the slope is greater or less than zero. (Use the .05 significance level.) Can we be sure that variation in a weld's shear strength does *not* cause variation in its diameter? Why, or why not?

*11.31 Based on the data in Exercise 11.25, Minitab is used to calculate the regression of the number of televisions in a country (in millions) on the number of telephones in the country (in millions). Interpret the results, shown on p. 501. (The dependent variable has been put in column 6; the independent variable has been put in column 5. See the answer to Exercise 11.25.)

* Exercise requiring some familiarity with Minitab.

```
REGRESS C6 ON 1 PREDICTOR IN COLUMN C5

THE REGRESSION EQUATION IS
Y = -2.24 + 1.07 X1

                                         ST. DEV.      T-RATIO =
           COLUMN     COEFFICIENT        OF COEF.      COEF/S.D.
             -          -2.243            2.724          -0.82
X1         C5           1.0688            0.1129          9.47

THE ST. DEV. OF Y ABOUT REGRESSION LINE IS
S = 5.337
WITH (10 - 2) = 8 DEGREES OF FREEDOM

R-SQUARED = 91.8 PERCENT
R-SQUARED = 90.8 PERCENT, ADJUSTED FOR D.F.
```

Appendix 11.1

RANK CORRELATION

In Chapter 9 we pointed out that nonparametric tests are often available as analogues to parametric tests. An important nonparametric technique used for measuring the degree of correlation between two variables Y and X is *rank correlation*. This technique is particularly useful when one or more of the variables in question cannot be measured in ordinary ways, but can only be ranked. For example, a psychologist may want to determine how closely a person's persistence is associated with his or her IQ score. Since there is no objective measure of persistence, the psychologist asks a professor to rank 12 students on the basis of his perception of their persistence. Each student's persistence rank and IQ score are shown in Table 11.3.

*The **rank correlation coefficient** measures the closeness of the relationship between the two sets of rankings—that is, between the rankings of the one variable and the rankings of the other variable.* For example, column (3) in Table 11.3 shows each person's rank with regard to IQ score and column (1) shows the corresponding rank with regard to persistence. Are these ranks correlated? In other words, do people who rank highly with regard to IQ score also tend to rank highly with regard to persistence? If so, is this relationship close? To help answer such questions, statisticians have devised the rank correlation coefficient, which is defined as follows:

$$r_r = 1 - \frac{6 \sum_{i=1}^{n} d_i^2}{n(n^2 - 1)},$$ *(11.16)*

Table 11.3
Results of Psychologist's
Tests

Person	(1) Persistence Rank	(2) IQ score	(3) IQ rank	(4) $d_i = (3) - (1)$	(5) d_i^2
A	1	120	8	7	49
B	9	130	3	−6	36
C	4	132	2	−2	4
D	3	127	4	1	1
E	8	126	5	−3	9
F	2	110	11	9	81
G	11	98	12	1	1
H	10	113	10	0	0
I	6	121	7	1	1
J	5	123	6	1	1
K	7	119	9	2	4
L	12	140	1	−11	121
Σd_i^2					308

$$r_r = 1 - \frac{6(308)}{12(144 - 1)} = 1 - \frac{1848}{1716} = -.077$$

where d_i is the difference between the two ranks of the ith observation, and n is the number of observations. As shown in Table 11.3, the rank correlation coefficient in the case of the psychologist's study equals −.077.

The value of the rank correlation coefficient can vary from +1 to −1. A positive value of the rank correlation coefficient suggests that the two variables are directly related; a negative value suggests that they are inversely related. If the rank correlation coefficient is +1, there is a perfect direct relationship between the two rankings. If the rank correlation coefficient is −1, there is a perfect inverse relationship between the two rankings. If the two variables are statistically independent, the population rank correlation coefficient is zero. (However, due to chance, the *sample* rank correlation coefficient may not be zero under these circumstances.)

To test whether the population rank correlation coefficient is zero, we compute the following test statistic:

$$t = \frac{r_r}{\sqrt{(1 - r_r^2)/(n - 2)}}.$$

If the null hypothesis (that the population rank correlation coefficient is zero) is true, this statistic has the t distribution wih $(n - 2)$ degrees of freedom. Thus, if the alternative hypothesis is that the population rank correlation coefficient is either greater or less than zero the decision rule is as follows.

Decision Rule: *Reject the null hypothesis that the population rank correlation coefficient is zero if* $t < -t_{\alpha/2}$ *or if* $t > t_{\alpha/2}$; *accept the null hypothesis if* $-t_{\alpha/2} \leq t \leq t_{\alpha/2}$. *(There are* $(n-2)$ *degrees of freedom, and* α *is the significance level.)*

To illustrate the application of this decision rule, let's return to the psychologist's results in Table 11.3. Since the sample rank correlation coefficient in this case equals $-.077$,

$$t = \frac{-.077}{\sqrt{.9941/10}} = \frac{-.077}{.315} = -.24.$$

If the significance level is set at .05, $t_{.025} = 2.228$ since there are 10 degrees of freedom. Since the observed value of t is not less than -2.228 or greater than 2.228, the psychologist should not reject the null hypothesis that persistence and IQ score are unrelated.

The rank correlation coefficient is often used instead of the ordinary correlation coefficient when the dependent variable is far from normally distributed. Wide departures from normality have no effect on the test described in the two previous paragraphs because this test does not assume normality. Also, the rank correlation coefficient is sometimes preferred to the ordinary correlation coefficient because the former is easier to compute. As you can see from Table 11.3, the computations are relatively simple.

12

OBJECTIVES

In most cases, analysts are interested in the way in which a dependent variable is related to more than one independent variable. The overall purpose of this chapter is to describe the nature and application of multiple regression and correlation techniques that are used when there is more than one independent variable. Among the specific objectives are:

1. To show how one can calculate and interpret A, B_1, and B_2, the intercept and slopes of a multiple regression with two independent variables.

2. To define the multiple coefficient of determination and indicate how it can be computed and used.

3. To discuss the role of the computer in calculating multiple regressions with any number of independent variables, and to indicate in detail how computer printouts should be interpreted and used.

Multiple Regression and Correlation

12.1 Introduction

In the previous chapter we discussed regression and correlation techniques in the case where there is only one independent variable. In practical applications of regression and correlation techniques, it frequently is necessary and desirable to include two or more independent variables. In this chapter we will extend our treatment of regression and correlation to the case in which there is more than one independent variable. In addition, we will indicate how tests can be carried out to check some of the assumptions underlying regression analysis, and we will describe how violations of these assumptions may affect the results.

12.2 Multiple Regression: Nature and Purposes

Whereas a *simple regression* includes only one independent variable, a *multiple regression* includes two or more independent variables. Basically, there are two important reasons why a multiple regression must often be used instead of a simple regression. First, *one frequently can predict the dependent variable more accurately if more than one independent variable is used.* In the case of the university secretaries (discussed in the previous chapter), the university administrator may feel that tests other than the aptitude test may be useful in predicting a secretary's performance rating. For example, a typing test is likely to be useful. Holding constant the score on the aptitude test, a secretary's expected performance rating may be a linear function of his or her score on the typing test. In other words, it may be reasonable to assume that

$$E(Y_i) = A + B_1 X_{1i} + B_2 X_{2i}, \qquad (12.1)$$

where Y_i is the performance rating for the ith secretary, X_{1i} is the aptitude test score for that secretary, and X_{2i} is the number of errors made by the ith secretary on the typing test. Of course, if increases in the number of errors are associated with lower performance ratings, B_2 is negative.

According to equation (12.1), the expected value of a secretary's performance rating is dependent on the secretary's aptitude test score and on the number of mistakes he or she makes on the typing test. The equation says that if the aptitude test score increases by 1 point, the expected value of the performance rating increases by B_1, and that if the number of mistakes on the typing test increases by one, the expected value of the performance rating increases by B_2. (Since B_2 is presumed to be negative, this means that the expected value of the performance rating will decrease if the number of mistakes on the typing test increases by one.) If it is true that a secretary's performance rating is associated both with his or her aptitude test score and his or her typing test score, then the university administrator can predict the performance rating more accurately by using a multiple regression—that is, an equation relating the dependent variable (performance rating) to both independent variables (aptitude test score and number of mistakes on typing test)—than by using the simple regression in the previous chapter. The latter relates the dependent variable (performance rating) to only one of the independent variables (aptitude test score).

A second reason for using multiple regression instead of simple regression is that *if the dependent variable depends on more than one independent variable, a simple regression of the dependent variable on a single independent variable may result in a biased estimate of the effect of this independent variable on the dependent variable.* For example, suppose that, in accord with equation (12.1), performance rating is related both to aptitude test score and typing score. Estimating the simple regression of performance rating on aptitude test score (as we did in the previous chapter) may result in a biased estimate of B_1, which measures the increase in performance rating associated with a 1-point increase in aptitude test score (when the typing score is held constant). To understand why such a bias may arise, suppose that the secretaries who score well on the aptitude test tend to have more formal schooling but less typing experience than those who do not do so well on the aptitude test. If this is the case, the performance ratings of secretaries with low aptitude test scores have been raised by the fact that they have tended to be good typists, and the performance ratings of secretaries with high aptitude test scores have been lowered by the fact that they have tended to be poor typists. Thus, the estimated simple regression will result in an estimate of B_1 that will be biased downward.[1]

When a dependent variable is a function of more than one independent variable, the observed relationship between the dependent variable and any one of the independent variables may be misleading because the observed relationship may reflect the variation in the other independent variables. Since these other independent variables are totally uncontrolled, they may be varying in such a way as to make it appear that this independent variable has more effect or less effect on the dependent variable than in fact is true. To estimate the true effects of this independent variable on the dependent variable, we must include all the independent variables in the regression; that is, we must construct a multiple regression.

12.3 The Multiple-Regression Model

As pointed out in Chapter 11, the basic model underlying simple regression is

$$Y_i = A + BX_i + e_i,$$

where Y_i is the ith observed value of the dependent variable, X_i is the ith observed value of the independent variable, and e_i is a normally distributed random variable with a mean of zero and a standard deviation of σ_e. Essentially, e_i is an *error term*—that is, a random amount that is added to $A + BX_i$ (or subtracted from it, if e_i is negative). The conditional mean of Y_i is assumed to be a linear function of X_i—namely, $A + BX_i$. And the values of e_i are assumed to be statistically independent.

The model underlying multiple regression is essentially the same as that above, the only difference being that the conditional mean of the dependent variable is assumed to be a linear function of more than

---

[1] Why will the estimate of B_1 be biased downward? Because the performance ratings of secretaries scoring well on the aptitude test are lower and the performance ratings of secretaries scoring poorly on the aptitude test are higher than if the typing test score for all secretaries were held constant at its average level. Thus, the increase in performance rating associated with a 1-point increase in aptitude test score is underestimated, which is the same as saying that B_1 is biased downward.

one independent variable. If there are two independent variables X_1 and X_2, the model is

$$Y_i = A + B_1 X_{1i} + B_2 X_{2i} + e_i, \qquad (12.2)$$

where e_i is an error term. As in the case of simple regression, it is assumed that the expected value of e_i is zero, that e_i is normally distributed, and that the standard deviation of e_i is the same, regardless of the value of X_{1i} or X_{2i}. Also, the values of e_i are assumed to be statistically independent. In contrast to the case of simple regression, the conditional mean of Y_i is a linear function of both X_{1i} and X_{2i}. Specifically, the conditional mean equals $A + B_1 X_{1i} + B_2 X_{2i}$.

12.4 Least-Squares Estimates of the Regression Coefficients

The first step in multiple-regression analysis is to identify the independent variables, and to specify the mathematical form of the equation relating the expected value of the dependent variable to these independent variables. In the case of the university secretaries, this step is carried out in equation (12.1), which indicates that the independent variables are the aptitude test score and typing test score. The relationship between the expected value of the dependent variable (performance rating) and these independent variables is linear. Having carried out this first step, *we next estimate the unknown constants* A, B_1, *and* B_2 *in the true regression equation.* Just as in the case of simple regression, these constants are estimated by finding the value of each that minimizes the sum of the squared deviations of the observed values of the dependent variable from the values of the dependent variable predicted by the regression equation.

To understand more precisely the nature of least-squares estimates of A, B_1, and B_2, suppose that a is an estimator of A, b_1 an estimator of B_1, and b_2 an estimator of B_2. Then the value of the dependent variable $\hat{Y}_i$ predicted by the estimated regression equation is

$$\hat{Y}_i = a + b_1 X_{1i} + b_2 X_{2i},$$

and the deviation of this predicted value from the actual value of the dependent variable is

$$Y_i - \hat{Y}_i = Y_i - a - b_1 X_{1i} - b_2 X_{2i}.$$

Just as in the case of simple regression, the closeness of fit of the estimated regression equation to the data is measured by the sum of squares of these deviations:

$$\sum_{i=1}^{n} (Y_i - \hat{Y}_i)^2 = \sum_{i=1}^{n} (Y_i - a - b_1 X_{1i} - b_2 X_{2i})^2, \qquad (12.3)$$

where n is the number of observations in the sample. The larger the sum of squares, the less closely the estimated regression equation fits; the smaller this sum of squares, the more closely it fits. Thus, it seems reasonable to choose the values of a, b_1, and b_2 that minimize the expression in equation (12.3). These estimates are least-squares estimates, as in the case of simple regression.

It can be shown that the values of a, b_1, and b_2 that minimize the sum of squared deviations in equation (12.3) must satisfy the following (so-called normal) equations:

$$\sum_{i=1}^{n} Y_i = na + b_1 \sum_{i=1}^{n} X_{1i} + b_2 \sum_{i=1}^{n} X_{2i} \qquad (12.4)$$

$$\sum_{i=1}^{n} X_{1i} Y_i = a \sum_{i=1}^{n} X_{1i} + b_1 \sum_{i=1}^{n} X_{1i}^2 + b_2 \sum_{i=1}^{n} X_{1i} X_{2i}$$

$$\sum_{i=1}^{n} X_{2i} Y_i = a \sum_{i=1}^{n} X_{2i} + b_1 \sum_{i=1}^{n} X_{1i} X_{2i} + b_2 \sum_{i=1}^{n} X_{2i}^2.$$

Solving these equations for a, b_1, and b_2, we obtain the following results:

$$b_1 = \frac{\sum_{i=1}^{n} (X_{2i} - \overline{X}_2)^2 \sum_{i=1}^{n} (X_{1i} - \overline{X}_1)(Y_i - \overline{Y}) - \sum_{i=1}^{n} (X_{1i} - \overline{X}_1)(X_{2i} - \overline{X}_2) \sum_{i=1}^{n} (X_{2i} - \overline{X}_2)(Y_i - \overline{Y})}{\sum_{i=1}^{n} (X_{1i} - \overline{X}_1)^2 \sum_{i=1}^{n} (X_{2i} - \overline{X}_2)^2 - \left[\sum_{i=1}^{n} (X_{1i} - \overline{X}_1)(X_{2i} - \overline{X}_2) \right]^2}$$

$$b_2 = \frac{\sum_{i=1}^{n} (X_{1i} - \overline{X}_1)^2 \sum_{i=1}^{n} (X_{2i} - \overline{X}_2)(Y_i - \overline{Y}) - \sum_{i=1}^{n} (X_{1i} - \overline{X}_1)(X_{2i} - \overline{X}_2) \sum_{i=1}^{n} (X_{1i} - \overline{X}_1)(Y_i - \overline{Y})}{\sum_{i=1}^{n} (X_{1i} - \overline{X}_1)^2 \sum_{i=1}^{n} (X_{2i} - \overline{X}_2)^2 - \left[\sum_{i=1}^{n} (X_{1i} - \overline{X}_1)(X_{2i} - \overline{X}_2) \right]^2}$$

$$a = \overline{Y} - b_1 \overline{X}_1 - b_2 \overline{X}_2.$$

Least-Squares Estimators of A, B_1, and B_2

$$(12.5)$$

To make these computations simpler, note that

$$\sum_{i=1}^{n} (X_{1i} - \overline{X}_1)^2 = \sum_{i=1}^{n} X_{1i}^2 - \frac{\left(\sum_{i=1}^{n} X_{1i}\right)^2}{n}$$

$$\sum_{i=1}^{n} (X_{2i} - \overline{X}_2)^2 = \sum_{i=1}^{n} X_{2i}^2 - \frac{\left(\sum_{i=1}^{n} X_{2i}\right)^2}{n}$$

$$\sum_{i=1}^{n} (X_{1i} - \overline{X}_1)(Y_i - \overline{Y}) = \sum_{i=1}^{n} X_{1i}Y_i - \frac{\left(\sum_{i=1}^{n} X_{1i}\right)\left(\sum_{i=1}^{n} Y_i\right)}{n}$$

$$\sum_{i=1}^{n} (X_{2i} - \overline{X}_2)(Y_i - \overline{Y}) = \sum_{i=1}^{n} X_{2i}Y_i - \frac{\left(\sum_{i=1}^{n} X_{2i}\right)\left(\sum_{i=1}^{n} Y_i\right)}{n}$$

$$\sum_{i=1}^{n} (X_{1i} - \overline{X}_1)(X_{2i} - \overline{X}_2) = \sum_{i=1}^{n} X_{1i}X_{2i} - \frac{\left(\sum_{i=1}^{n} X_{1i}\right)\left(\sum_{i=1}^{n} X_{2i}\right)}{n}$$

The following example illustrates how least-squares estimates of A, B_1, and B_2 are calculated. Although (as we shall see in subsequent sections) electronic computers are generally used for such calculations, it is worthwhile to work through at least one sample calculation of this sort by hand.

EXAMPLE 12.1 The university administrator feels that equation (12.1) is true, and wants to obtain least-squares estimates of A, B_1, and B_2. Data are obtained concerning performance rating, aptitude test score, and typing test score, the results being shown in Table 12.1. Calculate the least-squares estimates of A, B_1, and B_2. Compare the least-squares estimate of B_1 with the estimate of B that we obtained in the previous chapter, based on the same data concerning performance ratings and aptitude test scores. Why is the present estimate different from the latter? Which of the two estimates is likely to be better?

SOLUTION: Based on the data in Table 12.1,

$$\sum_{i=1}^{9} X_{1i}^2 = 340 \qquad \sum_{i=1}^{9} X_{1i} = 50 \qquad \sum_{i=1}^{9} Y_i^2 = 303$$

$$\sum_{i=1}^{9} X_{2i}^2 = 204 \qquad \sum_{i=1}^{9} X_{2i} = 36 \qquad \sum_{i=1}^{9} Y_i = 49$$

Table 12.1
Performance Ratings,
Aptitude Test Scores,
and Typing Test
Scores (Sample of Nine
Secretaries)

Performance Rating	Aptitude Test Score	Number of Errors on Typing Test
8	8	2
5	5	3
2	3	0
1	2	1
6	6	4
4	4	5
8	7	8
7	6	6
9	8	7

$$\sum_{i=1}^{9} X_{1i}X_{2i} = 245 \qquad \sum_{i=1}^{9} X_{1i}Y_i = 319 \qquad \sum_{i=1}^{9} X_{2i}Y_i = 225.$$

Inserting these figures into the formulas given prior to this example.

$$\sum_{i=1}^{9} (X_{1i} - \overline{X}_1)^2 = 340 - \frac{50^2}{9} = 340 - 277.78 = 62.22$$

$$\sum_{i=1}^{9} (X_{2i} - \overline{X}_2)^2 = 204 - \frac{36^2}{9} = 204 - 144 = 60$$

$$\sum_{i=1}^{9} (X_{1i} - \overline{X}_1)(Y_i - \overline{Y}) = 319 - \frac{(50)(49)}{9} = 319 - 272.22 = 46.78$$

$$\sum_{i=1}^{9} (X_{2i} - \overline{X}_2)(Y_i - \overline{Y}) = 225 - \frac{(36)(49)}{9} = 225 - 196 = 29$$

$$\sum_{i=1}^{9} (X_{1i} - \overline{X}_1)(X_{2i} - \overline{X}_2) = 245 - \frac{(50)(36)}{9} = 245 - 200 = 45.$$

Then, inserting these figures into the equations in (12.5), we obtain

$$b_1 = \frac{(60)(46.78) - (45)(29)}{(62.22)(60) - 45^2} = 0.88$$

$$b_2 = \frac{(62.22)(29) - (45)(46.78)}{(62.22)(60) - 45^2} = -0.18$$

$$a = \frac{49}{9} - .88 \left(\frac{50}{9}\right) + .18 \left(\frac{36}{9}\right) = 1.28.$$

Consequently, the estimated regression equation is

$$\hat{Y}_i = 1.28 + 0.88 \, X_{1i} - 0.18 X_{2i}.$$

The estimated value of B_1 is 0.88, as contrasted with the estimate of B in the previous chapter, which was 0.75. In other words, a 1-point increase in aptitude test score is associated with an increase in performance rating of 0.88, as contrasted with 0.75 in the previous chapter. The reason these estimates differ is that the present estimate holds constant the typing test score, whereas the earlier estimate did not hold this factor constant. Since this factor is related to the performance rating, the earlier estimate is likely to be a biased estimate of the relationship between aptitude test score and performance rating.[2]

---

[2] Of course, this regression is only supposed to be appropriate when X_{1i} and X_{2i} vary in a certain limited range. If X_{2i} is large and X_{1i} is small, the regression would predict a negative value of the performance rating, which obviously is inadmissable. But as long as the regression is not used to make predictions for values of X_{1i} and X_{2i} outside the range of the data given in Table 12.1, this is no problem. For simplicity, we assume in equation (12.1) that the relationship between the expected value of performance rating and the number of typing errors (holding aptitude test score constant) can be regarded as linear in the relevant range. Alternatively, we could have assumed that it was quadratic (or some other nonlinear forms could have been used). In Section 12.9 we discuss how multiple regression can be used to estimate quadratic equations.

GETTING DOWN TO CASES

DETERMINANTS OF THE STRENGTH OF COTTON YARN

The U.S. Department of Agriculture has presented data concerning the skein strength (in pounds) of cotton yarn, which is perhaps the most important single measure of spinning quality. For a sample of varieties of upland cotton grown in the United States, the Department showed the skein strength (in pounds) as well as the fiber length (in hundredths of an inch) and fiber tensile strength (in thousands of pounds per square inch). The results for a sample of 20 pieces of cotton yarn were as follows:[3]

[3] See U.S. Department of Agriculture, *Results of Fiber and Spinning Tests for Some Varieties of Upland Cotton Grown in the United States;* and A. J. Duncan, *Quality Control and Industrial Statistics.*

Skein strength	Fiber length	Fiber tensile strength
99	85	76
93	82	78
99	75	73
97	74	72
90	76	73
96	74	69
93	73	69
130	96	80
118	93	78
88	70	73
89	82	71
93	80	72
94	77	76
75	67	76
84	82	70
91	76	76
100	74	78
98	71	80
101	70	83
80	64	79

(a) Calculate the regression of skein strength on fiber length and fiber tensile strength. (That is, let skein strength be the dependent variable, and let fiber length and fiber tensil strength be the independent variables.)

(b) Based on your results in (a), what is the effect on the average strength of a piece of yarn of an increase in fiber length of .01 inches? What is the effect on the average strength of a piece of yarn of an increase in fiber tensile strength of 1,000 pounds per square inch?

(c) Based on your results in (a), estimate the strength of a piece of yarn if the fiber length is .80 inches and the fiber tensile strength is 75,000 pounds per square inch.

(d) Write a one-paragraph report summarizing your findings.

12.5 Confidence Intervals and Tests of Hypotheses Concerning B_1 and B_2

As in the case of simple regression, least-squares estimators have many statistically desirable properties. Specifically, a, b_1, *and* b_2 *are unbiased and consistent.* Moreover, the Gauss-Markov theorem tells us that *of all unbiased estimators that are linear functions of the dependent variables,* a, b_1, *and* b_2 *have the smallest standard deviation.* As in the case of simple regression, these desirable properties hold if the observations are independent and if the standard deviation of the conditional probability distribution is the same, regardless of the value of the independent variables. It is *not* necessary that the conditional probability distribution of the dependent variable be normal. Or, stating the same thing differently, it is *not* necessary that the probability distribution of e_i be normal.

If we are willing to assume that e_i is normally distributed, we can calculate a confidence interval for B_1 or B_2. This frequently is an important purpose of a multiple-regression analysis. For example, in the case of the university secretaries, an important purpose of the analysis may be to obtain a confidence interval for B_1, which is often called the **true regression coefficient of X_1.** (The least-squares estimator of B_1— that is, b_1—is often called the *estimated regression coefficient of* X_1). This true regression coefficient is of interest because it measures the effect of a one-point increase in X_1 (aptitude test score) on the expected value of Y (performance rating), when X_2 (the number of mistakes on the typing test) is held constant. Another important purpose of the analysis may be to obtain a confidence interval for B_2, which is often called the **true regression coefficient of X_2.** (The least-squares estimator of B_2— that is, b_2—is often called the *estimated regression coefficient of* X_2). This true regression coefficient is of interest because it measures the effect of a one-unit increase in X_2 (the number of mistakes on the typing test) on the expected value of Y (performance rating), when X_1 (aptitude test score) is held constant.[4]

As pointed out in the previous section, multiple regressions are generally carried out by electronic computer, not by hand. Because of the importance of multiple-regression techniques, standard programs have been formulated for calculation by computer of the least-squares estimates of A, B_1, and B_2. Since these estimates are sample statistics, they are obviously subject to sampling error. Besides calculating the

---

[4] Sometimes statisticians are also interested in obtaining a confidence interval for A, which is often called the intercept of the regression, and which measures the expected value of Y when both X_1 and X_2 are zero. (See note 5.)

values of the least-squares estimators of A, B_1, and B_2 (that is, a, b_1, and b_2), these programs provide an estimate of the standard deviation of b_1 (often called the *standard error of* b_1), and an estimate of the standard deviation of b_2 (often called the *standard error of* b_2).

Given the computer printout, it is relatively simple to construct a confidence interval for B_1 or B_2. As noted above, in any standard computer printout the standard error of b_1 and the standard error of b_2 are shown. If the confidence coefficient is set at $(1 - \alpha)$, *a confidence interval for* B_1 *is*

$$b_1 \pm t_{\alpha/2} s_{b_1},$$

(12.6)   *Confidence Interval for* B_1

where s_{b_1} is the standard error of b_1 and where t has $n - k - 1$ degrees of freedom (where k is the number of independent variables included in the regression—two in this case). If the confidence coefficient is set at $(1 - \alpha)$, *a confidence interval for* B_2 *is*

$$b_2 \pm t_{\alpha/2} s_{b_2},$$

(12.7)   *Confidence Interval for* B_2

where s_{b_2} is the standard error of b_2.[5]

Given the computer printout, it is also easy to test the null hypothesis that B_1 or B_2 equals zero, if we assume once again that e_i—in equation (12.2)—is normally distributed. The computer printout shows the t *statistic (or* t *value) for* b_1, this statistic being defined as $b_1 \div s_{b_1}$. If B_1 equals zero, this t statistic has the t distribution with $(n - k - 1)$ degrees of freedom. Thus, if the alternative hypothesis is two-sided and α is the significance level, the decision rule is as follows.

> **Decision Rule:** *Reject the null hypothesis that* B_1 *equals zero if the* t *statistic for* b_1 *exceeds* $t_{\alpha/2}$ *or is less than* $-t_{\alpha/2}$; *otherwise accept the null hypothesis. (The number of degrees of freedom is* $n - k - 1$.)   *Test that* $B_1 = 0$

Similarly, if B_2 equals zero, the t statistic (or t value) for b_2—defined as $b_2 \div s_{b_2}$—has the t distribution with $(n - k - 1)$ degrees of freedom; and the decision rule is as follows.

> **Decision Rule:** *Reject the null hypothesis that* B_2 *equals zero if the* t *statistic for* b_2 *exceeds* $t_{\alpha/2}$ *or is less than* $-t_{\alpha/2}$; *otherwise accept the null hypothesis. (The number of degrees of freedom is* $n - k - 1$.)   *Test that* $B_2 = 0$

---

[5] Some computer printouts also show s_a, the standard error of a. If the confidence coefficient is set at $(1 - \alpha)$, a confidence interval for A is $a \pm t_{\alpha/2} s_a$.

The following examples illustrate how, with a computer printout of the results of a multiple regression, one can construct confidence intervals for some of the true regression coefficients, and test whether some of these coefficients are zero.[6]

[6] It is also possible, if the printout shows the t statistic (or t value) for a, to test the null hypothesis that A equals zero. If A equals zero, the t statistic (or t value) for a—defined as $a \div s_a$—has the t distribution with $(n - k - 1)$ degrees of freedom; and the decision rule is: Reject the null hypothesis that A equals zero if the t statistic for a exceeds $t_{\alpha/2}$ or is less than $-t_{\alpha/2}$; otherwise, accept the null hypothesis. (The number of degrees of freedom is $n - k - 1$.)

EXAMPLE 12.2 The computer printout of the multiple regression of a secretary's performance rating on (1) aptitude test score and (2) the number of mistakes on the typing test is shown, in part, in Table 12.2. (The basic data were given in Table 12.1.) In this printout, aptitude test score is designated as variable 2, and the number of mistakes on the typing test is designated as variable 3. As you can see, the computer prints out the value of b_1, s_{b_1}, and the t statistic for b_1 in the first row, labeled 2 (for variable 2). In the second row, labeled 3 (for variable 3), it prints out the value of b_2, s_{b_2}, and the t statistic for b_2. (Because of rounding errors, the estimates of a, b_1, and b_2 obtained by the computer will differ slightly from those obtained by hand.) Use these results to calculate a 95 percent confidence interval for B_1; then use these results to test the hypothesis that B_2 equals zero. (The alternative hypothesis is two-sided, and the significance level should be set equal to .05.)

SOLUTION: Since the standard error of b_1 equals .0347, a 95 percent confidence interval for B_1 is $0.879 \pm t_{.025}(.0347)$. Because $n - k - 1 = 9 - 2 - 1$, or 6, the t distribution has 6 degrees of freedom, and $t_{.025} = 2.447$. Thus, the confidence interval for B_1 is $0.879 \pm .0849$. In other words, the 95 percent confidence interval for the increase in performance rating associated with a 1-point increase in aptitude test score is .794 to .964.

Table 12.2
Section of Computer Printout, Showing Results of Multiple Regression of Performance Rating on Aptitude Test Score and Number of Mistakes on the Typing Test

INDEPENDENT VARIABLE(S)	2	3
DEPENDENT VARIABLE	1	

VAR	COEFF	STD. ERROR	T-VALUE
2	0.8790	0.0347	25.3400
3	−0.1759	0.0353	−4.9800
INTERCEPT	1.26472		

If B_2 were zero, the probability would be .05 that the t statistic for b_2 would be greater than 2.447 or less than −2.447. (This is because $t_{.025} = 2.447$.) Since the t statistic for b_2 equals −4.98, it is less than −2.447. Thus, we must reject the null hypothesis that B_2 equals zero.

EXAMPLE 12.3 A professor at the University of Pennsylvania was interested in how accurately a student's grade on the final exam in a particular course could be predicted on the basis of his or her grades on the two hour exams. The computer printout in Table 12.3 indicates the results of a multiple regression where the dependent variable is a student's final exam grade in this course, and the independent variables are his or her grade on the first hour exam in this course (variable 2), and his or her grade on the second hour exam (variable 3). This multiple regression, which was calculated from actual data, is based on a sample of 45 students from a very large class.[7] Compute a 95 percent confidence interval for the true regression coefficient of the grade on the first hour exam. Using a two-tailed test, test the null hypothesis that the true regression coefficient of the grade on the second hour exam is zero. (Set the significance level at .05.)

SOLUTION: Since the estimated regression coefficient of the grade on the first hour exam is 0.8434, and the standard error of this estimated regression coefficient is .2358, the 95 percent confidence interval is $0.8434 \pm 2.02(.2358)$, since $t_{.025}$ is approximately 2.02 when there are $45 − 2 − 1$, or 42 degrees of freedom. In other words, the confidence interval is $0.8434 \pm .4763$, or .3671 to 1.3197. To interpret this result, suppose that a student's grade on the first hour exam were to increase by 1 point. Based on the regression, it appears that such an increase is associated, on the average, with an increase in his or her final exam grade of from .3671 to 1.3197 points, according to the 95 percent confidence interval.

The t value of the estimated regression coefficient of the grade on the second hour exam is 0.41, which is less than 2.02, the value of $t_{.025}$ when there are 42 degrees of freedom. Thus, we should not reject the null hypothesis that the true regression coefficient of the grade on the second hour exam is zero.

---

[7] Note that the highest possible grade on the final exam was 250, whereas the highest possible grade on each hour exam was 120.

```
INDEPENDENT VARIABLE(S)  2  3
DEPENDENT VARIABLE       1


VAR        COEFF       STD. ERROR      T-VALUE
 2        0.8434        0.2358          3.58
 3        0.1505        0.3705          0.41

INTERCEPT  107.39
```

EXERCISES

12.1 A sociologist wants to determine the relationship in San Francisco between the annual amount spent on clothing, on the one hand, and the
number of family members and family income, on the other. She obtains
the following data concerning 10 randomly selected San Francisco families:

Annual amount spent on clothing ($000)	Number of family members	Annual family income ($000)
0.8	1	21
1.4	2	25
0.3	1	10
2.3	2	37
3.8	2	48
2.1	3	27
3.3	4	35
2.9	3	35
3.0	5	25
4.0	3	51

(a) Calculate the sample multiple regression, clothing expenditure being
the dependent variable and number of family members and family income being the independent variables.
(b) What is the estimated effect of an additional family member on annual clothing expenditure?
(c) What is the estimated effect of an additional thousand dollars of income on annual clothing expenditure?
(d) Would you expect that this relationship would be linear for all values
of the independent variables?
(e) Predict a family's annual clothing expenditure if it contains two
members and its income is $25,000.
(f) Predict a family's annual clothing expenditure if it contains four
members and its income is $35,000.
(g) Do you think that the regression you calculated in (a) would be applicable to Minneapolis families? If not, what differences would you expect between this regression and that for Minneapolis families?

12.2 A college wants to estimate the relationship between a student's grade-point average, on the one hand, and his or her high school grade-point average and SAT score. For 12 students, these variables are as follows:

College grade-point average	High School grade-point average	SAT score
2.5	3.0	1,000
2.9	3.2	1,100
2.9	3.5	980
3.1	3.4	1,120
3.2	3.8	1,010
3.6	4.0	1,140
3.4	3.6	1,180
3.6	3.8	1,250
3.6	3.8	1,200
3.9	3.9	1,260
3.8	4.0	1,210
4.0	4.0	1,300

(a) Calculate the sample multiple regression, college grade-point average being the dependent variable and high school grade-point average and SAT score being the independent variables.

(b) What is the estimated effect of an additional 1.0 in high school grade-point average on college grade-point average?

(c) What is the estimated effect of an additional 100 points in SAT score on college grade-point average?

(d) Do colleges actually use formulas of this sort? For what reason?

(e) Predict a student's grade-point average at this college if he has a high school grade-point average of 3.0 and an SAT score of 1150.

(f) Predict a student's grade-point average at this college if she has a high school grade-point average of 3.8 and an SAT score of 1240.

(g) Do you think that the multiple regression you calculated in (a) can be applied to any college, or just this one? Explain.

12.6 Multiple Coefficient of Determination

In the previous chapter we described how the coefficient of determination can be used to measure how well a simple regression equation fits the data. When a multiple regression is calculated, the multiple coefficient of determination, rather than the simple coefficient of determination discussed in the previous chapter, is used for this purpose. The multiple coefficient of determination is defined as

Formula for R^2

$$R^2 = 1 - \frac{\sum_{i=1}^{n} (Y_i - \hat{Y}_i)^2}{\sum_{i=1}^{n} (Y_i - \overline{Y})^2},$$

(12.8)

where $\hat{Y}_i$ is the value of the dependent variable that is predicted from the regression equation.[8] Thus, as in the case of the simple coefficient of determination covered in the previous chapter,

$$R^2 = \frac{\text{variation explained by regression}}{\text{total variation}},$$

(12.9)

which means that R^2 *measures the proportion of the total variation in the dependent variable that is explained by the regression equation.* The positive square root of the multiple coefficient of determination is called the **multiple correlation coefficient** and is denoted by R. It, too, is sometimes used to measure how well a multiple-regression equation fits the data.

If there are only two independent variables in a multiple regression, as in equation (12.1), a relatively simple way to compute the multiple coefficient of determination is as follows:

Alternative Formula for R^2

$$R^2 = \frac{b_1 \sum_{i=1}^{n} (X_{1i} - \overline{X}_1)(Y_i - \overline{Y}) + b_2 \sum_{i=1}^{n} (X_{2i} - \overline{X}_2)(Y_i - \overline{Y})}{\sum_{i=1}^{n} Y_i^2 - \frac{\left(\sum_{i=1}^{n} Y_i \right)^2}{n}}.$$

(12.10)

[8] This is the *unadjusted* multiple coefficient of determination, which is biased away from zero. An unbiased estimate of the population multiple coefficient of determination is the *adjusted* multiple coefficient of determination, which is

$$\overline{R}^2 = 1 - \frac{\sum_{i=1}^{n} (Y_i - \hat{Y}_i)^2 \div (n - k - 1)}{\sum_{i=1}^{n} (Y_i - \overline{Y})^2 \div (n - 1)}$$

where n is the number of observations and k is the number of independent variables. (By adjusted, we mean adjusted for degrees of freedom.) See Section 12.4 for a more detailed definition of $\hat{Y}_i$.

If there are more than two independent variables, a multiple regression is almost always carried out on an electronic computer, which is programmed to print out the value of the multiple coefficient of determination (or of the multiple correlation coefficient).

The following are examples of the calculation and interpretation of the multiple coefficient of determination.

EXAMPLE 12.4 Use the data in Table 12.1 to calculate the multiple coefficient of determination between a secretary's performance rating, on the one hand, and his or her aptitude test score and number of mistakes on the typing test, on the other. Interpret your result.

SOLUTION: We know from Example 12.1 that $b_1 = 0.88$, $b_2 = -0.18$,

$$\sum_{i=1}^{9} (X_{1i} - \overline{X}_1)(Y_i - \overline{Y}) = 46.78, \sum_{i=1}^{9} (X_{2i} - \overline{X}_2)(Y_i - \overline{Y}) = 29, \text{ and}$$

$$\sum_{i=1}^{9} Y_i^2 - \frac{\left(\sum_{i=1}^{9} Y_i\right)^2}{9} = 303 - \frac{(49)^2}{9} = 303 - 266.78 = 36.22.$$

Thus it follows from equation (12.10) that

$$R^2 = \frac{.88(46.78) - .18(29)}{36.22} = .99.$$

This means that 99 percent of the variation in the performance ratings of the sample of secretaries can be explained by the multiple regression equation derived in Example 12.1.

EXAMPLE 12.5 Table 12.4 shows another part of the computer printout of the results of the multiple regression of a student's final exam grade on his or her first and second hour exam grades (described in Example 12.3). Interpret the figure labeled Multiple R-Squared in this printout.

Table 12.4
Another Section of the Computer Printout, Showing Results of Multiple Regression Concerning Final Exam Grades

```
S.E. OF ESTIMATE      28.35
F-VALUE                7.22
MULTIPLE R-SQUARED      .256
```

SOLUTION: This figure—.256—is the multiple coefficient of determination. It means that the regression equation shown in the printout explains 25.6 percent of the variation among these students in their final exam grades.

12.7 Geometrical Interpretation of Results of Multiple Regression and Correlation

As we have seen, the estimated multiple regression equation shows the average relationship between the dependent variable and the independent variables. If there are only two independent variables, it is possible to represent this average relationship by a plane rather than by a line (which was used for this purpose in the previous chapter). Figure 12.1 shows the plane corresponding to the regression equation relating

Figure 12.1
Plane Corresponding
to Regression Equation
Relating Performance
Rating to Aptitude and
Typing Test Scores

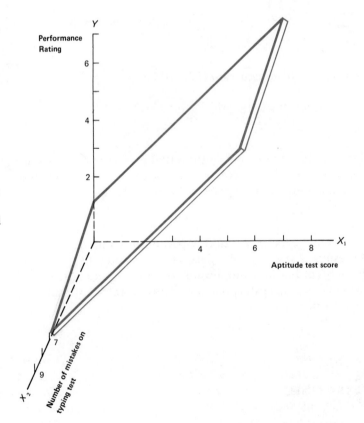

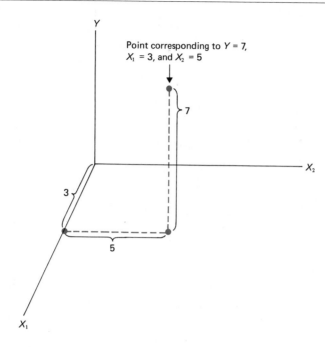

Figure 12.2
Geometrical Representation of an Observation (Y Is the Value of the Dependent Variable; X_1 and X_2 Are the Values of the Independent Variables)

the performance rating of a secretary to his or her aptitude test score and number of errors on the typing test:

$$\hat{Y}_i = 1.28 + 0.88X_{1i} - 0.18X_{2i}.$$

When both X_{1i} and X_{2i} are zero, Figure 12.1 shows that the predicted value of the performance rating is 1.28, which is the intercept of the regression. Figure 12.1 shows that when X_{1i} is held constant, if X_{2i} increases by one unit (one error in this case), the average value of the dependent variable decreases by an amount equal to the regression coefficient of X_{2i}, which is .18 in this case. Figure 12.1 also shows that when X_{2i} is held constant, if X_{1i} is increased by one unit (one point in this case), the average value of the dependent variable increases by an amount equal to the regression coefficient of X_{1i}, which is .88.

Just as each observation in a simple regression can be represented as a point in a scatter diagram, each observation in a multiple regression (with two independent variables) can be represented as a point in three-dimensional space. Figure 12.2 shows how we can represent an observation where the dependent variable equals 7 and the independent variables equal 3 and 5. *The multiple coefficient of determination is a measure of how well the plane representing the regression equation fits the points representing the individual observations.* For example, panel A of

A: R^2 relatively low

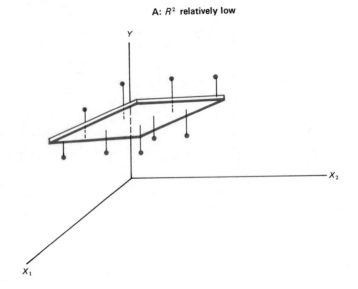

Figure 12.3
Closeness of Fit of
Regression Plane: Two
Cases

B: R^2 relatively high

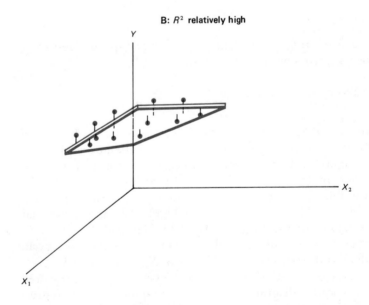

Figure 12.3 shows a case where the regression plane does not fit the points at all well; on the other hand, panel B of Figure 12.3 shows a case where the regression plane fits the points much more closely than in panel A. Clearly, the multiple coefficient of determination (as well as the multiple correlation coefficient) is higher in panel B than in panel A.

12.8 Analysis of Variance[9]

The analysis of variance, which we discussed at length in Chapter 10, is used to test the overall statistical significance of a regression equation. That is, it is used to test whether *all* the true regression coefficients in the equation equal zero. In the case of the performance rating of secretaries, we might want to test whether both B_1 and B_2 are zero in order to see whether there is any relationship between the dependent variable and all the independent variables taken together. The analysis of variance can be used in this way in simple as well as multiple regressions. In simple regressions, the result is precisely the same as in the test of $B = 0$, described in Section 11.13.

In our discussion of the analysis of variance in Chapter 10, we defined the total variation in Y as the sum of the squared deviations of the values of Y from the mean of Y. Thus, the total variation in Y equals

$$\sum_{i=1}^{n} (Y_i - \overline{Y})^2.$$

As pointed out in Section 11.10,

$$\sum_{i=1}^{n} (Y_i - \overline{Y})^2 = \sum_{i=1}^{n} (\hat{Y}_i - \overline{Y})^2 + \sum_{i=1}^{n} (Y_i - \hat{Y}_i)^2.$$

The first term on the right is the variation explained by the regression, and the second term on the right is the variation unexplained by the regression (also called the *error sum of squares* or the *residual sum of squares*). Thus

total variation = explained variation + unexplained variation.

To carry out the analysis of variance, we construct a table of the general form of Table 12.5. The first column shows the source or type of variation, and the second column shows the corresponding sum of squares (that is, the corresponding variation). The third column shows the number of degrees of freedom corresponding to each sum of squares. For the explained variation, the number of degrees of freedom equals the number of independent variables k. For the unexplained variation, the number of degrees of freedom equals $(n - k - 1)$.

[9] To understand this section, it is not essential that the reader be familiar with Chapter 10. However, it is essential that Section 10.4 (on the F distribution) be read prior to studying this section.

The fourth column shows the mean square, which is the sum of squares divided by the number of degrees of freedom. The fifth column shows the ratio of the two mean squares.[10]

If the null hypothesis is true (that is, if all the true regression coefficients are zero), this ratio has an F distribution with k and $(n - k - 1)$ degrees of freedom. Thus, to carry out the appropriate analysis of variance, the decision rule is as follows.

Test that All True Regression Coefficients = 0

Decision Rule: *Reject the null hypothesis that the true regression coefficients are all zero if the ratio of the explained mean square to the unexplained mean square exceeds* F_α, *where* α *is the desired significance level. Otherwise, accept the null hypothesis.*

Because multiple regressions are generally carried out on an electronic computer, it is seldom necessary to calculate the numbers in Table 12.5 by hand. The computer printout generally gives the analysis-of-variance table. Some printouts refer to the explained mean square as the *regression mean square* and to the unexplained mean square as the *error mean square* or the *residual mean square*. The ratio of

Table 12.5
Analysis of Variance
for Regression

Source of variation	Sum of squares	Degrees of freedom	Mean square	F ratio
Explained by regression	$\sum_{i=1}^{n} (\hat{Y}_i - \bar{Y})^2$	k	$\dfrac{\sum_{i=1}^{n} (\hat{Y}_i - \bar{Y})^2}{k}$	$\dfrac{\dfrac{\sum_{i=1}^{n} (\hat{Y}_i - \bar{Y})^2}{k}}{\dfrac{\sum_{i=1}^{n} (Y_i - \hat{Y}_i)^2}{(n - k - 1)}}$
Unexplained by regression	$\sum_{i=1}^{n} (Y_i - \hat{Y}_i)^2$	$n - k - 1$	$\dfrac{\sum_{i=1}^{n} (Y_i - \hat{Y}_i)^2}{(n - k - 1)}$	
Total	$\sum_{i=1}^{n} (Y_i - \bar{Y})^2$	$n - 1$		

[10] In calculating the sum of squares explained by the regression, it is convenient to note that

$$\sum_{i=1}^{n} (\hat{Y}_i - \bar{Y})^2 = b_1 \sum_{i=1}^{n} (X_{1i} - \bar{X}_1)(Y_i - \bar{Y}) + b_2 \sum_{i=1}^{n} (X_{2i} - \bar{X}_2)(Y_i - \bar{Y}).$$

The right-hand side of this equation can be evaluated relatively easily. (This assumes that $k = 2$.)

the explained to the unexplained mean square is often referred to as the F value or the F ratio. The way in which the analysis-of-variance table is displayed on the computer printout may vary from that in Table 12.5, but the information imparted is the same. In Section 12.10 we will examine the printout of one computer program in some detail.

The following is an example of the application of the analysis of variance.

EXAMPLE 12.6 Table 12.4 shows the F ratio for the multiple regression of a student's final exam grade on his or her grades on the first and second hour exams (described in Example 12.3). Use this F ratio to test the hypothesis that all the true regression coefficients in this multiple regression are zero. (Set $\alpha = .05$.)

SOLUTION: Table 12.4 shows that the F ratio equals 7.22. Since $n = 45$ and $k = 2$, the number of degrees of freedom are 2 and 22. (Why? Because $k = 2$ and $(n - k - 1) = 22$.) Thus, $F_{.05}$ equals 3.44. (See Appendix Table 9.) Since the F ratio exceeds 3.44, we should reject the null hypothesis that all the true regression coefficients in this multiple regression equal zero.

12.9 Coping with Departures from Linearity

In Chapter 11 we assumed that the conditional mean of the dependent variable is a linear function of the independent variable. If an examination of the data indicates that the relationship is curvilinear rather than linear, what can be done? One possibility is to fit a quadratic equation to the data. For example, suppose that a statistician is interested in the relationship between a family's income and the amount the family spends on food. To detect signs of nonlinearity, the statistician examines the scatter diagram and finds there is an obvious tendency for the relationship between these two variables to be curvilinear. He then assumes that

$$Y_i = A + B_1 X_i + B_2 X_i^2 + e_i,$$

where Y_i is the amount spent on food by the ith family and X_i is its income. Letting X_i be the first independent variable and X_i^2 be the second independent variable, A, B_1, and B_2 are estimated in the way described in Section 12.4. In this way, a quadratic relationship is fitted to the data.

In a case of this sort, the least-squares estimators of B_1, B_2, and A are as follows:

$$b_1 = \frac{\sum_{i=1}^{n} (u_i - \bar{u})^2 \sum_{i=1}^{n} (X_i - \bar{X})(Y_i - \bar{Y}) - \sum_{i=1}^{n} (X_i - \bar{X})(u_i - \bar{u}) \sum_{i=1}^{n} (Y_i - \bar{Y})(u_i - \bar{u})}{\sum_{i=1}^{n} (X_i - \bar{X})^2 \sum_{i=1}^{n} (u_i - \bar{u})^2 - \left[\sum_{i=1}^{n} (X_i - \bar{X})(u_i - \bar{u}) \right]^2}$$

$$b_2 = \frac{\sum_{i=1}^{n} (X_i - \bar{X})^2 \sum_{i=1}^{n} (u_i - \bar{u})(Y_i - \bar{Y}) - \sum_{i=1}^{n} (X_i - \bar{X})(u_i - \bar{u}) \sum_{i=1}^{n} (Y_i - \bar{Y})(X_i - \bar{X})}{\sum_{i=1}^{n} (X_i - \bar{X})^2 \sum_{i=1}^{n} (u_i - \bar{u})^2 - \left[\sum_{i=1}^{n} (X_i - \bar{X})(u_i - \bar{u}) \right]^2}$$

$$a = \bar{Y} - b_1 \bar{X} - b_2 \bar{u},$$

where $u_i = X_i^2$. To derive these formulas, one can substitute X_i for X_{1i} and u_i for X_{2i} in the equations in (12.5).

EXERCISES

12.3 Based on the data in Exercise 12.1, calculate the multiple coefficient of determination. Interpret your results.

12.4 Based on the data in Exercise 12.2, calculate the multiple coefficient of determination. Interpret your results.

12.5 A multiple regression is carried out where there are 4 independent variables and 32 observations. The multiple coefficient of determination (R^2) equals 0.63.
(a) What is $\bar{R}^2$, the multiple coefficient of determination adjusted for degrees of freedom?
(b) In what way is $\bar{R}^2$ preferable to R^2?

12.6 A multiple regression is carried out where there are three independent variables and 45 observations. The multiple coefficient of determination adjusted for degree of freedom, $\bar{R}^2$, equals 0.45.
(a) What is R^2?
(b) Must R^2 always be greater than $\bar{R}^2$?

12.7 A statistician at the Mayo Clinic carries out an analysis of variance, with the following results:

Source of variation	Sum of squares	Degrees of freedom	Mean square	F
Explained by regression	‾‾‾	3	42	‾‾‾
Unexplained by regression	‾‾‾	‾‾‾	‾‾‾	
Total	542	31		

(a) Fill in the blanks.
(b) How many independent variables are there?
(c) How many observations are included?

(d) What is the regression mean square?

(e) What is the error mean square?

(f) What is the residual mean square?

(g) Interpret the results of the analysis of variance.

12.8 Based on the data in Exercise 12.1, carry out an analysis of variance to test whether all of the true regression coefficients equal zero. (Let $\alpha = .05$.)

12.9 Based on the data in Exercise 12.2, carry out an analysis of variance to test whether all of the true regression coefficients equal zero. (Let $\alpha = .01$.)

12.10 A criminologist believes that there is a quadratic relationship between the unemployment rate in a particular city and the rate of occurrence of robberies. In other words, the criminologist believes that

$$E(R_i) = A + B_1 u_i + B_2 u_i^2,$$

where R_i is the number of robberies per 100,000 inhabitants in the ith city, and u_i is the unemployment rate in the ith city. Can the criminologist use multiple regression techniques to estimate A, B_1, and B_2? If so, what assumptions must be made?

12.10 Computer Programs and Multiple Regression

The advent of computer technology has caused a marked reduction in the amount of effort and expense required to calculate multiple regressions with large numbers of independent variables. Thirty years ago there was an enormous amount of drudgery in computing a multiple regression with more than a few independent variables; now such computations are relatively simple. It is important for you to be familiar with the kind of information printed out by computers and the form in which it appears. Since there is a wide variety of "canned" programs for calculating regressions, there is no single format or list of items which are printed out. However, the various sorts of computer printouts are sufficiently similar so that it is worthwhile looking at one illustration in some detail.

Table 12.6 shows the printout from a multiple regression of a student's final exam grade (variable 1) on his or her first and second hour exam grades (variables 2 and 3). We are already familiar with the top three horizontal rows of this printout, which were reproduced in Table 12.3. As we already know, these rows show the value, standard error, and t value of each of the estimated regression coefficients, as well as the estimated intercept of the regression equation.

MISTAKES TO AVOID

THE CASE OF THE UNAPPRECIATED RESIDUALS

A salesman who works for an ice cream firm regresses its daily sales on two independent variables (namely, the mean temperature during the day and the price it charges). He takes the results to the firm's statistician, who says that the residuals from the regression indicate that an important independent variable has been omitted from the analysis. The salesman asks what a residual is. The statistician says that, if Y_i is the ith value of the dependent variable, the residual corresponding to the ith observation is $Y_i - \hat{Y}_i$, where $\hat{Y}_i$ is the value of Y_i predicted by the sample regression. He goes on to say that a plot of the 61 residuals (corresponding to the 61 days in the sample) shows the following:

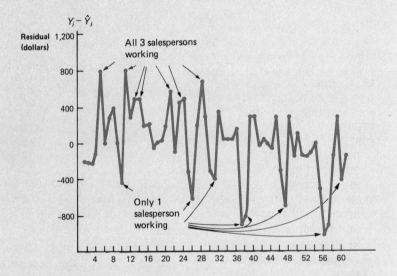

The salesman, after studying the above diagram, replies that he sees nothing wrong with his regression. Where does his mistake lie?

SOLUTION

Many of the days when the residuals were large and positive were days when the firm had all three of its salespersons on the job, whereas many of the days when the residuals were large and negative were days when only one of the firm's salespeople worked (because the others were ill or on vacation). This pattern, as pointed out by the statistician, suggests that the regression does not contain all the important independent variables. To decide whether the number of salespeople working on a particular day should be used as an additional independent variable in the regression, the salesman can use this number as a third independent variable, and see whether its regression coefficient is statistically significant.

Statisticians generally plot and analyze the residuals from their regression equations. In evaluating any model, it is useful to calculate the difference between each observation and what the model predicts this observation will be. These differences—or residuals—are very helpful in indicating whether the model excludes some important explanatory variables and whether its assumptions are valid.

```
INDEPENDENT VARIABLE(S)    2    3
DEPENDENT VARIABLE         1
VAR    COEFF    STD. ERROR    T-VALUE
 2    0.8434      0.2358       3.58
 3    0.1505      0.3705       0.41
INTERCEPT    107.39
REGRESSION
    DEGREES OF FREEDOM         2
    SUM OF SQUARES         11608.2
    MEAN SQUARE            5804.1
ERROR
    DEGREES OF FREEDOM         42
    SUM OF SQUARES         33758.1
    MEAN SQUARE            803.8
S.E. OF ESTIMATE           28.35
F-VALUE                     7.22
MULTIPLE R-SQUARED           .256
```

Table 12.6
Computer Printout of Multiple Regression Concerning Students' Final Exam Grades

The next six rows in Table 12.6 present the results of an analysis of variance for the multiple regression. The *variation explained by the regression* equals 11,608.2. The *error sum of squares* equals 33,758.1. The *regression mean square* is the variation explained by the regression divided by its degrees of freedom (which equal the number of independent variables). Thus, the regression mean square in this case equals 11,608.2 ÷ 2, or 5804.1. The *error mean square* here is the error sum of squares divided by its degrees of freedom (which is one less than the number of observations minus the number of independent variables). Thus, the error mean square in this case equals $33,758.1 \div 42$, or 803.8. The F value in Table 12.6 is shown in the next-to-last row in the table.

The only item left to be explained in Table 12.6 is the *standard error of estimate*. As in the case of simple regression, the standard error of estimate is an estimate of the standard deviation of the probability distribution of the dependent variable when the independent variables are all held constant. Thus, it is a measure of the amount of scatter of individual observations about the regression equation. In this case the standard error of estimate is 28.35, which means that in the population as a whole, the standard deviation of the differences between a student's final exam grade and that predicted by the true regression is estimated to be 28.35 points. Clearly, there is a considerable amount of scatter.[11]

12.11 Choosing the Best Form of a Multiple-Regression Equation

As emphasized throughout this chapter, the widespread availability and application of computer technology have facilitated greatly the use of multiple regression and correlation. Statisticians now can try various versions of a particular multiple regression equation to see which version fits best. These experiments can take various forms. For one thing, one can experiment with *various measures of a particular variable*. For example, the university administrator may want to try alternative types of aptitude and typing tests. For another thing, one can experiment with *various forms of the regression equation*. Thus, in the case of the university administrator, he might want to try the logarithm of the aptitude test score, rather than the aptitude test score, as the independent variable.

Still another way in which statisticians can experiment with a particular regression equation is to *add and omit various independent variables to see how the results are affected*. The university administrator might be interested in adding another independent variable, the secretary's IQ score, to the multiple regression to see whether it seems to

---

[11] The printout in Table 12.6 is not meant to be exhaustive. Besides the information shown there, the printout often shows the values of each of the dependent and independent variables (and their means and standard deviations); the simple correlation coefficient between the dependent variable and each independent variable and between each pair of independent variables; and the values of the residuals. Each of these items has been explained previously.

have a significant relationship to the performance rating. Also, he may want to determine (1) whether the estimated regression coefficients of the independent variables currently used are statistically significant when this new independent variable is introduced, and (2) whether the values of these regression coefficients are altered considerably by its introduction.

Statisticians sometimes use a procedure called *stepwise multiple regression* to specify which independent variables seem to provide the best explanation of the behavior of the dependent variable. Using commonly available programs, a computer can determine which of a set of independent variables is most highly correlated with the dependent variable Y. (The computer programs allow the inclusion of dozens of independent variables in the set.) Suppose that this independent variable is X. The computer then selects the independent variable from the remainder of the set that results in the greatest reduction in the variation unexplained by the regression of Y on X. Suppose that this added independent variable is R. Then the computer selects the independent variable from the remainder of this set that results in the greatest reduction in the variation unexplained by the regression of Y on X and R. If this independent variable is S, then the computer selects the independent variable from the remainder of this set that results in the greatest reduction in the variation unexplained by the regression of Y on X, R, and S. This process continues until all the independent variables in the set have been added to the regression, or until none of the remaining independent variables in the set reduces significantly the unexplained variation.

Given the relative ease with which one can calculate alternative forms and versions of a multiple regression equation, it is extremely important that the statistician have good a priori reasons for including each of the independent variables in the regression equation being used. Using stepwise multiple regression, it is not very difficult to select some independent variables that will explain much of the variation in practically any dependent variable, even if these independent variables really have little or no effect on the dependent variable. Why is this so? Because some variables are bound to be correlated with the dependent variable by chance, or they may be influenced by the same factors as the dependent variable, with the result that they are correlated with the dependent variable even though they do not influence the dependent variable. If one uses the computer to hunt around long enough, one is likely to find a combination of independent variables that explains much of the variation in the dependent variable. But if these variables are merely the result of an indiscriminate, mechanical quest for a good-fitting regression equation, the resulting equation is likely to be useless for purposes of prediction.

In choosing the best form of a multiple regression equation, the statistician sometimes must decide whether to include an independent variable which on *a priori* grounds is almost certain to influence the dependent variable, but which has an estimated regression coefficient that is not statistically significant. If there are very strong *a priori* grounds for believing that an independent variable affects the dependent variable, it is acceptable to include this variable in the regression equation, even if its regression coefficient is not statistically significant. After all, the fact that the estimated regression coefficient is not statistically significant does not prove that the true regression coefficient is zero. And the estimated regression coefficient would be likely to constitute a better estimate of the true regression coefficient than zero (which would be the value attributed to it if the independent variable were omitted from the equation).

12.12 Race and City Police Expenditures: A Case Study

To illustrate the use of multiple regression, let's consider once again a study by Pamela Jackson of Rhode Island College and Leo Carroll of the University of Rhode Island of the sociopolitical determinants of municipal police expenditures in 90 nonsouthern American cities.[12] Jackson and Carroll investigated how a city's per capita expenditures for police salaries and operations were related to a variety of socioeconomic characteristics of the city: (1) city population density, (2) percent black, (3) population size, (4) ratio of black to white median income, (5) percent below the poverty line, (6) total city revenue, (7) civil rights mobilization effort,[13] (8) crime rate, (9) number of riots, (10) (percent black)^2, and (11) (percent black)^3.

The results of this multiple regression are shown in Table 12.7. Based on the .05 significance level (and one-tailed tests), per capita expenditures for police salaries and operations seemed to be directly related to population density, population size, total city revenue, and (percent black)^2, and inversely related to the ratio of black to white median income and (percent black)^3. On the other hand, the percent below the poverty line, civil rights mobilization activity, crime rate, and number of riots appeared to have no statistically significant effect on expenditures for police.

---

[12] P. Jackson and L. Carroll, "Race and the War on Crime: The Sociopolitical Determinants of Municipal Police Expenditures in 90 Non-Southern U.S. Cities, *American Sociological Review*, June 1981.

[13] This variable is measured by the number of people involved in civil rights activities, among other things. See ibid.

Independent variable	Regression coefficient	Standard error	t value
Population density	.680	.267	2.55
Percent black	−.340	.443	−0.77
Population size	.002	.001	2.00
Ratio of black to white median income	−21.414	11.281	−1.90
Percent below the poverty level	−.305	.370	−0.82
Total city revenue	.021	.010	2.10
Civil rights mobilization activity	1.758	1.287	1.37
Crime rate	.167	.112	1.49
Number of riots	−.610	.991	−0.62
(Percent black)^2	.028	.016	1.75
(Percent black)^3	−.00035	*	*
Intercept	27.856		
Standard error of estimate	6.045		
R^2	.76		

Table 12.7
Results of Multiple Regression to Explain Variation Among 90 Nonsouthern American Cities in Per Capita Expenditures for Police Salaries and Operations

    * Jackson and Carroll do not give the standard error in this case. (They provide its value rounded to three decimal places, which is .000.) However, they indicate that the regression coefficient differs significantly from zero. For all other independent variables, we calculate the *t* value by dividing the regression coefficient by its standard error.

Source: Jackson and Carroll, op. cit.

    Jackson and Carroll were particularly interested in the relationship between percent black and expenditures for police. According to some sociologists, police expenditures are likely to be relatively high when minority groups constitute a large percent of the population. As noted above, the results indicate that the regression coefficients of (percent black)^2 and (percent black)^3 are both statistically significant at the 0.05 level (one-tailed test). Of course, it is difficult to sort out the reasons why this relationship exists. (Facts really do not speak for themselves.) But the fact that there is such a relationship is of considerable interest to sociologists and others.

12.13 Voter Registration: Another Case Study

As a further illustration of the use of multiple regression, let's return to voter registration, a topic discussed at length in Chapters 1 and 2. Political scientists and others are interested in the fact that the percentage of people of voting age who are registered to vote varies greatly from city to city. A classic study was carried out by Kelley, Ayres, and Bowen,[14] who calculated a multiple regression to estimate the effects on this percentage of six factors: the percent of the persons of voting age who were under 34 years old (X_1); the percent of nonwhites in the population (X_2); median school years completed by persons over 25 years of age (X_3); the difference between 50 and the average percentage of the two-party vote received by Democratic candidates in the late 1950s (X_4); a variable that measures roughly whether literacy tests were required for registration, and if so, how stringent they were (X_5);[15] and the number of days before the election at which registration was closed (X_6).

Based on data for 104 cities, Kelley, Ayres, and Bowen obtained the following regression equation:

$$\hat{Y}_i = 52.20 - 0.50X_{1i} - 0.17X_{2i} + 2.88X_{3i} - 0.41X_{4i} + 0.12X_{5i} + 0.15X_{6i}$$
$$\quad\;\;(0.18)\quad\;(0.07)\quad\;(0.72)\quad\;(0.07)\quad\;(0.04)\quad\;(0.03)$$

where $\hat{Y}_i$ is the predicted percent of the voting-age population registered to vote in the ith city, and $X_{1i}, X_{2i}, \cdots, X_{6i}$ are the values of the independent variables for the ith city. The standard error of each of the regression coefficients is shown in parentheses below the regression coefficient. The t value of each of the estimated regression coefficients can be obtained by dividing the regression coefficient by the standard error. Since each t value is well above 1.96, it is clear that each of the regression coefficients is statistically significant at the 0.05 level. The multiple coefficient of determination, R^2, equals 0.76. Thus, this regression equation can explain 76 percent of the variation among the cities in voter registration rates.

This study was a pioneering attempt to use statistical techniques to analyze voter registration. It provided some of the first quantitative evidence concerning the effects of the age, race, and educational characteristics of a city's population on voter registration. Also, it estimated the effects of factors affecting the difficulty of registration, namely, literacy tests and how quickly registration was closed. Further, it demonstrated that in cities where elections tend to be close (that is, where X_4 is small), registration tends to be high, due in part to the fact that potential voters feel that their votes will have more chance of influencing the results.

[14] S. Kelley, R. Ayres, and W. Bowen, "Registration and Voting: Putting First Things First," *American Political Science Review,* June 1967.

[15] Because of the way this variable is measured, low values of X_5 indicated very stringent tests, and a high value of X_5 (specifically, a value of 100), indicated no test at all.

EXERCISES

12.11 A statistician who works for the Department of Agriculture is interested in the effects on corn output of the amount of nitrogen fertilizer and of the amount of phosphate fertilizer. Using the number of bushels of corn per acre (variable 1) as the dependent variable, he regresses this variable on the number of pounds of nitrogen per acre (variable 2) and the number of pounds of phosphate per acre (variable 3). There are 25 observations. The computer printout is as follows:

```
Independent variable(s)   2   3
Dependent variable        1
Var.          Coeff.        Std. error        t value

  2            .29              .11               2.64
  3            .67              .23               2.91

Intercept     19
```

(a) Compute a 90 percent confidence interval for the true regression coefficient of the number of pounds of nitrogen per acre.
(b) Compute a 95 percent confidence interval for the true regression coefficient of the number of pounds of phosphate per acre.
(c) Suppose that the printout did not include the standard errors of the regression coefficients, but that the other figures shown above were included. Could you figure out the standard errors? If so, how?

12.12 A store wants to estimate the relationship between its sales, on the one hand, and the number of salespersons it hires and its advertising expenditures, on the other. Using its monthly sales (variable 1) as the dependent variable, the store's statistician regresses this variable on the number of salespersons employed by the store during the month (variable 2) and the store's monthly advertising expenditure (variable 3). There are 43 observations. Both variables 1 and 3 are measured in millions of dollars. The computer printout is as follows:

```
Independent variable(s)   2   3
Dependent variable        1
Var.          Coeff.        Std. error        t value

  2           0.052            .021              2.48
  3           5.621           2.012              2.79

Intercept     0.31
```

(a) Compute a 95 percent confidence interval for the true regression coefficient of the number of salespersons employed by the store.

(b) Compute a 90 percent confidence interval for the true regression coefficient of the store's advertising expenditure.

(c) If both variables 2 and 3 equal zero, wouldn't one expect that variable 1 should equal zero? If the standard error of the intercept is 0.19, can you test whether this is true? How? (Let $\alpha = .05$.)

The Chapter in a Nutshell

1. Whereas a *simple regression* includes only one independent variable, a *multiple regression* includes more than one independent variable. An advantage of multiple regression over simple regression is that one frequently can predict the dependent variable more accurately if more than one independent variable is used. Also, if the dependent variable is influenced by more than one independent variable, a simple regression of the dependent variable on a single independent variable may result in a biased estimate of the effect of this independent variable on the dependent variable.

2. The first step in *multiple regression analysis* is to identify the independent variables, and then to specify the mathematical form of the equation relating the expected value of the dependent variable to the independent variables. For example, if Y is the dependent variable and X_1 and X_2 are identified as the independent variables, one might specify that

$$Y_i = A + B_1 X_{1i} + B_2 X_{2i} + e_i,$$

where e_i is the difference between Y_i and $E(Y_i)$. To estimate B_1 and B_2 (called the true regression coefficients of X_1 and X_2) as well as A (the intercept of this true regression equation), we use the values that minimize the sum of squared deviations of Y_i from $\hat{Y}_i$, the value of the dependent variable predicted by the estimated regression equation.

3. Multiple regressions are generally calculated by computers rather than by hand. The standard programs print out the *estimated standard deviations* of the least-squares estimators of B_1 and B_2 (these estimated standard deviations being called *standard errors*). Using the value of the least-squares estimator of B_1 or B_2 together with these standard errors, one can obtain a confidence interval for B_1 or B_2. For example, a confidence interval for B_1 is $b_1 \pm t_{\alpha/2} s_{b_1}$, where b_1 is the least-squares estimator of B_1 and s_{b_1} is the standard error of b_1. The standard programs also print out the t value for each estimated regression coefficient, the t value being the estimated regression coefficient divided by its standard error. To test whether a true regression coefficient equals zero, one should see whether its t value exceeds $t_{\alpha/2}$ or is

less than $-t_{\alpha/2}$; if so, one should reject the hypothesis that the true regression coefficient is zero. (This is a two-tailed test.)

4. The *multiple coefficient of determination* R^2 equals the ratio of the variation explained by the multiple regression to the total variation in the dependent variable. The positive square root of the multiple coefficient of determination is called the *multiple correlation coefficient* and is denoted by R. Both R^2 and R are measures of how well the regression equation fits the data: The closer they are to zero, the poorer the fit; the closer they are to 1, the better the fit.

5. The advent of computer technology has caused a marked reduction in the amount of effort and expense required to calculate multiple regressions. We have described and explained the items generally printed out by "canned" programs, including the F test for determining whether all true regression coefficients are zero. With existing computer technology, one can experiment with various forms of a regression equation. It is important that the statistician have good a priori reasons for including each of the independent variables in the equation. If independent variables are chosen indiscriminately in a mechanical quest for a good-fitting equation, the resulting equation is likely to be useless for purposes of prediction.

Chapter Review Exercises

12.13 A government official asks a research assistant to calculate a multiple regression on an electronic computer. The dependent variable is a particular industry's rate of productivity increase during 1948–66, and the independent variables are (1) the percent of the industry's employees belonging to unions in 1953 (variable 2); (2) the percent of the industry's value added spent on applied research and development in 1958 (variable 4); and (3) the percentage of the firms in the industry reporting in 1958 that they expected their R and D (research and development) expenditures to pay out in no less than six years (variable 8). Only 17 industries can be included because data are lacking concerning the third independent variable for 3 of the industries. The research assistant comes back with the printout, which is as follows:

INDEPENDENT VARIABLE(S)	2	4	8
DEPENDENT VARIABLE	1		

VAR	COEFF	STD. ERROR	T-VALUE
2	−0.0622	0.0090	−6.9400
4	0.0963	0.0154	6.2700
8	0.0704	0.0098	7.2000

INTERCEPT 4.65948

(a) What is the estimated regression equation?

(b) Calculate a 95 percent confidence interval for the true regression coefficient of the percent of the industry's value added spent on applied research and development.

(c) Calculate a 90 percent confidence interval for the true regression coefficient of the percent of the industry's employees belonging to unions.

(d) The official wants to test the hypothesis that whether an industry's R and D consists mainly of long-term or short-term projects has no effect on its rate of productivity increase. If the significance level is set at .05, and if the alternative hypothesis is two-sided, use the printout to test this hypothesis.

(e) Another part of the computer printout is shown below. What is the multiple coefficient of determination in this case? What does this result mean?

```
S.E. OF ESTIMATE      0.3458
F-VALUE              29.01
MULTIPLE R-SQUARED    .87
```

(f) Test the null hypothesis that all the true regression coefficients in the multiple regression are zero. (Set $\alpha = .05$.)

12.14 The Rotunda Corporation believes that its annual sales depend on disposable income in its city of location and on the price of its product. Data concerning these variables are given below:

Year	Number of units sold annually (millions)	Disposable income (billions of dollars)	Price of product (dollars per unit)
1986	8	4	9
1985	8	4	8
1984	5	3	8
1983	4	3	9
1982	6	3	7
1981	4	3	10
1980	2	2	8
1979	3	2	6
1978	4	2	5
1977	2	2	7
1976	2	2	8
1975	1	1	6
1974	1	1	5
1973	1	1	7
1972	1	1	5

(a) Calculate the multiple regression equation, if Rotunda's sales are the dependent variable and disposable income and price of product are the independent variables.

(b) Based on your results in (a), predict the volume of annual sales for the Rotunda Corporation under each of the following sets of circumstances:

   (i) disposable income equals $3 billion and price equals $7;

   (ii) disposable income equals $1 billion and price equals $6;

   (iii) disposable income equals $2 billion and price equals $10.

(c) What is the estimated effect on the expected value of Rotunda's annual sales of an increase of $1 billion in disposable income? What is the estimated effect of an increase of $1 in price?

(d) The president of the Rotunda Corporation uses the multiple regression equation you derived in (a) to predict the firm's annual sales for next year. Since he believes that disposable income will be $6 billion and product price will be $11, he bases his prediction on these values of the independent variables. What objections might legitimately be raised against these procedures?

(e) Use the data to calculate the multiple coefficient of determination between Rotunda's sales, on the one hand, and disposable income and price, on the other. Interpret your results. Does it appear that the regression equation you derived in (a) fits the data well? Why, or why not?

(f) Between 1972 and 1986, the population of the city in which the Rotunda Corporation is located grew steadily. By 1986, it was about four times its size in 1972. The marketing director of the Rotunda Corporation objects to the equation that you derived in (a) on the grounds that population, not disposable income, has been the variable that has influenced Rotunda's sales.

   (i) Is the marketing director correct in saying that if population and disposable income are perfectly correlated, the observed effect of disposable income in your equation may be due to population?

   (ii) Can you suggest some ways of altering the multiple regression so that the marketing director's hypothesis can be tested?

12.15 For 20 working women (with children) in a particular community, their age at the birth of their first child, their score on an IQ test, and their current annual income are as follows:

Age at birth of first child (years)	IQ test score	Current annual income ($1000)
18	105	16
17	108	17
20	110	17
22	109	18
24	110	17
26	120	21
27	125	20
19	111	17
18	85	9
20	96	11
21	94	10
24	130	26
30	134	21
28	103	11
25	102	12
24	89	8
22	92	12
20	95	13
19	100	14
17	91	9

Use Minitab (or some other computer package) to calculate the regression of income on IQ test score and age at birth of first child, as well as the multiple correlation coefficient. If no computer package is available, do the calculations by hand.

Appendix

Appendix Table 1

BINOMIAL PROBABILITY DISTRIBUTION

This table shows the value of

$$P(x) = \frac{n!}{(n-x)!x!} \Pi^x (1 - \Pi)^{n-x}$$

for selected values of Π and for $n = 1$ to 20. For values of Π exceeding 0.5, the value of $P(x)$ can be obtained by substituting $(1 - \Pi)$ for Π and by finding $P(n - x)$. (See Section 4.6.)

| | | | | | | Π | | | | | |
n	x	.05	.10	.15	.20	.25	.30	.35	.40	.45	.50
1	0	.9500	.9000	.8500	.8000	.7500	.7000	.6500	.6000	.5500	.5000
	1	.0500	.1000	.1500	.2000	.2500	.3000	.3500	.4000	.4500	.5000
2	0	.9025	.8100	.7225	.6400	.5625	.4900	.4225	.3600	.3025	.2500
	1	.0950	.1800	.2550	.3200	.3750	.4200	.4550	.4800	.4950	.5000
	2	.0025	.0100	.0225	.0400	.0625	.0900	.1225	.1600	.2025	.2500
3	0	.8574	.7290	.6141	.5120	.4219	.3430	.2746	.2160	.1664	.1250
	1	.1354	.2430	.3251	.3840	.4219	.4410	.4436	.4320	.4084	.3750
	2	.0071	.0270	.0574	.0960	.1406	.1890	.2389	.2880	.3341	.3750
	3	.0001	.0010	.0034	.0080	.0156	.0270	.0429	.0640	.0911	.1250
4	0	.8145	.6561	.5220	.4096	.3164	.2401	.1785	.1296	.0915	.0625
	1	.1715	.2916	.3685	.4096	.4219	.4116	.3845	.3456	.2995	.2500
	2	.0135	.0486	.0975	.1536	.2109	.2646	.3105	.3456	.3675	.3750
	3	.0005	.0036	.0115	.0256	.0469	.0756	.1115	.1536	.2005	.2500
	4	.0000	.0001	.0005	.0016	.0039	.0081	.0150	.0256	.0410	.0625

Appendix Table 1 (Continued)

		Π									
n	x	.05	.10	.15	.20	.25	.30	.35	.40	.45	.50
5	0	.7738	.5905	.4437	.3277	.2373	.1681	.1160	.0778	.0503	.0312
	1	.2036	.3280	.3915	.4096	.3955	.3602	.3124	.2592	.2059	.1562
	2	.0214	.0729	.1382	.2048	.2637	.3087	.3364	.3456	.3369	.3125
	3	.0011	.0081	.0244	.0512	.0879	.1323	.1811	.2304	.2757	.3125
	4	.0000	.0004	.0022	.0064	.0146	.0284	.0488	.0768	.1128	.1562
	5	.0000	.0000	.0001	.0003	.0010	.0024	.0053	.0102	.0185	.0312
6	0	.7351	.5314	.3771	.2621	.1780	.1176	.0754	.0467	.0277	.0156
	1	.2321	.3543	.3993	.3932	.3560	.3025	.2437	.1866	.1359	.0938
	2	.0305	.0984	.1762	.2458	.2966	.3241	.3280	.3110	.2780	.2344
	3	.0021	.0146	.0415	.0819	.1318	.1852	.2355	.2765	.3032	.3125
	4	.0001	.0012	.0055	.0154	.0330	.0595	.0951	.1382	.1861	.2344
	5	.0000	.0001	.0004	.0015	.0044	.0102	.0205	.0369	.0609	.0938
	6	.0000	.0000	.0000	.0001	.0002	.0007	.0018	.0041	.0083	.0516
7	0	.6983	.4783	.3206	.2097	.1335	.0824	.0490	.0280	.0152	.0078
	1	.2573	.3720	.3960	.3670	.3115	.2471	.1848	.1306	.0872	.0547
	2	.0406	.1240	.2097	.2753	.3115	.3177	.2985	.2613	.2140	.1641
	3	.0036	.0230	.0617	.1147	.1730	.2269	.2679	.2903	.2918	.2734
	4	.0002	.0026	.0109	.0287	.0577	.0972	.1442	.1935	.2388	.2734
	5	.0000	.0002	.0012	.0043	.0115	.0250	.0466	.0774	.1172	.1641
	6	.0000	.0000	.0001	.0004	.0013	.0036	.0084	.0172	.0320	.0547
	7	.0000	.0000	.0000	.0000	.0001	.0002	.0006	.0016	.0037	.0078
8	0	.6634	.4305	.2725	.1678	.1001	.0576	.0319	0168	.0084	.0039
	1	.2793	.3826	.3847	.3355	.2670	.1977	.1373	.0896	.0548	.0312
	2	.0515	.1488	.2376	.2936	.3115	.2965	.2587	.2090	.1569	.1094
	3	.0054	.0331	.0839	.1468	.2076	.2541	.2786	.2787	.2568	.2188
	4	.0004	.0046	.0815	.0459	.0865	.1361	.1875	.2322	.2627	.2734
	5	.0000	.0004	.0026	.0092	.0231	.0467	.0808	.1239	.1719	.2188
	6	.0000	.0000	.0002	.0011	.0038	.0100	.0217	.0413	.0703	.1094
	7	.0000	.0000	.0000	.0001	.0004	.0012	.0033	.0079	.0164	.0312
	8	.0000	.0000	.0000	.0000	.0000	.0001	.0002	.0007	.0017	.0039
9	0	.6302	.3874	.2316	.1342	.0751	.0404	.0207	.0101	.0046	.0020
	1	.2985	.3874	.3679	.3020	.2253	.1556	.1004	.0605	.0339	.0176
	2	.0629	.1722	.2597	.3020	.3003	.2668	.2162	.1612	.1110	.0703
	3	.0077	.0446	.1069	.1762	.2336	.2668	.2716	.2508	.2119	.1641
	4	.0006	.0074	.0283	.0661	.1168	.1715	.2194	.2508	.2600	.2461

Appendix Table 1 (Continued)

n	x	.05	.10	.15	.20	.25	.30	.35	.40	.45	.50
9	5	.0000	.0008	.0050	.0165	.0389	.0735	.1181	.1672	.2128	.2461
	6	.0000	.0001	.0006	.0028	.0087	.0210	.0424	.0743	.1160	.1641
	7	.0000	.0000	.0000	.0003	.0012	.0039	.0098	.0212	.0407	.0703
	8	.0000	.0000	.0000	.0000	.0001	.0004	.0013	.0035	.0083	.0716
	9	.0000	.0000	.0000	.0000	.0000	.0000	.0001	.0003	.0008	.0020
10	0	.5987	.3487	.1969	.1074	.0563	.0282	.0135	.0060	.0025	.0010
	1	.3151	.3874	.3474	.2684	.1877	.1211	.0725	.0403	.0207	.0098
	2	.0746	.1937	.2759	.3020	.2816	.2335	.1757	.1209	.0763	.0439
	3	.0105	.0574	.1298	.2013	.2503	.2668	.2522	.2150	.1665	.1172
	4	.0010	.0112	.0401	.0881	.1460	.2001	.2377	.2508	.2384	.2051
	5	.0001	.0015	.0085	.0264	.0584	.1029	.1536	.2007	.2340	.2461
	6	.0000	.0001	.0012	.0055	.0162	.0368	.0689	.1115	.1596	.2051
	7	.0000	.0000	.0001	.0008	.0031	.0090	.0212	.0425	.0746	.1172
	8	.0000	.0000	.0000	.0001	.0004	.0014	.0043	.0106	.0229	.0439
	9	.0000	.0000	.0000	.0000	.0000	.0001	.0005	.0016	.0042	.0098
	10	.0000	.0000	.0000	.0000	.0000	.0000	.0000	.0001	.0003	.0010
11	0	.5688	.3138	.1673	.0859	.0422	.0198	.0088	.0036	.0014	.0005
	1	.3293	.3835	.3248	.2362	.1549	.0932	.0518	.0266	.0125	.0054
	2	.0867	.2131	.2866	.2953	.2581	.1998	.1395	.0887	.0513	.0269
	3	.0137	.0710	.1517	.2215	.2581	.2568	.2254	.1774	.1259	.0806
	4	.0014	.0158	.0536	.1107	.1721	.2201	.2428	.2365	.2060	.1611
	5	.0001	.0025	.0132	.0388	.0803	.1321	.1830	.2207	.2360	.2256
	6	.0000	.0003	.0023	.0097	.0268	.0566	.0985	.1471	.1931	.2256
	7	.0000	.0000	.0003	.0017	.0064	.0173	.0379	.0701	.1128	.1611
	8	.0000	.0000	.0000	.0002	.0011	.0037	.0102	.0234	.0462	.0806
	9	.0000	.0000	.0000	.0000	.0001	.0005	.0018	.0052	.0126	.0269
	10	.0000	.0000	.0000	.0000	.0000	.0000	.0002	.0007	.0021	.0054
	11	.0000	.0000	.0000	.0000	.0000	.0000	.0000	.0000	.0002	.0005
12	0	.5404	.2824	.1422	.0687	.0317	.0138	.0057	.0022	.0008	.0002
	1	.3413	.3766	.3012	.2062	.1267	.0712	.0368	.0174	.0075	.0029
	2	.0988	.2301	.2924	.2835	.2323	.1678	.1088	.0639	.0339	.0161
	3	.0173	.0852	.1720	.2362	.2581	.2397	.1954	.1419	.0923	.0537
	4	.0021	.0213	.0683	.1329	.1936	.2311	.2367	.2128	.1700	.1208
	5	.0002	.0038	.0193	.0532	.1032	.1585	.2039	.2270	.2225	.1934
	6	.0000	.0005	.0040	.0155	.0401	.0792	.1281	.1766	.2124	.2256
	7	.0000	.0000	.0006	.0033	.0115	.0291	.0591	.1009	.1489	.1934
	8	.0000	.0000	.0001	.0005	.0024	.0078	.0199	.0420	.0762	.1208
	9	.0000	.0000	.0000	.0001	.0004	.0015	.0048	.0125	.0277	.0537

Appendix Table 1 (Continued)

						Π					
n	x	.05	.10	.15	.20	.25	.30	.35	.40	.45	.50
12	10	.0000	.0000	.0000	.0000	.0000	.0002	.0008	.0025	.0068	.0161
	11	.0000	.0000	.0000	.0000	.0000	.0000	.0001	.0003	.0010	.0029
	12	.0000	.0000	.0000	.0000	.0000	.0000	.0000	.0000	.0001	.0002
13	0	.5133	.2542	.1209	.0550	.0238	.0097	.0037	.0013	.0004	.0001
	1	.3512	.3672	.2774	.1787	.1029	.0540	.0259	.0113	.0045	.0016
	2	.1109	.2448	.2937	.2680	.2059	.1388	.0836	.0453	.0220	.0095
	3	.0214	.0997	.1900	.2457	.2517	.2181	.1651	.1107	.0660	.0349
	4	.0028	.0277	.0838	.1535	.2097	.2337	.2222	.1845	.1350	.0873
	5	.0003	.0055	.0266	.0691	.1258	.1803	.2154	.2214	.1989	.1571
	6	.0000	.0008	.0063	.0230	.0559	.1030	.1546	.1968	.2169	.2095
	7	.0000	.0001	.0011	.0058	.0186	.0442	.0833	.1312	.1775	.2095
	8	.0000	.0000	.0001	.0011	.0047	.0142	.0336	.0656	.1089	.1571
	9	.0000	.0000	.0000	.0001	.0009	.0034	.0101	.0243	.0495	.0873
	10	.0000	.0000	.0000	.0000	.0001	.0006	.0022	.0065	.0162	.0349
	11	.0000	.0000	.0000	.0000	.0000	.0001	.0003	.0012	.0036	.0095
	12	.0000	.0000	.0000	.0000	.0000	.0000	.0000	.0001	.0005	.0016
	13	.0000	.0000	.0000	.0000	.0000	.0000	.0000	.0000	.0000	.0001
14	0	.4877	.2288	.1028	.0440	.0178	.0068	.0024	.0008	.0002	.0001
	1	.3593	.3559	.2539	.1539	.0832	.0407	.0181	.0073	.0027	.0009
	2	.1229	.2570	.2912	.2501	.1802	.1134	.0634	.0317	.0141	.0056
	3	.0259	.1142	.2056	.2501	.2402	.1943	.1366	.0845	.0462	.0222
	4	.0037	.0348	.0998	.1720	.2202	.2290	.2022	.1549	.1040	.0611
	5	.0004	.0078	.0352	.0860	.1468	.1963	.2178	.2066	.1701	.1222
	6	.0000	.0013	.0093	.0322	.0734	.1262	.1759	.2066	.2088	.1833
	7	.0000	.0002	.0019	.0092	.0280	.0618	.1082	.1574	.1952	.2095
	8	.0000	.0000	.0003	.0020	.0082	.0232	.0510	.0918	.1398	.1833
	9	.0000	.0000	.0000	.0003	.0018	.0066	.0183	.0408	.0762	.1222
	10	.0000	.0000	.0000	.0000	.0003	.0014	.0049	.0136	.0312	.0611
	11	.0000	.0000	.0000	.0000	.0000	.0002	.0010	.0033	.0093	.0222
	12	.0000	.0000	.0000	.0000	.0000	.0000	.0001	.0005	.0019	.0056
	13	.0000	.0000	.0000	.0000	.0000	.0000	.0000	.0001	.0002	.0009
	14	.0000	.0000	.0000	.0000	.0000	.0000	.0000	.0000	.0000	.0001
15	0	.4633	.2059	.0874	.0352	.0134	.0047	.0016	.0005	.0001	.0000
	1	.3658	.3432	.2312	.1319	.0668	.0305	.0126	.0047	.0016	.0005
	2	.1348	.2669	.2856	.2309	.1559	.0916	.0476	.0219	.0090	.0032
	3	.0307	.1285	.2184	.2501	.2252	.1700	.1110	.0634	.0318	.0139
	4	.0049	.0428	.1156	.1876	.2252	.2186	.1792	.1268	.0780	.0417

Appendix Table 1 (Continued)

n	x	.05	.10	.15	.20	.25	.30	.35	.40	.45	.50
						Π					
15	5	.0006	.0105	.0449	.1032	.1651	.2061	.2123	.1859	.1404	.0916
	6	.0000	.0019	.0132	.0430	.0917	.1472	.1906	.2066	.1914	.1527
	7	.0000	.0003	.0030	.0138	.0393	.0811	.1319	.1771	.2013	.1964
	8	.0000	.0000	.0005	.0035	.0131	.0348	.0710	.1181	.1647	.1964
	9	.0000	.0000	.0001	.0007	.0034	.0116	.0298	.0612	.1048	.1527
	10	.0000	.0000	.0000	.0001	.0007	.0030	.0096	.0245	.0515	.0916
	11	.0000	.0000	.0000	.0000	.0001	.0006	.0024	.0074	.0191	.0417
	12	.0000	.0000	.0000	.0000	.0000	.0001	.0004	.0016	.0052	.0139
	13	.0000	.0000	.0000	.0000	.0000	.0000	.0001	.0003	.0010	.0032
	14	.0000	.0000	.0000	.0000	.0000	.0000	.0000	.0000	.0001	.0005
	15	.0000	.0000	.0000	.0000	.0000	.0000	.0000	.0000	.0000	.0000
16	0	.4401	.1853	.0743	.0281	.0100	.0033	.0010	.0003	.0001	.0000
	1	.3706	.3294	.2097	.1126	.0535	.0228	.0087	.0030	.0009	.0002
	2	.1463	.2745	.2775	.2111	.1336	.0732	.0353	.0150	.0056	.0018
	3	.0359	.1423	.2285	.2463	.2079	.1465	.0888	.0468	.0215	.0085
	4	.0061	.0514	.1311	.2001	.2252	.2040	.1553	.1014	.0572	.0278
	5	.0008	.0137	.0555	.1201	.1802	.2099	.2008	.1623	.1123	.0667
	6	.0001	.0028	.0180	.0550	.1101	.1649	.1982	.1983	.1684	.1222
	7	.0000	.0004	.0045	.0197	.0524	.1010	.1524	.1889	.1969	.1746
	8	.0000	.0001	.0009	.0055	.0197	.0487	.0923	.1417	.1812	.1964
	9	.0000	.0000	.0001	.0012	.0058	.0185	.0442	.0840	.1318	.1746
	10	.0000	.0000	.0000	.0002	.0014	.0056	.0167	.0392	.0755	.1222
	11	.0000	.0000	.0000	.0000	.0002	.0013	.0049	.0142	.0337	.0667
	12	.0000	.0000	.0000	.0000	.0000	.0002	.0011	.0040	.0115	.0278
	13	.0000	.0000	.0000	.0000	.0000	.0000	.0002	.0008	.0029	.0085
	14	.0000	.0000	.0000	.0000	.0000	.0000	.0000	.0001	.0005	.0018
	15	.0000	.0000	.0000	.0000	.0000	.0000	.0000	.0000	.0001	.0002
	16	.0000	.0000	.0000	.0000	.0000	.0000	.0000	.0000	.0000	.0000
17	0	.4181	.1668	.0631	.0225	.0075	.0023	.0007	.0002	.0000	.0000
	1	.3741	.3150	.1893	.0957	.0426	.0169	.0060	.0019	.0005	.0001
	2	.1575	.2800	.2673	.1914	.1136	.0581	.0260	.0102	.0035	.0010
	3	.0415	.1556	.2359	.2393	.1893	.1245	.0701	.0341	.0144	.0052
	4	.0076	.0605	.1457	.2093	.2209	.1868	.1320	.0796	.0411	.0182

Appendix Table 1 (Continued)

						Π					
n	x	.05	.10	.15	.20	.25	.30	.35	.40	.45	.50
17	5	.0010	.0175	.0668	.1361	.1914	.2081	.1849	.1379	.0875	.0472
	6	.0001	.0039	.0236	.0680	.1276	.1784	.1991	.1839	.1432	.0944
	7	.0000	.0007	.0065	.0267	.0668	.1201	.1685	.1927	.1841	.1484
	8	.0000	.0001	.0014	.0084	.0279	.0644	.1134	.1606	.1883	.1855
	9	.0000	.0000	.0003	.0021	.0093	.0276	.0611	.1070	.1540	.1855
	10	.0000	.0000	.0000	.0004	.0025	.0095	.0263	.0571	.1008	.1484
	11	.0000	.0000	.0000	.0001	.0005	.0026	.0090	.0242	.0525	.0944
	12	.0000	.0000	.0000	.0000	.0001	.0006	.0024	.0021	.0215	.0472
	13	.0000	.0000	.0000	.0000	.0000	.0001	.0005	.0021	.0068	.0182
	14	.0000	.0000	.0000	.0000	.0000	.0000	.0001	.0004	.0016	.0052
	15	.0000	.0000	.0000	.0000	.0000	.0000	.0000	.0001	.0003	.0010
	16	.0000	.0000	.0000	.0000	.0000	.0000	.0000	.0000	.0000	.0001
	17	.0000	.0000	.0000	.0000	.0000	.0000	.0000	.0000	.0000	.0000
18	0	.3972	.1501	.0536	.0180	.0056	.0016	.0004	.0001	.0000	.0000
	1	.3763	.3002	.1704	.0811	.0338	.0126	.0042	.0012	.0003	.0001
	2	.1683	.2835	.2556	.1723	.0958	.0458	.0190	.0069	.0022	.0006
	3	.0473	.1680	.2406	.2297	.1704	.1046	.0547	.0246	.0095	.0031
	4	.0093	.0700	.1592	.2153	.2130	.1681	.1104	.0614	.0291	.0117
	5	.0014	.0218	.0787	.1507	.1988	.2017	.1664	.1146	.0666	.0327
	6	.0002	.0052	.0301	.0816	.1436	.1873	.1941	.1655	.1181	.0708
	7	.0000	.0010	.0091	.0350	.0820	.1376	.1792	.1892	.1657	.1214
	8	.0000	.0002	.0022	.0120	.0376	.0811	.1327	.1734	.1864	.1669
	9	.0000	.0000	.0004	.0033	.0139	.0386	.0794	.1284	.1694	.1855
	10	.0000	.0000	.0001	.0008	.0042	.0149	.0385	.0771	.1248	.1669
	11	.0000	.0000	.0000	.0001	.0010	.0046	.0151	.0374	.0742	.1214
	12	.0000	.0000	.0000	.0000	.0002	.0012	.0047	.0145	.0354	.0708
	13	.0000	.0000	.0000	.0000	.0000	.0002	.0012	.0044	.0134	.0327
	14	.0000	.0000	.0000	.0000	.0000	.0000	.0002	.0011	.0039	.0117
	15	.0000	.0000	.0000	.0000	.0000	.0000	.0000	.0002	.0009	.0031
	16	.0000	.0000	.0000	.0000	.0000	.0000	.0000	.0000	.0001	.0006
	17	.0000	.0000	.0000	.0000	.0000	.0000	.0000	.0000	.0000	.0001
	18	.0000	.0000	.0000	.0000	.0000	.0000	.0000	.0000	.0000	.0000
19	0	.3774	.1351	.0456	.0144	.0042	.0011	.0003	.0001	.0000	.0000
	1	.3774	.2852	.1529	.0685	.0268	.0093	.0029	.0008	.0002	.0000
	2	.1787	.2852	.2428	.1540	.0803	.0358	.0138	.0046	.0013	.0003
	3	.0533	.1796	.2428	.2182	.1517	.0869	.0422	.0175	.0062	.0018
	4	.0112	.0798	.1714	.2182	.2023	.1491	.0909	.0467	.0203	.0074

Appendix Table 1 (Continued)

						Π					
n	x	.05	.10	.15	.20	.25	.30	.35	.40	.45	.50
19	5	.0018	.0266	.0907	.1636	.2023	.1916	.1468	.0933	.0497	.0222
	6	.0002	.0069	.0374	.0955	.1574	.1916	.1844	.1451	.0949	.0518
	7	.0000	.0014	.0122	.0443	.0974	.1525	.1844	.1797	.1443	.0961
	8	.0000	.0002	.0032	.0166	.0487	.0981	.1489	.1797	.1771	.1442
	9	.0000	.0000	.0007	.0051	.0198	.0514	.0980	.1464	.1771	.1762
	10	.0000	.0000	.0001	.0013	.0066	.0220	.0528	.0976	.1449	.1762
	11	.0000	.0000	.0000	.0003	.0018	.0077	.0233	.0532	.0970	.1442
	12	.0000	.0000	.0000	.0000	.0004	.0022	.0083	.0237	.0529	.0961
	13	.0000	.0000	.0000	.0000	.0001	.0005	.0024	.0085	.0233	.0518
	14	.0000	.0000	.0000	.0000	.0000	.0001	.0006	.0024	.0082	.0222
	15	.0000	.0000	.0000	.0000	.0000	.0000	.0001	.0005	.0022	.0074
	16	.0000	.0000	.0000	.0000	.0000	.0000	.0000	.0001	.0005	.0018
	17	.0000	.0000	.0000	.0000	.0000	.0000	.0000	.0000	.0001	.0003
	18	.0000	.0000	.0000	.0000	.0000	.0000	.0000	.0000	.0000	.0000
	19	.0000	.0000	.0000	.0000	.0000	.0000	.0000	.0000	.0000	.0000
20	0	.3585	.1216	.0388	.0115	.0032	.0008	.0002	.0000	.0000	.0000
	1	.3774	.2702	.1368	.0576	.0211	.0068	.0020	.0005	.0001	.0000
	2	.1887	.2852	.2293	.1369	.0669	.0278	.0100	.0031	.0008	.0002
	3	.0596	.1901	.2428	.2054	.1339	.0716	.0323	.0123	.0040	.0011
	4	.0133	.0898	.1821	.2182	.1897	.1304	.0738	.0350	.0139	.0046
	5	.0022	.0319	.1028	.1746	.2023	.1789	.1272	.0746	.0365	.0148
	6	.0003	.0089	.0454	.1091	.1686	.1916	.1712	.1244	.0746	.0370
	7	.0000	.0020	.0160	.0545	.1124	.1643	.1844	.1659	.1221	.0739
	8	.0000	.0004	.0046	.0222	.0609	.1144	.1614	.1797	.1623	.1201
	9	.0000	.0001	.0011	.0074	.0271	.0654	.1158	.1597	.1771	.1602
	10	.0000	.0000	.0002	.0020	.0099	.0308	.0686	.1171	.1593	.1762
	11	.0000	.0000	.0000	.0005	.0030	.0120	.0336	.0710	.1185	.1602
	12	.0000	.0000	.0000	.0001	.0008	.0039	.0136	.0355	.0727	.1201
	13	.0000	.0000	.0000	.0000	.0002	.0010	.0045	.0146	.0366	.0739
	14	.0000	.0000	.0000	.0000	.0000	.0002	.0012	.0049	.0150	.0370
	15	.0000	.0000	.0000	.0000	.0000	.0000	.0003	.0013	.0049	.0148
	16	.0000	.0000	.0000	.0000	.0000	.0000	.0000	.0003	.0013	.0046
	17	.0000	.0000	.0000	.0000	.0000	.0000	.0000	.0000	.0002	.0011
	18	.0000	.0000	.0000	.0000	.0000	.0000	.0000	.0000	.0000	.0002
	19	.0000	.0000	.0000	.0000	.0000	.0000	.0000	.0000	.0000	.0000
	20	.0000	.0000	.0000	.0000	.0000	.0000	.0000	.0000	.0000	.0000

SOURCE: This table is taken from National Bureau of Standards, *Tables of the Binomal Probability Distribution*, Applied Mathematics Series. (U.S. Department of Commerce, 1950).

Appendix Table 2

AREAS UNDER THE STANDARD NORMAL CURVE

This table shows the area between zero (the mean of a standard normal variable) and z. For example, if $z = 1.50$, this is the shaded area shown below which equals .4332.

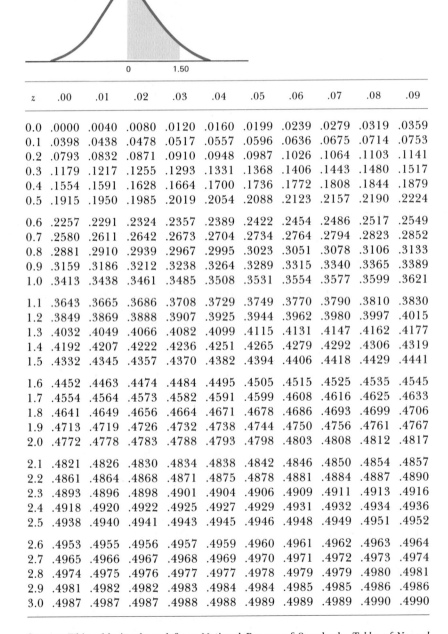

z	.00	.01	.02	.03	.04	.05	.06	.07	.08	.09
0.0	.0000	.0040	.0080	.0120	.0160	.0199	.0239	.0279	.0319	.0359
0.1	.0398	.0438	.0478	.0517	.0557	.0596	.0636	.0675	.0714	.0753
0.2	.0793	.0832	.0871	.0910	.0948	.0987	.1026	.1064	.1103	.1141
0.3	.1179	.1217	.1255	.1293	.1331	.1368	.1406	.1443	.1480	.1517
0.4	.1554	.1591	.1628	.1664	.1700	.1736	.1772	.1808	.1844	.1879
0.5	.1915	.1950	.1985	.2019	.2054	.2088	.2123	.2157	.2190	.2224
0.6	.2257	.2291	.2324	.2357	.2389	.2422	.2454	.2486	.2517	.2549
0.7	.2580	.2611	.2642	.2673	.2704	.2734	.2764	.2794	.2823	.2852
0.8	.2881	.2910	.2939	.2967	.2995	.3023	.3051	.3078	.3106	.3133
0.9	.3159	.3186	.3212	.3238	.3264	.3289	.3315	.3340	.3365	.3389
1.0	.3413	.3438	.3461	.3485	.3508	.3531	.3554	.3577	.3599	.3621
1.1	.3643	.3665	.3686	.3708	.3729	.3749	.3770	.3790	.3810	.3830
1.2	.3849	.3869	.3888	.3907	.3925	.3944	.3962	.3980	.3997	.4015
1.3	.4032	.4049	.4066	.4082	.4099	.4115	.4131	.4147	.4162	.4177
1.4	.4192	.4207	.4222	.4236	.4251	.4265	.4279	.4292	.4306	.4319
1.5	.4332	.4345	.4357	.4370	.4382	.4394	.4406	.4418	.4429	.4441
1.6	.4452	.4463	.4474	.4484	.4495	.4505	.4515	.4525	.4535	.4545
1.7	.4554	.4564	.4573	.4582	.4591	.4599	.4608	.4616	.4625	.4633
1.8	.4641	.4649	.4656	.4664	.4671	.4678	.4686	.4693	.4699	.4706
1.9	.4713	.4719	.4726	.4732	.4738	.4744	.4750	.4756	.4761	.4767
2.0	.4772	.4778	.4783	.4788	.4793	.4798	.4803	.4808	.4812	.4817
2.1	.4821	.4826	.4830	.4834	.4838	.4842	.4846	.4850	.4854	.4857
2.2	.4861	.4864	.4868	.4871	.4875	.4878	.4881	.4884	.4887	.4890
2.3	.4893	.4896	.4898	.4901	.4904	.4906	.4909	.4911	.4913	.4916
2.4	.4918	.4920	.4922	.4925	.4927	.4929	.4931	.4932	.4934	.4936
2.5	.4938	.4940	.4941	.4943	.4945	.4946	.4948	.4949	.4951	.4952
2.6	.4953	.4955	.4956	.4957	.4959	.4960	.4961	.4962	.4963	.4964
2.7	.4965	.4966	.4967	.4968	.4969	.4970	.4971	.4972	.4973	.4974
2.8	.4974	.4975	.4976	.4977	.4977	.4978	.4979	.4979	.4980	.4981
2.9	.4981	.4982	.4982	.4983	.4984	.4984	.4985	.4985	.4986	.4986
3.0	.4987	.4987	.4987	.4988	.4988	.4989	.4989	.4989	.4990	.4990

SOURCE: This table is adapted from National Bureau of Standards, *Tables of Normal Probability Functions*, Applied Mathematics Series 23 (U.S. Department of Commerce, 1953).

Appendix Table 3

POISSON PROBABILITY DISTRIBUTION

This table shows the value of

$$P(x) = \frac{\mu^x e^{-\mu}}{x!}$$

for selected values of x and for $\mu = .005$ to 8.0.

x	.005	.01	.02	.03	.04	.05	.06	.07	.08	.09
0	.9950	.9900	.9802	.9704	.9608	.9512	.9418	.9324	.9231	.9139
1	.0050	.0099	.0192	.0291	.0384	.0476	.0565	.0653	.0738	.0823
2	.0000	.0000	.0002	.0004	.0008	.0012	.0017	.0023	.0030	.0037
3	.0000	.0000	.0000	.0000	.0000	.0000	.0000	.0001	.0001	.0001

x	0.1	0.2	0.3	0.4	0.5	0.6	0.7	0.8	0.9	1.0
0	.9048	.8187	.7408	.6703	.6065	.5488	.4966	.4493	.4066	.3679
1	.0905	.1637	.2222	.2681	.3033	.3293	.3476	.3595	.3659	.3679
2	.0045	.0164	.0333	.0536	.0758	.0988	.1217	.1438	.1647	.1839
3	.0002	.0011	.0033	.0072	.0126	.0198	.0284	.0383	.0494	.0613
4	.0000	.0001	.0002	.0007	.0016	.0030	.0050	.0077	.0111	.0153
5	.0000	.0000	.0000	.0001	.0002	.0004	.0007	.0012	.0020	.0031
6	.0000	.0000	.0000	.0000	.0000	.0000	.0001	.0002	.0003	.0005
7	.0000	.0000	.0000	.0000	.0000	.0000	.0000	.0000	.0000	.0001

x	1.1	1.2	1.3	1.4	1.5	1.6	1.7	1.8	1.9	2.0
0	.3329	.3012	.2725	.2466	.2231	.2019	.1827	.1653	.1496	.1353
1	.3662	.3614	.3543	.3452	.3347	.3230	.3106	.2975	.2842	.2707
2	.2014	.2169	.2303	.2417	.2510	.2584	.2640	.2678	.2700	.2707
3	.0738	.0867	.0998	.1128	.1255	.1378	.1496	.1607	.1710	.1804
4	.0203	.0260	.0324	.0395	.0471	.0551	.0636	.0723	.0812	.0902
5	.0045	.0062	.0084	.0111	.0141	.0176	.0216	.0260	.0309	.0361
6	.0008	.0012	.0018	.0026	.0035	.0047	.0061	.0078	.0098	.0120
7	.0001	.0002	.0003	.0005	.0008	.0011	.0015	.0020	.0027	.0034
8	.0000	.0000	.0001	.0001	.0001	.0002	.0003	.0005	.0006	.0009
9	.0000	.0000	.0000	.0000	.0000	.0000	.0001	.0001	.0001	.0002

Appendix Table 3 (Continued)

μ

x	2.1	2.2	2.3	2.4	2.5	2.6	2.7	2.8	2.9	3.0
0	.1225	.1108	.1003	.0907	.0821	.0743	.0672	.0608	.0550	.0498
1	.2572	.2438	.2306	.2177	.2052	.1931	.1815	.1703	.1596	.1494
2	.2700	.2681	.2652	.2613	.2565	.2510	.2450	.2384	.2314	.2240
3	.1890	.1966	.2033	.2090	.2138	.2176	.2205	.2225	.2237	.2240
4	.0992	.1082	.1169	.1254	.1336	.1414	.1488	.1557	.1622	.1680
5	.0417	.0476	.0538	.0602	.0668	.0735	.0804	.0872	.0940	.1008
6	.0146	.0174	.0206	.0241	.0278	.0319	.0362	.0407	.0455	.0504
7	.0044	.0055	.0068	.0083	.0099	.0118	.0139	.0163	.0188	.0216
8	.0011	.0015	.0019	.0025	.0031	.0038	.0047	.0057	.0068	.0081
9	.0003	.0004	.0005	.0007	.0009	.0011	.0014	.0018	.0022	.0027
10	.0001	.0001	.0001	.0002	.0002	.0003	.0004	.0005	.0006	.0008
11	.0000	.0000	.0000	.0000	.0000	.0001	.0001	.0001	.0002	.0002
12	.0000	.0000	.0000	.0000	.0000	.0000	.0000	.0000	.0000	.0001

μ

x	3.1	3.2	3.3	3.4	3.5	3.6	3.7	3.8	3.9	4.0
0	.0450	.0408	.0369	.0334	.0302	.0273	.0247	.0224	.0202	.0183
1	.1397	.1304	.1217	.1135	.1057	.0984	.0915	.0850	.0789	.0733
2	.2165	.2087	.2008	.1929	.1850	.1771	.1692	.1615	.1539	.1465
3	.2237	.2226	.2209	.2186	.2158	.2125	.2087	.2046	.2001	.1954
4	.1734	.1781	.1823	.1858	.1888	.1912	.1931	.1944	.1951	.1954
5	.1075	.1140	.1203	.1264	.1322	.1377	.1429	.1477	.1522	.1563
6	.0555	.0608	.0662	.0716	.0771	.0826	.0881	.0936	.0989	.1042
7	.0246	.0278	.0312	.0348	.0385	.0425	.0466	.0508	.0551	.0595
8	.0095	.0111	.0129	.0148	.0169	.0191	.0215	.0241	.0269	.0298
9	.0033	.0040	.0047	.0056	.0066	.0076	.0089	.0102	.0116	.0132
10	.0010	.0013	.0016	.0019	.0023	.0028	.0033	.0039	.0045	.0053
11	.0003	.0004	.0005	.0006	.0007	.0009	.0011	.0013	.0016	.0019
12	.0001	.0001	.0001	.0002	.0002	.0003	.0003	.0004	.0005	.0006
13	.0000	.0000	.0000	.0000	.0001	.0001	.0001	.0001	.0002	.0002
14	.0000	.0000	.0000	.0000	.0000	.0000	.0000	.0000	.0000	.0001

μ

x	4.1	4.2	4.3	4.4	4.5	4.6	4.7	4.8	4.9	5.0
0	.0166	.0150	.0136	.0123	.0111	.0101	.0091	.0082	.0074	.0067
1	.0679	.0630	.0583	.0540	.0500	.0462	.0427	.0395	.0365	.0337
2	.1393	.1323	.1254	.1188	.1125	.1063	.1005	.0948	.0894	.0842
3	.1904	.1852	.1798	.1743	.1687	.1631	.1574	.1517	.1460	.1404
4	.1951	.1944	.1933	.1917	.1898	.1875	.1849	.1820	.1789	.1755

Appendix Table 3 (Continued)

μ									
x	4.1	4.2	4.3	4.4	4.5	4.6	4.7	4.8	4.9	5.0
5	.1600	.1633	.1662	.1687	.1708	.1725	.1738	.1747	.1753	.1755
6	.1093	.1143	.1191	.1237	.1281	.1323	.1362	.1398	.1432	.1462
7	.0640	.0686	.0732	.0778	.0824	.0869	.0914	.0959	.1002	.1044
8	.0328	.0360	.0393	.0428	.0463	.0500	.0537	.0575	.0614	.0653
9	.0150	.0168	.0188	.0209	.0232	.0255	.0280	.0307	.0334	.0363
10	.0061	.0071	.0081	.0092	.0104	.0118	.0132	.0147	.0164	.0181
11	.0023	.0027	.0032	.0037	.0043	.0049	.0056	.0064	.0073	.0082
12	.0008	.0009	.0011	.0014	.0016	.0019	.0022	.0026	.0030	.0034
13	.0002	.0003	.0004	.0005	.0006	.0007	.0008	.0009	.0011	.0013
14	.0001	.0001	.0001	.0001	.0002	.0002	.0003	.0003	.0004	.0005
15	.0000	.0000	.0000	.0000	.0001	.0001	.0001	.0001	.0001	.0002

μ									
x	5.1	5.2	5.3	5.4	5.5	5.6	5.7	5.8	5.9	6.0
0	.0061	.0055	.0050	.0045	.0041	.0037	.0033	.0030	.0027	.0025
1	.0311	.0287	.0265	.0244	.0225	.0207	.0191	.0176	.0162	.0149
2	.0793	.0746	.0701	.0659	.0618	.0580	.0544	.0509	.0477	.0446
3	.1348	.1293	.1239	.1185	.1133	.1082	.1033	.0985	.0938	.0892
4	.1719	.1681	.1641	.1600	.1558	.1515	.1472	.1428	.1383	.1339
5	.1753	.1748	.1740	.1728	.1714	.1697	.1678	.1656	.1632	.1606
6	.1490	.1515	.1537	.1555	.1571	.1584	.1594	.1601	.1605	.1606
7	.1086	.1125	.1163	.1200	.1234	.1267	.1298	.1326	.1353	.1377
8	.0692	.0731	.0771	.0810	.0849	.0887	.0925	.0962	.0998	.1033
9	.0392	.0423	.0454	.0486	.0519	.0552	.0586	.0620	.0654	.0688
10	.0200	.0220	.0241	.0262	.0285	.0309	.0334	.0359	.0386	.0413
11	.0093	.0104	.0116	.0129	.0143	.0157	.0173	.0190	.0207	.0225
12	.0039	.0045	.0051	.0058	.0065	.0073	.0082	.0092	.0102	.0113
13	.0015	.0018	.0021	.0024	.0028	.0032	.0036	.0041	.0046	.0052
14	.0006	.0007	.0008	.0009	.0011	.0013	.0015	.0017	.0019	.0022
15	.0002	.0002	.0003	.0003	.0004	.0005	.0006	.0007	.0008	.0009
16	.0001	.0001	.0001	.0001	.0001	.0002	.0002	.0002	.0003	.0003
17	.0000	.0000	.0000	.0000	.0000	.0001	.0001	.0001	.0001	.0001

μ									
x	6.1	6.2	6.3	6.4	6.5	6.6	6.7	6.8	6.9	7.0
0	.0022	.0020	.0018	.0017	.0015	.0014	.0012	.0011	.0010	.0009
1	.0137	.0126	.0116	.0106	.0098	.0090	.0082	.0076	.0070	.0064
2	.0417	.0390	.0364	.0340	.0318	.0296	.0276	.0258	.0240	.0223
3	.0848	.0806	.0765	.0726	.0688	.0652	.0617	.0584	.0552	.0521
4	.1294	.1249	.1205	.1162	.1118	.1076	.1034	.0992	.0952	.0912

Appendix Table 3 (Continued)

μ

x	6.1	6.2	6.3	6.4	6.5	6.6	6.7	6.8	6.9	7.0
5	.1579	.1549	.1519	.1487	.1454	.1420	.1385	.1349	.1314	.1277
6	.1605	.1601	.1595	.1586	.1575	.1562	.1546	.1529	.1511	.1490
7	.1399	.1418	.1435	.1450	.1462	.1472	.1480	.1486	.1489	.1490
8	.1066	.1099	.1130	.1160	.1188	.1215	.1240	.1263	.1284	.1304
9	.0723	.0757	.0791	.0825	.0858	.0891	.0923	.0954	.0985	.1014
10	.0441	.0469	.0498	.0528	.0558	.0588	.0618	.0649	.0679	.0710
11	.0245	.0265	.0285	.0307	.0330	.0353	.0377	.0401	.0426	.0452
12	.0124	.0137	.0150	.0164	.0179	.0194	.0210	.0227	.0245	.0264
13	.0058	.0065	.0073	.0081	.0089	.0098	.0108	.0119	.0130	.0142
14	.0025	.0029	.0033	.0037	.0041	.0046	.0052	.0058	.0064	.0071
15	.0010	.0012	.0014	.0016	.0018	.0020	.0023	.0026	.0029	.0033
16	.0004	.0005	.0005	.0006	.0007	.0008	.0010	.0011	.0013	.0014
17	.0001	.0002	.0002	.0002	.0003	.0003	.0004	.0004	.0005	.0006
18	.0000	.0001	.0001	.0001	.0001	.0001	.0001	.0002	.0002	.0002
19	.0000	.0000	.0000	.0000	.0000	.0000	.0000	.0001	.0001	.0001

μ

x	7.1	7.2	7.3	7.4	7.5	7.6	7.7	7.8	7.9	8.0
0	.0008	.0007	.0007	.0006	.0006	.0005	.0005	.0004	.0004	.0003
1	.0059	.0054	.0049	.0045	.0041	.0038	.0035	.0032	.0029	.0027
2	.0208	.0194	.0180	.0167	.0156	.0145	.0134	.0125	.0116	.0107
3	.0492	.0464	.0438	.0413	.0389	.0366	.0345	.0324	.0305	.0286
4	.0874	.0836	.0799	.0764	.0729	.0696	.0663	.0632	.0602	.0573
5	.1241	.1204	.1167	.1130	.1094	.1057	.1021	.0986	.0951	.0916
6	.1468	.1445	.1420	.1394	.1367	.1339	.1311	.1282	.1252	.1221
7	.1489	.1486	.1481	.1474	.1465	.1454	.1442	.1428	.1413	.1396
8	.1321	.1337	.1351	.1363	.1373	.1382	.1388	.1392	.1395	.1396
9	.1042	.1070	.1096	.1121	.1144	.1167	.1187	.1207	.1224	.1241
10	.0740	.0770	.0800	.0829	.0858	.0887	.0914	.0941	.0967	.0993
11	.0478	.0504	.0531	.0558	.0585	.0613	.0640	.0667	.0695	.0722
12	.0283	.0303	.0323	.0344	.0366	.0388	.0411	.0434	.0457	.0481
13	.0154	.0168	.0181	.0196	.0211	.0227	.0243	.0260	.0278	.0296
14	.0078	.0086	.0095	.0104	.0113	.0123	.0134	.0145	.0157	.0169
15	.0037	.0041	.0046	.0051	.0057	.0062	.0069	.0075	.0083	.0090
16	.0016	.0019	.0021	.0024	.0026	.0030	.0033	.0037	.0041	.0045
17	.0007	.0008	.0009	.0010	.0012	.0013	.0015	.0017	.0019	.0021
18	.0003	.0003	.0004	.0004	.0005	.0006	.0006	.0007	.0008	.0009
19	.0001	.0001	.0001	.0002	.0002	.0002	.0003	.0003	.0003	.0004
20	.0000	.0000	.0001	.0001	.0001	.0001	.0001	.0001	.0001	.0002
21	.0000	.0000	.0000	.0000	.0000	.0000	.0000	.0000	.0001	.0001

Appendix Table 4

VALUES OF e^{-x}

x	e^{-x}	x	e^{-x}	x	e^{-x}
.0	1.000	1.5	.223	3.0	.050
.1	.905	1.6	.202	3.1	.045
.2	.819	1.7	.183	3.2	.041
.3	.741	1.8	.165	3.3	.037
.4	.670	1.9	.150	3.4	.033
.5	.607	2.0	.135	3.5	.030
.6	.549	2.1	.122	3.6	.027
.7	.497	2.2	.111	3.7	.025
.8	.449	2.3	.100	3.8	.022
.9	.407	2.4	.091	3.9	.020
1.0	.368	2.5	.082	4.0	.018
1.1	.333	2.6	.074	4.5	.011
1.2	.301	2.7	.067	5.0	.007
1.3	.273	2.8	.061	6.0	.002
1.4	.247	2.9	.055	7.0	.001

Appendix Table 5

RANDOM NUMBERS

```
11850 53535 04260 77609 93799 92171 45524 10968 30231 70864 29908
11851 41292 15201 66342 59155 46163 69248 31029 62034 21855 27863
11852 07320 22682 09595 44805 54593 53350 61354 14029 10195 18644
11853 77676 67772 45072 08940 02592 45976 82099 90739 77072 42081
11854 43227 20568 16309 23841 53173 39475 27282 82699 00022 96419

11855 90712 41695 67474 27567 93269 10163 94190 36188 41491 71217
11856 88103 21514 60787 33170 58215 89951 01634 98155 05154 08971
11857 72252 35791 84125 31962 81093 93068 41197 57779 88515 48002
11858 51702 49516 69510 19678 47298 11355 68459 96360 13436 66314
11859 63055 86998 22187 59898 96371 61370 35937 34292 00678 33505

11860 32373 57889 85880 66515 37489 37854 72926 23437 62233 38651
11861 71996 16525 25618 56577 69130 25035 93551 54394 81572 90624
11862 26912 70619 22576 22780 99118 18487 58801 36063 32886 60453
11863 74589 82677 13353 67658 17080 43212 34585 17179 86980 81899
11864 56041 53072 19912 47466 32585 41414 07564 80712 27286 07966

11865 09286 68067 84883 10023 78195 84711 85988 31545 39904 14984
11866 33610 84843 07145 38437 06148 06094 89601 96751 49124 55092
11867 14113 06396 59084 02534 09360 81918 77118 91640 92978 24815
11868 56302 89765 63857 42747 28592 41784 00822 60356 96389 11728
11869 06362 94540 29532 09994 55277 43897 63268 40481 00312 46039

11870 48568 34412 84939 54850 84317 92032 60430 49071 68962 28953
11871 65975 60965 77679 95782 67541 50654 09482 56111 98710 35803
11872 66686 32977 48472 30226 54226 72490 18395 37338 88279 79089
11873 51610 13000 73849 46654 30324 78000 72852 28934 83197 59003
11874 47600 86103 25788 08774 72020 04543 25849 88887 41159 30131

11875 34860 67572 83116 99579 81303 41889 56577 64142 51596 25329
11876 76649 50908 67006 29332 29689 68786 98987 34815 53512 20620
11877 78321 54309 85956 04976 37863 06711 72679 03405 28770 08515
11878 35775 21295 39621 02339 16537 42246 06571 81193 94930 05376
11879 06783 21338 89886 78826 02303 37886 70453 11021 62887 36855

11880 25887 53024 71881 51208 95739 98572 01903 68043 62661 71273
11881 37784 42100 70838 78963 10927 05448 25759 74051 47577 30196
11882 02120 59536 82996 22671 89267 65924 46725 69179 15182 59158
11883 55292 03836 28883 71134 08547 93204 09656 11671 29735 59573
11884 66186 43648 97926 80469 66412 73647 36779 84688 96862 51937
```

Appendix Table 5 (Continued)

```
11885 55010 11479 55036 82146 37120 62328 56276 28906 45311 61818
11886 02322 18679 18478 30052 05666 84405 47513 09244 78978 91819
11887 78056 67836 82582 25809 20198 37222 62629 75733 77420 58746
11888 69812 88260 83519 10062 60865 35038 14665 18163 59351 25794
11889 84904 66864 26982 37928 32988 87652 81415 24416 93778 20391

11890 83143 47631 79772 08576 10311 17597 71049 63326 47168 05737
11891 44423 71197 91081 40781 72403 76245 31881 55716 89255 71997
11892 59882 58479 59609 80115 91569 23152 51781 85744 78640 80172
11893 74890 90405 75945 31645 61008 24448 42249 84909 29013 12529
11894 52174 64334 77631 19855 17723 02897 80427 20700 92210 92091

11895 41361 24347 53420 33639 83765 97935 83630 33765 21502 15589
11896 94585 84798 98480 08335 08728 60428 22282 76784 37316 08624
11897 36020 71966 61443 12554 67446 08676 46177 22422 87471 27283
11898 08112 59807 28404 60316 49676 52901 90604 48379 85233 52060
11899 05853 69681 52034 77617 78644 75321 14162 01849 94684 14628
```

SOURCE: The Rand Corporation, *A Million Random Digits with 100,000 Normal Deviates* (New York: Free Press, 1955), p. 238.

Appendix Table 6

VALUES OF *t* THAT WILL BE EXCEEDED WITH SPECIFIED PROBABILITIES

This table shows the value of *t* where the area under the *t* distribution exceeding this value of *t* equals the specified amount. For example, the probability that a *t* variable with 14 degrees of freedom will exceed 1.345 equals .10.

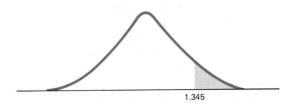

Degrees of freedom	Probability						
	0.40	0.25	0.10	0.05	0.025	0.01	0.005
1	0.325	1.000	3.078	6.314	12.706	31.821	63.657
2	.289	0.816	1.886	2.920	4.303	6.965	9.925
3	.277	.765	1.638	2.353	3.182	4.541	5.841
4	.271	.741	1.533	2.132	2.776	3.747	4.604
5	0.267	0.727	1.476	2.015	2.571	3.365	4.032
6	.265	.718	1.440	1.943	2.447	3.143	3.707
7	.263	.711	1.415	1.895	2.365	2.998	3.499
8	.262	.706	1.397	1.860	2.306	2.896	3.355
9	.261	.703	1.383	1.833	2.262	2.821	3.250
10	0.260	0.700	1.372	1.812	2.228	2.764	3.169
11	.260	.697	1.363	1.796	2.201	2.718	3.106
12	.259	.695	1.356	1.782	2.179	2.681	3.055
13	.259	.694	1.350	1.771	2.160	2.650	3.012
14	.258	.692	1.345	1.761	2.145	2.624	2.977
15	0.258	0.691	1.341	1.753	2.131	2.602	2.947
16	.258	.690	1.337	1.746	2.120	2.583	2.921
17	.257	.689	1.333	1.740	2.110	2.567	2.898
18	.257	.688	1.330	1.734	2.101	2.552	2.878
19	.257	.688	1.328	1.729	2.093	2.539	2.861

Appendix Table 6 (Continued)

Degrees of freedom	0.40	0.25	0.10	Probability 0.05	0.025	0.01	0.005
20	0.257	0.687	1.325	1.725	2.086	2.528	2.845
21	.257	.686	1.323	1.721	2.080	2.518	2.831
22	.256	.686	1.321	1.717	2.074	2.508	2.819
23	.256	.685	1.319	1.714	2.069	2.500	2.807
24	.256	.685	1.318	1.711	2.064	2.492	2.797
25	0.256	0.684	1.316	1.708	2.060	2.485	2.787
26	.256	.684	1.315	1.706	2.056	2.479	2.779
27	.256	.684	1.314	1.703	2.052	2.473	2.771
28	.256	.683	1.313	1.701	2.048	2.467	2.763
29	.256	.683	1.311	1.699	2.045	2.462	2.756
30	0.256	0.683	1.310	1.697	2.042	2.457	2.750
40	.255	.681	1.303	1.684	2.021	2.423	2.704
60	.254	.679	1.296	1.671	2.000	2.390	2.660
120	.254	.677	1.289	1.658	1.980	2.358	2.617
∞	.253	.674	1.282	1.645	1.960	2.326	2.576

SOURCE: *Biometrika Tables for Statisticians* (Cambridge, England: Cambridge University, 1954).

Appendix Table 7a

CHART PROVIDING 95 PERCENT CONFIDENCE INTERVAL FOR POPULATION PROPORTION, BASED ON SAMPLE PROPORTION*

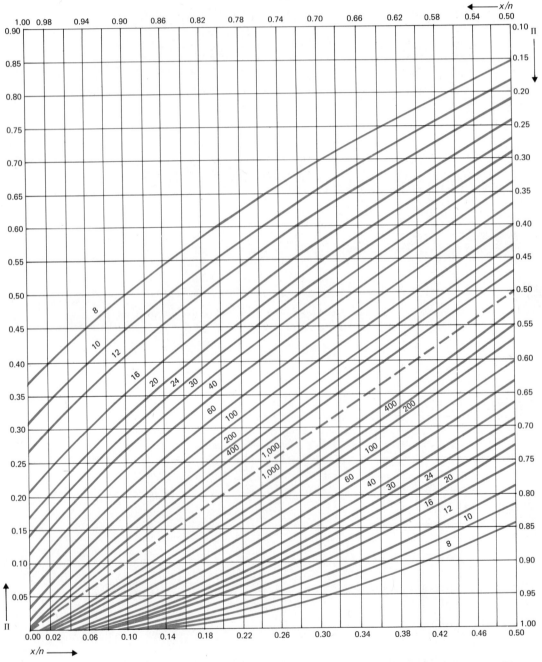

* The sample proportion equals x/n; the population proportion equals Π.

Appendix Table 7b

CHART PROVIDING 99 PERCENT CONFIDENCE INTERVAL FOR POPULATION PROPORTION, BASED ON SAMPLE PROPORTION*

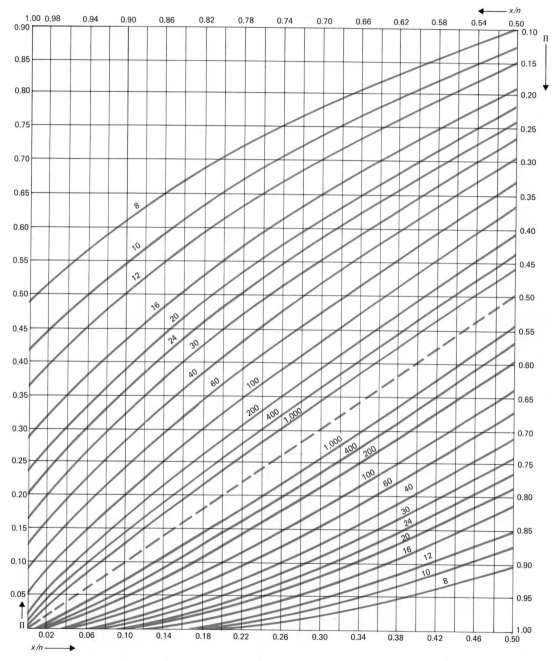

Appendix Table 8

VALUES OF χ^2 THAT WILL BE EXCEEDED WITH SPECIFIED PROBABILITIES

This table shows the value of χ^2 where the area under the χ^2 distribution exceeding this value of χ^2 equals the specified amount. For example, the probability that a χ^2 variable with 8 degrees of freedom will exceed 13.3616 equals 0.10.

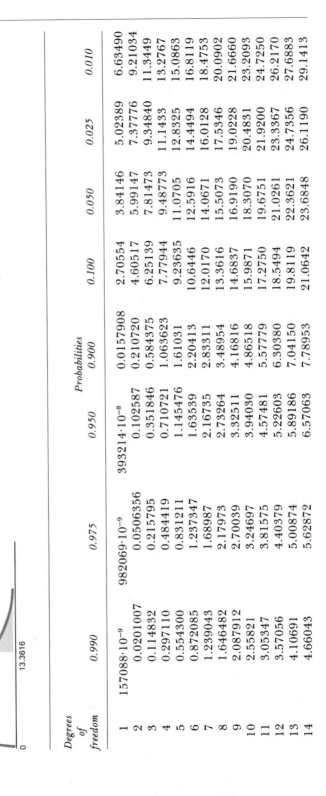

Degrees of freedom	Probabilities							
	0.990	0.975	0.950	0.900	0.100	0.050	0.025	0.010
1	$157088 \cdot 10^{-9}$	$982069 \cdot 10^{-9}$	$393214 \cdot 10^{-8}$	0.0157908	2.70554	3.84146	5.02389	6.63490
2	0.0201007	0.0506356	0.102587	0.210720	4.60517	5.99147	7.37776	9.21034
3	0.114832	0.215795	0.351846	0.584375	6.25139	7.81473	9.34840	11.3449
4	0.297110	0.484419	0.710721	1.063623	7.77944	9.48773	11.1433	13.2767
5	0.554300	0.831211	1.145476	1.61031	9.23635	11.0705	12.8325	15.0863
6	0.872085	1.237347	1.63539	2.20413	10.6446	12.5916	14.4494	16.8119
7	1.239043	1.68987	2.16735	2.83311	12.0170	14.0671	16.0128	18.4753
8	1.646482	2.17973	2.73264	3.48954	13.3616	15.5073	17.5346	20.0902
9	2.087912	2.70039	3.32511	4.16816	14.6837	16.9190	19.0228	21.6660
10	2.55821	3.24697	3.94030	4.86518	15.9871	18.3070	20.4831	23.2093
11	3.05347	3.81575	4.57481	5.57779	17.2750	19.6751	21.9200	24.7250
12	3.57056	4.40379	5.22603	6.30380	18.5494	21.0261	23.3367	26.2170
13	4.10691	5.00874	5.89186	7.04150	19.8119	22.3621	24.7356	27.6883
14	4.66043	5.62872	6.57063	7.78953	21.0642	23.6848	26.1190	29.1413

Appendix Table 8 (Continued)

Degrees of freedom				Probabilities				
	0.990	0.975	0.950	0.900	0.100	0.050	0.025	0.010
15	5.22935	6.26214	7.26094	8.54675	22.3072	24.9958	27.4884	30.5779
16	5.81221	6.90766	7.96164	9.31223	23.5418	26.2962	28.8454	31.9999
17	6.40776	7.56418	8.67176	10.0852	24.7690	27.5871	30.1910	33.4087
18	7.01491	8.23075	9.39046	10.8649	25.9894	28.8693	31.5264	34.8053
19	7.63273	8.90655	10.1170	11.6509	27.2036	30.1435	32.8523	36.1908
20	8.26040	9.59083	10.8508	12.4426	28.4120	31.4104	34.1696	37.5662
21	8.89720	10.28293	11.5913	13.2396	29.6151	32.6705	35.4789	38.9321
22	9.54249	10.9823	12.3380	14.0415	30.8133	33.9244	36.7807	40.2894
23	10.19567	11.6885	13.0905	14.8479	32.0069	35.1725	38.0757	41.6384
24	10.8564	12.4011	13.8484	15.6587	33.1963	36.4151	39.3641	42.9798
25	11.5240	13.1197	14.6114	16.4734	34.3816	37.6525	40.6465	44.3141
26	12.1981	13.8439	15.3791	17.2919	35.5631	38.8852	41.9232	45.6417
27	12.8786	14.5733	16.1513	18.1138	36.7412	40.1133	43.1944	46.9630
28	13.5648	15.3079	16.9279	18.9392	37.9159	41.3372	44.4607	48.2782
29	14.2565	16.0471	17.7083	19.7677	39.0875	42.5569	45.7222	49.5879
30	14.9535	16.7908	18.4926	20.5992	40.2560	43.7729	46.9792	50.8922
40	22.1643	24.4331	26.5093	29.0505	51.8050	55.7585	59.3417	63.6907
50	29.7067	32.3574	34.7642	37.6886	63.1671	67.5048	71.4202	76.1539
60	37.4848	40.4817	43.1879	46.4589	74.3970	79.0819	83.2976	88.3794
70	45.4418	48.7576	51.7393	55.3290	85.5271	90.5312	95.0231	100.425
80	53.5400	57.1532	60.3915	64.2778	96.5782	101.879	106.629	112.329
90	61.7541	65.6466	69.1260	73.2912	107.565	113.145	118.136	124.116
100	70.0648	74.2219	77.9295	82.3581	118.498	124.342	129.561	135.807

SOURCE: See Appendix Table 6.

Appendix Table 9

VALUE OF AN *F* VARIABLE THAT IS EXCEEDED WITH
PROBABILITY EQUAL TO .05

		Degrees of freedom for numerator							
	1	*2*	*3*	*4*	*5*	*6*	*7*	*8*	*9*
1	161.4	199.5	215.7	224.6	230.2	234.0	236.8	238.9	240.5
2	18.51	19.00	19.16	19.25	19.30	19.33	19.35	19.37	19.38
3	10.13	9.55	9.28	9.12	9.01	8.94	8.89	8.85	8.81
4	7.71	6.94	6.59	6.39	6.26	6.16	6.09	6.04	6.00
5	6.61	5.79	5.41	5.19	5.05	4.95	4.88	4.82	4.77
6	5.99	5.14	4.76	4.53	4.39	4.28	4.21	4.15	4.10
7	5.59	4.74	4.35	4.12	3.97	3.87	3.79	3.73	3.68
8	5.32	4.46	4.07	3.84	3.69	3.58	3.50	3.44	3.39
9	5.12	4.26	3.86	3.63	3.48	3.37	3.29	3.23	3.18
10	4.96	4.10	3.71	3.48	3.33	3.22	3.14	3.07	3.02
11	4.84	3.98	3.59	3.36	3.20	3.09	3.01	2.95	2.90
12	4.75	3.89	3.49	3.26	3.11	3.00	2.91	2.85	2.80
13	4.67	3.81	3.41	3.18	3.03	2.92	2.83	2.77	2.71
14	4.60	3.74	3.34	3.11	2.96	2.85	2.76	2.70	2.65
15	4.54	3.68	3.29	3.06	2.90	2.79	2.71	2.64	2.59
16	4.49	3.63	3.24	3.01	2.85	2.74	2.66	2.59	2.54
17	4.45	3.59	3.20	2.96	2.81	2.70	2.61	2.55	2.49
18	4.41	3.55	3.16	2.93	2.77	2.66	2.58	2.51	2.46
19	4.38	3.52	3.13	2.90	2.74	2.63	2.54	2.48	2.42
20	4.35	3.49	3.10	2.87	2.71	2.60	2.51	2.45	2.39
21	4.32	3.47	3.07	2.84	2.68	2.57	2.49	2.42	2.37
22	4.30	3.44	3.05	2.82	2.66	2.55	2.46	2.40	2.34
23	4.28	3.42	3.03	2.80	2.64	2.53	2.44	2.37	2.32
24	4.26	3.40	3.01	2.78	2.62	2.51	2.42	2.36	2.30
25	4.24	3.39	2.99	2.76	2.60	2.49	2.40	2.34	2.28
26	4.23	3.37	2.98	2.74	2.59	2.47	2.39	2.32	2.27
27	4.21	3.35	2.96	2.73	2.57	2.46	2.37	2.31	2.25
28	4.20	3.34	2.95	2.71	2.56	2.45	2.36	2.29	2.24
29	4.18	3.33	2.93	2.70	2.55	2.43	2.35	2.28	2.22
30	4.17	3.32	2.92	2.69	2.53	2.42	2.33	2.27	2.21
40	4.08	3.23	2.84	2.61	2.45	2.34	2.25	2.18	2.12
60	4.00	3.15	2.76	2.53	2.37	2.25	2.17	2.10	2.04
120	3.92	3.07	2.68	2.45	2.29	2.17	2.09	2.02	1.96
∞	3.84	3.00	2.60	2.37	2.21	2.10	2.01	1.94	1.88

Degrees of freedom for denominator

Appendix Table 9 (Continued)

	Degrees of freedom for numerator									
	10	12	15	20	24	30	40	60	120	∞
1	241.9	243.9	245.9	248.0	249.1	250.1	251.1	252.2	253.3	254.3
2	19.40	19.41	19.43	19.45	19.45	19.46	19.47	19.48	19.49	19.50
3	8.79	8.74	8.70	8.66	8.64	8.62	8.59	8.57	8.55	8.53
4	5.96	5.91	5.86	5.80	5.77	5.75	5.72	5.69	5.66	5.63
5	4.74	4.68	4.62	4.56	4.53	4.50	4.46	4.43	4.40	4.36
6	4.06	4.00	3.94	3.87	3.84	3.81	3.77	3.74	3.70	3.67
7	3.64	3.57	3.51	3.44	3.41	3.38	3.34	3.30	3.27	3.23
8	3.35	3.28	3.22	3.15	3.12	3.08	3.04	3.01	2.97	2.93
9	3.14	3.07	3.01	2.94	2.90	2.86	2.83	2.79	2.75	2.71
10	2.98	2.91	2.85	2.77	2.74	2.70	2.66	2.62	2.58	2.54
11	2.85	2.79	2.72	2.65	2.61	2.57	2.53	2.49	2.45	2.40
12	2.75	2.69	2.62	2.54	2.51	2.47	2.43	2.38	2.34	2.30
13	2.67	2.60	2.53	2.46	2.42	2.38	2.34	2.30	2.25	2.21
14	2.60	2.53	2.46	2.39	2.35	2.31	2.27	2.22	2.18	2.13
15	2.54	2.48	2.40	2.33	2.29	2.25	2.20	2.16	2.11	2.07
16	2.49	2.42	2.35	2.28	2.24	2.19	2.15	2.11	2.06	2.01
17	2.45	2.38	2.31	2.23	2.19	2.15	2.10	2.06	2.01	1.96
18	2.41	2.34	2.27	2.19	2.15	2.11	2.06	2.02	1.97	1.92
19	2.38	2.31	2.23	2.16	2.11	2.07	2.03	1.98	1.93	1.88
20	2.35	2.28	2.20	2.12	2.08	2.04	1.99	1.95	1.90	1.84
21	2.32	2.25	2.18	2.10	2.05	2.01	1.96	1.92	1.87	1.81
22	2.30	2.23	2.15	2.07	2.03	2.98	1.94	1.89	1.84	1.78
23	2.27	2.20	2.13	2.05	2.01	1.96	1.91	1.86	1.81	1.76
24	2.25	2.18	2.11	2.03	1.98	1.94	1.89	1.84	1.79	1.73
25	2.24	2.16	2.09	2.01	1.96	1.92	1.87	1.82	1.77	1.71
26	2.22	2.15	2.07	1.99	1.95	1.90	1.85	1.80	1.75	1.69
27	2.20	2.13	2.06	1.97	1.93	1.88	1.84	1.79	1.73	1.67
28	2.19	2.12	2.04	1.96	1.91	1.87	1.82	1.77	1.71	1.65
29	2.18	2.10	2.03	1.94	1.90	1.85	1.81	1.75	1.70	1.64
30	2.16	2.09	2.01	1.93	1.89	1.84	1.79	1.74	1.68	1.62
40	2.08	2.00	1.92	1.84	1.79	1.74	1.69	1.64	1.58	1.5i
60	1.99	1.92	1.84	1.75	1.70	1.65	1.59	1.53	1.47	1.39
120	1.91	1.83	1.75	1.66	1.61	1.55	1.50	1.43	1.35	1.25
∞	1.83	1.75	1.67	1.57	1.52	1.46	1.39	1.32	1.22	1.00

Degrees of freedom for denominator

SOURCE: See Appendix Table 6.

Appendix Table 10

VALUE OF AN *F* VARIABLE EXCEEDED WITH PROBABILITY EQUAL TO .01

	Degrees of freedom for numerator								
	1	2	3	4	5	6	7	8	9
1	4052	4999.5	5403	5625	5764	5859	5928	5982	6022
2	98.50	99.00	99.17	99.25	99.30	99.33	99.36	99.37	99.39
3	34.12	30.82	29.46	28.71	28.24	27.91	27.67	27.49	27.35
4	21.20	18.00	16.69	15.98	15.52	15.21	14.98	14.80	14.66
5	16.26	13.27	12.06	11.39	10.97	10.67	10.46	10.29	10.16
6	13.75	10.92	9.78	9.15	8.75	8.47	8.26	8.10	7.98
7	12.25	9.55	8.45	7.85	7.46	7.19	6.99	6.84	6.72
8	11.26	8.65	7.59	7.01	6.63	6.37	6.18	6.03	5.91
9	10.56	8.02	6.99	6.42	6.06	5.80	5.61	5.47	5.35
10	10.04	7.56	6.55	5.99	5.64	5.39	5.20	5.06	4.94
11	9.65	7.21	6.22	5.67	5.32	5.07	4.89	4.74	4.63
12	9.33	6.93	5.95	5.41	5.06	4.82	4.64	4.50	4.39
13	9.07	6.70	5.74	5.21	4.86	4.62	4.44	4.30	4.19
14	8.86	6.51	5.56	5.04	4.69	4.46	4.28	4.14	4.03
15	8.68	6.36	5.42	4.89	4.56	4.32	4.14	4.00	3.89
16	8.53	6.23	5.29	4.77	4.44	4.20	4.03	3.89	3.78
17	8.40	6.11	5.18	4.67	4.34	4.10	3.93	3.79	3.68
18	8.29	6.01	5.09	4.58	4.25	4.01	3.84	3.71	3.60
19	8.18	5.93	5.01	4.50	4.17	3.94	3.77	3.63	3.52
20	8.10	5.85	4.94	4.43	4.10	3.87	3.70	3.56	3.46
21	8.02	5.78	4.87	4.37	4.04	3.81	3.64	3.51	3.40
22	7.95	5.72	4.82	4.31	3.99	3.76	3.59	3.45	3.35
23	7.88	5.66	4.76	4.26	3.94	3.71	3.54	3.41	3.30
24	7.82	5.61	4.72	4.22	3.90	3.67	3.50	3.36	3.26
25	7.77	5.57	4.68	4.18	3.85	3.63	3.46	3.32	3.22
26	7.72	5.53	4.64	4.14	3.82	3.59	3.42	3.29	3.18
27	7.68	5.49	4.60	4.11	3.78	3.56	3.39	3.26	3.15
28	7.64	5.45	4.57	4.07	3.75	3.53	3.36	3.23	3.12
29	7.60	5.42	4.54	4.04	3.73	3.50	3.33	3.20	3.09
30	7.56	5.39	4.51	4.02	3.70	3.47	3.30	3.17	3.07
40	7.31	5.18	4.31	3.83	3.51	3.29	3.12	2.99	2.89
60	7.08	4.98	4.13	3.65	3.34	3.12	2.95	2.82	2.72
120	6.85	4.79	3.95	3.48	3.17	2.96	2.79	2.66	2.56
∞	6.63	4.61	3.78	3.32	3.02	2.80	2.64	2.51	2.41

Degrees of freedom for denominator

Appendix Table 10 (Continued)

	Degrees of freedom for numerator									
	10	12	15	20	24	30	40	60	120	∞
1	6056	6106	6157	6209	6235	6261	6287	6313	6339	6366
2	99.40	99.42	99.43	99.45	99.46	99.47	99.47	99.48	99.49	99.50
3	27.23	27.05	26.87	26.69	26.60	26.50	26.41	26.32	26.22	26.13
4	14.55	14.37	14.20	14.02	13.93	13.84	13.75	13.65	13.56	13.46
5	10.05	9.89	9.72	9.55	9.47	9.38	9.29	9.20	9.11	9.02
6	7.87	7.72	7.56	7.40	7.31	7.23	7.14	7.06	6.97	6.88
7	6.62	6.47	6.31	6.16	6.07	5.99	5.91	5.82	5.74	5.65
8	5.81	5.67	5.52	5.36	5.28	5.20	5.12	5.03	4.95	4.86
9	5.26	5.11	4.96	4.81	4.73	4.65	4.57	4.48	4.40	4.31
10	4.85	4.71	4.56	4.41	4.33	4.25	4.17	4.08	4.00	3.91
11	4.54	4.40	4.25	4.10	4.02	3.94	3.86	3.78	3.69	3.60
12	4.30	4.16	4.01	3.86	3.78	3.70	3.62	3.54	3.45	3.36
13	4.10	3.96	3.82	3.66	3.59	3.51	3.43	3.34	3.25	3.17
14	3.94	3.80	3.66	3.51	3.43	3.35	3.27	3.18	3.09	3.00
15	3.80	3.67	3.52	3.37	3.29	3.21	3.13	3.05	2.96	2.87
16	3.69	3.55	3.41	3.26	3.18	3.10	3.02	2.93	2.84	2.75
17	3.59	3.46	3.31	3.16	3.08	3.00	2.92	2.83	2.75	2.65
18	3.51	3.37	3.23	3.08	3.00	2.92	2.84	2.75	2.66	2.57
19	3.43	3.30	3.15	3.00	2.92	2.84	2.76	2.67	2.58	2.49
20	3.37	3.23	3.09	2.94	2.86	2.78	2.69	2.61	2.52	2.42
21	3.31	3.17	3.03	2.88	2.80	2.72	2.64	2.55	2.46	2.36
22	3.26	3.12	2.98	2.83	2.75	2.67	2.58	2.50	2.40	2.31
23	3.21	3.07	2.93	2.78	2.70	2.62	2.54	2.45	2.35	2.26
24	3.17	3.03	2.89	2.74	2.66	2.58	2.49	2.40	2.31	2.21
25	3.13	2.99	2.85	2.70	2.62	2.54	2.45	2.36	2.27	2.17
26	3.09	2.96	2.81	2.66	2.58	2.50	2.42	2.33	2.23	2.13
27	3.06	2.93	2.78	2.63	2.55	2.47	2.38	2.29	2.20	2.10
28	3.03	2.90	2.75	2.60	2.52	2.44	2.35	2.26	2.17	2.06
29	3.00	2.87	2.73	2.57	2.49	2.41	2.33	2.23	2.14	2.03
30	2.98	2.84	2.70	2.55	2.47	2.39	2.30	2.21	2.11	2.01
40	2.80	2.66	2.52	2.37	2.29	2.20	2.11	2.02	1.92	1.80
60	2.63	2.50	2.35	2.20	2.12	2.03	1.94	1.84	1.73	1.60
120	2.47	2.34	2.19	2.03	1.95	1.86	1.76	1.66	1.53	1.38
∞	2.32	2.18	2.04	1.88	1.79	1.70	1.59	1.47	1.32	1.00

Degrees of freedom for denominator (left margin label)

SOURCE: See Appendix Table 6.

Appendix Table 11

COMMON LOGARITHMS

N	0	1	2	3	4	5	6	7	8	9
10	0000	0043	0086	0128	0170	0212	0253	0294	0334	0374
11	0414	0453	0492	0531	0569	0607	0645	0682	0719	0755
12	0792	0828	0864	0899	0934	0969	1004	1038	1072	1106
13	1139	1173	1206	1239	1271	1303	1335	1367	1399	1430
14	1461	1492	1523	1553	1584	1614	1644	1673	1703	1732
15	1761	1790	1818	1847	1875	1903	1931	1959	1987	2014
16	2041	2068	2095	2122	2148	2175	2201	2227	2253	2279
17	2304	2330	2355	2380	2405	2430	2455	2480	2504	2529
18	2553	2577	2601	2625	2648	2672	2695	2718	2742	2765
19	2788	2810	2833	2856	2878	2900	2923	2945	2967	2989
20	3010	3032	3054	3075	3096	3118	3139	3160	3181	3201
21	3222	3243	3263	3284	3304	3324	3345	3365	3385	3404
22	3424	3444	3464	3483	3502	3522	3541	3560	3579	3598
23	3617	3636	3655	3674	3692	3711	3729	3747	3766	3784
24	3802	3826	3838	3856	3874	3892	3909	3927	3945	3962
25	3979	3997	4014	4031	4048	4065	4082	4099	4116	4133
26	4150	4166	4183	4200	4216	4232	4249	4205	4281	4298
27	4314	4330	4346	4362	4378	4393	4409	4425	4440	4456
28	4472	4487	4502	4518	4533	4548	4564	4579	4594	4609
29	4624	4639	4654	4669	4683	4698	4713	4728	4742	4757
30	4771	4786	4800	4814	4829	4843	4857	4871	4886	4900
31	4914	4928	4942	4955	4969	4983	4997	5011	5024	5038
32	5051	5065	5079	5092	5105	5119	5132	5145	5159	5172
33	5185	5198	5211	5224	5237	5250	5263	5276	5289	5302
34	5315	5328	5340	5353	5366	5378	5391	5403	5416	5423
35	5441	5453	5465	5478	5490	5502	5514	5527	5539	5551
36	5563	5575	5587	5599	5611	5623	5635	5647	5658	5670
37	5682	5694	5705	5717	5729	5740	5752	5763	5775	5786
38	5798	5809	5821	5832	5843	5855	5866	5877	5888	5899
39	5911	5922	5933	5944	5955	5966	5977	5988	5999	6010
40	6021	6031	6042	6053	6064	6075	6085	6096	6107	6117
41	6128	6138	6149	6160	6170	6180	6191	6201	6212	6222
42	6232	6243	6253	6263	6274	6284	6294	6304	6314	6325
43	6335	6345	6355	6365	6375	6385	6395	6405	6415	6425
44	6435	6444	6454	6464	6474	6484	6493	6503	6513	6522

Appendix Table 11 (Continued)

N	0	1	2	3	4	5	6	7	8	9
45	6532	6542	6551	6561	6571	6580	6590	6599	6609	6618
46	6628	6637	6646	6656	6665	6675	6684	6693	6702	6712
47	6721	6730	6739	6749	6758	6767	6776	6785	6794	6803
48	6812	6821	6830	6839	6848	6857	6866	6875	6884	6893
49	6902	6911	6920	6928	6937	6946	6955	6964	6972	6981
50	6990	6998	7007	7016	7024	7033	7042	7050	7059	7067
51	7076	7084	7093	7101	7110	7118	7126	7135	7143	7152
52	7160	7168	7177	7185	7193	7202	7210	7218	7226	7235
53	7243	7251	7259	7267	7275	7284	7292	7300	7308	7316
54	7324	7332	7340	7348	7356	7364	7372	7380	7388	7396
55	7404	7412	7419	7427	7435	7443	7451	7459	7466	7474
56	7482	7490	7497	7505	7513	7520	7528	7536	7543	7551
57	7559	7566	7574	7582	7589	7597	7604	7612	7619	7627
58	7634	7642	7649	7657	7664	7672	7679	7686	7694	7701
59	7709	7716	7723	7731	7738	7745	7752	7760	7767	7774
60	7782	7789	7796	7803	7810	7818	7825	7832	7839	7846
61	7853	7860	7868	7875	7882	7889	7896	7903	7910	7917
62	7924	7931	7938	7945	7952	7959	7966	7973	7980	7987
63	7993	8000	8007	8014	8021	8028	8035	8041	8048	8055
64	8062	8069	8075	8082	8089	8096	8102	8109	8116	8122
65	8129	8136	8142	8149	8156	8162	8169	8176	8182	8189
66	8195	8202	8209	8215	8222	8228	8235	8241	8248	8254
67	8261	8267	8274	8280	8287	8293	8299	8306	8312	8319
68	8325	8331	8338	8344	8351	8357	8363	8370	8376	8382
69	8388	8395	8401	8407	8414	8420	8426	8432	8439	8445
70	8451	8457	8463	8470	8476	8482	8488	8494	8500	8506
71	8513	8519	8525	8531	8537	8543	8549	8555	8561	8567
72	8573	8579	8585	8591	8597	8603	8609	8615	8621	8627
73	8633	8639	8645	8651	8637	8663	8669	8675	8681	8686
74	8692	8698	8704	8710	8716	8722	8727	8733	8739	8745
75	8751	8756	8762	8768	8774	8779	8785	8791	8797	8802
76	8808	8814	8820	8825	8831	8837	8842	8848	8854	8859
77	8865	8871	8876	8882	8887	8893	8899	8904	8910	8915
78	8921	8927	8932	8938	8943	8949	8954	8960	8965	8971
79	8976	8982	8987	8993	8998	9004	9009	9015	9020	9025

Appendix Table 11 (Continued)

N	0	1	2	3	4	5	6	7	8	9
80	9031	9036	9042	9047	9053	9058	9063	9069	9074	9079
81	9085	9090	9096	9101	9106	9112	9117	9122	9128	9133
82	9138	9143	9149	9154	9159	9165	9170	9175	9180	9186
83	9191	9190	9201	9206	9212	9217	9222	9227	9232	9238
84	9243	9248	9253	9258	9263	9269	9274	9279	9284	9289
85	9294	9299	9304	9309	9315	9320	9325	9330	9335	9340
86	9345	9350	9355	9360	9365	9370	9375	9380	9385	9390
87	9395	9400	9405	9410	9415	9420	9425	9430	9435	9440
88	9445	9450	9455	9460	9465	9469	9474	9479	9484	9489
89	9494	9499	9504	9509	9513	9518	9523	9528	9533	9538
90	9542	9547	9552	9557	9562	9566	9571	9576	9581	9586
91	9590	9595	9600	9605	9609	9614	9619	9624	9628	9633
92	9638	9643	9647	9652	9657	9661	9666	9671	9675	9680
93	9685	9689	9694	9699	9703	9708	9713	9717	9722	9727
94	9731	9736	9741	9745	9750	9754	9759	9763	9768	9773
95	9777	9782	9786	9791	9795	9800	9805	9808	9814	9818
96	9823	9827	9832	9836	9841	9845	9850	9854	9859	9863
97	9868	9872	9877	9881	9886	9890	9894	9899	9903	9908
98	9912	9917	9921	9926	9930	9934	9939	9943	9948	9952
99	9956	9961	9965	9969	9974	9978	9983	9987	9991	9996

SOURCE: Adapted from National Bureau of Standards, *Tables of 10^x*, Applied Mathematics Series 27 (U.S. Department of Commerce, 1953).

Appendix Table 12

SQUARES AND SQUARE ROOTS

n	n^2	$\sqrt{n}$	$\sqrt{10n}$	n	n^2	$\sqrt{n}$	$\sqrt{10n}$
				35	1 225	5.916 080	18.708 29
1	1	1.000 00	3.162 278	36	1 296	6.000 000	18.973 67
2	4	1.414 214	4.472 136	37	1 369	6.082 763	19.235 38
3	9	1.732 051	5.477 226	38	1 444	6.164 414	19.493 59
4	16	2.000 000	6.324 555	39	1 521	6.244 998	19.748 42
5	25	2.236 068	7.071 068	40	1 600	6.324 555	20.000 00
6	36	2.449 490	7.745 967	41	1 681	6.403 124	20.248 46
7	49	2.645 751	8.366 600	42	1 764	6.480 741	20.493 90
8	64	2.828 427	8.944 272	43	1 849	6.557 439	20.736 44
9	81	3.000 000	9.486 833	44	1 936	6.633 250	20.976 18
10	100	3.162 278	10.000 00	45	2 025	6.708 204	21.213 20
11	121	3.316 625	10.488 09	46	2 116	6.782 330	21.447 61
12	144	3.464 102	10.954 45	47	2 209	6.855 655	21.679 48
13	169	3.605 551	11.401 75	48	2 304	6.928 203	21.908 90
14	196	3.741 657	11.832 16	49	2 401	7.000 000	22.135 94
15	225	3.872 983	12.247 45	50	2 500	7.071 068	22.360 68
16	256	4.000 000	12.649 11	51	2 601	7.141 428	22.583 18
17	289	4.123 106	13.038 40	52	2 704	7.211 103	22.803 51
18	324	4.242 641	13.416 41	53	2 809	7.280 110	23.021 73
19	361	4.358 899	13.784 05	54	2 916	7.348 469	23.237 90
20	400	4.472 136	14.142 14	55	3 025	7.416 198	23.452 08
21	441	4.582 576	14.491 38	56	3 136	7.483 315	23.664 32
22	484	4.690 416	14.832 40	57	3 249	7.549 834	23.874 67
23	529	4.795 832	15.165 75	58	3 364	7.615 773	24.083 19
24	576	4.898 979	15.491 93	59	3 481	7.681 146	24.289 92
25	625	5.000 000	15.811 39	60	3 600	7.745 967	24.494 90
26	676	5.099 020	16.124 52	61	3 721	7.810 250	24.698 18
27	729	5.196 152	16.431 68	62	3 844	7.874 008	24.899 80
28	784	5.291 503	16.733 20	63	3 969	7.937 254	25.099 80
29	841	5.385 165	17.029 39	64	4 096	8.000 000	25.298 22
30	900	5.477 226	17.320 51	65	4 225	8.062 258	25.495 10
31	961	5.567 764	17.606 82	66	4 356	8.124 038	25.690 47
32	1 024	5.656 854	17.888 54	67	4 489	8.185 353	25.884 36
33	1 089	5.744 563	18.165 90	68	4 624	8.246 211	26.076 81
34	1 156	5.830 952	18.439 09	69	4 761	8.306 624	26.267 85

Appendix Table 12 (Continued)

n	n^2	$\sqrt{n}$	$\sqrt{10n}$	n	n^2	$\sqrt{n}$	$\sqrt{10n}$
70	4 900	8.366 600	26.457 51	110	12 100	10.488 09	33.166 25
71	5 041	8.426 150	26.645 83	111	12 321	10.535 65	33.316 66
72	5 184	8.485 281	26.832 82	112	12 544	10.583 01	33.466 40
73	5 329	8.544 004	27.018 51	113	12 769	10.630 15	33.615 47
74	5 476	8.602 325	27.202 94	114	12 996	10.677 08	33.763 89
75	5 625	8.660 254	27.386 13	115	13 225	10.723 81	33.911 65
76	5 776	8.717 798	27.568 10	116	13 456	10.770 33	34.058 77
77	5 929	8.774 964	27.748 87	117	13 689	10.816 65	34.205 26
78	6 084	8.831 761	27.928 48	118	13 924	10.862 78	34.351 13
79	6 241	8.888 194	28.106 94	119	14 161	10.908 71	34.496 38
80	6 400	8.944 272	28.284 27	120	14 400	10.954 45	34.641 02
81	6 561	9.000 000	28.460 50	121	14 641	11.000 00	34.785 05
82	6 724	9.055 385	28.635 64	122	14 884	11.045 36	34.928 50
83	6 889	9.110 434	28.809 72	123	15 129	11.090 54	35.071 36
84	7 056	9.165 151	28.982 75	124	15 376	11.135 53	35.213 63
85	7 225	9.219 544	29.154 76	125	15 625	11.180 34	35.355 34
86	7 396	9.273 618	29.325 76	126	15 876	11.224 97	35.496 48
87	7 569	9.327 379	29.495 76	127	16 129	11.269 43	35.637 06
88	7 744	9.380 832	29.664 79	128	16 384	11.313 71	35.777 09
89	7 921	9.433 981	29.832 87	129	16 641	11.357 82	35.916 57
90	8 100	9.486 833	30.000 00	130	16 900	11.401 75	36.055 51
91	8 281	9.539 392	30.166 21	131	17 161	11.445 52	36.193 92
92	8 464	9.591 663	30.331 50	132	17 424	11.489 13	36.331 80
93	8 649	9.643 651	30.495 90	133	17 689	11.532 56	36.469 17
94	8 836	9.695 360	30.659 42	134	17 956	11.575 84	36.606 01
95	9 025	9.746 794	30.822 07	135	18 225	11.618 95	36.742 35
96	9 216	9.797 959	30.983 87	136	18 496	11.661 90	36.878 18
97	9 409	9.848 858	31.144 82	137	18 769	11.704 70	37.013 51
98	9 604	9.899 495	31.304 95	138	19 044	11.747 34	37.148 35
99	9 801	9.949 874	31.464 27	139	19 321	11.789 83	37.282 70
100	10 000	10.000 00	31.622 78	140	19 600	11.832 16	37.416 57
101	10 201	10.049 88	31.780 50	141	19 881	11.874 34	37.549 97
102	10 404	10.099 50	31.937 44	142	20 164	11.916 38	37.682 89
103	10 609	10.148 89	32.093 61	143	20 449	11.958 26	37.815 34
104	10 816	10.198 04	32.249 03	144	20 736	12.000 00	37.947 33
105	11 025	10.246 95	32.403 70	145	21 025	12.041 59	38.078 87
106	11 236	10.295 63	32.557 64	146	21 316	12.083 05	38.209 95
107	11 449	10.344 08	32.710 85	147	21 609	12.124 36	38.340 58
108	11 664	10.392 30	32.863 35	148	21 904	12.165 53	38.470 77
109	11 881	10.440 31	33.015 15	149	22 201	12.206 56	38.600 52

Appendix Table 12 (Continued)

n	n^2	$\sqrt{n}$	$\sqrt{10n}$	n	n^2	$\sqrt{n}$	$\sqrt{10n}$
150	22 500	12.247 45	38.729 83	190	36 100	13.784 05	43.588 99
151	22 801	12.288 21	38.858 72	191	36 481	13.820 27	43.703 55
152	23 104	12.328 83	38.987 18	192	36 864	13.856 41	43.817 80
153	23 409	12.369 32	39.115 21	193	37 249	13.892 44	43.931 77
154	23 716	12.409 67	39.242 83	194	37 636	13.928 39	44.045 43
155	24 025	12.449 90	39.370 04	195	38 025	13.964 24	44.158 80
156	24 336	12.490 00	39.496 84	196	38 416	14.000 00	44.271 89
157	24 649	12.529 96	39.623 23	197	38 809	14.035 67	44.384 68
158	24 964	12.569 81	39.749 21	198	39 204	14.071 25	44.497 19
159	25 281	12.609 52	39.874 80	199	39 601	14.106 74	44.609 42
160	25 600	12.649 11	40.000 00	200	40 000	14.142 14	44.721 36
161	25 921	12.688 58	40.124 81	201	40 401	14.177 45	44.833 02
162	26 244	12.727 92	40.249 22	202	40 804	14.212 67	44.944 41
163	26 569	12.767 15	40.373 26	203	41 209	14.247 81	45.055 52
164	26 896	12.806 25	40.496 91	204	41 616	14.282 86	45.166 36
165	27 225	12.845 23	40.620 19	205	42 025	14.317 82	45.276 93
166	27 556	12.884 10	40.743 10	206	42 436	14.352 70	45.387 22
167	27 889	12.922 85	40.865 63	207	42 849	14.387 49	45.497 25
168	28 224	12.961 48	40.987 80	208	43 264	14.422 21	45.607 02
169	28 561	13.000 00	41.109 61	209	43 681	14.456 83	45.716 52
170	28 900	13.038 40	41.231 06	210	44 100	14.491 38	45.825 76
171	29 241	13.076 70	41.352 15	211	44 521	14.525 84	45.934 74
172	29 584	13.114 88	41.472 88	212	44 944	14.560 22	46.043 46
173	29 929	13.152 95	41.593 27	213	45 369	14.594 52	46.151 92
174	30 276	13.190 91	41.713 31	214	45 796	14.628 74	46.260 13
175	30 625	13.228 76	41.833 00	215	46 225	14.662 88	46.368 09
176	30 976	13.266 50	41.952 35	216	46 656	14.696 94	46.475 80
177	31 329	13.304 13	42.071 37	217	47 089	14.730 92	46.583 26
178	31 684	13.341 66	42.190 05	218	47 524	14.764 82	46.690 47
179	32 041	13.379 09	42.308 39	219	47 961	14.798 65	46.797 44
180	32 400	13.416 41	42.426 41	220	48 400	14.832 40	46.904 16
181	32 761	13.453 62	42.544 09	221	48 841	14.866 07	47.010 64
182	33 124	13.490 74	42.661 46	222	49 284	14.899 66	17.116 88
183	33 489	13.527 75	42.778 50	223	49 729	14.933 18	47.222 88
184	33 856	13.564 66	42.895 22	224	50 176	14.966 63	47.328 64
185	34 225	13.601 47	43.011 63	225	50 625	15.000 00	47.434 16
186	34 596	13.638 18	43.127 72	226	51 076	15.033 30	47.539 46
187	·34 969	13.674 79	43.243 50	227	51 529	15.066 52	47.644 52
188	35 344	13.711 31	43.358 97	228	51 984	15.099 67	47.749 35
189	35 721	13.747 73	43.474 13	229	52 441	15.132 75	47.853 94

Appendix Table 12 (Continued)

n	n^2	$\sqrt{n}$	$\sqrt{10n}$	n	n^2	$\sqrt{n}$	$\sqrt{10n}$
230	52 900	15.165 75	47.958 32	270	72 900	16.431 68	51.961 52
231	53 361	15.198 68	48.062 46	271	73 441	16.462 08	52.057 66
232	53 824	15.231 55	48.166 38	272	73 984	16.492 42	52.153 62
233	54 289	15.264 34	48.270 07	273	74 529	16.522 71	52.249 40
234	54 756	15.297 06	48.373 55	274	75 076	16.552 95	52.345 01
235	55 225	15.329 71	48.476 80	275	75 625	16.583 12	52.440 44
236	55 696	15.362 29	48.579 83	276	76 176	16.613 25	52.535 70
237	56 169	15.394 80	48.682 65	277	76 729	16.643 32	52.630 79
238	56 644	15.427 25	48.785 24	278	77 284	16.673 33	52.725 71
239	57 121	15.459 62	48.887 63	279	77 841	16.703 29	52.820 45
240	57 600	15.491 93	48.989 79	280	78 400	16.733 20	52.915 03
241	58 081	15.524 17	49.091 75	281	78 961	16.763 05	53.009 43
242	58 564	15.556 35	49.193 50	282	79 524	16.792 86	53.103 67
243	59 049	15.588 46	49.295 03	283	80 089	16.822 60	53.197 74
244	59 536	15.620 50	49.396 36	284	80 656	16.852 30	53.291 65
245	60 025	15.652 48	49.497 47	285	81 225	16.881 94	53.385 39
246	60 516	15.684 39	49.598 39	286	81 796	16.911 53	53.478 97
247	61 009	15.716 23	49.699 09	287	82 369	16.941 07	53.572 38
248	61 504	15.748 02	49.799 60	288	82 944	16.970 56	53.665 63
249	62 001	15.779 73	49.899 90	289	83 521	17.000 00	53.758 72
250	62 500	15.811 39	50.000 00	290	84 100	17.029 39	53.851 65
251	63 001	15.842 98	50.099 90	291	84 681	17.058 72	53.944 42
252	63 504	15.874 51	50.199 60	292	85 264	17.088 01	54.037 02
253	64 009	15.905 97	50.299 11	293	85 849	17.117 24	54.129 47
254	64 516	15.937 38	50.398 41	294	86 436	17.146 43	54.221 77
255	65 025	15.968 72	50.497 52	295	87 025	17.175 56	54.313 90
256	65 536	16.000 00	50.596 44	296	87 616	17.204 65	54.405 88
257	66 049	16.031 22	50.695 17	297	88 209	17.233 69	54.497 71
258	66 564	16.062 38	50.793 70	298	88 804	17.262 68	54.589 38
259	67 081	16.093 48	50.892 04	299	89 401	17.291 62	54.680 89
260	67 600	16.124 52	50.990 20	300	90 000	17.320 51	54.772 26
261	68 121	16.155 49	51.088 16	301	90 601	17.349 35	54.863 47
262	68 644	16.186 41	51.185 94	302	91 204	17.378 15	54.954 53
263	69 169	16.217 27	51.283 53	303	91 809	17.406 90	55.045 44
264	69 696	16.248 08	51.380 93	304	92 416	17.435 60	55.136 20
265	70 225	16.278 82	51.478 15	305	93 025	17.464 25	55.226 81
266	70 756	16.309 51	51.575 19	306	93 636	17.492 86	55.317 27
267	71 289	16.340 13	51.672 04	307	94 249	17.521 42	55.407 58
268	71 824	16.370 71	51.768 72	308	94 864	17.549 93	55.497 75
269	72 361	16.401 22	51.865 21	309	95 481	17.578 40	55.587 77

Appendix Table 12 (Continued)

n	n^2	$\sqrt{n}$	$\sqrt{10n}$	n	n^2	$\sqrt{n}$	$\sqrt{10n}$
310	96 100	17.606 82	55.677 64	350	122 500	18.708 29	59.160 80
311	96 721	17.635 19	55.767 37	351	123 201	18.734 99	59.245 25
312	97 344	17.663 52	55.856 96	352	123 904	18.761 66	59.329 59
313	97 969	17.691 81	55.946 40	353	124 609	18.788 29	59.413 80
314	98 596	17.720 05	56.035 70	354	125 316	18.814 89	59.497 90
315	99 225	17.748 24	56.124 86	355	126 025	18.841 44	59.581 88
316	99 856	17.776 39	56.213 88	356	126 736	18.867 96	59.665 74
317	100 489	17.804 49	56.302 75	357	127 449	18.894 44	59.749 48
318	101 124	17.832 55	56.391 49	358	128 164	18.920 89	59.833 10
319	101 761	17.860 57	56.480 08	359	128 881	18.947 30	59.916 61
320	102 400	17.888 54	56.568 54	360	129 600	18.973 67	60.000 00
321	103 041	17.916 47	56.656 86	361	130 321	19.000 00	60.083 28
322	103 684	17.944 36	56.745 04	362	131 044	19.026 30	60.166 44
323	104 329	17.972 20	56.833 09	363	131 769	19.052 56	60.249 48
324	104 976	18.000 00	56.921 00	364	132 496	19.078 78	60.332 41
325	105 625	18.027 76	57.008 77	365	133 225	19.104 97	60.415 23
326	106 276	18.055 47	57.096 41	366	133 956	19.131 13	60.497 93
327	106 929	18.083 14	57.183 91	367	134 689	19.157 24	60.580 52
328	107 584	18.110 77	57.271 28	368	135 424	19.183 33	60.663 00
329	108 241	18.138 36	57.358 52	369	136 161	19.209 37	60.745 37
330	108 900	18.165 90	57.445 63	370	136 900	19.235 38	60.827 63
331	109 561	18.193 41	57.532 60	371	137 641	19.261 36	60.909 77
332	110 224	18.220 87	57.619 44	372	138 384	19.287 30	60.991 80
333	110 889	18.248 29	57.706 15	373	139 129	19.313 21	61.073 73
334	111 556	18.275 67	57.792 73	374	139 876	19.339 08	61.115 54
335	112 225	18.303 01	57.879 18	375	140 625	19.364 92	61.237 24
336	112 896	18.330 30	57.965 51	376	141 376	19.390 72	61.318 84
337	113 569	18.357 56	58.051 70	377	142 129	19.416 49	61.400 33
338	114 244	18.384 78	58.137 77	378	142 884	19.442 22	61.481 70
339	114 921	18.411 95	58.223 71	379	143 641	19.467 92	61.562 98
340	115 600	18.439 09	58.309 52	380	144 400	19.493 59	61.644 14
341	116 281	18.466 19	58.395 21	381	145 161	19.519 22	61.725 20
342	116 964	18.943 24	58.480 77	382	145 924	19.544 82	61.806 15
343	117 649	18.520 26	58.566 20	383	146 689	19.570 39	61.886 99
344	118 336	18.547 24	58.651 51	384	147 456	19.595 92	61.967 73
345	119 025	18.574 18	58.736 70	385	148 225	19.621 42	62.048 37
346	119 716	18.601 08	58.821 76	386	148 996	19.646 88	62.128 90
347	120 409	18.627 94	58.906 71	387	149 769	19.672 32	62.209 32
348	121 104	18.654 76	58.991 52	388	150 544	19.697 72	62.289 65
349	121 801	18.681 54	59.076 22	389	151 321	19.723 08	62.369 86

Appendix Table 12 (Continued)

n	n^2	$\sqrt{n}$	$\sqrt{10n}$	n	n^2	$\sqrt{n}$	$\sqrt{10n}$
390	152 100	19.748 42	62.449 98	430	184 900	20.736 44	65.574 39
391	152 881	19.773 72	62.529 99	431	185 761	20.760 54	65.650 59
392	153 664	19.798 99	62.609 90	432	186 624	20.784 61	65.726 71
393	154 449	19.824 23	62.689 71	433	187 489	20.808 65	65.802 74
394	155 236	19.849 43	62.769 42	434	188 356	20.832 67	65.878 68
395	156 025	19.874 61	62.849 03	435	189 225	20.856 65	65.954 53
396	156 816	19.899 75	62.928 53	436	190 096	20.880 61	66.030 30
397	157 609	19.924 86	63.007 94	437	190 969	20.904 54	66.105 98
398	158 404	19.949 94	63.087 24	438	191 844	20.928 45	66.181 57
399	159 201	19.974 98	63.166 45	439	192 721	20.952 33	66.257 08
400	160 000	20.000 00	63.245 55	440	193 600	20.976 18	66.332 50
401	160 801	20.024 98	63.324 56	441	194 481	21.000 00	66.407 83
402	161 604	20.049 94	63.403 47	442	195 364	21.023 80	66.483 08
403	162 409	20.074 86	63.482 28	443	196 249	21.047 57	66.558 25
404	163 216	20.099 75	63.560 99	444	197 136	21.071 31	66.633 32
405	164 025	20.124 61	63.639 61	445	198 025	21.095 02	66.708 32
406	164 836	20.149 44	63.718 13	446	198 916	21.118 71	66.783 23
407	165 649	20.174 24	63.796 55	447	199 809	21.142 37	66.858 06
408	166 464	20.199 01	63.874 88	448	200 704	21.166 01	66.932 80
409	167 281	20.223 75	63.953 11	449	201 601	21.189 62	67.007 46
410	168 100	20.248 46	64.031 24	450	202 500	21.213 20	67.082 04
411	168 921	20.273 13	64.109 28	451	203 401	21.236 76	67.156 53
412	169 744	20.297 78	64.187 23	452	204 304	21.260 29	67.230 95
413	170 569	20.322 40	64.265 08	453	205 209	21.283 80	67.305 27
414	171 396	20.346 99	64.342 83	454	206 116	21.307 28	67.379 52
415	172 225	20.371 55	64.420 49	455	207 025	21.330 73	67.453 69
416	173 056	20.396 08	64.498 06	456	207 936	21.354 16	67.527 77
417	173 889	20.420 58	64.575 54	457	208 849	21.377 56	67.601 78
418	174 724	20.445 05	64.652 92	458	209 764	21.400 93	67.675 70
419	175 561	20.469 49	64.730 21	459	210 681	21.424 29	67.749 54
420	176 400	20.493 90	64.807 41	460	211 600	21.447 61	67.823 30
421	177 241	20.518 28	64.884 51	461	212 521	21.470 91	67.896 98
422	178 084	20.542 64	64.961 53	462	213 444	21.494 19	67.970 58
423	178 929	20.566 96	65.038 45	463	214 369	21.517 43	68.044 10
424	179 776	20.591 26	65.115 28	464	215 296	21.540 66	68.117 55
425	180 625	20.615 53	65.192 02	465	216 225	21.563 86	68.190 91
426	181 476	20.639 77	65.268 68	466	217 156	21.587 03	68.264 19
427	182 329	20.663 98	65.345 24	467	218 089	21.610 18	68.337 40
428	183 184	20.688 16	65.421 71	468	219 024	21.633 31	68.410 53
429	184 041	20.712 32	65.498 09	469	219 961	21.656 41	68.483 57

Appendix Table 12 (Continued)

n	n^2	$\sqrt{n}$	$\sqrt{10n}$	n	n^2	$\sqrt{n}$	$\sqrt{10n}$
470	220 900	21.679 48	68.556 55	510	260 100	22.583 18	71.414 28
471	221 841	21.702 53	68.629 44	511	261 121	22.605 31	71.484 26
472	222 784	21.725 56	68.702 26	512	262 144	22.627 42	71.554 18
473	223 729	21.748 56	68.775 00	513	263 169	22.649 50	71.624 02
474	224 676	21.771 54	68.847 66	514	264 196	22.671 57	71.693 79
475	225 625	21.794 49	68.920 24	515	265 225	22.693 61	71.763 50
476	226 576	21.817 42	68.992 75	516	266 256	22.715 63	71.833 14
477	227 529	21.840 33	69.065 19	517	267 289	22.737 63	71.902 71
478	228 484	21.863 21	69.137 54	518	268 324	22.759 61	71.972 22
479	229 441	21.886 07	69.209 83	519	269 361	22.781 57	72.041 65
480	230 400	21.908 90	69.282 03	520	270 400	22.803 51	72.111 03
481	231 361	21.931 71	69.354 16	521	271 441	22.825 42	72.180 33
482	232 324	21.954 50	69.426 22	522	272 484	22.847 32	72.249 57
483	233 289	21.977 26	69.498 20	523	273 529	22.869 19	72.318 74
484	234 256	22.000 00	69.570 11	524	274 576	22.891 05	72.387 84
485	235 225	22.022 72	69.641 94	525	275 625	22.912 88	72.456 88
486	236 196	22.045 41	69.713 70	526	276 676	22.934 69	72.525 86
487	237 169	22.068 08	69.785 39	527	277 729	22.956 48	72.594 77
488	238 144	22.090 72	69.857 00	528	278 784	22.978 25	72.663 61
489	239 121	22.113 34	69.928 53	529	279 841	23.000 00	72.732 39
490	240 100	22.135 94	70.000 00	530	280 900	23.021 73	72.801 10
491	241 081	22.158 52	70.071 39	531	281 961	23.043 44	72.869 75
492	242 064	22.181 07	70.142 71	532	283 024	23.065 13	72.938 33
493	243 049	22.203 60	70.213 96	533	284 089	23.086 79	73.006 85
494	244 036	22.226 11	70.285 13	534	285 156	23.108 44	73.075 30
495	245 025	22.248 60	70.356 24	535	286 225	23.130 07	73.143 69
496	246 016	22.271 06	70.427 27	536	287 296	23.151 67	73.212 02
497	247 009	22.293 50	70.498 23	537	288 369	23.173 26	73.280 28
498	248 004	22.315 91	70.569 12	538	289 444	23.194 83	73.348 48
499	249 001	22.338 31	70.639 93	539	290 521	23.216 37	73.416 62
500	250 000	22.360 68	70.710 68	540	291 600	23.237 90	73.484 69
501	251 001	22.383 03	70.781 35	541	292 681	23.259 41	73.552 70
502	252 004	22.405 36	70.851 96	542	293 764	23.280 89	73.620 65
503	253 009	22.427 66	70.922 49	543	294 849	23.302 36	73.688 53
504	254 016	22.449 94	70.992 96	544	295 936	23.323 81	73.756 36
505	255 025	22.472 21	71.063 35	545	297 025	23.345 24	73.824 12
506	256 036	22.494 44	71.133 68	546	298 116	23.366 64	73.891 81
507	257 049	22.516 66	71.203 93	547	299 209	23.388 03	73.959 45
508	258 064	22.538 86	71.274 12	548	300 304	23.409 40	74.027 02
509	259 081	22.561 03	71.344 24	549	301 401	23.430 75	74.094 53

Appendix Table 12 (Continued)

n	n^2	$\sqrt{n}$	$\sqrt{10n}$	n	n^2	$\sqrt{n}$	$\sqrt{10n}$
550	302 500	23.452 08	74.161 98	590	348 100	24.289 92	76.811 46
551	303 601	23.473 39	74.229 37	591	349 281	24.310 49	76.876 52
552	304 704	23.494 68	74.296 70	592	350 464	24.331 05	76.941 54
553	305 809	23.515 95	74.363 97	593	351 649	24.351 59	77.006 49
554	306 916	23.537 20	74.431 18	594	352 836	24.372 12	77.071 40
555	308 025	23.558 44	74.498 32	595	354 025	24.392 62	77.136 24
556	309 136	23.579 65	74.565 41	596	355 216	24.413 11	77.201 04
557	310 249	23.600 85	74.632 43	597	356 409	24.433 58	77.265 78
558	311 364	23.622 02	74.699 40	598	357 604	24.454 04	77.330 46
559	312 481	23.643 18	74.766 30	599	358 801	24.474 48	77.395 09
560	313 600	23.664 32	74.833 15	600	360 000	24.494 90	77.459 67
561	314 721	23.685 44	74.899 93	601	361 201	24.515 30	77.524 19
562	315 844	23.706 54	74.966 66	602	362 404	24.535 69	77.588 66
563	316 969	23.727 62	75.033 33	603	363 609	24.556 06	77.653 07
564	318 096	23.748 68	75.099 93	604	364 816	24.576 41	77.717 44
565	319 225	23.769 73	75.166 48	605	366 025	24.596 75	77.781 75
566	320 356	23.790 75	75.232 97	606	367 236	24.617 07	77.846 00
567	321 489	23.811 76	75.299 40	607	368 449	24.637 37	77.910 20
568	322 624	23.832 75	75.365 77	608	369 664	24.657 66	77.974 35
569	323 761	23.853 72	75.432 09	609	370 881	24.677 93	78.038 45
570	324 900	23.874 67	75.498 34	610	372 100	24.698 18	78.102 50
571	326 041	23.895 61	75.564 54	611	373 321	24.718 41	78.166 49
572	327 184	23.916 52	75.630 68	612	374 544	24.738 63	78.230 43
573	328 329	23.937 42	75.696 76	613	375 769	24.758 84	78.294 32
574	329 476	23.958 30	75.762 79	614	376 996	24.779 02	78.358 15
575	330 625	23.979 16	75.828 75	615	378 225	24.799 19	78.421 94
576	331 776	24.000 00	75.894 66	616	379 456	24.819 35	78.485 67
577	332 929	24.020 82	75.960 52	617	380 689	24.839 48	78.549 35
578	334 084	24.041 63	76.026 31	618	381 924	24.859 61	78.612 98
579	335 241	24.062 42	76.092 05	619	383 161	24.879 71	78.676 55
580	336 400	24.083 19	76.157 73	620	384 400	24.899 80	78.740 08
581	337 561	24.103 94	76.223 36	621	385 641	24.919 87	78.803 55
582	338 724	24.124 68	76.288 92	622	386 884	24.939 93	78.866 98
583	339 889	24.145 39	76.354 44	623	388 129	24.959 97	78.930 35
584	341 056	24.166 09	76.419 89	624	389 376	24.979 99	78.993 67
585	342 225	24.186 77	76.485 29	625	390 625	25.000 00	79.056 94
586	343 396	24.207 44	76.550 64	626	391 876	25.019 99	79.120 16
587	344 569	24.228 08	76.615 93	627	393 129	25.039 97	79.183 33
588	345 744	24.248 71	76.681 16	628	394 384	25.059 93	79.246 45
589	346 921	24.269 32	76.746 34	629	395 641	25.079 87	79.309 52

Appendix Table 12 (Continued)

n	n^2	$\sqrt{n}$	$\sqrt{10n}$	n	n^2	$\sqrt{n}$	$\sqrt{10n}$
630	396 900	25.099 80	79.372 54	670	448 900	25.884 36	81.853 53
631	398 161	25.119 71	79.435 51	671	450 241	25.903 67	81.914 59
632	399 424	25.139 61	79.498 43	672	451 584	25.922 96	81.975 61
633	400 689	25.159 49	79.561 30	673	452 929	25.942 24	82.036 58
634	401 956	25.179 36	79.624 12	674	454 276	25.961 51	82.097 50
635	403 225	25.199 21	79.686 89	675	455 625	25.980 76	82.158 38
636	404 496	25.219 04	79.749 61	676	456 976	26.000 00	82.219 22
637	405 769	25.238 86	79.812 28	677	458 329	26.019 22	82.280 01
638	407 044	25.258 66	79.874 90	678	459 684	26.038 43	82.340 76
639	408 321	25.278 45	79.937 48	679	461 041	26.057 63	82.401 46
640	409 600	25.298 22	80.000 00	680	462 400	26.076 81	82.462 11
641	410 881	25.317 98	80.062 48	681	463 761	26.095 98	82.522 72
642	412 164	25.337 72	80.124 90	682	465 124	26.115 13	82.583 29
643	413 449	25.357 44	80.187 28	683	466 489	26.134 27	82.643 81
644	414 736	25.377 16	80.249 61	684	467 856	26.153 39	82.704 29
645	416 025	25.396 85	80.311 89	685	469 225	26.172 50	82.764 73
646	417 316	25.416 53	80.374 13	686	470 596	26.191 60	82.825 12
647	418 609	25.436 19	80.436 31	687	471 969	26.210 68	82.885 46
648	419 904	25.455 84	80.498 45	688	473 344	26.229 75	82.945 77
649	421 201	25.475 48	80.560 54	689	474 721	26.248 81	83.006 02
650	422 500	25.495 10	80.622 58	690	476 100	26.267 85	83.066 24
651	423 801	25.514 70	80.684 57	691	477 481	26.286 88	83.126 41
652	425 104	25.534 29	80.746 52	692	478 864	26.305 89	83.186 54
653	426 409	25.553 86	80.808 42	693	480 249	26.324 89	83.246 62
654	427 716	25.573 42	80.870 27	694	481 636	26.343 88	83.306 66
655	429 025	25.592 97	80.932 07	695	483 025	26.362 85	83.366 66
656	430 336	25.612 50	80.993 83	696	484 416	26.381 81	83.426 61
657	431 649	25.632 01	81.055 54	697	485 809	26.400 76	83.486 53
658	432 964	25.651 51	81.117 20	698	487 204	26.419 69	83.546 39
659	434 281	25.671 00	81.178 81	699	488 601	26.438 61	83.606 22
660	435 600	25.690 47	81.240 38	700	490 000	26.457 51	83.666 00
661	436 921	25.709 92	81.301 91	701	491 401	26.476 40	83.725 74
662	438 244	25.729 36	81.363 38	702	492 804	26.495 28	83.785 44
663	439 569	25.748 79	81.424 81	703	494 209	26.514 15	83.845 10
664	440 896	25.768 20	81.486 20	704	495 616	26.533 00	83.904 71
665	442 225	25.787 59	81.547 53	705	497 025	26.551 84	83.964 28
666	443 556	25.806 98	81.608 82	706	498 436	26.570 66	84.023 81
667	444 889	25.826 34	81.670 07	707	499 849	26.589 47	84.083 29
668	446 224	25.845 70	81.731 27	708	501 264	26.608 27	84.142 74
669	447 561	25.865 03	81.792 42	709	502 681	26.627 05	84.202 14

Appendix Table 12 (Continued)

n	n^2	$\sqrt{n}$	$\sqrt{10n}$	n	n^2	$\sqrt{n}$	$\sqrt{10n}$
710	504 100	26.645 83	84.261 50	750	562 500	27.386 13	86.602 54
711	505 521	26.664 58	84.320 82	751	564 001	27.404 38	86.660 26
712	506 944	26.683 33	84.380 09	752	565 504	27.422 62	86.717 93
713	508 369	26.702 06	84.439 33	753	567 009	27.440 85	86.775 57
714	509 796	26.720 78	84.498 52	754	568 516	27.459 06	86.833 17
715	511 225	26.739 48	84.557 67	755	570 025	27.477 26	86.890 74
716	512 656	26.758 18	84.616 78	756	571 536	27.495 45	86.948 26
717	514 089	26.776 86	84.675 85	757	573 049	27.513 63	87.005 75
718	515 524	26.795 52	84.734 88	758	574 564	27.531 80	87.063 20
719	516 961	26.814 18	84.793 87	759	576 081	27.549 95	87.120 61
720	518 400	26.832 82	84.852 81	760	577 600	27.568 10	87.177 98
721	519 841	26.851 44	84.911 72	761	579 121	27.586 23	87.235 31
722	521 284	26.870 06	84.970 58	762	580 644	27.604 35	87.292 61
723	522 729	26.888 66	85.029 41	763	582 169	27.622 45	87.349 87
724	524 176	26.907 25	85.088 19	764	583 696	27.640 55	87.407 09
725	525 625	26.925 82	85.146 93	765	585 225	27.658 63	87.464 28
726	527 076	26.944 39	85.205 63	766	586 756	27.676 71	87.521 43
727	528 529	26.962 94	85.264 29	767	588 289	27.694 76	87.578 54
728	529 984	26.981 48	85.322 92	768	589 824	27.712 81	87.635 61
729	531 441	27.000 00	85.381 50	769	591 361	27.730 85	87.692 65
730	532 900	27.018 51	85.440 04	770	592 900	27.748 87	87.749 64
731	534 361	27.037 01	85.498 54	771	594 441	27.766 89	87.806 61
732	535 824	27.055 50	85.557 00	772	595 984	27.784 89	87.863 53
733	537 289	27.073 97	85.615 42	773	597 529	27.802 88	87.920 42
734	538 756	27.092 43	85.673 80	774	599 076	27.820 86	87.977 27
735	540 225	27.110 88	85.732 14	775	600 625	27.838 82	88.034 08
736	541 696	27.129 32	85.790 44	776	602 176	27.856 78	88.090 86
737	543 169	27.147 74	85.848 70	777	603 729	27.874 72	88.147 60
738	544 644	27.166 16	85.906 93	778	605 284	27.892 65	88.204 31
739	546 121	27.184 55	85.965 11	779	606 841	27.910 57	88.260 98
740	547 600	27.202 94	86.023 25	780	608 400	27.928 48	88.317 61
741	549 081	27.221 32	86.081 36	781	609 961	27.946 38	88.374 20
742	550 564	27.239 68	86.139 42	782	611 524	27.964 26	88.430 76
743	552 049	27.258 03	86.197 45	783	613 089	27.982 14	88.487 29
744	553 536	27.276 36	86.255 43	784	614 656	28.000 00	88.543 77
745	555 025	27.294 69	86.313 38	785	616 225	28.017 85	88.600 23
746	556 516	27.313 00	86.371 29	786	617 796	28.035 69	88.656 64
747	558 009	27.331 30	86.429 16	787	619 369	28.053 52	88.713 02
748	559 504	27.349 59	86.486 99	788	620 944	28.071 34	88.769 36
749	561 001	27.367 86	86.544 79	789	622 521	28.089 14	88.825 67

Appendix Table 12 (Continued)

n	n^2	$\sqrt{n}$	$\sqrt{10n}$	n	n^2	$\sqrt{n}$	$\sqrt{10n}$
790	624 100	28.106 94	88.881 94	830	688 900	28.809 72	91.104 34
791	625 681	28.124 72	88.938 18	831	690 561	28.827 07	91.159 20
792	627 264	28.142 49	88.994 38	832	692 224	28.844 41	91.214 03
793	628 849	28.160 26	89.050 55	833	693 889	28.861 74	91.268 83
794	630 436	28.178 01	89.106 68	834	695 556	28.879 06	91.323 60
795	632 025	28.195 74	89.162 77	835	697 225	28.896 37	91.378 33
796	633 616	28.213 47	89.218 83	836	698 896	28.913 66	91.433 04
797	635 209	28.231 19	89.274 86	837	700 569	28.930 95	91.487 70
798	636 804	28.248 89	89.330 85	838	702 244	28.948 23	91.542 34
799	638 401	28.266 59	89.386 80	839	703 921	28.965 50	91.596 94
800	640 000	28.284 27	89.442 72	840	705 600	28.982 75	91.651 51
801	641 601	28.301 94	89.498 60	841	707 281	29.000 00	91.706 05
802	643 204	28.319 60	89.554 45	842	708 964	29.017 24	91.760 56
803	644 809	28.337 25	89.610 27	843	710 649	29.034 46	91.815 03
804	646 416	28.354 89	89.666 05	844	712 336	29.051 68	91.869 47
805	648 025	28.372 52	89.721 79	845	714 025	29.068 88	91.923 88
806	649 636	28.390 14	89.777 50	846	715 716	29.086 08	91.978 26
807	651 249	28.407 75	89.833 18	847	717 409	29.103 26	92.032 60
808	652 864	28.425 34	89.888 82	848	719 104	29.120 44	92.086 92
809	654 481	28.442 93	89.944 43	849	720 801	29.137 60	92.141 20
810	656 100	28.460 50	90.000 00	850	722 500	29.154 76	92.195 44
811	657 721	28.478 06	90.055 54	851	724 201	29.171 90	92.249 66
812	659 344	28.495 61	90.111 04	852	725 904	29.189 04	92.303 85
813	660 969	28.513 15	90.166 51	853	727 609	29.206 16	92.358 00
814	662 596	28.530 69	90.221 95	854	729 316	29.223 28	92.412 12
815	664 225	28.548 20	90.277 35	855	731 025	29.240 38	92.466 21
816	665 856	28.565 71	90.332 72	856	732 736	29.257 48	92.520 27
817	667 489	28.583 21	90.388 05	857	734 449	29.274 56	92.574 29
818	669 124	28.600 70	90.443 35	858	736 164	29.291 64	92.628 29
819	670 761	28.618 18	90.498 62	859	737 881	29.308 70	92.682 25
820	672 400	28.635 64	90.553 85	860	739 600	29.325 76	92.736 18
821	674 041	28.653 10	90.609 05	861	741 321	29.342 80	92.790 09
822	675 684	28.670 54	90.664 22	862	743 044	29.359 84	92.843 96
823	677 329	28.687 98	90.719 35	863	744 769	29.376 86	92.897 79
824	678 976	28.705 40	90.774 45	864	746 496	29.393 88	92.951 60
825	680 625	28.722 81	90.829 51	865	748 225	29.410 88	93.005 38
826	682 276	28.740 22	90.884 54	866	749 956	29.427 88	93.059 12
827	683 929	28.757 61	90.939 54	867	751 689	29.444 86	93.112 83
828	685 584	28.774 99	90.994 51	868	753 424	29.461 84	93.166 52
829	687 241	28.792 36	91.049 44	869	755 161	29.478 81	93.220 17

Appendix Table 12 (Continued)

n	n^2	$\sqrt{n}$	$\sqrt{10n}$	n	n^2	$\sqrt{n}$	$\sqrt{10n}$
870	756 900	29.495 76	93.273 79	910	828 100	30.166 21	95.393 92
871	758 641	29.512 71	93.327 38	911	829 921	30.182 78	95.446 32
872	760 384	29.529 65	93.380 94	912	831 744	30.199 34	95.498 69
873	762 129	29.546 57	93.434 47	913	833 569	30.215 89	95.551 03
874	763 876	29.563 49	93.487 97	914	835 396	30.232 43	95.603 35
875	765 625	29.580 40	93.541 43	915	837 225	30.248 97	95.655 63
876	767 376	29.597 30	93.594 87	916	839 056	30.265 49	95.707 89
877	769 129	29.614 19	93.648 28	917	840 889	30.282 01	95.760 12
878	770 884	29.631 06	93.701 65	918	842 724	30.298 51	95.812 32
879	772 641	29.647 93	93.755 00	919	844 561	30.315 01	95.864 49
880	774 400	29.664 79	93.808 32	920	846 400	30.331 50	95.916 63
881	776 161	29.681 64	93.861 60	921	848 241	30.347 98	95.968 74
882	777 924	29.698 48	93.914 86	922	850 084	30.364 45	96.020 83
883	779 689	29.715 32	93.968 08	923	851 929	30.380 92	96.072 89
884	781 456	29.732 14	94.021 27	924	853 776	30.397 37	96.124 92
885	783 225	29.748 95	94.074 44	925	855 625	30.413 81	96.176 92
886	784 996	29.765 75	94.127 57	926	857 476	30.430 25	96.228 89
887	786 769	29.782 55	94.180 68	927	859 329	30.446 67	96.280 84
888	788 544	29.799 33	94.233 75	928	861 184	30.463 09	96.332 76
889	790 321	29.816 10	94.286 80	929	863 041	30.479 50	96.384 65
890	792 100	29.832 87	94.339 81	930	864 900	30.495 90	96.436 51
891	793 881	29.849 62	94.392 80	931	866 761	30.512 29	96.488 34
892	795 664	29.866 37	94.445 75	932	868 624	30.528 68	96.540 15
893	797 449	29.883 11	94.498 68	933	870 489	30.545 05	96.591 93
894	799 236	29.899 83	94.551 57	934	872 356	30.561 41	96.643 68
895	801 025	29.916 55	94.604 44	935	874 225	30.577 77	96.695 40
896	802 816	29.933 26	94.657 28	936	876 096	30.594 12	96.747 09
897	804 609	29.949 96	94.710 08	937	877 969	30.610 46	96.798 76
898	806 404	29.966 65	94.762 86	938	879 844	30.626 79	96.850 40
899	808 201	29.983 33	94.815 61	939	881 721	30.643 11	96.902 01
900	810 000	30.000 00	94.868 33	940	883 600	30.659 42	96.953 60
901	811 801	30.016 66	94.921 02	941	885 481	30.675 72	97.005 15
902	813 604	30.033 31	94.973 68	942	887 364	30.692 02	97.056 68
903	815 409	30.049 96	95.026 31	943	889 249	30.708 31	97.108 19
904	817 216	30.066 59	95.078 91	944	891 136	30.724 58	97.159 66
905	819 025	30.083 22	95.131 49	945	893 025	30.740 85	97.211 11
906	820 836	30.099 83	95.184 03	946	894 916	30.757 11	97.262 53
907	822 649	30.116 44	95.236 55	947	896 809	30.773 37	97.313 93
908	824 464	30.133 04	95.289 03	948	898 704	30.789 61	97.365 29
909	826 281	30.149 63	95.341 49	949	900 601	30.805 84	97.416 63

Appendix Table 12 (Continued)

n	n^2	$\sqrt{n}$	$\sqrt{10n}$	n	n^2	$\sqrt{n}$	$\sqrt{10n}$
950	902 500	30.822 07	97.467 94	975	950 625	31.224 99	98.742 09
951	904 401	30.838 29	97.519 23	976	952 576	31.241 00	98.792 71
952	906 304	30.854 50	97.570 49	977	954 529	31.257 00	98.843 31
953	908 209	30.870 70	97.621 72	978	956 484	31.272 99	98.893 88
954	910 116	30.886 89	97.672 92	979	958 441	31.288 98	98.944 43
955	912 025	30.903 07	97.724 10	980	960 400	31.304 95	98.994 95
956	913 936	30.919 25	97.775 25	981	962 361	31.320 92	99.045 44
957	915 849	30.935 42	97.826 38	982	964 324	31.336 88	99.095 91
958	917 764	30.951 58	97.877 47	983	966 289	31.352 83	99.146 36
959	919 681	30.967 73	97.928 55	984	968 256	31.368 77	99.196 77
960	921 600	30.983 87	97.979 59	985	970 225	31.384 71	99.247 17
961	923 521	31.000 00	98.030 61	986	972 196	31.400 64	99.297 53
962	925 444	31.016 12	98.081 60	987	974 169	31.416 56	99.347 87
963	927 369	31.032 24	98.132 56	988	976 144	31.432 47	99.398 19
964	929 296	31.048 35	98.183 50	989	978 121	31.448 37	99.448 48
965	931 225	31.064 45	98.234 41	990	980 100	31.464 27	99.498 74
966	933 156	31.080 54	98.285 30	991	982 081	31.480 15	99.548 98
967	935 089	31.096 62	98.336 16	992	984 064	31.496 03	99.599 20
968	937 024	31.112 70	98.386 99	993	986 049	31.511 90	99.649 39
969	938 961	31.128 76	98.437 80	994	988 036	31.527 77	99.699 55
970	940 900	31.144 82	98.488 58	995	990 025	31.543 62	99.749 69
971	942 841	31.160 87	98.539 33	996	992 016	31.559 47	99.799 80
972	944 784	31.176 91	98.590 06	997	994 009	31.575 31	99.849 89
973	946 729	31.192 95	98.640 76	998	996 004	31.591 14	99.899 95
974	948 676	31.208 97	98.691 44	999	998 001	31.606 96	99.949 99

Answers to Odd-Numbered Exercises

Chapter 1

1.1 (a) The population consists of a set of observations indicating whether each child of a drug-addicted mother in Boston has this particular form of learning disability.

(b) A frame can be obtained by getting a list of all children of drug-addicted mothers in Boston, but it may, of course, be difficult to get such a list.

(c) Finite.

(d) Qualitative.

1.3 (a) No. It would be a sample of the opinions of those on picket lines.

(b) Finite.

(c) Qualitative.

(d) Teachers on picket lines may be less likely to believe that the strike would be settled on unfavorable terms.

1.5 (a) Correlation may not imply causation. Even if the neighboring community had not allowed bars to stay open after midnight, the crime rate might have increased.

(b) It would be difficult, but possible. Some communities might allow longer hours for bars, and their crime rates could be compared with communities that maintained existing hours for bars. But it would be difficult to randomize such an experiment properly, because communities are not likely to allow their laws to be dictated by the experimental design.

(c) Perhaps. For example, one might see whether communities that lengthened the hours for bars in recent years have experienced greater increases in crime rates than those that have not lengthened them. However, the former communities may differ from the latter communities in many other ways, and it may be difficult to estimate the extent to which the results are due to these differences, rather than to the change in hours for bars.

1.7 The first class interval might be: zero and under 2 inches. The last class interval might be: 18 and under 20 inches. The class mark for the first class interval would be 1 inch; for the last class interval, it would be 19 inches.

1.9

Amount of insurance	*Number of males*
Under $40,000	_____
$40,000 and under $80,000	_____
$80,000 and under $120,000	_____
. . .	
$360,000 and under $400,000	_____

The class mark in each interval would be $20,000, $60,000,..., $340,000, and $380,000.

1.11 (a)

Score	*Number of adults*
40 and under 50	2
50 and under 60	4
60 and under 70	6
70 and under 80	16
80 and under 90	7
90 and under 100	5

(b)

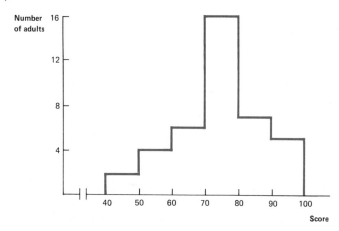

1.13 (a)

Weight	*Number of soldiers*
Less than 120 pounds	0
Less than 140 pounds	205
Less than 160 pounds	576
Less than 180 pounds	979
Less than 200 pounds	1502
Less than 220 pounds	1814
Less than 240 pounds	2000

(b)

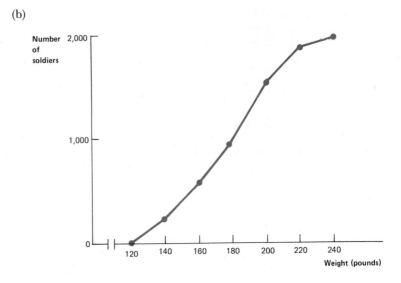

1.15

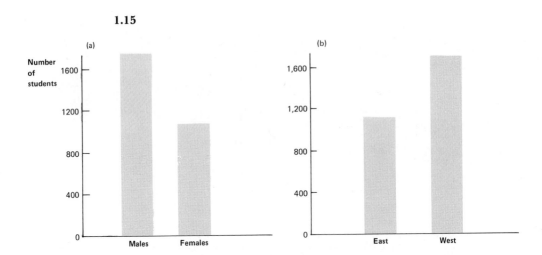

1.17

```
4 | 2 3
5 | 0 1 3 8
6 | 0 1 2 3 5 9
7 | 1 1 1 2 3 4 4 5 5 6 6 6 8 8 8 9
8 | 0 1 1 1 2 3 6
9 | 2 3 5 8 8
```

1.19 (a) 0.50
(b) It is unlikely that the proportion of cases will be exactly equal to 0.50.

1.21 No, because the smallest bill may not be at the lower limit of the lowest class interval and the largest bill may not be at the upper limit of the highest class interval.

1.23 (a)

Age of first consumption of alcohol (years)	*Number of alcoholics*
6 and under 8	2
8 and under 10	7
10 and under 12	9
12 and under 14	4
14 and under 16	6
16 and under 18	2
Total	30

(b)

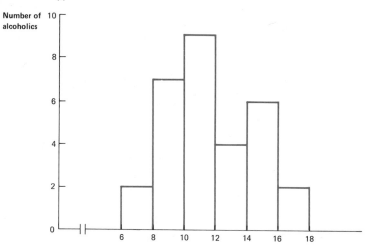

(c)

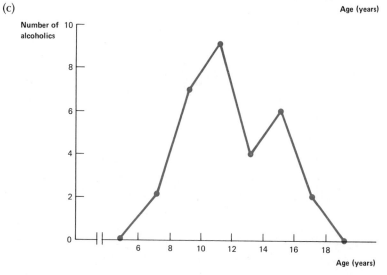

1.25 (a) They are −$1.00 to −$0.75, −$0.75 to −$0.50, and so forth.
(b) Yes
(c) The fourth class interval: −$0.25 and under $0.00
(d)

Error (dollars)	Number of customers
Less than −1.00	0
Less than −0.75	1
Less than −0.50	3
Less than −0.25	7
Less than 0.00	37
Less than +0.25	43
Less than +0.50	45
Less than +0.75	47
Less than +1.00	49
Less than +1.25	50

(e) The ogive is as follows:

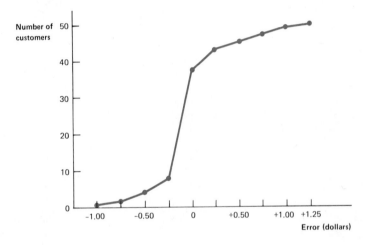

Chapter 2

2.1 (a) $\dfrac{28.3}{18} = 1.572$

(b) $(1.56 + 1.68) \div 2 = 1.62$
(c) If one is interested in only these 18 students, it is the population. If these students are only part of those one is interested in, it is a sample.
(d) If this is the population, they are parameters; if it is a sample, they are statistics.

2.3 $\dfrac{\$6.05(4{,}000) + \$7.39(8{,}000)}{12{,}000} = \dfrac{83{,}320}{12{,}000} = \$6.943$

2.5 (a) $\dfrac{29(.5) + 32(2) + 29(4) + 10(6)}{100} = \$2.545$

(b) $\left(\dfrac{100/2 - 29}{32}\right) \$2 + \$1 = \$2.312$

2.7 (a) 98.
(b) Yes. To the left.
(c) The mode.

2.9 (a) $\dfrac{191}{6} = 31.83$

(b) $\dfrac{113}{5} = 22.60$

(c) The average number of successful launchings per year was lower in 1977–81 than in 1971–76.
(d) The median in 1971–76 was 31.5; in 1977–81, it was 18. The median, like the mean, was lower in 1977–81 than in 1971–76.

2.11 $\displaystyle\sum_{i=1}^{5} X_i = 10.1$

$\displaystyle\sum_{i=1}^{5} X_i^2 = 3.24 + 3.61 + 4.41 + 5.29 + 4.00 = 20.55$

$s^2 = \dfrac{20.55 - \dfrac{1}{5}(10.1)^2}{4} = \dfrac{20.55 - 20.402}{4} = 0.037$

$s = 0.19$ pounds

2.13 (a) $\displaystyle\sum_{i=1}^{5} X_i = 150.11$ $s^2 = \dfrac{4506.6027 - \dfrac{1}{5}(150.11)^2}{4} = .00007$

$\displaystyle\sum_{i=1}^{5} X_i^2 = 4506.6027$ $s = .0084$ inches

(b) $\displaystyle\sum_{i=1}^{5} X_i = .11$ $s^2 = \dfrac{.0027 - \dfrac{1}{5}(.11)^2}{4} = .00007$

$\displaystyle\sum_{i=1}^{5} X_i^2 = .0027$ $s = .0084$ inches

(c) They are equal because the standard deviation remains the same if all observations are decreased by the same amount.

2.15 $\displaystyle\sum_{j=1}^{k} f_j X_j' = 254.5$

$$\sum_{j=1}^{k} f_j X_j'^2 = 29(.25) + 32(4) + 29(16) + 10(36) = 959.25$$

$$s^2 = \frac{959.25 - \dfrac{1}{100}(254.5)^2}{99} = \frac{959.25 - 647.7025}{99} = \frac{311.5475}{99} = 3.1469$$

$s = \$1.77$

If the 100 workers were the population, 100 would have to be substituted for 99. Thus, $s^2 = 3.1155$, and $s = \$1.765$.

2.17 Mary scores $\dfrac{95 - 80}{10} = 1.5$ standard deviations above the mean. Annie scores $\dfrac{94 - 75}{12} = 1.58$ standard deviations above the mean. Thus, Annie seems to have done better (relative to others that have taken the same test) than Mary.

2.19 Statements (b) and (c) are true.

2.21 (a) $45 - 23 = 22$
(b) $34 - 16 = 18$
(c) $45 - 16 = 29$
(d) no

2.23 These figures do not prove that civilians were more likely to suffer amputations. There were many more civilians than military personnel during the war. To estimate the likelihood that a civilian would suffer an amputation, the 120,000 would have to be divided by the number of civilians. To estimate the likelihood that a member of the military would suffer an amputation, the 18,000 would have to be divided by the number of military personnel.

2.25 No. Perhaps divorce is more common among relatively young people who have not had a chance to have many children. Other possible explanations could be given too. Correlation does not prove causation.

2.27 No. The difference between the means is only about $200. This is a very small difference when compared to the variation about each average. It might have been more meaningful to present the standard deviation (or some other measure of dispersion) of income of farms in each state as well as averages.

2.29 No. The pitfall here is that the variation about the average is neglected.

2.31 Yes, because the frequency distribution of income tax payments is likely to be skewed to the right.

2.33 (a) Zero
(b) If the frequency distribution is symmetrical, the median equals the mean, so the mean must also equal 3. Thus, the sum equals 1,000(3), or 3,000.

2.35 (a) 6.84
   (b) 6.86
   (c) 8.25

2.37 Since we know from Exercise 2.34 that the mean is $20, the standard deviation (if the data in Exercise 2.34 are the entire population) equals

$$\sqrt{\frac{60(10-20)^2 + 30(30-20)^2 + 10(50-20)^2}{100}} = \sqrt{180} = \$13.42.$$

If we calculate the standard deviation in cents, we have

$$\sqrt{\frac{60(1000-2000)^2 + 30(3000-2000)^2 + 10(5000-2000)^2}{100}}$$

$$= \sqrt{1,800,000}$$

$$= 1,342 \text{ cents.}$$

The ratio of the latter standard deviation to the former is 100, since each observation expressed in cents is 100 times the same observation expressed in dollars.

2.39 (a) Not necessarily, as explained in (b).
   (b) Yes, it is consistent with them. Not unless the female applicants were better qualified than the male applicants, or unless admission was deliberately reduced in majors that have high female application rates.

2.41 The Minitab computer printout is as follows:

```
SET FOLLOWING DATA INTO COL C1

73 65 74 46 71 60 73 62 54 73 72 72 47 46 73 76 60 41 48 71 70 73
64 75 61 57 73 63 70 39

HISTOGRAM OF DATA IN COLUMN C1

C1

MIDDLE OF      NUMBER OF
INTERVAL       OBSERVATIONS
   40.         1        *
   45.         3        ***
   50.         1        *
   55.         2        **
   60.         4        ****
   65.         2        **
   70.         5        *****
   75.         9        *********

AVERAGE THE VALUES IN COLUMN C1

AVERAGE =    64.074
STANDARD DEVIATION OF COLUMN C1

ST.DEV. =    10.795
```

Chapter 3

3.1
heads	heads	heads
heads	heads	tails
heads	tails	heads
tails	heads	heads

tails	tails	heads
tails	heads	tails
heads	tails	tails
tails	tails	tails

The probability is 1/8 that each of these outcomes will occur.

3.3 (a) 1/6   (b) 1/3   (c) zero   (d) 1/2   (e) 1/3

3.5 (a) 8/36   (b) 2/36

3.7 0.25

3.9 $0.75 + 0.25 - 0.1875 = 0.8125$

3.11 (a) No, because P (A and B) does not equal zero.
(b) No. If A and B were statistically independent, $P(A$ and $B)$ would equal $P(A)$ times $P(B)$.

3.13 (a) From Figure 3.1 it is clear that there are five points in the sample space where the sum equals six: (5,1), (4,2), (3,3), (2,4), (1,5). Thus, the probability is 5/36.
(b) From Figure 3.1 it is clear that there are ten points in the sample space where the sum is less than six: (1,1), (1,2), (1,3), (1,4), (2,1), (2,2), (2,3), (3,1), (3,2) (4,1). Thus, the probability equals 10/36.
(c) Since $P(7$ or more$) = 1 - P(6) - P($less than 6$)$, it follows from (a) and (b) that $P(7$ or more$) = 1 - 5/36 - 10/36 = 21/36$.

3.15 $2/3(2/3) = 4/9$

3.17 (a) $0.4(0.4)(0.6) = .096$
(b) $3(.096) + 0.4^3 = .352$

3.19 1/3

3.21 5/6

3.23 The probability that he will get an A in either subject (or both) is $0.20 + 0.25 - 0.05 = 0.40$. Thus the probability that he will get an A in neither subject is $1 - 0.40 = 0.60$.

3.25 (a) .70
(b) .40

3.27 2/3

3.29 (a) $400 \div 1400$
(b) $1200 \div 4900$
(c) No. For example, 300 out of 1,100 motors received by plant 1 come from supplier III, whereas 400 out of 1,400 motors received by plant 2 come from supplier III.

(d) (i) $100 \div 900$
  (ii) $100 \div 1100$
  (iii) $400 \div 2100$
  (iv) $300 \div 2500$
(e) $P(\text{defective}|\text{supplier I}) = P(\text{defective and supplier I}) \div P(\text{supplier I}) = 0.1 \div (800/4900) = 0.1 \div 0.163 = .61.$
(f) $P(\text{supplier I}|\text{defective}) = P(\text{defective and supplier I}) \div P(\text{defective}) = 0.1 \div .20 = 0.50.$

3.31 Since $\binom{n}{x} = \dfrac{n!}{(n-x)!x!}$

it follows that

$$\binom{n}{n-x} = \frac{n!}{[n-(n-x)]!(n-x)!} = \frac{n!}{x!(n-x)!} = \binom{n}{x}.$$

3.33 $8(7)(6)(5)(4)(3)(2)(1) = 40{,}320$

3.35 $6(5)(4) \div 3(2)(1) = 20$ years

3.37 There are $5! \div (3!2!)$ different pairs of men that can be included with each of the two women. Thus, the answer is

$$(2)\left(\frac{5!}{3!2!}\right) = \frac{(2)(5)(4)}{(2)(1)} = 20.$$

3.39 $5(4)(3)(2)(1) = 120$ hours

3.41 $2^8 = 256$

$$\frac{1}{256} = .0039$$

Chapter 4

4.1 (a) $P(x) = 1/3$, for $x = 7, 8, 9$
  (b) $1/3$ $2/3$

4.3 None is a probability distribution. In the case of (a) and (b) $\Sigma P(x) \neq 1$. In the case of (c), $P(x)$ is not always nonnegative.

4.5 (a) Yes
  (b) 0, 1, 2, 3, or 4
  (c)

Number of bicycles	Probability
0	1/5
1	1/5
2	1/5
3	1/5
4	1/5

(d)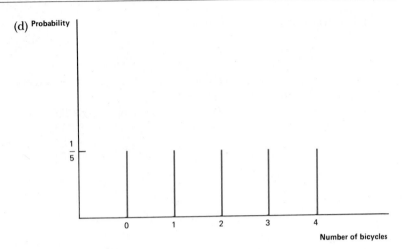

$$P(x) = 1/5, \text{ for } x = 0, 1, 2, 3, 4$$

(e) 1/5
    3/5
    4/5
    1

(f) Yes
    0, $20, $50, $90

Income	Probability
0	2/5
$20	1/5
50	1/5
90	1/5

(g)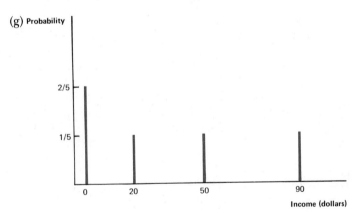

(h) 2/5
    2/5
    2/5

(i)

Number of bicycles	Probability
0	1/25
1	2/25
2	3/25
3	4/25
4	5/25
5	4/25
6	3/25
7	2/25
8	1/25

19/25
10/25
3/25
0

(j) The salesman can make \$100 or more in the two-day period if he sells
    six or more bicycles in this period. As shown in (i) the probability of this
    occurring is 6/25.

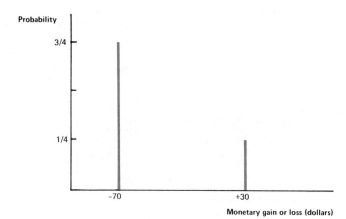

4.7 (a) 28
    (b) −6

4.9 (a) Discrete
    (b) The last digit in a randomly-selected two-digit number would be ex-
        pected to have this probability distribution.
    (c) $E(X) = 0(1/10) + 1(1/10) + \cdots + 9(1/10) = 4.5$
        $\sigma^2(X) = (0 - 4.5)^2(1/10) + (1 - 4.5)^2(1/10) + (2 - 4.5)^2(1/10) +$
        $\qquad (3 - 4.5)^2(1/10) + (4 - 4.5)^2(1/10) + (5 - 4.5)^2(1/10) +$
        $\qquad (6 - 4.5)^2(1/10) + (7 - 4.5)^2(1/10) + (8 - 4.5)^2(1/10) +$
        $\qquad (9 - 4.5)^2(1/10) = 8.25$
        $\sigma(X) = 2.87$

4.11 (a) The expected value equals $(1/5)(0) + (1/5)(1) + (1/5)(2) + (1/5)(3)$
$+ (1/5)(4) = 2$

(b) Yes

(c) $\frac{1}{5}(0-2)^2 + \frac{1}{5}(1-2)^2 + \frac{1}{5}(2-2)^2 + \frac{1}{5}(3-2)^2 + \frac{1}{5}(4-2)^2 = \frac{1}{5}(10) =$
$2(\text{bicycles})^2$

(d) $\sqrt{2} = 1.414$ bicycles

4.13 (a) $\$20(10/16) + \$50(5/16) + \$100(1/16) = \$34.375$

(b) $\sqrt{(20-34.375)^2(10/16) + (50-34.375)^2(5/16) + (100-34.375)^2 1/16}$

$= \sqrt{129.15 + 76.29 + 269.16} = \sqrt{474.60} = \$21.79$

(c) No

(d) No

4.15 5
5
Yes

4.17 (a) $(.20)^{12} = .0000$

(b) $.0155 + .0033 + .0005 + .0001 = .0194$. See Appendix Table 1.

(c) $.0687$

(d) $.0687 + .2062 = .2749$

(e) The mean equals 12 (.2), or 2.4. Thus, the answer is $.0687 + .2062 + .2835 = .5584$.

4.19 (a) Yes. No.

(b) $\$5,000(.0038) + \$2,000(.9962) = \$2,011$

(c) $\sqrt{(5,000 - 2,011)^2(.0038) + (2,000 - 2,011)^2(.9962)}$

$= \sqrt{(8,934,121)(.0038) + (121)(.9962)} = \sqrt{34,070} = \$185$

4.21 (a) From Appendix Table 1, $.6302 + .2985 = .9287$

(b) $.3874 + .3874 = .7748$

(c) $.2316 + .3679 = .5995$

4.23 (a) Since $n = 20$ and $\Pi = .20$, Appendix Table 1 indicates that this probability equals $.0005 + .0001 + .0000 = .0006$.

(b) According to Appendix Table 1 this probability equals $.0115 + .0576 + .1369 = 0.206$.

(c) According to Appendix Table 1 the probability that the number expressing dissatisfaction is more than 20 percent of the sample equals $1 - (.0115 + .0576 + .1369 + .2054 + .2182) = .3704$. Thus the expected value of the gamble to the director of dining services is $(.3704)(-\$100) + (.6296)(\$100) = \$25.92$. This is not a fair bet since the expected value does not equal zero.

(d) The expected number of students is 20 (0.20) = 4. The standard deviation is $\sqrt{20 (0.20)(0.80)} = \sqrt{3.2} = 1.79$ students.

4.25 None. All are known constants, not quantities to be determined by chance.

4.27 The Minitab computer printout is as follows:

```
BINOMIAL PROBABILITIES FOR N=21 AND P=0.4
  K     P( X = K)     P( X LESS OR = K )
  0     0.0000        0.0000
  1     0.0003        0.0003
  2     0.0020        0.0024
  3     0.0086        0.0110
  4     0.0259        0.0370
  5     0.0588        0.0957
  6     0.1045        0.2002
  7     0.1493        0.3495
  8     0.1742        0.5237
  9     0.1677        0.6914
 10     0.1342        0.8256
 11     0.0895        0.9151
 12     0.0497        0.9648
 13     0.0229        0.9877
 14     0.0087        0.9964
 15     0.0027        0.9992
 16     0.0007        0.9998
 17     0.0001        1.0000
```

4.29 The expected gain is $(.993)(\$25) + (.007)(-\$975) = \$18$. Thus, the standard deviation of the gain equals

$$\sqrt{(25 - 18)^2(.993) + (-975 - 18)^2(.007)} = \sqrt{48.657 + 6,902.343}$$

$$= \sqrt{6,951} = \$83.37.$$

4.31 If she debates her opponent, the expected value of the number of votes she will gain is $0.55\,(100,000) + 0.45\,(-150,000) = -12,500$. Since this expected value is negative, she should not debate.

Chapter 5

5.1 (a) $.5 - .4893 = .0107$
  (b) $.5 - .4987 = .0013$
  (c) $.5 - .2580 = .2420$
  (d) $.4772 - .3413 = .1359$
  (e) $.3413 + .4772 = .8185$

5.3 No. This is true only if the variable's mean equals zero.

5.5 (a) 0.52
  (b) 0.59
  (c) 0.23
  (d) 0.86
  (e) 0.23

5.7 (a) $.5000 - .3413 = .1587$
(b) $.5000 - .4332 = .0668$
(c) $.1915 + .1915 = .3830$
(d) $.4938 - .3413 = .1525$

5.9 According to Appendix Table 2, the point on the standard normal distribution that is exceeded with a probability of .05 is 1.64. The weight that corresponds to this point on the standard normal distribution is $\mu + 1.64\sigma$, or $170 + (1.64)(20)$, or 202.8 pounds.

5.11 The number of times he hits the target is a binomial random variable with mean equal to $(50)(1/3)$, or 16.67 and a standard deviation equal to

$$\sqrt{(50)(1/3)(2/3)}, \text{ or } \sqrt{11.11} = 3.333.$$

We want the probability that this variable is 9 or less, which can be approximated by the probability that the standard normal variable is less than $(9\frac{1}{2} - 16.67) \div 3.333 = -7.17 \div 3.333$, or -2.15. Using Appendix Table 2, we find that this probability equals .0158.

5.13 (a) $Pr\{Z < (7.5 - 10) \div 2.828\} = Pr\{Z < -.88\} = .1894$
(b) $Pr\{Z > (20.5 - 10) \div 2.828\} = Pr\{Z > 3.71\}$, which is less than .001
(c) $Pr\{(10.5 - 10) \div 2.828 < Z < (18.5 - 10) \div 2.828\}$

$$= Pr\{.18 < Z < 3.01\} = .4987 - .0714 = .4273$$

(d) $Pr\{(8.5 - 10) \div 2.828 < Z < (11.5 - 10) \div 2.828\}$

$$= Pr\{-.53 < Z < .53\} = .2019 + .2019 = .4038$$

5.15 Let X be the number of voters in the sample that are Republicans. Obviously, X has the binomial distribution with mean ($n\Pi$) equal to $(1/3)(300) = 100$ and with standard deviation equal to

$$\sqrt{300(1/3)(2/3)} = \sqrt{66.67}, \text{ or } 8.165.$$

The probability that between 84½ and 115½ voters are Republicans equals $Pr\{84\ 1/2 < X < 115\ 1/2\}$.

The point on the standard normal distribution corresponding to 84 1/2 is

$$\frac{84\ 1/2 - 100}{8.165} = \frac{-15.5}{8.165} = -1.90.$$

The point on the standard normal distribution corresponding to 115 1/2 is

$$\frac{115\ 1/2 - 100}{8.165} = \frac{15.5}{8.165} = +1.90.$$

Using Appendix Table 2, the probability that a standard normal variable lies between -1.90 and $+1.90$ is $.4713 + .4713 = .9426$. Thus, the probability that either less than 85 or more than 115 voters in the sample are Republicans is $1 - .9426$, or .0574. (Of course, this is only an approximation.)

5.17 (a) $n\Pi = 20(.25) = 5$

$$\sqrt{n\Pi(1 - \Pi)} = \sqrt{20(.25)(.75)} = \sqrt{3.75} = 1.9365$$

$$Pr\{(3.5 - 5) \div 1.9365 < Z < (7.5 - 5) \div 1.9365\}$$

$$= Pr\{-.77 < Z < 1.29\} = .2794 + .4015 = .6809$$

(b) $.1897 + .2023 + .1686 + .1124 = .6730$

(c) The difference is only .0079.

5.19 (a) Since the standard deviation is .02 inches, the point on the standard normal distribution corresponding to a width that exceeds the mean by .03 inches is 1.5. Using Appendix Table 2, the probability that the standard normal variable exceeds 1.5 is .0668.

(b) The point on the standard normal distribution that corresponds to a width that falls short of the mean by .05 inches is -2.5. Using Appendix Table 2, the probability that the standard normal variable is less than -2.5 is .0062.

(c) The point on the standard normal distribution that corresponds to a width that is .015 inches below the mean is -0.75. The point on the standard normal distribution that corresponds to a width that is .015 inches above the mean is 0.75. Using Appendix Table 2, the probability that the standard normal variable lies between -0.75 and 0.75 is $.2734 + .2734 = .5468$.

(d) $1/2$

(e) Since the mean is .01 inches greater than the design, all pedals of widths greater than .01 inches less than the mean exceeded the design. The point on the standard normal distribution corresponding to .01 inches less than the mean is $-1/2$ since the standard deviation equals .02 inches. Using Appendix Table 2, the probability that the standard normal variable exceeds $-1/2$ equals $.1915 + .5000$, or .6915.

(f) Since the average width equaled the design, if a pedal was .04 inches wider than the design, its width was .04 inches above the mean. Since the standard deviation was .02 inches, the point on the standard normal distribution corresponding to .04 inches above the mean is 2.0. Using Appendix Table 2, the probability that the standard normal variable exceeds 2.0 is .0228.

(g) Since the average width was .01 inches greater than the design, if a pedal was .04 inches wider than the design, its width was .03 inches above the mean. The point on the standard normal distribution corresponding to .03 inches above the mean is 1.5. Using Appendix Table 2, the probability that the standard normal variable exceeds 1.5 is .0668.

5.21 The normal distribution

5.23 3.0

5.25 Using Appendix Table 1, the binomial probability is .3585. Using Appendix Table 3, the Poisson probability is .3679 since $\mu = (20)(.05) = 1$.

5.27 Since the standard deviation is 1.732, the mean must be $(1.732)^2$, or 3.0. Using Appendix Table 3, we find that the probability of no accidents, given that the mean is 3.0, equals .0498.

5.29 (a) $P(2) = .0842$
(b) $P(3) = .1404$
(c) $P(0) = .0067$

5.31 $\sigma = \sqrt{3} = 1.732$

$Pr\{1.268 < X < 4.732\} = P(2) + P(3) + P(4) = .2240 + .2240 + .1680 = .6160$

Thus, the desired probability equals $1 - .6160 = .3840$.

5.33 (a) $1 - (.0025 + .0149 + .0446) = 1 - .0620 = .9380$
(b) .0892

5.35 If $\Pi = .001$, the probability that none of the sample is defective is $(.999)^n$. If n is at least 20, the Poisson approximation should be useful. According to Appendix Table 3, the probability that $X = 0$ (that is, that there are no defectives) equals .9512 (which is very close to .95) when $\mu = .05$. Since $\mu = n\Pi$, n must equal 50 because $\Pi = .001$. Thus, 50 motors should be examined from each day's output.

5.37 The Minitab computer output is as follows:

POISSON PROBABILITIES FOR MEAN=1.45

K	P(X = K)	P(X LESS OR = K)
0	0.2346	0.2346
1	0.3401	0.5747
2	0.2466	0.8213
3	0.1192	0.9405
4	0.0432	0.9837
5	0.0125	0.9962
6	0.0030	0.9992
7	0.0006	0.9999
8	0.0001	1.0000

Chapter 6

6.1 (a) *AB, AC, AD, AE, BC, BD, BE, CD, CE, DE.* Of course, this assumes that sampling is without replacement.
(b) *CDE, BDE, BCE, BCD, ADE, ACE, ACD, ABE, ABD, ABC.*
(c) 6/10
3/10

6.5 No. Not every guest has the same probability of filling out a questionnaire. For example, guests that have complaints are more likely than others to do so.

6.7 (a) Yes.
(b) No.

(c) The probability that each voter will be included in the sample is unknown and uncontrolled.

(d) He might choose a random sample of voters in each part of his district and ask each person in the sample whether he should vote for or against the proposed legislation.

6.9 (a) No.

(b) The population of Philadelphia residents with children attending the public schools.

(c) Philadelphia residents with no children currently in the public schools may tend to have different opinions on this score than those with children in the public schools.

6.11 (a) 2/7 from stratum A, and 5/7 from stratum B.

(b) 2/3 from stratum A, and 1/3 from stratum B.

(c) The allocation scheme in (a).

6.13 The standard deviation of the sample mean is $100 \div \sqrt{100}$, or 10 points. Thus, if the sample mean differs by more than 15 points from the population mean, it differs by more than 1.5 of its standard deviations from its mean. Since the sample size is large, the sample mean is approximately normally distributed. The probability of a normal variable lying more than 1.5 standard deviations from its mean is $(2)(.0668)$, or .1336 according to Appendix Table 2.

6.15 (a) $\sigma_{\bar{x}} = \dfrac{\sigma}{\sqrt{n}} \sqrt{\dfrac{N-n}{N-1}}$

$$= \frac{\sigma}{\sqrt{n}} \sqrt{\left(1 - \frac{n}{N}\right) \div \left(1 - \frac{1}{N}\right)}$$

If N is large, $(1 - 1/N)$ is approximately equal to one. Thus,

$$\sigma_{\bar{x}} \div \frac{\sigma}{\sqrt{n}} \sqrt{1 - \frac{n}{N}}.$$

(b) $\dfrac{2}{\sqrt{16}} \sqrt{1 - \dfrac{16}{100}} = \dfrac{2}{4} \sqrt{.84} = (1/2)(.9165) = .458$

$\dfrac{2}{\sqrt{16}} \sqrt{1 - \dfrac{16}{10,000}} = \dfrac{2}{4} \sqrt{.9984} = (1/2)(.9992) = .500$

6.17 (a) $950,000

(b) $310,000 \div \sqrt{16} = \$77,500$

(c) No. The central limit theorem assumes that the sample size is at least 30. But if the distribution of the sales of the 3,000 gas stations is close to normal, the sampling distribution of the sample mean will be close to normal, even though the sample size is only 16.

6.19 (a) If $\bar{X} < 78$, $(\bar{X} - 80) \div 0.9 < (78 - 80) \div 0.9 = -2.22$. The probability that a standard normal variable is less than -2.22 equals $0.5 - 0.4868 = 0.0132$.

(b) Since $(78 - 80) \div 9/7 = -1.56$, and since the probability that a standard normal variable is less than -1.56 equals .0594, the desired probability is .0594.

(c) Because the standard error of the sample mean is inversely related to the sample size.

6.21 (a) $\sigma_{\bar{x}} = \frac{10}{\sqrt{50}} \sqrt{\frac{85-50}{85-1}} = 10 \sqrt{\frac{35}{4200}} = 10 \sqrt{.008333} = .913.$

Since $(120 - 115) \div .913 = 5.48$, this probability is less than .001.

(b) If there had been 170 students,

$$\sigma_{\bar{x}} = \frac{10}{\sqrt{50}} \sqrt{\frac{170-50}{170-1}} = 10 \sqrt{\frac{120}{8450}} = 10 \sqrt{.0142} = 1.19.$$

Since $(120 - 115) \div 1.19 = 4.20$, this probability is less than .001. Thus, the doubling of the class size would have little effect on the answer to (a).

6.23 In this case, $\Pi = \frac{1,000}{3,000} = 1/3.$

If the sample proportion differs from the population proportion by more than .02, the number in the sample having relatives in the United States must be other than 32, 33, 34, or 35. Using the normal approximation to the binomial distribution, the probability that this is the case equals:

$$1 - Pr \left\{ \frac{31.5 - 33\ 1/3}{\sqrt{200/9}} < Z < \frac{35.5 - 33\ 1/3}{\sqrt{200/9}} \right\}$$

$$= 1 - Pr \left\{ -\frac{1.83}{4.71} < Z < \frac{2.17}{4.71} \right\}$$

$$= 1 - Pr \{ -0.39 < Z < 0.46 \} = 1 - (.1517 + .1772) = 0.6711.$$

6.25 Yes. There is some evidence that the candidate named first is at an advantage.

6.27 (a) It can number the farms on the list from 1 to 3,000, and it can use a table of random numbers to draw the sample.

(b) No, it is a systematic sample.
Yes.

(c) Yes. Holding sample size constant, the sampling error would tend to be smaller if stratification of this sort occurred.

(d) If proportional allocation is used, the number sampled in each stratum is proportional to the total number in the stratum. Since the total sample size is 100, 1/30 of the farms in each stratum will be chosen. Thus the number chosen from each stratum is as follows:

Farm size(acres)	Number of farms in sample
0–50	1,000/30 = 33
51–100	500/30 = 17
101–150	500/30 = 17
151–200	400/30 = 13
201–250	400/30 = 13
Over 250	200/30 = 7
Total	100

(e) The product of number of farms and standard deviation in each stratum is shown in column (3) below:

Farm size (acres)	(1) Number of farms	(2) Standard deviation	(3) (1) × (2)	Column (3) as percent of total	Number in sample
0–50	1,000	10	10,000	24	24
51–100	500	10	5,000	12	12
101–150	500	10	5,000	12	12
151–200	400	15	6,000	14	14
201–250	400	15	6,000	14	14
Over 250	200	50	10,000	24	24
Total			42,000		

In the next column, this product in each stratum is divided by the sum of these products (42,000). Finally, the result is multiplied by 100 to get the sample size in each stratum.

(f) No, because the farms contained in a particular cluster are not close together (unless people with names beginning with the same initial live close together, which is unlikely).

6.29 Suppose that the median is based on a sample size of n_1, and the mean is based on a sample size of n_2. If

$$\sqrt{\frac{\pi}{2}}\frac{\sigma}{\sqrt{n_1}} = \frac{\sigma}{\sqrt{n_2}},$$

it follows that $n_1/n_2 = \pi/2$. Since $\pi = 3.1416$, $\pi/2 = 1.57$. Thus, n_1 must be 1.57 times n_2.

6.31 (a) Its sampling distribution is normal with a mean of 3.00 inches and a standard deviation of .05 inches.
(b) The point on the standard normal distribution corresponding to 3.01 inches is $(3.01 - 3.00) \div .05 = 0.20$. According to Appendix Table 2, the probability that the standard normal variable will exceed 0.20 is .4207.
(c) 100

Chapter 7

7.1 (a) Estimate. Point estimate.
(b) Estimator.
(c) Estimate. Interval estimate.

7.3 The sample mean is 4.448.

7.5 (a) 8/60, or .13
(b) A sample of this size need not be unreliable; whether it is unreliable or not depends on the required accuracy. The sample percentage is not a biased estimate of the population percentage.

7.7 (a) .01

(b) .005

(c) .025

7.9 (a) $\overline{X} - 1.64 \dfrac{s}{\sqrt{n}} < \mu < \overline{X} + 1.64 \dfrac{s}{\sqrt{n}}$

$810 - 1.64 \dfrac{85}{9.487} < \mu < 810 + 164 \dfrac{85}{9.487}$

$810 - 14.69 < \mu < 810 + 14.69$

$\$795.31 < \mu < \$824.69$

(b) $810 - 1.96(8.96) < \mu < 810 + 1.96(8.96)$

$810 - 17.56 < \mu < 810 + 17.56$

$\$792.44 < \mu < \$827.56$

(c) $810 - 2.576(8.96) < \mu < 810 + 2.576(8.96)$

$810 - 23.08 < \mu < 810 + 23.08$

$\$786.92 < \mu < \$833.08$

7.11 (a) $8.4 - 1.64 \dfrac{(1.8)}{\sqrt{90}} < \mu < 8.4 + 1.64 \dfrac{(1.8)}{\sqrt{90}}$

$8.4 - 1.64(.19) < \mu < 8.4 + 1.64(.19)$

$8.4 - .31 < \mu < 8.4 + .31$

$8.1 \text{ years} < \mu < 8.7 \text{ years}$

(b) $8.4 - 2.33(.19) < \mu < 8.4 + 2.33(.19)$

$8.4 - .44 < \mu < 8.4 + .44$

$8.0 \text{ years} < \mu < 8.8 \text{ years}$

7.13 (a) $31.15 - 1.96 \dfrac{(.08)}{\sqrt{80}} < \mu < 31.15 + 1.96 \dfrac{(.08)}{\sqrt{80}}$

$31.15 - 1.96(.00895) < \mu < 31.15 + 1.96(.00895)$

$31.15 - .018 < \mu < 31.15 + .018$

$31.13 \text{ ounces} < \mu < 31.17 \text{ ounces}$

(b) No. The sample size is greater than 30, so we can be sure that the sample mean is approximately normally distributed whether or not the population is normal.

7.15 (a) $\Sigma X_i = 67$

$\Sigma X_i^2 = 143$

$s^2 = \dfrac{143 - \dfrac{67^2}{36}}{35} = \dfrac{143 - 124.7}{35} = \dfrac{18.3}{35} = .523$

$s = .72$

$$1.86 - 1.96 \left(\frac{.72}{6}\right) \sqrt{\frac{60-36}{60-1}} < \mu < 1.86 + 1.96 \left(\frac{.72}{6}\right) \sqrt{\frac{60-36}{60-1}}$$

$1.86 - 1.96(.12)(.64) < \mu < 1.86 + 1.96(.12)(.64)$

$1.86 - .15 < \mu < 1.86 + .15$

1.7 persons $< \mu < 2.0$ persons

   (b) No. The central limit theorem says that the sample mean is normally distributed if the sample size exceeds 30.

7.17 -0.735 to $-.265$ miles

7.19 (a) $.36 - 1.64 \sqrt{\dfrac{(.36)(.64)}{100}} < \Pi < .36 + 1.64 \sqrt{\dfrac{(.36)(.64)}{100}}$

       $.36 - 1.64(.048) < \Pi < .36 + 1.64(.048)$

       $.36 - .08 < \Pi < .36 + .08$

       $.28 < \Pi < .44$

   (b) $.36 - 1.96(.048) < \Pi < .36 + 1.96(.048)$

       $.36 - .09 < \Pi < .36 + .09$

       $.27 < \Pi < .45$

7.21 (a) Based on Appendix Table 7a,

       $0.24 < \Pi < 0.57.$

   (b) Based on Appendix Table 7b,

       $0.21 < \Pi < 0.62.$

7.23 (a) $58.15 - 56.35 - 1.96 \sqrt{\dfrac{3.42^2}{50} + \dfrac{4.13^2}{50}} < \mu_1 - \mu_2 < 58.15 - 56.35 +$

      $1.96 \sqrt{\dfrac{3.42^2}{50} + \dfrac{4.13^2}{50}}$

      $1.80 - 1.96 \sqrt{\dfrac{11.6964 + 17.0659}{50}} < \mu_1 - \mu_2 < 1.80 +$

      $1.96 \sqrt{\dfrac{11.6964 + 17.0659}{50}}$

      $1.80 - 1.96\,(.758) < \mu_1 - \mu_2 < 1.80 + 1.96\,(.758)$

      $1.80 - 1.49 < \mu_1 - \mu_2 < 1.80 + 1.49$

      $\$0.31 < \mu_1 - \mu_2 < \$3.29$

   (b) No. Each sample size is large enough so that the central limit theorem tells us that the sample mean will be normally distributed.

7.25 (a) $.59 - .52 - 1.64 \sqrt{\frac{(.59)(.41)}{200} + \frac{(.52)(.48)}{200}} < \Pi_1 - \Pi_2 < .59 - .52 +$

$1.64 \sqrt{\frac{(.59)(.41)}{200} + \frac{(.52)(.48)}{200}}$

$0.07 - 1.64 \sqrt{\frac{.2419 + .2496}{200}} < \Pi_1 - \Pi_2 < 0.07 +$

$1.64 \sqrt{\frac{.2419 + .2496}{200}}$

$0.07 - 1.64(.0496) < \Pi_1 - \Pi_2 < 0.07 + 1.64(.0496)$

$0.07 - .08 < \Pi_1 - \Pi_2 < 0.07 + .08$

$-0.01 < \Pi_1 - \Pi_2 < 0.15$

(b) $0.07 - 2.576(.0496) < \Pi_1 - \Pi_2 < 0.07 + 2.576(.0496)$

$0.07 - .13 < \Pi_1 - \Pi_2 < 0.07 + .13$

$-0.06 < \Pi_1 - \Pi_2 < 0.20$

7.27 $n = \left[\frac{(1.96)(15,000)}{5,000} \right]^2 = 5.88^2 = 35$

7.29 Yes, because part or all of any apparent difference between the half-pieces treated with the chlorinating agent and those that were untreated could be due to the difference in the machines on which they were evaluated.

7.31 $n = \left(\frac{2.576}{.02} \right)^2 (.4)(.6) = 128.8^2 (.24) = 3,981$

7.33 (a) 4,376 to 4,624 points.
(b) The width of such a confidence interval equals $(2)(1.64)(\sigma/\sqrt{n})$. If $\sigma = 500$ and $n = 36$, the width equals $(2)(1.64)(500/6)$, or $1,640/6$, or 273.33 points. Yet the width of the confidence interval he gives is 300 points (that is, $4800 - 4500$). Thus, there is a contradiction. He has made a mistake somewhere.
(c) The width of a 95 percent confidence interval is $(2)(1.96)(\sigma/\sqrt{n})$, whereas the width of a 90 percent confidence interval is $(2)(1.64)(\sigma/\sqrt{n})$. Thus, the ratio of the former to the latter is $1.96/1.64$, or 1.20. Thus, he is right.

7.35 If s is (approximately) distributed normally with mean equal to σ and standard deviation equal to $\sigma \div \sqrt{2n}$, then $(s - \sigma) \div \sigma/\sqrt{2n}$ has (approximately) the standard normal distribution. Thus,

$$Pr \left\{ -z_{\alpha/2} < \frac{s - \sigma}{\sigma/\sqrt{2n}} < z_{\alpha/2} \right\} = 1 - \alpha$$

$$Pr \left\{ \frac{-z_{\alpha/2}}{\sqrt{2n}} < \frac{s}{\sigma} - 1 < \frac{z_{\alpha/2}}{\sqrt{2n}} \right\} = 1 - \alpha$$

$$Pr\left\{\frac{-z_{\alpha/2}}{\sqrt{2n}} + 1 < \frac{s}{\sigma} < \frac{z_{\alpha/2}}{\sqrt{2n}} + 1\right\} = 1 - \alpha$$

$$Pr\left\{\frac{1}{1 + \frac{z_{\alpha/2}}{\sqrt{2n}}} < \frac{\sigma}{s} < \frac{1}{1 - \frac{z_{\alpha/2}}{\sqrt{2n}}}\right\} = 1 - \alpha$$

$$Pr\left\{\frac{s}{1 + \frac{z_{\alpha/2}}{\sqrt{2n}}} < \sigma < \frac{s}{1 - \frac{z_{\alpha/2}}{\sqrt{2n}}}\right\} = 1 - \alpha.$$

7.37 (a) .262 to .418

(b) The confidence interval is

$$\frac{8.7}{1 + \frac{1.64}{\sqrt{200}}} < \sigma < \frac{8.7}{1 - \frac{1.64}{\sqrt{200}}}$$

$$\frac{8.7}{1 + \frac{1.64}{14.14}} < \sigma < \frac{8.7}{1 - \frac{1.64}{14.14}}$$

$$\frac{8.7}{1.116} < \sigma < \frac{8.7}{.884}$$

$$7.80 < \sigma < 9.84$$

Thus, the answer is \$7.80 to \$9.84.

7.39 $n = \left(\frac{(1.96)(60)}{10}\right)^2 = 11.76^2 = 138$

7.41 $\left(\frac{1.96 \times 400}{20}\right)^2 = 39.2^2 = 1537$

7.43 The 95 percent confidence interval is 19.46 to 26.94 years, as indicated in the last line of the printout.

Chapter 8

8.1 (a) The null hypothesis is that the death rate with holothane is the same as with other anesthetics like pentothal or cyclopropane. The alternative hypothesis is that the death rate with holothane is higher than with these other anesthetics.

(b) A Type I error means that one concludes that the death rate with holothane is higher than with these other anesthetics when there really is no difference in this regard. A Type II error means that one concludes that the death rate is the same with holothane as with these other anesthetics, when in fact the death rate with holothane is higher. Both types of error may result in incorrect decisions by physicians and others.

8.3 (a) No. With a fixed sample size, the probability of Type I error can be reduced only by increasing the probability of Type II error. Thus, these probabilities should be chosen to reflect the relative costs of Type I and Type II errors.

(b) No.

8.5 (a) Reject the hypothesis that $\mu = 12$ if $\bar{x} < 12 - 1.64(.09) \div \sqrt{60}$. In other words, reject H_0 if $\bar{x} < 11.98$.

(b) If $\mu = 11.9$, the probability that $\bar{X}$ is greater than or equal to 11.98 equals the probability that a standard normal variable is greater than or equal to $(11.98 - 11.9) \div .09/\sqrt{60} = .08 \div .0116 = 6.90$. This probability is less than .001.

(c) Since $\bar{x} = 11.95$, the hypothesis that $\mu = 12$ should be rejected.

8.7 (a) Reject the hypothesis that $\mu = 120$ if $\bar{x} > 120 +$
$$\frac{2.05(10.9)}{\sqrt{36}} \sqrt{\frac{(60 - 36)}{(60 - 1)}}.$$ In other words, reject H_0 if $\bar{x} > 120 + 2.38 = 122.38$.

(b) If $\mu = 121$, the probability that $\bar{X} \leq 122.38$ equals the probability that a standard normal variable is less than or equal to $(122.38 - 121) \div 1.159 = 1.19$. This probability is .883.

(c) Since $\bar{x} = 122.8$, the hypothesis that $\mu = 120$ should be rejected.

8.9 (a) The test procedure in the text is: reject H_0 if $\bar{X} < 50 - 2.33(10)/\sqrt{64}$. In other words, H_0 is accepted if $\bar{X} \geq 47.09$ years. Thus, if $\mu = 47$, the probability of a Type II error equals the probability that $Z \geq (47.09 - 47)/1.25$, or that $Z \geq .07$; this probability equals .47. If $\mu = 48$, the probability of a Type II error equals the probability that $Z \geq (47.09 - 48)/1.25$, or that $Z \geq -.73$; this probability equals .77. If $\mu = 49$, the probability of a Type II error equals the probability that $Z \geq (47.09 - 49)/1.25$, or that $Z \geq -1.53$; this probability equals .94. If $\mu = 50$, the probability of a Type II error equals the probability that $Z \geq (47.09 - 50)/1.25$, or that $Z \geq -2.33$; this probability equals .99.

(b)

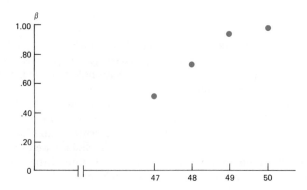

(c)

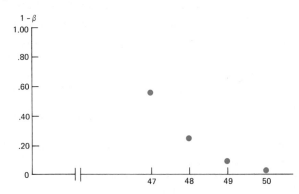

8.11 (a) Reject the hypothesis that the clerk is performing adequately if one or more out of twenty randomly chosen forms per day is handled incorrectly.

(b) $1 - (.99)^{20} = 1 - (.904)^2 = 1 - .82 = .18$

(c) .3585

8.13 (a) The null hypothesis is that $\Pi = .40$. It should be rejected if $p < .40 - 2.576 \sqrt{[.4(.6)]/120}$ or if $p > .40 + 2.576 \sqrt{[.4(.6)]/120}$. In other words, H_0 should be rejected if $p < .285$ or if $p > .515$.

(b) This probability is approximately equal to the probability that a standard normal variable's value lies between $(.285 - .42) \div \sqrt{(.42)(.58)/120}$ and $(.515 - .42) \div \sqrt{(.42)(.58)/120}$, which is the probability that $-3.00 < Z < 2.11$. This probability equals .98.

(c) Do not reject the null hypothesis that no change has occurred (and $\Pi = .40$).

8.15 (a) If the percentage is the same as last year (81 percent), the probability that 317 or less would be filed on time is approximately equal to the probability that a standard normal variable is less than $(.7925 - .81) \div \sqrt{(.81)(.19)/400} = -0.89$. This probability is .19. Thus, the Service cannot be reasonably sure (if "reasonably sure" is defined as the probability being less than .19).

(b) No.

(c) The test in the text is: Reject the hypothesis that $\Pi = .81$ if $p < .81 - 1.64 \sqrt{(.81)(.19)/400} = .81 - .032 = .778$. Thus, if $\Pi = .79$, the probability of accepting the hypothesis that $\Pi = .81$ is approximately equal to the probability that a standard normal variable is greater than or equal to $(.778 - .79) \div \sqrt{(.79)(.21)/400} = -.012 \div .020 = -0.60$. This probability is .73.

8.17 (a) The null hypothesis (that $\Pi = .50$) should be rejected if $p > .50 + 1.64 \sqrt{[(.5)(.5)]/100}$ or if $p < .50 - 1.64 \sqrt{[(.5)(.5)]/100}$. In other words, it should be rejected if $p > .582$ or if $p < .418$. Since the observed value of p is .34, it should be rejected.

(b) The null hypothesis should be rejected if $p < .50 -$ $1.28 \sqrt{[(.5)(.5)]/100}$; that is, if $p < .436$. Since the observed value of p is .34, it should be rejected.

8.19 (a) (i) The null hypothesis is that the population mean score of the mathematics students equals that of the history students. The alternative hypothesis is that these population mean scores are not equal.

(ii) Reject the null hypothesis if $(\bar{x}_1 - \bar{x}_2) \div \sqrt{(100/100) + (100/100)}$ is greater than $z_{\alpha/2}$ or less than $-z_{\alpha/2}$. Otherwise accept the null hypothesis.

(iii) Since $(\bar{x}_1 - \bar{x}_2) \div \sqrt{2} = (80 - 78) \div 1.414 = 1.414$, it follows that it does not exceed $z_{.05}$, which equals 1.64. Thus the null hypothesis should not be rejected.

(iv) Since $z_{.025} = 1.96$, the null hypothesis should not be rejected.

(v) Since $z_{.005} = 2.576$, the null hypothesis should not be rejected.

(b) (i) The null hypothesis is that the population mean score of the mathematics students equals that of the history students. The alternative hypothesis is that the former exceeds the latter.

(ii) Reject the null hypothesis if $(\bar{x}_1 - \bar{x}_2) \div \sqrt{(100/100) + (100/100)}$ $> z_\alpha$, where the scores of the mathematics students are population 1 and those of the history students are population 2. Otherwise accept the null hypothesis.

(iii) From above we know that $(\bar{x}_1 - \bar{x}_2) \div \sqrt{(100/100) + (100/100)}$ $= 1.414$. Since $z_{.10} = 1.28$, the null hypothesis should be rejected.

(iv) Since $z_{.05} = 1.64$, which exceeds 1.414, the null hypothesis should not be rejected.

(v) Since $z_{.01} = 2.33$, which exceeds 1.414, the null hypothesis should not be rejected.

(c) (i) The null hypothesis is that the population mean score of the mathematics students equals that of the history students. The alternative hypothesis is that the latter exceeds the former.

(ii) Reject the null hypothesis if $(\bar{x}_1 - \bar{x}_2) \div \sqrt{(100/100) + (100/100)}$ $< -z_\alpha$, where the scores of the mathematics students are population 1 and those of the history students are population 2. Otherwise accept the null hypothesis.

(iii) Since 1.414 is not less than -1.28, the null hypothesis should not be rejected.

(iv) Since 1.414 is not less than -1.64, the null hypothesis should not be rejected.

(v) Since 1.414 is not less than -2.33, the null hypothesis should not be rejected.

8.21 (a) $z = (60.8 - 58.4) \div \sqrt{(9.9^2/100) + (8.7^2/100)} = 2.4 \div \sqrt{1.737} = 1.82$. Since $z_{.05} = 1.64$, the evidence does indicate that men are better than women at this task.

(b) No.

8.23 (a) $p = (71 + 56) \div 200 = .635$
$z = (.71 - .56) \div \sqrt{(.635)(.365)/50} = .15/ \sqrt{.004636} = 2.21$

Since $z_{.01} = 2.33$, the evidence does not indicate that the null hypothesis (that there is no regional difference) should be rejected.

(b) Since $z_{.025} = 1.96$, the evidence indicates that the null hypothesis (that there is no regional difference) should be rejected.

8.25 $t = (2.3 - 2.0) \div 1.4/\sqrt{16} = .3/.35 = .86$.

Since this value of t is less than $t_{.05}$ (which is 1.753), there is no reason to reject the hypothesis that $\mu = \$2.0$ million.

8.27 (a) $t = (11.3 - 8.9) \div \sqrt{15.625(1/6)} = 2.4/1.614 = 1.49$,

since $s^2 = \dfrac{11(4.1^2) + 11(3.8^2)}{22} = \dfrac{11(16.81) + 11(14.44)}{22} = 15.625$.

Since the value of t is less than $t_{.025}$ (which is 2.074), the data seem to be consistent with the hypothesis that the average effect is the same in both groups.

(b) No.

(c) We are assuming that the weight loss in each group is normally distributed and that the variance is equal in the two groups.

8.29 (a) $t = (2.7 - 2.0) \div 1.1/\sqrt{12} = .7/.318 = 2.20$. Since this value of t is less than $t_{.01}$ (which is 2.718), there is no reason to reject the belief that $\mu = 2.0$ ounces.

(b) $z = (2.7 - 2.0) \div 1.1/\sqrt{120} = .7/.100 = 7.0$. Since this value of z is far greater than $z_{.01}$ (which is 2.33), one would reject the belief that $\mu = 2.0$ ounces.

8.31 (a) Reject the null hypothesis if $\bar{x} > 28 + 1.96 \, (6/\sqrt{100})$ or if $\bar{x} < 28 - 1.96 \, (6/\sqrt{100})$; otherwise, accept the null hypothesis. In other words, reject the null hypothesis if $\bar{x} > 29.176$ miles per gallon or if $\bar{x} < 26.824$ miles per gallon; otherwise, accept the null hypothesis.

(b) Since $\bar{x}$ is less than 26.824 miles per gallon, the company should reject the hypothesis that the population mean is 28 miles per gallon, based on the test given in Section 8.4.

(c) Reject the null hypothesis if $\bar{x} < 28 - 1.64 \, (6/\sqrt{100})$, that is, if $\bar{x} < 27.016$ miles per gallon. Otherwise, accept the null hypothesis.

(d) Yes, because the sample mean is less than 27.016 miles per gallon.

8.33 (a) Reject H_0 if $p < .40 - 2.576 \sqrt{(.4)(.6)/200}$ or if $p > .40 + 2.576 \sqrt{(.4)(.6)/200}$. In other words, accept H_0 if $.311 < p < .489$. Since $p = .445$, the difference is not statistically significant.

(b) The probability of accepting the hypothesis that $\Pi = 0.4$ when $\Pi = .38$ is approximately equal to the probability that $(.311 - .38) \div \sqrt{(.38)(.62)/200} \le Z \le (.489 - .38) \div \sqrt{(.38)(.62)/200}$. Since this is the probability that $-2.01 \le Z \le 3.18$, it equals .98. The probability of accepting the hypothesis that $\Pi = 0.4$ when $\Pi = .40$ is .99. The probability of accepting the hypothesis that $\Pi = 0.4$ when $\Pi = .42$ is approximately equal to the probability that $(.311 - .42) \div \sqrt{(.42)(.58)/200} \le Z \le (.489 - .42) \div \sqrt{(.42)(.58)/200}$. Since this is the probability that $-3.12 \le Z \le 1.98$, it equals 0.98.

(c)

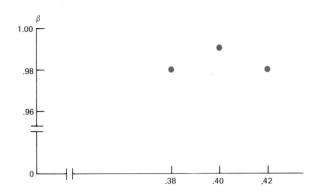

8.35 (a) $p = (121 + 109) \div 400 = .575$

$z = (.605 - .545) \div \sqrt{(.575)(.425)/100} = .06/.0494 = 1.21$

Since $z_{.05} = 1.64$, the evidence does not indicate that the null hypothesis (that the proportion is the same) should be rejected.

(b) The results are the same as in (a).

8.37 (a) Since $n = 36$, $\bar{x} = \$15,561$, and $s = \$9,010$,

$$t = \frac{15,561 - 15,000}{9,010 \div 6} = \frac{561}{1502} = 0.37.$$

Since $t_{.025}$ equals about 2.03, the null hypothesis should not be rejected since the observed value of t is not greater than 2.03.

(b) Since $t_{.005}$ equals about 2.73, the null hypothesis should not be rejected since the observed value of t is not greater than 2.73.

8.39 (a) Since $s^2 = \dfrac{(35)(9,010)^2 + (8)(5,624)^2}{36 + 9 - 2}$

$= 71,961,361,$

it follows that,

$$t = \frac{15,561 - 15,078}{\sqrt{71,961,361 \left(\dfrac{1}{36} + \dfrac{1}{9} \right)}}$$

$$= \frac{483}{\sqrt{(71,961,361)(.1389)}} = \frac{483}{\sqrt{9,995,433}} = \frac{483}{3162} = .153.$$

Since $t_{.025}$ equals about 2.02, the null hypothesis should not be rejected since the observed value of t is not greater than 2.02.

(b) Since $t_{.005}$ equals about 2.70, the null hypothesis should not be rejected since the observed value of t is not greater than 2.70.

Chapter 9

9.1 (a) 11.0705
(b) 18.3070
(c) 31.4104

9.3 Since the mean equals v, and the standard deviation equals $\sqrt{2v}$, the coefficient of variation equals $\sqrt{2v} \div v$, or $\sqrt{2/v}$, where v is the number of degrees of freedom. If $\sqrt{2/v}$ equals 1, v must equal 2.

9.5 (a) It has the χ^2 distribution with 15 degrees of freedom.
(b) 15
(c) 30

9.7 The proportion containing falsified information is $(6 + 8 + 9 + 12) \div 800$ $= .04375$. Thus, the expected number containing falsified information at each university is 8.75, and the expected number containing no falsified information at each university is 191.25. Thus,

$$\Sigma \frac{(f-e)^2}{e} = \frac{(6-8.75)^2}{8.75} + \frac{(194-191.25)^2}{191.25} + \frac{(8-8.75)^2}{8.75}$$

$$+ \frac{(192-191.25)^2}{191.25} + \frac{(9-8.75)^2}{8.75} + \frac{(191-191.25)^2}{191.25} + \frac{(12-8.75)^2}{8.75}$$

$$+ \frac{(188-191.25)^2}{191.25}$$

$$= (1/8.75)(7.5625 + .5625 + .0625 + 10.5625)$$

$$+ (1/191.25)(7.5625 + .5625 + .0625 + 10.5625)$$

$$= (.1143 + .0052)(18.7500)$$

$$= (.1195)(18.7500) = 2.2406.$$

Since $\chi^2_{.05} = 7.81473$, the null hypothesis (that the proportion was the same at each university) should not be rejected.

9.9 (a) Summing up the data for all communities, 65 of the 320 children watch TV more than 5 hours per day. Thus, if the null hypothesis is true, the expected number of children in each sample who watch TV for more than 5 hours per day would be 16.25. Thus,

$$\Sigma \frac{(f-e)^2}{e} = \frac{(9-16.25)^2}{16.25} + \frac{(15-16.25)^2}{16.25} + \frac{(20-16.25)^2}{16.25} +$$

$$\frac{(21-16.25)^2}{16.25} + \frac{(71-63.75)^2}{63.75} + \frac{(65-63.75)^2}{63.75}$$

$$+ \frac{(60-63.75)^2}{63.75} + \frac{(59-63.75)^2}{63.75}$$

$$= \frac{1}{16.25}(52.5625 + 1.5625 + 14.0625 + 22.5625)$$

$$+ \frac{1}{63.75}(52.5625 + 1.5625 + 14.0625 + 22.5625)$$

$$= (.06154 + .01569)(90.75) = (.07723)(90.75) = 7.01.$$

Since $\chi^2_{.05} = 7.81473$, the observed differences can be attributed to chance.

(b) Yes.

9.11 (a) The expected frequencies are shown below in parentheses (next to the actual frequencies):

	California	Illinois	New York	Total
First cover	$81 \left(\dfrac{129,200}{1200} = 107.67 \right)$	60 (107.67)	182 (107.67)	323
Second cover	$78 \left(\dfrac{106,400}{1200} = 88.67 \right)$	93 (88.67)	95 (88.67)	266
Third cover	$241 \left(\dfrac{244,400}{1200} = 203.67 \right)$	247 (203.67)	123 (203.67)	611
Total	400	400	400	1200

$$\sum \frac{(f-e)^2}{e} = \frac{1}{107.67}\{(81 - 107.67)^2 + (60 - 107.67)^2 + (182 - 107.67)^2\}$$

$$+ \frac{1}{88.67}\{(78 - 88.67)^2 + (93 - 88.67)^2 + (95 - 88.67)^2\}$$

$$+ \frac{1}{203.67}\{(241 - 203.67)^2 + (247 - 203.67)^2 + (123 - 203.67)^2\}$$

$$= (.00929)(8508.67) + (.01128)(172.67) + (.00491)(9778.67)$$

$$= 79.05 + 1.95 + 48.01 = 129.01$$

Since $\chi^2_{.05} = 9.48773$, the data indicate that there are regional differences.

(b) Since $\chi^2_{.01} = 13.2767$, the data indicate that there are regional differences.

9.13 The expected frequency of each number is 16 2/3. Thus,

$$\sum \frac{(f-e)^2}{e} = \frac{1}{16\,2/3}\left\{ \left(\frac{5}{3}\right)^2 + \left(\frac{4}{3}\right)^2 + \left(\frac{10}{3}\right)^2 + \left(\frac{1}{3}\right)^2 + \left(\frac{11}{3}\right)^2 + \left(\frac{1}{3}\right)^2 \right\}$$

$$= \frac{1}{16\,2/3}\left(\frac{25 + 16 + 100 + 1 + 121 + 1}{9} \right)$$

$$= 264/150 = 1.76.$$

Since $\chi^2_{.05} = 11.0705$, the null hypothesis (that the die is true) should not be rejected.

9.15 (a) The actual and expected frequencies are:

	A	B	C	D	F
Actual	5	22	50	27	6
Expected	8.8	27.5	37.4	27.5	8,8

Thus,

$$\sum \frac{(f-e)^2}{e} = \frac{(5-8.8)^2}{8.8} + \frac{(22-27.5)^2}{27.5} + \frac{(50-37.4)^2}{37.4} + \frac{(27-27.5)^2}{27.5}$$

$$+ \frac{(6-8.8)^2}{8.8}$$

$$= 1.64 + 1.10 + 4.24 + 0.01 + 0.89 = 7.88.$$

Since $\chi^2_{.05} = 9.48773$, the data seem to be consistent with her claim.
(b) The conclusion is the same as in (a).

9.17 (a) $(n-1)\dfrac{s^2}{\sigma_0^2} = \dfrac{(29)(10)}{20} = 14.5$

Since $\chi^2_{.025} = 45.7222$ and $\chi^2_{.975} = 16.0471$, the null hypothesis (that $\sigma^2 = 20$) should be rejected since 14.5 is less than 16.0471.
(b) The confidence interval is

$$\frac{(29)(10)}{\chi^2_{.01}} < \sigma^2 < \frac{(29)(10)}{\chi^2_{.99}},$$

or

$$\frac{290}{49.5879} < \sigma^2 < \frac{290}{14.2565},$$

or

$$5.85 < \sigma^2 < 20.34.$$

9.19 (a) $\dfrac{17(.000009)}{30.1910} < \sigma^2 < \dfrac{17(.000009)}{7.56418}$

$.00000507 < \sigma^2 < .00002023$

$.0023 < \sigma < .0045$

(b) $\dfrac{17(.000009)}{27.5871} < \sigma^2 < \dfrac{17(.000009)}{8.67176}$

$.00000555 < \sigma^2 < .00001764$

$.0024 < \sigma < .0042$

(c) We assume that the population is normal.

9.21 (a) There are 15 black workers in the sample with incomes exceeding $15,000.

$$\frac{n}{2} - z_{\alpha/2}\sqrt{\frac{n}{4}} = 18 - (1.96)(3) = 12.12$$

$$\frac{n}{2} + z_{\alpha/2}\sqrt{\frac{n}{4}} = 18 + (1.96)(3) = 23.88$$

Since 15 is not less than 12.12 or greater than 23.88, we should not reject the null hypothesis that the median income equals $15,000.

(b) $$\frac{n}{2} - 2.576\sqrt{\frac{n}{4}} = 18 - (2.576)(3) = 10.27$$

$$\frac{n}{2} + 2.576\sqrt{\frac{n}{4}} = 18 + (2.576)(3) = 25.73$$

Since 15 is not less than 10.27 or greater than 25.73, we should not reject the null hypothesis that the median income equals $15,000.

(c) Yes.

No. The present results test a hypothesis about the median whereas Exercise 8.37 tests a hypothesis about the mean.

No. The test in Exercise 8.37 assumes normality whereas this test does not. However, since $n > 30$, the test in Exercise 8.37 should be dependable even if the population is not normal.

9.23 (a) $R_1 = 96$

$$\frac{U - E_u}{\sigma_u} = \frac{11(11) + \dfrac{11(12)}{2} - 96 - \dfrac{11(11)}{2}}{\sqrt{11(11)(23)/12}} = \frac{121 + 66 - 96 - 60.5}{\sqrt{2783/12}}$$

$$= \frac{30.5}{\sqrt{231.92}} = 2.00.$$

Since $(U - E_u) \div \sigma_u$ is greater than 1.96, the data do not seem to be consistent with the psychologist's claim.

(b) Since $(U - E_u) \div \sigma_u$ is less than 2.576, the psychologist's claim is not rejected.

9.25 The sample consists of matched pairs, and the sign test can be used (as in Example 9.3). The number of persons that made fewer errors under the new procedure than the old was nine. Since n is not large, the binomial distribution must be used. If the null hypothesis is true (and there is no difference between the old and new procedures), the probability is .5 that a person will make fewer errors under the new procedure than the old. If this is the case, Appendix Table 1 shows that the probability of nine or more people making fewer errors with the new technique than the old is .0098 + .0010 = .0108. Thus, the sign test results in a rejection of the null hypothesis.

9.27 (a) The sequence is

$$\underline{E}\ \underline{OO}\ \underline{EEE}\ \underline{OO}\ \underline{E}\ \underline{O}\ \underline{E}\ \underline{O}\ \underline{EE}\ \underline{O}\ \underline{E}\ \underline{O}\ \underline{E}$$

$$\underline{O}\ \underline{E}\ \underline{O}\ \underline{E}\ \underline{OO}\ \underline{E}\ \underline{O}\ \underline{E}\ \underline{O}\ \underline{EE}\ \underline{OO}\ \underline{E}\ \underline{O}\ \underline{EE}$$

$$\underline{O}\ \underline{E}\ \underline{O}\ \underline{E}\ \underline{O}\ \underline{E}\ \underline{O}\ \underline{E}\ \underline{O}\ \underline{E}\ \underline{OO}\ \underline{E}\ \underline{O}\ \underline{E}\ \underline{O}$$

$$\underline{EE}\ \underline{OO}\ \underline{E}\ \underline{O}\ \underline{E}\ \underline{OO}\ \underline{E}\ \underline{O}\ \underline{EEE}\ \underline{O}$$

Thus, there are 52 runs.

(b) Since n_1 (the number of even numbers, including zeroes) equals 34 and n_2 (the number of odd numbers) equals 33,

$$E_r = \frac{2(34)(33)}{67} + 1 = \frac{2244}{67} + 1 = 33.49 + 1 = 34.49$$

$$\sigma_r = \sqrt{\frac{2(34)(33)[2(34)(33) - 67]}{67^2(66)}} = \sqrt{\frac{2244(2244 - 67)}{4489(66)}}$$
$$= 4.06.$$

We should reject the null hypothesis (of randomness) if the number of runs exceeds $34.49 + 1.96\ (4.06) = 34.49 + 7.96 = 42.45$. Since the number of runs is 52, the null hypothesis should be rejected.

(c) The sequence is

$$\underline{AAAA}\ \underline{BBBBBB}\ \underline{AAAA}\ \underline{BBBBB}\ \underline{AAAAAAA}$$

$$\underline{BBBBBB}\ \underline{A}\ \underline{B}\ \underline{AA}\ \underline{BBBBB}\ \underline{AAAAA}$$

$$\underline{BBBB}\ \underline{AAAAAAA}\ \underline{BBBBBB}\ \underline{AAA}\ \underline{B}$$

Thus, there are 16 runs.

(d) Since n_1 (the number of As) equals 33 and n_2 (the number of Bs) equals 34,

$$E_r = \frac{2(33)(34)}{67} + 1 = 34.49$$

$$\sigma_r = \sqrt{\frac{2(33)(34)[2(33)(34) - 67]}{67^2(66)}} = 4.06.$$

We reject the null hypothesis (of randomness) if the number of runs is less than $34.49 - 2.576\ (4.06) = 34.49 - 10.46 = 24.03$, or greater than $34.49 + 10.46 = 44.95$. Since the number of runs equals 16, the null hypothesis should be rejected.

9.29 (a) The number of runs is 15.

$$E_r = \frac{2(13)(11)}{24} + 1 = \frac{286}{24} + 1 = 12.92$$

$$\sigma_r = \sqrt{\frac{2(13)(11)[2(13)(11) - 13 - 11]}{24^2(23)}} = \frac{1}{24}\sqrt{\frac{286(286 - 24)}{23}}$$

$$= \frac{1}{24} \sqrt{\frac{74{,}932}{23}} = \frac{1}{24} \sqrt{3257.91} = 57.08/24 = 2.38$$

Thus, $E_r - 2.33\sigma_r = 7.37$ and $E_r + 2.33\sigma_r = 18.47$. Since the number of runs is between 7.37 and 18.47, there is no evidence, based on the runs test, of a departure from randomness.

(b) No.

9.31 Summing up the data for all types of cars, 162 of the 600 cars stalled. Thus, if the null hypothesis is true, the expected number of cars in each sample that would stall is 27. Thus,

$$\sum \frac{(f-e)^2}{e} = \frac{1}{27}\{(31-27)^2 + (22-27)^2 + (29-27)^2$$
$$+ (33-27)^2 + (23-27)^2 + (24-27)^2\}$$

$$+ \frac{1}{73}\{(69-73)^2 + (78-73)^2 + (71-73)^2$$
$$+ (67-73)^2 + (77-73)^2 + (76-73)^2\}$$

$$= (.03704 + .01370)(16 + 25 + 4 + 36 + 16 + 9)$$
$$= (.05074)(106) = 5.38.$$

Since $\chi^2_{.10} = 9.23635$, there is a (considerably) better than 10 percent chance that the observed differences could have been due to chance.

9.33 (a) Summing up the data for all districts, 212 voters support an increase. Thus, if the null hypothesis is true, the expected number in each sample that would support an increase is 53. Thus,

$$\sum \frac{(f-e)^2}{e} = \frac{1}{53}\{(38-53)^2 + (43-53)^2 + (78-53)^2$$
$$+ (53-53)^2\} + \frac{1}{47}\{(62-47)^2 + (57-47)^2$$
$$+ (22-47)^2 + (47-47)^2\} =$$
$$(.01887 + .02128)(225 + 100 + 625) =$$
$$(.04015)(950) = 38.14.$$

Since $\chi^2_{.10} = 6.25139$, it does not appear that the differences can be attributed to chance.

(b) and (c) The conclusion is the same as in (a).

9.35 (a) $\dfrac{(n-1)s^2}{\sigma_0^2} = \dfrac{14(64)}{100} = 8.96$

Since $\chi^2_{.05} = 23.6848$, there is no reason to reject the hypothesis that $\sigma = 10$.

(b) Since $\chi^2_{.99} = 4.66043$, there is no reason to reject the hypothesis that $\sigma = 10$.

9.37 (a) $\dfrac{n}{2} - z_{.025}\sqrt{\dfrac{n}{4}} = 15 - 1.96\sqrt{7.5} = 15 - 1.96(2.74) = 9.63$

$$\frac{n}{2} + z_{.025}\sqrt{\frac{n}{4}} = 15 + 1.96\sqrt{7.5} = 15 + 1.96(2.74) = 20.37$$

Since x (which equals 19) does not exceed 20.37 or fall below 9.63, we cannot reject the hypothesis that there is no difference between the alloys.

(b) No.

9.39 The Minitab computer output is as follows:

```
READ THE FOLLOWING DATA INTO COLUMNS C1,C2,C3

20 12 19
24 28 21
16 20 20
16 20 20

CHISQUARE ANALYSIS OF THE TABLE IN COLUMNS C1,C2,C3

       EXPECTED FREQUENCIES ARE PRINTED
       BELOW OBSERVED FREQUENCIES

          I   C1   I   C2   I   C3   I TOTALS
- - - - - -I- - - - - -I- - - - - -I- - - - - -I- - - - - -
     1    I   20   I   12   I   19   I   51
          I  16.4  I  17.3  I  17.3  I
- - - - - -I- - - - - -I- - - - - -I- - - - - -I- - - - - -
     2    I   24   I   28   I   21   I   73
          I  23.5  I  24.7  I  24.7  I
- - - - - -I- - - - - -I- - - - - -I- - - - - -I- - - - - -
     3    I   16   I   20   I   20   I   56
          I  18.0  I  19.0  I  19.0  I
- - - - - -I- - - - - -I- - - - - -I- - - - - -I- - - - - -
     4    I   16   I   20   I   20   I   56
          I  18.0  I  19.0  I  19.0  I
- - - - - -I- - - - - -I- - - - - -I- - - - - -I- - - - - -
 TOTALS I  76   I   80   I   80   I  236

TOTAL CHI SQUARE =
```

$$0.78 + 1.62 + 0.17 +$$
$$0.01 + 0.43 + 0.57 +$$
$$0.23 + 0.05 + 0.05 +$$
$$0.23 + 0.05 + 0.05$$
$$= 4.25$$

```
DEGREES OF FREEDOM - (4-1) × (3-1) = 6
```

Since $\sum \frac{(f-e)^2}{e} = 4.25$, which is less than $\chi^2_{.05} = 12.5916$, we should not reject the null hypothesis that student opinion on this issue is independent of whether a student belongs to a fraternity or sorority or is unaffiliated.

9.41 As indicated in the last two lines of the printout, we cannot reject the null hypothesis that the median time for the two ambulances is the same. Only if the significance level were set at 0.1190 or higher would we reject this hypothesis.

Chapter 10

10.1 One possible explanation is that, in experiments without controls, there may be a tendency to use the surgical technique on patients that are in relatively good physical condition. This biases the results.

10.3 (a) No, because it may perform no better than a placebo.
  (b) As pointed out in Exercise 10.2, placebos can make people feel better.

10.5 (a) It would not tell anything about how these men would have performed in the absence of vitamin supplementation.
  (b) Soldiers with prior use of vitamin supplements all receive the vitamins, not the placebo. This can bias the results. It is better to allocate the soldiers at random to the two groups.
  (c) It would be better not to tell the soldiers whether they are receiving the vitamins or the placebo.
  (d) (i) Yes.
     (ii) The blocks are the platoons, and the treatments are the vitamins or the placebo.
     (iii) Yes.

10.7 (a) .05
  (b) .01

10.9 (a) 2.75
  (b) 4.30

10.11 (a) Using the formulas in Appendix 10.2, we find that

$$BSS = \frac{1}{4}(74^2 + 87^2 + 106^2 + 106^2) - \frac{1}{16}(373)^2$$

$$= 8{,}879.25 - 8{,}695.5625 = 183.6875.$$

$$TSS = 324 + 400 + 361 + 289 + 484 + 441 + 576 + 400$$

$$+ 625 + 729 + 676 + 784 + 841 + 784 + 576 + 625$$

$$- \frac{1}{16}(373)^2$$

$$= 8{,}915 - 8{,}695.5625 = 219.4375.$$

$$WSS = 219.4375 - 183.6875 = 35.75.$$

$$F = \frac{BSS/3}{WSS/12} = \frac{61.229}{2.979} = 20.55.$$

According to Appendix Table 9, $F_{.05}$ equals 3.49. Since the observed value of F exceeds 3.49, we should reject the null hypothesis that the mean number of miles per gallon is the same for all four firms' cars.

(b) The table is as follows:

Source of Variation	Sum of squares	Degrees of freedom	Mean square	F
Between groups	183.6875	3	61.229	20.55
Within groups	35.7500	12	2.979	
Total	219.4375	15		

(c) Let μ_1 be the mean for U.S. firm I, μ_2 be the mean for U.S. firm II, μ_3 be the mean for the German firm, and μ_4 be the mean for the Japanese firm. Then

$$\sqrt{3.49\ (1.726)}\ \sqrt{\frac{(3)(2)}{4}} = 3.95.$$

Consequently,

$$-8.0 - 3.95 < \mu_1 - \mu_4 < -8.0 + 3.95$$

$$\boxed{-11.95 < \mu_1 - \mu_4 < -4.05}$$

$$18.5 - 21.75 - 3.95 < \mu_1 - \mu_2 < 18.5 - 21.75 + 3.95$$

$$\boxed{-7.20 < \mu_1 - \mu_2 < 0.70}$$

$$18.5 - 26.5 - 3.95 < \mu_1 - \mu_3 < 18.5 - 26.5 + 3.95$$

$$\boxed{-11.95 < \mu_1 - \mu_3 < -4.05}$$

$$21.75 - 26.5 - 3.95 < \mu_2 - \mu_3 < 21.75 - 26.5 + 3.95$$

$$\boxed{-8.70 < \mu_2 - \mu_3 < -0.80}$$

$$21.75 - 26.5 - 3.95 < \mu_2 - \mu_4 < 21.75 - 26.5 + 3.95$$

$$\boxed{-8.70 < \mu_2 - \mu_4 < -0.80}$$

$$26.5 - 26.5 - 3.95 < \mu_3 - \mu_4 < 26.5 - 26.5 + 3.95$$

$$\boxed{-3.95 < \mu_3 - \mu_4 < 3.95}$$

10.13 (a) Method B seems fastest, and method D seems slowest.

(b) Using Appendix 10.2,

$$BSS = \frac{1}{4}(78^2 + 63^2 + 79^2 + 86^2) - \frac{1}{16}(306^2) = \frac{23,690}{4} - \frac{93,636}{16}$$

$$= 5922.5 - 5852.25 = 70.25$$

$$TSS = 5952 - 5852.25 = 99.75$$

$$WSS = 99.75 - 70.25 = 29.50$$

Source of Variation	Sum of squares	Degrees of freedom	Mean square	F
Between groups	70.25	3	23.42	9.52
Within groups	29.50	12	2.46	
Total	99.75	15		

(c) Since $F_{.01} = 5.95$, the answer seems to be no.

(d) Let μ_1 be the mean for method A, μ_2 be the mean for method B, μ_3 be the mean for method C, and μ_4 be the mean for method D.

$$3.75 - \sqrt{(5.95)(2.46)\frac{(2)(3)}{4}} < \mu_1 - \mu_2 < 3.75 +$$

$$\sqrt{(5.95)(2.46)\frac{(2)(3)}{4}}$$

$$-.94 < \mu_1 - \mu_2 < 8.44$$

$$-4.94 < \mu_1 - \mu_3 < 4.44$$

$$-6.69 < \mu_1 - \mu_4 < 2.69$$

$$-8.69 < \mu_2 - \mu_3 < 0.69$$

$$-10.44 < \mu_2 - \mu_4 < -1.06$$

$$-6.44 < \mu_3 - \mu_4 < 2.94$$

(e) There would be advantages if the skill and experience of the workers were controlled.

(f) No. Even though this method is *faster*, others may be *cheaper* or *better*.

10.15 (a) Catalyst 3. Catalyst 1.

(b) Plant C.

(c) Using Appendix 10.2,

$$BSS = \frac{1}{8}(294^2 + 309^2 + 318^2 + 310^2) - \frac{1}{32}(1231)^2$$

$$= \frac{379,141}{8} - \frac{1,515,361}{32}$$

$= 47{,}392.62 - 47{,}355.03 = 37.59$

$RSS = \dfrac{1}{4}(156^2 + 153^2 + 161^2 + 160^2 + 159^2 + 150^2 + 147^2 +$

$145^2) - 47{,}355.03$

$= \dfrac{189{,}681}{4} - 47{,}355.03 = 47{,}420.25 - 47{,}355.03 = 65.22$

$TSS = 47{,}493 - 47{,}355.03 = 137.97$

$WSS = 137.97 - 37.59 - 65.22 = 35.16$

Source of Variation	Sum of squares	Degrees of freedom	Mean square	F
Treatments (catalysts)	37.59	3	12.53	7.50
Blocks (plants)	65.22	7	9.32	5.58
Error	35.16	21	1.67	
Total	137.97	31		

(d) Since $F_{.01} = 4.87$, the answer seems to be no.

(e) Since $F_{.01} = 3.64$, the answer seems to be no.

(f) Let μ_1 be the mean for catalyst 1, μ_2 be the mean for catalyst 2, μ_3 be the mean for catalyst 3, and μ_4 be the mean for catalyst 4.

$$-1.875 - \sqrt{(4.87)(1.67)\dfrac{(2)(3)}{8}} < \mu_1 - \mu_2 < -1.875 +$$

$$\sqrt{(4.87)(1.67)\dfrac{(2)(3)}{8}}$$

$$-4.3 < \mu_1 - \mu_2 < 0.6$$

$$-5.5 < \mu_1 - \mu_3 < -0.5$$

$$-4.5 < \mu_1 - \mu_4 < 0.5$$

$$-3.6 < \mu_2 - \mu_3 < 1.3$$

$$-2.6 < \mu_2 - \mu_4 < 2.3$$

$$-1.5 < \mu_3 - \mu_4 < 3.5$$

10.17 (a) It seems highest in the 150–190 lb class, and lowest in the over-190 lb class.

(b) $BSS = 15[(15.1 - 14.9)^2 + (16.2 - 14.9)^2 + (13.4 - 14.9)^2] = 15(3.98) = 59.7$

$WSS = 14[3.2^2 + 4.0^2 + 3.1^2] = 14(35.85) = 501.9$

Source of Variation	Sum of squares	Degrees of freedom	Mean square	F
Between groups	59.7	2	29.85	2.50
Within groups	501.9	42	11.95	
Total	561.6	44		

(c) Since $F_{.05}$ is about 3.23, the answer seems to be yes.

(d) The times are assumed to be normally distributed in each weight class, and the variance is assumed to be the same in each of these weight classes.

(e) Let μ_1 be the mean in the under-150 lb class, μ_2 be the mean in the 150–190 lb class, and μ_3 be the mean in the over-190 lb class.

$$-1.1 - \sqrt{(3.23)(11.95)\frac{(2)(2)}{15}} < \mu_1 - \mu_2 < -1.1 +$$

$$\sqrt{(3.23)(11.95)\frac{(2)(2)}{15}}$$

$$-4.3 < \mu_1 - \mu_2 < 2.1$$

$$-1.5 < \mu_1 - \mu_3 < 4.9$$

$$-0.4 < \mu_2 - \mu_3 < 6.0$$

10.19 No. There is no way to compare this percentage with what would have occurred if a similar class had used another textbook. In other words, there is no control group.

10.21 (a) Brand B seems to wear best; brand C seems to wear most poorly.

(b) They seem to wear best in Type II cars and worst in Type I cars.

(c) Using Appendix 10.2,

$$BSS = \frac{1}{5}(140^2 + 150^2 + 130^2) - \frac{1}{15}(420^2) = \frac{59,000}{5} - \frac{176,400}{15}$$

$$= 11,800 - 11,760 = 40$$

$$RSS = \frac{1}{3}(79^2 + 89^2 + 83^2 + 81^2 + 88^2) - 11,760 = \frac{35,356}{3} -$$

$$11,760$$

$$= 11,785.33 - 11,760 = 25.33$$

$$TSS = 11,830 - 11,760 = 70$$

$$WSS = 70 - 40 - 25.33 = 4.67$$

Source of Variation	Sum of squares	Degrees of freedom	Mean square	F
Treatments (tires)	40.00	2	20.00	34
Blocks (cars)	25.33	4	6.33	11
Error	4.67	8	0.58	
Total	70.00	14		

(d) Since $F_{.05} = 4.46$, the differences are statistically significant at the .05 probability level.

(e) Since $F_{.05} = 3.84$, the differences are statistically significant at the .05 probability level.

(f) Let μ_1 be the mean for brand A, μ_2 be the mean for brand B, and μ_3 be the mean for brand C.

$$-2.0 - \sqrt{(4.46)(0.58)\frac{(2)(2)}{5}} < \mu_1 - \mu_2 < -2.0 +$$

$$\sqrt{(4.46)(0.58)\frac{(2)(2)}{5}}$$

$$-3.4 < \mu_1 - \mu_2 < -0.6$$

$$0.6 < \mu_1 - \mu_3 < 3.4$$

$$2.6 < \mu_2 - \mu_3 < 5.4$$

10.23 The F ratio is 1.32. Since there are 4 and 20 degrees of freedom, $F_{.05} = 2.87$. (See Appendix Table 9.) Thus, we should not reject the null hypothesis that the observed differences among the typewriters are due to chance.

Chapter 11

11.1

(a)

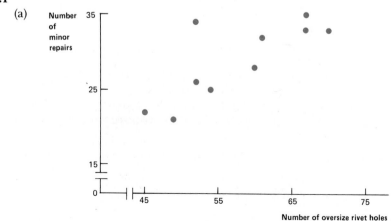

(b) Let Y be the number of minor repairs, X be the number of oversize rivet holes. $\Sigma X = 577$; $\Sigma Y = 289$; $\Sigma X^2 = 33{,}949$; $\Sigma XY = 16{,}987$; $n = 10$.

$$b = \frac{10(16{,}987) - (577)(289)}{10(33{,}949) - 577^2} = \frac{169{,}870 - 166{,}753}{339{,}490 - 332{,}929} = \frac{3117}{6561} = 0.475$$

$$a = 28.9 - (.475)(57.7) = 28.9 - 27.4 = 1.5$$

The regression line is $\hat{Y} = 1.5 + .475X$.

(c) $1.5 + .475(50)$. Thus, the answer is 25.2 repairs.

(d) $1.5 + .475(70)$. Thus, the answer is 34.7 repairs.

11.3

(b) $\Sigma X = 1560$; $\Sigma Y = 989$; $\Sigma X^2 = 176{,}928$; $\Sigma XY = 113{,}355$; $n = 14$

$$b = \frac{(14)(113{,}355) - (1560)(989)}{14(176{,}928) - 1560^2} = \frac{1{,}586{,}970 - 1{,}542{,}840}{2{,}476{,}992 - 2{,}433{,}600} =$$

$$\frac{44{,}130}{43{,}392} = 1.017$$

$$a = 70.64 - (1.017)(111.43) = 70.64 - 113.32 = -42.68$$

The regression line is $\hat{Y} = -42.68 + 1.017X$.

(c) $-42.68 + (1.017)(130) = -42.68 + 132.21 = 89.53$

(d) No. For values of IQ above 140, the relationship must become closer to a horizontal line. Otherwise the predicted test score would exceed 100.

11.5 (a) $\Sigma X = 158$; $\Sigma Y = 28$; $\Sigma X^2 = 3912$; $\Sigma XY = 653$; $n = 8$; $\Sigma Y^2 = 112$

$$b = \frac{8(653) - (158)(28)}{8(3912) - 158^2} = \frac{5224 - 4424}{31{,}296 - 24{,}964} = \frac{800}{6332} = 0.1263$$

$a = 3.5 - (0.1263)(19.75) = 3.5 - 2.494 = 1.006$

The regression line is $\hat{Y} = 1.006 + 0.1263X$.

(b) $s_e = \sqrt{\dfrac{112 - 1.006(28) - (.1263)(653)}{6}} =$

$\sqrt{\dfrac{112 - 28.168 - 82.474}{6}}$

$= \sqrt{\dfrac{1.358}{6}} = \sqrt{.2263} = .476$ thousands of dollars

(c) $1.006 + .1263(20) \pm 1.943(.476) \sqrt{.125 + \dfrac{.25^2}{3912 - 158(19.75)}}$

$1.006 + 2.526 \pm .925 \sqrt{.125 + \dfrac{.0625}{791.5}}$

$3.532 \pm (.925)(.354)$

$3.532 \pm .327$

Thus, the confidence interval is 3.205 to 3.859 thousand of dollars.

(d) $3.532 \pm .925 \sqrt{1.125}$

$3.532 \pm .981$

Thus, the confidence interval is 2.551 to 4.513 thousands of dollars.

11.7 (a) $s_e = \sqrt{\dfrac{1.1117 + .030(1.83) - .04427(12.9643)}{13}}$

$= \sqrt{\dfrac{1.1117 + .0549 - .5739}{13}} = \sqrt{\dfrac{.5927}{13}} = \sqrt{.04559}$

$= .214$ percentage points.

(b) $0.236 \pm 0.694(.214) \sqrt{\dfrac{16}{15} + \dfrac{(6 - 3.431)^2}{327.5882 - (51.46)(3.431)}}$

$.236 \pm .149 \sqrt{1.067 + \dfrac{6.600}{151.029}}$

$.236 \pm .149 \sqrt{1.111}$

$.236 \pm .157$

Thus, the confidence interval is 0.079 to .393 percent.

11.9 (a)

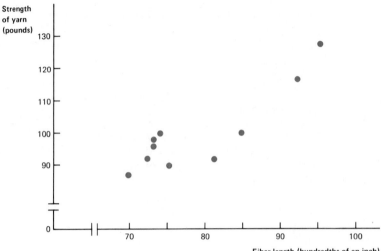

(b) Direct.

Yes, since one would expect longer fibers to be associated with greater strength.

It is hard to tell, but it may be linear.

(c)

	X	Y	X^2	Y^2	XY
	85	99	7,225	9,801	8,415
	82	93	6,724	8,649	7,626
	75	99	5,625	9,801	7,425
	74	97	5,476	9,409	7,178
	76	90	5,776	8,100	6,840
	74	96	5,476	9,216	7,104
	73	93	5,329	8,649	6,789
	96	130	9,216	16,900	12,480
	93	118	8,649	13,924	10,974
	70	88	4,900	7,744	6,160
Total	798	1,003	64,396	102,193	80,991
Mean	79.8	100.3			

$$b = \frac{10(80,991) - (798)(1003)}{10(64,396) - 798^2} = \frac{9516}{7156} = 1.330$$

$$a = 100.3 - (1.33)(79.8) = 100.3 - 106.1 = -5.8$$

(d) The sample regression is

$\hat{Y} = -5.8 + 1.33X.$

$\hat{Y} = -5.8 + (1.33)(80) = -5.8 + 106.4 = 100.6$ pounds.

$\hat{Y} = -5.8 + (1.33)(90) = -5.8 + 119.7 = 113.9$ pounds.

(e) $s_e = \sqrt{\dfrac{102,193 + (5.818)(1003) - (1.3298)(80,991)}{8}}$

$= \sqrt{40.8} = 6.387$ pounds

(Note that a and b are taken out to more decimal places than in (c).) The standard deviation of the conditional probability distribution of the strength of the yarn is estimated to be 6.387 pounds.

(f) The confidence interval is

$$-5.818 + 1.3298(80) \pm (1.860)(6.387) \sqrt{\frac{1}{10} + \frac{(80 - 79.8)^2}{715.6}}$$

or

$$100.566 \pm (11.8798) \sqrt{.10 + .000056}$$

or

$$100.566 \pm 3.758.$$

That is, it is 96.8 to 104.3 pounds.

(g) The confidence interval is

$$100.566 \pm 11.8798 \sqrt{1.100056}$$

or

$$100.566 \pm 12.461.$$

That is, it is 88.1 to 113.0 pounds. This confidence interval is wider than that in (f) because the sampling error in predicting the strength of a particular piece of yarn is greater than the sampling error in estimating the conditional mean strength.

11.11 (a) $\Sigma X = 190$; $\Sigma Y = 451$; $\Sigma X^2 = 5700$; $\Sigma XY = 11390$; $\Sigma Y^2 = 27,503$; $n = 8$.

$$r^2 = \frac{[8(11390) - (190)(451)]^2}{[8(5700) - 190^2][8(27,503) - 451^2]}$$

$$= \frac{(91,120 - 85,690)^2}{(45,600 - 36,100)(220,024 - 203,401)}$$

$$= \frac{5430^2}{9500(16,623)} = \frac{29,484,900}{157,918,500} = .1867$$

$r = 0.43$

(b) $t = \dfrac{.43}{\sqrt{.81/6}} = \dfrac{.43}{\sqrt{.135}} = \dfrac{.43}{.37} = 1.16$

Since $t_{.025} = 2.447$, we cannot reject the hypothesis that the population correlation coefficient is zero.

11.13 (a) Using equation (11.12b),

$$r^2 = \frac{1.5(289) + .475(16{,}987) - \dfrac{1}{10}(289)^2}{8593 - \dfrac{1}{10}(289^2)}$$

$$= \frac{433.5 + 8068.82 - 8352.1}{240.9} = \frac{150.22}{240.99} = 0.62$$

(b) $s_b = \dfrac{3.366}{\sqrt{33{,}949 - 577(57.7)}} = \dfrac{3.366}{\sqrt{33{,}949 - 33{,}292.9}} = 0.131$

$b \div s_b = .475/.131 = 3.63$

Since $t_{.025} = 2.306$, the null hypothesis that the true regression coefficient is zero should be rejected.

11.15 (a) $s_b = \dfrac{1.1726}{\sqrt{3{,}145{,}400 - 5240(524)}} = \dfrac{1.1726}{\sqrt{3{,}145{,}400 - 2{,}745{,}760}}$

$= .001855$

(b) $.02424 \pm 1.86\,(.001855)$
$.02424 \pm .00345$

The confidence interval is .02079 to .02769.

11.17 (a) $r^2 = \dfrac{9{,}516^2}{(7{,}156)(15{,}921)} = .7948$. Thus, $r = .89$.

(b) 79.48 percent of the variation can be explained.

11.19 This estimate involves extrapolation of the regression line beyond the range of the data. For reasons given in section 11.14, this is a very hazardous procedure.

11.21 (a) One might expect the standard deviation of the amount saved to increase as income increases.
(b) At higher levels of output, the standard deviation of the error term may be greater than at lower levels of output.

11.23 (a)

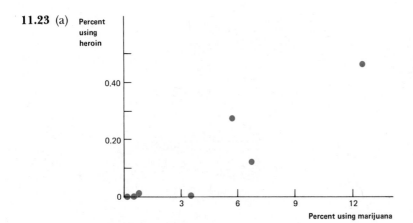

(b) $\Sigma X = 30.0$; $\Sigma Y = 0.94$; $\Sigma X^2 = 248.72$; $\Sigma XY = 8.307$; $n = 7$; $\Sigma Y^2 = .3030$

$$b = \frac{7(8.307) - (30)(0.94)}{7(248.72) - 30^2} = \frac{58.149 - 28.200}{1741.04 - 900} = \frac{29.949}{841.04} = .0356$$

$$a = .134 - (.0356)(4.286) = .134 - .153 = -.019$$

The regression line is $\hat{Y} = -0.019 + .0356X$.

(c) $-0.019 + .0356(2)$. Thus, the answer is about .05 percent.

11.25 The Minitab computer output is as follows:

```
READ THE FOLLOWING DATA INTO COLUMNS C5,C6
2.6    5.1
3.6    3.5
16.5  11.3
24.7  19.0
28.6  20.8
19.3  22.0
53.6  63.0
11.9   9.4
1.9    3.3
26.7  22.6
```

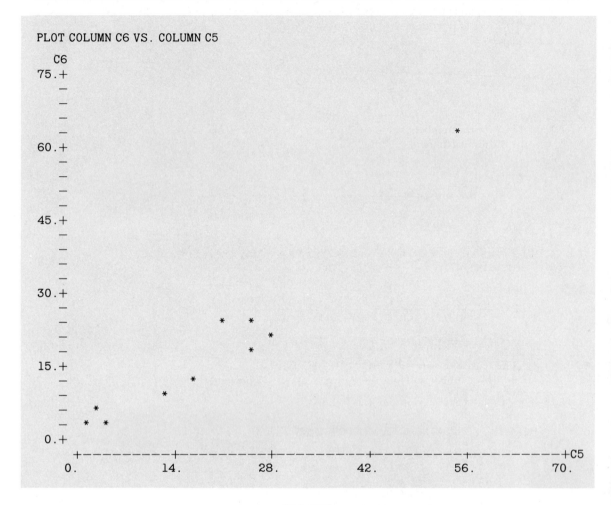

PLOT COLUMN C6 VS. COLUMN C5

11.27 (a) $r^2 = \dfrac{363,450^2}{(52,689)(3,305,500)} = (6.898)(.10995)$

$= .7584$. Thus, $r = .87$

(b) 75.84 percent of the variation is explained.

11.29 $s_b = \dfrac{6.387}{\sqrt{715.6}} = .239$

Thus, $b \div s_b = 1.330/.239 = 5.565$. Since $t_{.01} = 2.896$, the null hypothesis (that $B = 0$) should be rejected because $b \div s_b$ exceeds 2.896.

No. Correlation does not prove causation.

11.31 The regression equation is $\hat{Y} = -2.24 + 1.07\,X_1$, where Y equals the number of televisions in a country (in millions) and X_1 equals the number of telephones in the country (in millions). The standard error of estimate is 5.337 million televisions. The coefficient of determination equals 91.8 percent. Since $b \div s_b = 9.47$, and $t_{.025} = 2.306$ (because there are 8 degrees of freedom), the null hypothesis that B equals zero should be rejected.

Chapter 12

12.1 (a) Let X_1 be the number of family members, X_2 be family income, and Y be the annual amount spent on clothing.

$$\Sigma(X_{1i} - \overline{X}_1)^2 = 82 - \frac{1}{10}(26^2) = 82 - 67.6 = 14.4$$

$$\Sigma(X_{2i} - \overline{X}_2)^2 = 11{,}244 - \frac{1}{10}(314^2) = 11{,}244 - 9859.6 = 1384.4$$

$$\Sigma(Y_i - \overline{Y})^2 = 71.13 - \frac{1}{10}(23.9^2) = 71.13 - 57.121 = 14.009$$

$$\Sigma(X_{1i} - \overline{X}_1)(Y_i - \overline{Y}) = 71.3 - \frac{1}{10}(26)(23.9) = 71.3 - 62.14 = 9.16$$

$$\Sigma(X_{2i} - \overline{X}_2)(Y_i - \overline{Y}) = 875 - \frac{1}{10}(314)(23.9) = 875 - 750.46 = 124.54$$

$$\Sigma(X_{1i} - \overline{X}_1)(X_{2i} - \overline{X}_2) = 855 - \frac{1}{10}(26)(314) = 855 - 816.4 = 38.6$$

$$b_1 = \frac{(1384.4)(9.16) - (38.6)(124.54)}{(14.4)(1384.4) - 38.6^2} = \frac{12{,}681.104 - 4{,}807.244}{19{,}935.36 - 1{,}489.96} =$$

$$\frac{7{,}873.86}{18{,}445.4} = .427$$

$$b_2 = \frac{(14.4)(124.54) - (38.6)(9.16)}{(14.4)(1384.4) - 38.6^2} = \frac{1793.376 - 353.576}{18{,}445.4} =$$

$$\frac{1{,}439.8}{18{,}445.4} = .078$$

$$a = 2.39 - (.427)(2.6) - (.078)(31.4) = 2.39 - 1.1102 - 2.4492 =$$

$$-1.169$$

Thus, the multiple regression is $\hat{Y} = -1.169 + .427X_1 + .078X_2$

(b) Annual clothing expenditure increases by .427 thousands of dollars.

(c) Annual clothing expenditure increases by .078 thousands of dollars.

(d) No. As family income increases, one might expect that the extra clothing expenditure resulting from an extra thousand dollars of income might decrease eventually.

(e) $-1.169 + .427(2) + .078(25) = 1.635$ thousands of dollars.

(f) $-1.169 + .427(4) + .078(35) = 3.269$ thousands of dollars.

(g) No. Minneapolis families may spend more on clothing because the climate is colder than in San Francisco.

12.3 $\dfrac{.427(9.16) + .078(124.54)}{14.009} = \dfrac{13.625}{14.009} = .97$

This means that 97 percent of the variation in the dependent variable can be explained by the regression equation.

12.5 (a) $\bar{R}^2 = 1 - \left(\dfrac{n-1}{n-k-1}\right)(1 - R^2)$

$= 1 - \left(\dfrac{32-1}{32-4-1}\right)0.37 = 1 - \left(\dfrac{31}{27}\right).37 = 1 - .42 = .58$

(b) $\bar{R}^2$ is an unbiased estimate of the population multiple coefficient of determination.

12.7 (a) The complete table is

Sum of squares	Degrees of freedom	Mean square	F
126	3	42	2.83
416	28	14.86	
542	31		

(b) Three.

(c) 32.

(d) 42.

(e) 14.86.

(f) 14.86.

(g) Since $F_{.05} = 2.95$, there is no reason to reject the null hypothesis that the true regression coefficients are all zero.

12.9 $\Sigma(\hat{Y}_i - \bar{Y})^2 = .831(1.54) + .00214(477.8) = 2.302$

$\Sigma(Y_i - \hat{Y}_i)^2 = 2.322 - 2.302 = .020$

The analysis of variance table is

Source of variation	Sum of squares	Degrees of freedom	Mean square	F
Explained by regression	2.302	2	1.151	523
Unexplained by regression	.020	9	.0022	
Total	2.322	11		

Since $F_{.01} = 8.02$, we must reject the null hypothesis that the true regression coefficients are all zero.

12.11 (a) $.29 \pm (1.717)(.11)$, or .10 to .48

(b) $.67 \pm (2.074)(.23)$, or .19 to 1.15

(c) Yes. Divide the regression coefficient (Coeff) by the t value.

12.13 (a) $\hat{Y} = 4.659 - 0.062X_2 + 0.096X_4 + 0.070X_8$

(b) $.0963 \pm (2.16)(.0154)$. In other words, the confidence interval is .063 to .130.

(c) $-0.062 \pm (1.771)(.009)$. In other words, the confidence interval is $-.078$ to $-.046$.

(d) The t value for the regression coefficient of X_8 equals 7.20. Since $t_{.025} = 2.16$, the null hypothesis (that the population regression coefficient of X_8 equals zero) should be rejected since the t value exceeds 2.16.

(e) 0.87. It means that the regression equation explains 87 percent of the variation in the dependent variable.

(f) With 3 and 13 degrees of freedom, $F_{.05}$ equals 3.41. Thus the null hypothesis should be rejected.

12.15 The Minitab computer output is as follows:

```
READ THE FOLLOWING DATA INTO COLUMNS C1,C2,C3
18 105 16
17 108 17
20 110 17
22 109 18
24 110 17
26 120 21
27 125 20
19 111 17
18 85 9
20 96 11
21 94 10
24 130 26
30 134 21
28 103 11
25 102 12
24 89 8
22 92 12
20 95 13
19 100 14
17 91 9
REGRESS C3 ON 2 PREDICTORS C1 AND C2

THE REGRESSION EQUATION IS
Y = - 18.2 - 0.339 X1 + 0.385 X2
```

	COLUMN	COEFFICIENT	ST.DEV. OF COEF.	T-RATIO = COEF/S.D.
	–	−18.181	2.586	−7.03
X1	C1	−0.3394	0.1047	−3.24
X2	C2	0.38516	0.02909	13.24

THE ST. DEV. OF Y ABOUT REGRESSION LINE IS
S = 1.419
WITH (20− 3) = 17 DEGREES OF FREEDOM

R-SQUARED = 92.3 PERCENT
R-SQUARED = 91.4 PERCENT, ADJUSTED FOR D.F.

ANALYSIS OF VARIANCE

DUE TO	DF	SS	MS=SS/DF
REGRESSION	2	410.721	205.361
RESIDUAL	17	34.229	2.013
TOTAL	19	444.950	

As indicated above, the regression equation is: $\hat{Y} = -18.2 - 0.339\,X_1 + 0.385\,X_2$, where Y is income (in thousands of dollars), X_1 is age at birth of first child (in years), and X_2 is IQ score. The multiple correlation coefficient is $\sqrt{.923} = .96$.

Answers to *Getting Down to Cases*

How Often Are Psychologists' Numbers Wrong? (Chapter 2)

(a)

Percent of observations that are incorrect	*Number of experiments*
Zero and under 0.5	3
0.5 and under 1.0	6
1.0 and under 1.5	1
1.5 and under 2.0	2
2.0 and under 2.5	0
2.5 and under 3.0	1
3.0 and under 3.5	1
3.5 and over	1

(b)

```
0 | .00  .23  .41  .62  .67  .69  .72  .82  .95
1 | .13  .59  .69
2 | .50
3 | .17
4 | .17
```

(c) It costs money and time to eliminate errors of this sort. Although one obviously should try to keep such errors to a reasonable minimum, it can sometimes cost more than it is worth to eliminate them completely.

The Reliability of the Apollo Space Mission (Chapter 3)

(a) $(.99)^5 = .9510$
(b) $1 - (.01)^2 = .9999$
(c) $(.99)^4 [1 - (.01)^2] = (.9606)(.9999) = .9605$
(d) The probability is higher in (c) than in (a) because of the smaller chance that the engine will fail to function properly.
(e) It is expensive to add redundant components, and it adds to the weight of the system.

Quality Control in the Manufacture of Railway-Car Side Frames (Chapter 4)

(a) Since $n = 10$ and $\Pi = .20$ the probability that $x = 0$, 1, or 2 equals $.1074 + .2684 + .3020$, or $.6778$. Thus, the probability that the process would be stopped is $1 - .6778 = .3222$.
(b) Since $n = 10$ and $\Pi = .40$ the probability that $x = 0$, 1, or 2 equals $.0060 + .0403 + .1209$, or $.1672$. Thus, the probability that the productive process would be stopped is $1 - .1672$, or $.8328$.
(c) If 20 percent are defective there is a 32 percent probability that the firm will stop the productive process (even though this seems to have been the normal defective rate in the past). If the percent defective jumps to 40 percent, there is a 17 percent probability that the productive process will not be stopped (even though this seems to be a very high defective rate). Based on these considerations, the adequacy of this sampling plan seems to be questionable.
(d) According to Appendix Table 1, since $n = 5$ and $\Pi = .40$, the probability distribution of the number of defectives is

Number of defectives	Probability	Payment(dollars)
0	.0778	0
1	.2592	100
2	.3456	200
3	.2304	300
4	.0768	400
5	.0102	500

Thus, the graph is

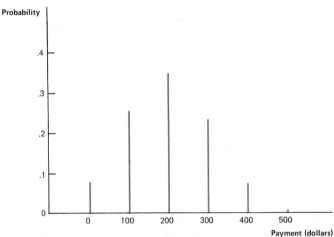

(e) The expected value equals

$$(.0778)(0) + (.2592)(\$100) + (.3456)(\$200) + (.2304)(\$300)$$

$$+ (.0768)(\$400) + (.0102)(\$500) = \$25.92 + \$69.12 +$$

$$\$69.12$$

$$+ \$30.72 + \$5.10 = \$200.$$

(f) The variance equals

$$(.0778)(0 - 200)^2 + (.2592)(100 - 200)^2 + (.3456)(200 - 200)^2 +$$

$$(.2304)(300 - 200)^2 + (.0768)(400 - 200)^2 + (.0102)(500 - 200)^2$$

$$= (.0778)(40,000) + (.2592)(10,000) + (.3456)(0) +$$

$$(.2304)(10,000)$$

$$+ (.0768)(40,000) + (.0102)(90,000) = 3112 + 2592 +$$

$$2304 + 3072 + 918 = 11,998.$$

Thus, the standard deviation equals about 110 dollars. (Actually, the variance equals 12,000; the slight difference is due to rounding errors.)

How Ma Bell Solved an Engineering Problem (Chapter 5)

(a) and (b) This is a situation where there are 2,000 Bernoulli trials with $\Pi = 1/30$. Thus, the number of subscribers wanting a trunkline to B has the binomial distribution with $n = 2,000$ and $\Pi = 1/30$. Using the normal approximation, the probability that L or more subscribers want a trunkline to B is the probability that a normal variable with mean of $(2,000)(1/30) = 66.667$ and a standard deviation of $\sqrt{2,000(1/30)(29/30)} = 8.027$ exceeds $(L - 1/2)$. (Because of the continuity correction, $(L - 1/2)$, rather than L, is correct.) If this probability is to equal .01, $(L - 1/2)$ must correspond to 2.33 on the standard normal distribution. That is,

$$(L - 1/2 - 66.667) \div 8.027 = 2.33$$

and

$$L = (2.33)(8.027) + 67.167 = 85.9.$$

Thus, the probability that about 85.9 or more subscribers will want a trunkline to B is approximately equal to .01. Consequently, about 85 or 86 trunklines should be installed. (This solution is given in W. Feller, *Probability Theory and Applications* (New York: Wiley, 1950), pp. 143–144. For further discussion, see this reference.)

The Decennial Census of Population (Chapter 6)

(a) 20 seconds × 200 million = 200 million hours ÷ 180, or approximately 1 million hours. $4 million.

(b) If a probability sample is used, one can calculate measures of how "far off" the results are likely to be.

(c) To a greater extent, respondents were given the questionnaires and asked to fill them out and mail them themselves. This resulted in fewer errors introduced by enumerators. In the 1960 census, this method was regarded as a considerable success, both in terms of cost and accuracy.

(d) Such a sample would not apportion representatives in the state legislatures in the same way as a complete census. Also, there could be substantial differences in the distribution of large sums of government money to many small areas. To obtain data for small areas, it is argued that a census is worthwhile. (For example, see Morris Hansen, "How to Count Better," in J. Tanur et al., eds., *Statistics: A Guide to the Unknown*.)

The Effect of a New Enzyme on a Pharmaceutical Process (Chapter 7)

(a) Since $\bar{x} = 1.268$ and $s = .228$, the confidence interval is

$$1.268 - 1.64\left(\frac{.228}{6}\right) < \mu < 1.268 + 1.64\left(\frac{.228}{6}\right)$$

$$1.268 - 1.64(.038) < \mu < 1.268 + 1.64(.038)$$

$$1.268 - .062 < \mu < 1.268 + .062$$

$$1.206 < \mu < 1.330.$$

Thus, the answer is 1.206 to 1.330.

(b) The confidence interval is

$$1.268 - 1.96(0.38) < \mu < 1.268 + 1.96(.038)$$

$$1.268 - .074 < \mu < 1.268 + .074$$

$$1.194 < \mu < 1.342.$$

Thus, the answer is 1.194 to 1.342.

(c) The confidence interval is

$$1.268 - 2.576(.038) < \mu < 1.268 + 2.576(.038)$$

$$1.268 - .098 < \mu < 1.268 + .098$$

$$1.170 < \mu < 1.366.$$

Thus, the answer is 1.170 to 1.366.

(d) It is assumed that the sample is random.

(e) 1.268

(f) Yes.

(g) Yes, since the 99 percent confidence interval is entirely above one.

(h) The mistake lies in comparing the difference between the sample mean and 1.00 with the sample standard deviation. It makes more sense to compare this difference with the standard error of the sample mean, as we shall see in Chapter 8.

Who Is Prone to Alcoholism? (Chapter 8)

(a) Yes.

(b) The null hypothesis is that the proportion of men with several alcohol-dependent relatives who themselves are dependent on alcohol is the same as the proportion of men with no such relatives who are dependent on alcohol.

(c) The alternative hypothesis is that the proportion of men with several alcohol-dependent relatives who themselves are dependent on alcohol is higher than the proportion of men with no such relatives who are dependent on alcohol.

(d) Yes.

$$z = (.479 - .056) \div \sqrt{.177(.823)\left(\frac{1}{178} + \frac{1}{71}\right)}$$

$$= .423 \div \sqrt{.1457(.005618 + .014084)}$$

$$= .423 \div \sqrt{.1457(.0197)} = .423 \div \sqrt{.00287} = .423 \div .054 = 7.83.$$

(Note that $p = \dfrac{10 + 34}{178 + 71} = \dfrac{44}{249} = .177$.)

Since $z_{.05} = 1.64$, the null hypothesis should be rejected because $z > 1.64$.

(e) No. For example, it could be due to cultural factors.

(f) No. For example, it could be due to hereditary factors.

Testing for Normality at the American Stove Company (Chapter 9)

(a) and (b)

Upper limit	$\dfrac{\text{Upper limit} - .8314}{.0059}$	Area under standard normal curve to the left of point in column 2	Theoretical proportion in class interval	Theoretical frequency
.8215	−1.68	.0465	.0465	6.7
.8245	−1.17	.1210	.0745	10.8
.8275	−0.66	.2546	.1336	19.4
.8305	−0.15	.4404	.1858	26.9
.8335	0.36	.6406	.2002	29.0
.8365	0.86	.8051	.1645	23.9
.8395	1.37	.9147	.1096	15.9
.8425	1.88	.9699	.0552 ⎫	
∞	∞	1.0000	.0301 ⎭	12.4

The second column in the table on p. A100 expresses the upper limit of each class interval (shown in column 1) as a deviation from the sample mean divided by the sample standard deviation. Using Appendix Table 2, we find the probability that the standard normal variable is less than each number in column 2; this probability is shown in column 3. Column 4 shows the probability that a normal variable (with mean equal to the sample mean and standard deviation equal to the sample standard deviation) will fall in each class interval. To obtain these numbers, we subtract the figure in column 3 for the previous class interval from the figure in column 3 for this class interval. Finally, multiplying the numbers in column 4 by 145, we get the theoretical frequencies in each class interval, shown in column 5. (The last two class intervals are combined so that the theoretical frequency will exceed 5.)

(c)

$$\sum \frac{(f-e)^2}{e} = \frac{(9-6.7)^2}{6.7} + \frac{(5-10.8)^2}{10.8} + \frac{(14-19.4)^2}{19.4} + \frac{(21-26.9)^2}{26.9}$$

$$+ \frac{(55-29.0)^2}{29.0} + \frac{(23-23.9)^2}{23.9} + \frac{(7-15.9)^2}{15.9} +$$

$$\frac{(11-12.4)^2}{12.4}$$

$$= 0.79 + 3.11 + 1.50 + 1.29 + 23.31 + 0.03 + 4.98 +$$

$$0.16 = 35.17.$$

(d) and (e) Since there are 2 parameters estimated from the sample, there are $8 - 3$, or 5 degrees of freedom. Thus, $\chi^2_{.05} = 11.0705$. Since the observed value (35.07) exceeds 11.0705, the null hypothesis (of normality) should be rejected.

(f) The evidence indicates that the heights of these metal pieces are not normally distributed. In particular, there were far more pieces with heights of .8305 and under .8335 inches than would be expected if they were normally distributed.

Smoking and Lung Cancer (**Chapter 10**)

(a) No. The persons who were "comparable" to the lung-cancer patients could only be chosen somewhat arbitrarily, and it is possible that the selection might tend to omit, or fail to recognize, smokers because of conscious or unconscious bias.

(b) No. Because it was not a randomized experiment, people were not assigned at random to smoking and nonsmoking groups. Thus, there was the possibility of confounding of the sort suggested by R. A. Fisher in part (c) of the case.

(c) Studies might be carried out with twins. Since identical twins share the same genetic endowment, one might see whether the cancer rate is higher among twins that smoke heavily than among twins that do not smoke.[1]

[1] D. Reid, "Does Inheritance Matter in Disease? The Use of Twin Studies in Medical Research," in J. Tanur, et al. (eds.), *Statistics: A Guide to the Unknown*.

Admissions Testing on Trial (Chapter 11)

(a) $.51^2 = .26$, or 26 percent.

(b) The coefficient of correlation averages about 0.34. Thus, the proportion of the variation that is explained is about $.34^2 = .12$, or 12 percent.

(c) It is more closely related to a student's cumulative law school grade point average.

(d) 88 percent of the variation in grade ranks is unexplained by SAT scores.

Determinants of the Strength of Cotton Yarn (Chapter 12)

(a) In the table on p. A103, we show the sums, the sums of squares, and the sums of cross products of Y (skein strength), X_1 (fiber length), and X_2 (tensile strength). Using these results, we find that

$$b_1 = \frac{[113,104 - (1502)(75.1)][148,344 - (1541)(95.4)] - [115,754 - (1541)(75.1)][143,626 - (1908)(75.1)]}{[119,951 - (1541)(77.05)][113,104 - (1502)(75.1)] - [115,754 - (1541)(75.1)]^2}$$

$$= \frac{(303.8)(1332.6) - (24.9)(335.2)}{(1,216.95)(303.8) - (24.9)^2} = \frac{404,843.88 - 8346.48}{369,709.4 - 620.01} = \frac{396,497.4}{369,089.4} = 1.074.$$

$$b_2 = \frac{[119,951 - (1541)(77.05)][143,626 - (1908)(75.1)] - [115,754 - (1541)(75.1)][148,344 - (1541)(95.4)]}{369,089.4}$$

$$= \frac{(1216.95)(335.2) - (24.9)(1,332.6)}{369,089.4} = \frac{407,921.64 - 33,181.74}{369,089.4} = 1.015$$

$$a = 95.4 - 1.074(77.05) - 1.015(75.1) = 95.4 - 82.75 - 76.23 = -63.58.$$

Thus, the regression is

$$\hat{Y} = -63.58 + 1.074X_1 + 1.015\,X_2.$$

(b) An increase in fiber length of .01 inches is estimated to increase the average strength by 1.074 pounds. An increase in fiber tensile strength of 1,000 pounds per square inch is estimated to increase the average strength by 1.015 pounds.

(c) $-63.58 + (1.074)(80) + (1.015)(75) = -63.58 + 85.92 + 76.125 = 98.46$ pounds. Thus, the answer is 98.46 pounds.

Y	X_1	X_2	X_1Y	X_2Y	X_1X_2	Y^2	X_1^2	X_2^2
99	85	76	8,415	7,524	6,460	9,801	7,225	5,776
93	82	78	7,626	7,254	6,396	8,649	6,724	6,084
99	75	73	7,425	7,227	5,475	9,801	5,625	5,329
97	74	72	7,178	6,984	5,328	9,409	5,476	5,184
90	76	73	6,840	6,570	5,548	8,100	5,776	5,329
96	74	69	7,104	6,624	5,106	9,216	5,476	4,761
93	73	69	6,789	6,417	5,037	8,649	5,329	4,761
130	96	80	12,480	10,400	7,680	16,900	9,216	6,400
118	93	78	10,974	9,204	7,254	13,924	8,649	6,084
88	70	73	6,160	6,424	5,110	7,744	4,900	5,329
89	82	71	7,298	6,319	5,822	7,921	6,724	5,041
93	80	72	7,440	6,696	5,760	8,649	6,400	5,184
94	77	76	7,238	7,144	5,852	8,836	5,929	5,776
75	67	76	5,025	5,700	5,092	5,625	4,489	5,776
84	82	70	6,888	5,880	5,740	7,056	6,724	4,900
91	76	76	6,916	6,916	5,776	8,281	5,776	5,776
100	74	78	7,400	7,800	5,772	10,000	5,476	6,084
98	71	80	6,958	7,840	5,680	9,604	5,041	6,400
101	70	83	7,070	8,383	5,810	10,201	4,900	6,889
80	64	79	5,120	6,320	5,056	6,400	4,096	6,241

Total
1,908 1,541 1,502 148,344 143,626 115,754 184,766 119,951 113,104

Mean
95.4   77.05 75.1

Index